Series in Real Analysis – Vol. 15

Kurzweil–Stieltjes Integral

Theory and Applications

SERIES IN REAL ANALYSIS

ISSN: 1793-1134

Published

Vol. 15: *Kurzweil–Stieltjes Integral: Theory and Applications*
Giselle Antunes Monteiro, Antonín Slavík & Milan Tvrdý

Vol. 14: *Nonabsolute Integration on Measure Spaces*
Wee Leng Ng

Vol. 13: *Theories of Integration: The Integrals of Riemann, Lebesgue, Henstock–Kurzweil, and McShane (Second Edition)*
Douglas S Kurtz & Charles W Swartz

Vol. 12: *Henstock–Kurzweil Integration on Euclidean Spaces*
Tuo Yeong Lee

Vol. 11: *Generalized Ordinary Differential Equations: Not Absolutely Continuous Solutions*
Jaroslav Kurzweil

Vol. 10: *Topics in Banach Space Integration*
Štefan Schwabik & Ye Guoju

Vol. 9: *Theories of Integration: The Integrals of Riemann, Lebesgue, Henstock–Kurzweil, and McShane*
Douglas S Kurtz & Charles W Swartz

Vol. 8: *Integration between the Lebesgue Integral and the Henstock–Kurzweil Integral: Its Relation to Local Convex Vector Spaces*
Jaroslav Kurzweil

Vol. 7: *Henstock–Kurzweil Integration: Its Relation to Topological Vector Spaces*
Jaroslav Kurzweil

Vol. 6: *Uniqueness and Nonuniqueness Criteria for Ordinary Differential Equations*
R P Agarwal & V Lakshmikantham

Vol. 5: *Generalized ODE*
Š Schwabik

Vol. 4: *Linear Functional Analysis*
W Orlicz

Vol. 3: *The Theory of the Denjoy Integral & Some Applications*
V G Celidze & A G Dzvarseisvili
translated by P S Bullen

Vol. 2: *Lanzhou Lectures on Henstock Integration*
Lee Peng Yee

Vol. 1: *Lectures on the Theory of Integration*
R Henstock

Series in Real Analysis – Vol. 15

Kurzweil–Stieltjes Integral
Theory and Applications

Giselle Antunes Monteiro
Czech Academy of Sciences, Czech Republic

Antonín Slavík
Charles University, Czech Republic

Milan Tvrdý
Czech Academy of Sciences, Czech Republic

World Scientific

NEW JERSEY • LONDON • SINGAPORE • BEIJING • SHANGHAI • HONG KONG • TAIPEI • CHENNAI • TOKYO

Published by

World Scientific Publishing Co. Pte. Ltd.
5 Toh Tuck Link, Singapore 596224
USA office: 27 Warren Street, Suite 401-402, Hackensack, NJ 07601
UK office: 57 Shelton Street, Covent Garden, London WC2H 9HE

Library of Congress Cataloging-in-Publication Data
Names: Monteiro, Giselle Antunes, author. | Slavík, Antonín, 1980– author. |
Tvrdý, Milan, 1944– author.
Title: Kurzweil-Stieltjes integral : theory and applications / by Giselle Antunes Monteiro
(Academy of Sciences of the Czech Republic, Czech Republic),
Antonín Slavík (Charles University, Czech Republic),
Milan Tvrdý (Academy of Sciences of the Czech Republic, Czech Republic).
Description: New Jersey : World Scientific, 2018. | Series: Series in real analysis : volume 15 |
Includes bibliographical references and index.
Identifiers: LCCN 2018035629 | ISBN 9789814641777 (hardcover : alk. paper)
Subjects: LCSH: Integrals. | Integrals, Stieltjes. | Integral equations. |
Theory of distributions (Functional analysis) | Functional analysis.
Classification: LCC QA311 .M545 2018 | DDC 515/.43--dc23
LC record available at https://lccn.loc.gov/2018035629

British Library Cataloguing-in-Publication Data
A catalogue record for this book is available from the British Library.

For any available supplementary material, please visit
https://www.worldscienti ic.com/worldscibooks/10.1142/9432#t=suppl

Preface

Il semble donc que les sommes de Riemann-Stieltjes aient encore un bel avenir devant elles en calcul intégral, et qu'elles pourront réserver encore, dans les mains d'habiles analystes, d'intéressantes surprises.

Jean Mawhin

This text is a free continuation of Š. Schwabik's book [Schwabik (1999a)], a Czech textbook devoted to integration theory on the real line, which begins with the Newton and Riemann integrals, but its main emphasis is on a more general integral known as the Kurzweil-Henstock integral. The definition of this integral was discovered by J. Kurzweil and published

Thomas Joannes Stieltjes

Jaroslav Kurzweil

Štefan Schwabik

in his landmark paper [Kurzweil (1957)] on generalized ordinary differential equations as an alternative definition of the Perron integral. In the 1960s,

Ralph Henstock, an expert in integration theory, independently rediscovered the same definition of integral, and developed it into a systematic theory (see e.g. [Henstock (1968)] and the references there).

It is remarkable that Kurzweil's simple definition, obtained by a slight modification of Riemann's definition, leads to a nonabsolutely convergent integral which is more powerful than the Lebesgue integral. Another great advantage of the Kurzweil-Henstock integral is that there is no need to introduce improper integrals; according to Hake's theorem, the existence of a limit of integrals with respect to the upper and/or lower bound is equivalent to the existence of the integral over the limiting domain.

The Kurzweil-Henstock integral can be generalized in various ways; for example, one can consider integrals over n-dimensional intervals, or integrals of functions taking values in a Banach space. Although the definitions remain simple, the resulting integrals are rather sophisticated, and their study led to many interesting and deep results.

Another possible direction is to study Stieltjes-type integrals; this is the main topic of the present book. We consider integrals of the form $\int_a^b f\,\mathrm{d}g$, in which a function $f:[a,b]\to\mathbb{R}$ (called the integrand) is integrated with respect to another function $g:[a,b]\to\mathbb{R}$ (referred to as the integrator). The simplest integral of this type is the Riemann-Stieltjes integral, whose definition is based on integral sums having the form

$$\sum_{j=1}^{m} f(\xi_j)\,(g(\alpha_j)-g(\alpha_{j-1})), \tag{1}$$

where $a=\alpha_0<\alpha_1<\cdots<\alpha_m=b$, and $\xi_j\in[\alpha_{j-1},\alpha_j]$ for each $j\in\{1,\dots,m\}$. Such integrals appeared for the first time in a famous treatise [Stieltjes (1894)] by T. J. Stieltjes. His investigations of continued fractions led to a surprising connection with the so-called moment problem, which asks for the distribution of mass on a half-line provided that the moments of all orders are given. To be able to describe both continuous and discrete masses, Stieltjes considered a distribution function g such that $g(x)$ equals the mass of the interval $[0,x)$. The calculation of the k-th moment then leads to the integral sum (1) with $f(x)=x^k$ (see also Section 1.1).

The Riemann-Stieltjes integral received further attention thanks to a fundamental result by F. Riesz, who discovered in [Riesz (1909)] that every continuous linear functional on the space of continuous functions can be represented by a Stieltjes integral with a suitable integrator g having bounded variation (see Section 8.1).

In the next year, H. L. Lebesgue showed in [Lebesgue (1910)] that the Riemann-Stieltjes integral of a continuous function $f:[a,b]\to\mathbb{R}$ with respect to a function $g:[a,b]\to\mathbb{R}$ having bounded variation can be expressed as an ordinary Lebesgue integral by means of the formula

$$\int_a^b f(x)\,\mathrm{d}g(x) = \int_0^{v(b)} f(w(t))\,\frac{\mathrm{d}g}{\mathrm{d}t}(w(t))\,\mathrm{d}t,$$

where $v(x)$ denotes the variation of g over $[a,x]$, and w is a generalized inverse to v given by $w(t)=\inf\{s\in[a,b]:v(s)=t\}$ for $t\in[a,b]$. Lebesgue then noted that the integral on the right-hand side can serve as the definition of the Stieltjes integral on the left-hand side in cases when f is discontinuous. An alternative definition of a Stieltjes integral was proposed by W. H. Young in [W.H. Young (1914)] (see Section 6.13). The general concept of a Lebesgue-Stieltjes integral, which is now standard (see Section 6.12), is due to J. Radon, who showed in [Radon (1913)] that a nondecreasing function g gives rise to a nonnegative additive set function (the Lebesgue-Stieltjes measure generated by g); a straightforward modification of Lebesgue's definition then gives the Lebesgue-Stieltjes integral.

During the next decades, Stieltjes integrals were the subject of investigation of numerous mathematicians, e.g. C. J. de la Vallée Poussin [de la Vallée Poussin (1916)], E. B. van Vleck [van Vleck (1917)], T. H. Hildebrandt [Hildebrandt (1917–18)], L. C. Young [L.C. Young (1927, 1936)], and others. The Perron-Stieltjes integral (see Section 6.11), a Stieltjes-type generalization of the Perron integral, was introduced by A. J. Ward in [Ward (1936)], and in the subsequent year it was already included in the second edition of S. Saks's famous monograph [Saks (1937)]. This integral is equivalent to the Kurzweil-Stieltjes integral, which is the central concept of the present book. More information on the history of Stieltjes integrals can be found in [Hawkins (2001)], [Mawhin (2001)], [Medvedev (1991)], or [Pesin (1970)].

Despite the fact that Kurzweil-Stieltjes integration theory was already well developed in the 1990s (see e.g. [Schwabik, Tvrdý, Vejvoda (1979)]), this topic did not fit into Š. Schwabik's textbook. The aim of our monograph is to fill this gap and provide an elementary introduction to Stieltjes integrals with an emphasis on the Kurzweil-Stieltjes integral, including selected applications.

The book consist of eight chapters. Chapter 1 has an introductory character, and describes some simple practical problems leading to Stieltjes integrals. Chapters 2–4 provide an overview of three classes of functions, which play an important role in integration theory: functions of bounded variation, absolutely continuous functions, and regulated functions. Readers who are already familiar with these concepts may skip this part and return later if necessary. The study of Stieltjes integrals begins in Chapter 5, which is devoted to Riemann-Stieltjes integrals. Chapter 6, which is the core of the whole book, focuses on the Kurzweil-Stieltjes integral and its properties, as well as relations to other Stieltjes-type integrals. Chapter 7 describes one important application of the Kurzweil-Stieltjes integral, the theory of generalized linear differential equations. Chapter 8 covers additional topics, such as representation of continuous linear functionals on various function spaces by means of Stieltjes integrals, applications of the Kurzweil-Stieltjes integral in the time scale calculus, and more.

A major part of the book assumes only a basic knowledge of real analysis; no previous familiarity with the Kurzweil-Henstock integral is needed. Those parts which require some notions and results from functional analysis or Lebesgue's integration theory can be skipped if necessary.

Let us conclude by providing several recommendations for further reading. The monographs [Hildebrandt (1963)] and [McLeod (1980)] are particularly close to the material of the present book. Other useful resources are the monographs [Dudley, Norvaiša (2011)], [Kolmogorov, Fomin (2012)], [Schechter E. (1997)], [Rudin (1973)], or the lecture notes [Lukeš, Malý (2017)]. Kurzweil's two recent monographs [Kurzweil (2000, 2002)], which deal with topological problems in integration theory, are not directly concerned with Stieltjes integration. On the other hand, his latest book [Kurzweil (2012)] covers both Kurzweil-Stieltjes integrals and generalized differential equations; another standard reference in this area is the monograph [Schwabik (1992)].

A very preliminary version of this book was written in Czech and published as lecture notes [Tvrdý (2012)] by Palacký University in Olomouc. Since then the manuscript has been considerably extended and the presentation improved. Some results presented in here were previously discussed in the *Seminar on differential equations and integration theory* in the Institute of Mathematics of the Czech Academy of Sciences. We are grateful to all participants of this seminar, in particular to J. Kurzweil, P. Krejčí and I. Vrkoč, for their valuable comments. We are also indebted to J. Jarník for a careful proofreading of the whole text and numerous corrections.

Giselle Antunes Monteiro (email: `gam@math.cas.cz`), Institute of Mathematics, Czech Academy of Sciences, Praha, Czech Republic,

Antonín Slavík (email: `slavik@karlin.mff.cuni.cz`), Charles University, Faculty of Mathematics and Physics, Praha, Czech Republic,

Milan Tvrdý (email: `tvrdy@math.cas.cz`), Institute of Mathematics, Czech Academy of Sciences, Praha, Czech Republic.

Acknowledgments. The first author was supported by the Institutional Research Plan RVO 67985840 of the Czech Academy of Sciences, and by the SASPRO Programme, Marie Curie Actions (European Union's 7th Framework Programme under REA grant agreement No. 609427) during her stay at the Mathematical Institute of the Slovak Academy of Sciences. The third author was supported by the grant No. 14-06958S of the Grant Agency of the Czech Republic and by the Institutional Research Plan RVO 6798584 of the Czech Academy of Sciences.

Contents

Conventions and notation

(i) $\mathbb{N}$ is the set of natural numbers (excluding zero), $\mathbb{Q}$ is the set of rational numbers, $\mathbb{R}$ is the set of real numbers, and $\mathbb{C}$ is the set of complex numbers. $\mathbb{R}^m$ is the m-dimensional Euclidean space consisting of real m-vectors (m-tuples of real numbers). If $x \in \mathbb{R}^m$, then its i-th component is denoted by x_i. For a Banach space X, the norm of its element x is denoted by $\|x\|_X$. If $X = \mathbb{R}^m$ for some $m \in \mathbb{N}$, we write $|x|$ instead of $\|x\|_{\mathbb{R}^m}$ and define $|x| = \sum_{i=1}^{m} |x_i|$.

(ii) $\{x \in A : B(x)\}$ stands for the set of all elements x of the set A which satisfy the condition $B(x)$.
For given sets P, Q, the symbol $P \setminus Q$ represents the set

$$P \setminus Q = \{x \in P : x \notin Q\}.$$

As usual, $P \subset Q$ means that P is a subset of Q (every element of set P is also an element of set Q). Unless it may cause a misunderstanding, we write $\{x_n\}$ instead of $\{x_n : n \in \mathbb{N}\}$. A sequence $\{x_n\}$ is called *non-repeating* if $x_k \neq x_n$ whenever $k \neq n$.

(iii) For each $a \in \mathbb{R}$, we write $a^+ = \max\{a, 0\}$ and $a^- = \max\{-a, 0\}$. (Let us recall that $a^+ + a^- = |a|$ and $a^+ - a^- = a$ for every $a \in \mathbb{R}$.) Furthermore,

$$\operatorname{sgn}(a) = \begin{cases} 1 & \text{if } a > 0, \\ -1 & \text{if } a < 0, \\ 0 & \text{if } a = 0. \end{cases}$$

(iv) If $-\infty < a < b < \infty$, then $[a,b]$ is the *closed* interval $\{t \in \mathbb{R} : a \le t \le b\}$ and (a,b) is the *open* interval $\{t \in \mathbb{R} : a < t < b\}$. The corresponding *half-closed*, or *half-open* intervals are denoted by $[a,b)$ and $(a,b]$. In all cases, the points a, b are called the endpoints of the interval. If $a = b \in \mathbb{R}$, we say that the interval $[a,b]$ *degenerates* to the singleton set $\{a\}$, which we also denote by $[a]$. If I is an interval (closed or open or half-open) with endpoints a, b, then $|I| = |b-a|$ stands for its length. Of course, $|[a]| = 0$.

(v) A finite set of points $\boldsymbol{\alpha} = \{\alpha_0, \alpha_1, \dots, \alpha_m\}$ of the interval $[a,b]$ is called a *division of the interval*, $[a,b]$ if $a = \alpha_0 < \alpha_1 < \dots < \alpha_m = b$. The set of all divisions of the interval $[a,b]$ is denoted by $\mathscr{D}[a,b]$. If $\boldsymbol{\alpha} \in \mathscr{D}[a,b]$, then, unless otherwise stated, its elements will be denoted by α_j, $|\boldsymbol{\alpha}|$ is the length of the largest subinterval $[\alpha_{j-1}, \alpha_j]$, and $\nu(\boldsymbol{\alpha})$ is the number of subintervals, i.e.,

$$\alpha_{\nu(\boldsymbol{\alpha})} = b \quad \text{and} \quad |\boldsymbol{\alpha}| = \max_{j=1,\dots,\nu(\boldsymbol{\alpha})} (\alpha_j - \alpha_{j-1}) \quad \text{for } \boldsymbol{\alpha} \in \mathscr{D}[a,b].$$

If $\boldsymbol{\alpha}' \supset \boldsymbol{\alpha}$, then we say that $\boldsymbol{\alpha}'$ is a *refinement* of $\boldsymbol{\alpha}$.

(vi) For a given set $M \subset \mathbb{R}$ the symbol χ_M denotes the characteristic function of M, i.e.,

$$\chi_M(t) = \begin{cases} 1 & \text{if } t \in M, \\ 0 & \text{if } t \notin M. \end{cases}$$

(vii) The supremum (or the infimum) of a set $M \subset \mathbb{R}$ is denoted by $\sup M$ (or $\inf M$). If $m = \sup M$ is an element of M (or $m = \inf M$ is an element of M), i.e., if m is a maximum (or a minimum) of M, we write $m = \max M$ (or $m = \min M$). If M is the set of all values $F(x)$ of a function F over the domain B (i.e., if $M = \{F(x) : x \in B\}$), we write $\sup_{x \in B} F(x)$, and similarly for the infimum, maximum or minimum, respectively.

(viii) Let X be a Banach space. The notation $f : [a,b] \to X$ means that f is a function from the interval $[a,b]$ into X. For functions $f : [a,b] \to X$ and $g : [a,b] \to X$ and a real number λ, we define the functions $f + g$ and λf by

$$(f+g)(x) = f(x) + g(x) \quad \text{for } x \in [a,b]$$

and

$$(\lambda f)(x) = \lambda f(x) \quad \text{for} \quad x \in [a, b].$$

For a given function $f : [a, b] \to X$, we set

$$\|f\| = \sup_{t \in [a,b]} \|f(t)\|_X.$$

(Of course, if f is unbounded on $[a, b]$, then $\|f\| = \infty$.)

(ix) If $\{x_n\}$ is a sequence of real numbers which has a limit

$$\lim_{n \to \infty} x_n = x \in \mathbb{R} \cup \{-\infty\} \cup \{\infty\},$$

we will write also shortly $x_n \to x$. Given a sequence of functions $\{f_n\}$ defined on $[a, b]$, the symbol $f_n \to f$ stands for the pointwise convergence, i.e., it means that $f_n(x) \to f(x)$ for each $x \in [a, b]$. If $\{f_n\}$ converges uniformly to f on $[a, b]$, i.e. $\lim_{n \to \infty} \|f_n - f\| = 0$, we write also $f_n \rightrightarrows f$ on $[a, b]$.

(x) If $f : [a, b] \to \mathbb{R}$, $t \in [a, b)$, $s \in (a, b]$ and the one-sided limits

$$\lim_{\tau \to t+} f(\tau) \quad \text{and} \quad \lim_{\tau \to s-} f(\tau)$$

exist and are finite, we denote

$$f(t+) = \lim_{\tau \to t+} f(\tau), \quad f(s-) = \lim_{\tau \to s-} f(\tau),$$

$$\Delta^+ f(t) = f(t+) - f(t), \quad \Delta^- f(s) = f(s) - f(s-),$$

$$\Delta f(t) = f(t+) - f(t-) \quad \text{for} \quad t \in (a, b).$$

The following convention is often used:

$$f(a-) = f(a), \quad f(b+) = f(b), \quad \Delta^- f(a) = \Delta^+ f(b) = 0.$$

(xi) $\mathrm{C}([a, b])$ is the space of all continuous real functions on the interval $[a, b]$ with the supremum norm defined in (viii). The space of all continuous vector-valued functions from $[a, b]$ to a Banach space X is denoted by $\mathrm{C}([a, b], X)$.

(xii) If M is a subset of a linear space X, then $\mathrm{Lin}(M)$ denotes the linear span of M, i.e., the set of all elements $x \in X$ of the form

$$x = \sum_{j=1}^{m} c_j\, x_j,$$

where $m \in \mathbb{N}$, $c_1, c_2, \dots, c_m \in \mathbb{R}$ and $x_1, x_2, \dots, x_m \in M$.

(xiii) The set of all continuous linear mappings from a Banach space X to a Banach space Y is denoted by $\mathscr{L}(X,Y)$. It is known that $\mathscr{L}(X,Y)$ is a Banach space when equipped with the operator norm

$$\|A\|_{\mathscr{L}(X,Y)} = \sup_{\substack{x\in X\\ \|x\|_X \le 1}} \|A\,x\|_Y \quad \text{for} \quad A \in \mathscr{L}(X,Y).$$

If $Y = X$ or $Y = \mathbb{R}$, we write simply $\mathscr{L}(X) = \mathscr{L}(X,X)$ or $X^* = \mathscr{L}(X,\mathbb{R})$, respectively. In particular, $\mathscr{L}(\mathbb{R}^m,\mathbb{R}^n)$ is the space of $m \times n$ real matrices and $\mathscr{L}(\mathbb{R}^m,\mathbb{R})$ is the space of column m-vectors, which coincides with the space $\mathbb{R}^m$.

(xiv) Given a matrix $A \in \mathscr{L}(\mathbb{R}^m,\mathbb{R}^n)$, the element in the i-th line and j-th column is denoted by $a_{i,j}$. We write

$$A = \big(a_{i,j}\big)_{\substack{i=1,\dots,m\\ j=1,\dots,n}}.$$

The symbol I stands for the identity matrix, i.e.,

$$I = \big(e_{i,j}\big)_{\substack{i=1,\dots,n\\ j=1,\dots,n}}, \qquad \text{where } e_{i,j} = \begin{cases} 1 & \text{if } i = j,\\ 0 & \text{if } i \ne j. \end{cases}$$

The size of the identity matrix will be always clear from the context. The norm in $\mathscr{L}(\mathbb{R}^m,\mathbb{R}^n)$ is defined by

$$|A| = \max_{j=1,\dots,m} \sum_{i=1}^{n} |a_{i,j}| \quad \text{for} \quad A = \big(a_{i,j}\big)_{\substack{i=1,\dots,m\\ j=1,\dots,n}} \in \mathscr{L}(\mathbb{R}^m,\mathbb{R}^n).$$

In particular, for $x \in \mathbb{R}^m = \mathscr{L}(\mathbb{R}^m,\mathbb{R})$, we have $|x| = \sum_{i=1}^{m} |x_i|$, which agrees with the definition in (i). Furthermore, for $A \in \mathscr{L}(\mathbb{R}^m,\mathbb{R}^n)$, we have

$$|A| = \sup\big\{|A\,x| : |x| \le 1\big\},$$

i.e., the norm of a matrix coincides with the operator norm on $\mathscr{L}(\mathbb{R}^m,\mathbb{R}^n)$ (provided that the norm introduced in (i) is used on $\mathbb{R}^n$).

Chapter 1

Introduction

Stieltjes integrals were originally introduced in order to describe the moments for a given distribution of mass. However, there exist many other problems in mathematics and physics that lead to Stieltjes integrals. Some of them are treated in the present chapter. The exposition here is rather informal; precise statements of definitions and theorems will be given in Chapters 5 and 6.

1.1 Areas of planar regions

Consider a nonnegative continuous function $f:[a,b]\to\mathbb{R}$ and the plane region M bounded by the graph of f and the lines $y=0$, $x=a$, $x=b$, i.e.,

$$M=\{(x,y)\in\mathbb{R}^2 : a\le x\le b,\ 0\le y\le f(x)\}.$$

It is well known that the area of M equals the value of the Riemann integral $\int_a^b f$. An intuitive explanation of this fact is as follows:

Suppose that $a=\alpha_0<\alpha_1<\cdots<\alpha_m=b$, i.e., $\boldsymbol{\alpha}=\{\alpha_0,\alpha_1,\dots,\alpha_m\}$ is a division of $[a,b]$. For every $j\in\{1,\dots,m\}$, choose a point $\xi_j\in[\alpha_{j-1},\alpha_j]$, referred to as the tag of $[\alpha_{j-1},\alpha_j]$, and denote $\boldsymbol{\xi}=(\xi_1,\dots,\xi_m)$. The area of the region M can be approximated by the sum

$$S(\boldsymbol{\alpha},\boldsymbol{\xi})=\sum_{j=1}^{m} f(\xi_j)(\alpha_j-\alpha_{j-1}),$$

where the j-th term corresponds to the area of the rectangle whose base is the interval $[\alpha_{j-1},\alpha_j]$, and whose height is $f(\xi_j)$; see Figure 1.1. Obviously, finer divisions of $[a,b]$ lead to a better approximation. Thus, we can expect

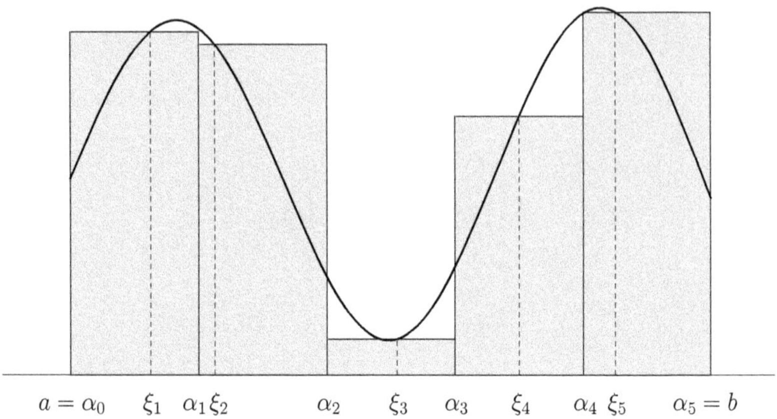

Fig. 1.1 Geometric meaning of Riemann sums

that if we consider finer and finer divisions of $[a,b]$ (a precise description of this limit process will be given later), the sums $S(\boldsymbol{\alpha},\boldsymbol{\xi})$ will approach a certain real number. In this way, we obtain the Riemann integral of the function $f:[a,b]\to\mathbb{R}$, which is denoted by $\int_a^b f$, and whose value gives the area of M.

Next, consider a nonnegative continuous function $f:[a,b]\to\mathbb{R}$ and a nondecreasing continuous function $g:[a,b]\to\mathbb{R}$. Then the mapping $\varphi(t)=(g(t),f(t))$ for $t\in[a,b]$ is a parametric curve in $\mathbb{R}^2$. In fact, if g is increasing, then this curve coincides with the graph of the function $y=f(g^{-1}(x))$. If g is merely nondecreasing, then φ may contain vertical line segments, and need not correspond to a graph. In any case, we can consider the plane region M bounded by φ and the lines $y=0$, $x=g(a)$, $x=g(b)$, i.e.,

$$M=\{(x,y)\in\mathbb{R}^2 : x=g(t),\ 0\le y\le f(t)\ \text{ for } t\in[a,b]\}.$$

We now choose a division $\boldsymbol{\alpha}=\{\alpha_0,\alpha_1,\dots,\alpha_m\}$ of $[a,b]$ and a vector of tags $\boldsymbol{\xi}=(\xi_1,\dots,\xi_m)$ satisfying $\xi_j\in[\alpha_{j-1},\alpha_j]$ for each $j\in\{1,\dots,m\}$. Then the area of M may be approximated by the sum

$$S(\boldsymbol{\alpha},\boldsymbol{\xi})=\sum_{j-1}^{m} f(\xi_j)\,(g(\alpha_j)-g(\alpha_{j-1})),$$

where the j-th term corresponds to the area of the rectangle whose base

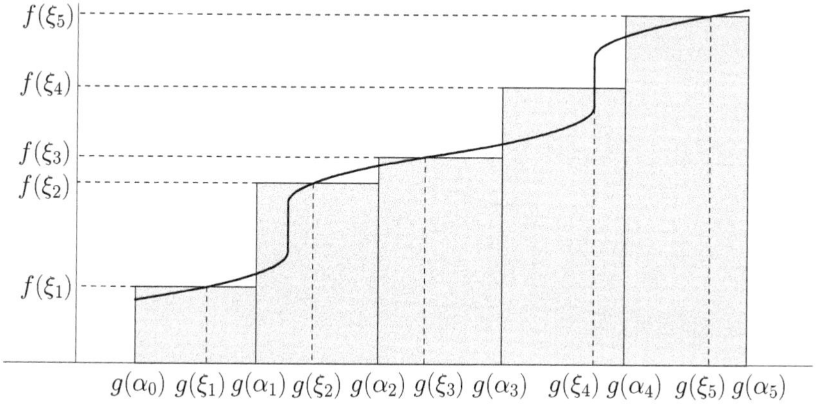

Fig. 1.2 Geometric meaning of Riemann-Stieltjes sums

is the interval $[g(\alpha_{j-1}), g(\alpha_j)]$, and whose height is $f(\xi_j)$; see Figure 1.2. As before, finer divisions lead to a better approximation, and one can show that the sums $S(\boldsymbol{\alpha}, \boldsymbol{\xi})$ tend to a certain real number called the Riemann-Stieltjes integral of f with respect to g. It is denoted by $\int_a^b f\,\mathrm{d}g$, and gives the area of M.

1.1.1 Exercise. Assume that $f:[a,b]\to\mathbb{R}$, $g:[a,b]\to\mathbb{R}$ are continuous functions, and consider the parametric curve $\varphi(t)=(g(t), f(t))$, $t\in[a,b]$.

(i) Observe that if f is nonnegative and g is nonincreasing, then $\int_a^b f\,\mathrm{d}g$ equals minus the area of the region between φ and the x-axis.

(ii) Assume that φ is a closed curve (i.e., $\varphi(a)=\varphi(b)$) with no self-intersections. Moreover, suppose that $[a,b]$ can be written as a union of finitely many intervals on which g is monotone. If φ has clockwise orientation, what is the geometric meaning of $\int_a^b f\,\mathrm{d}g$?

Suppose now that both f and g are nondecreasing and nonnegative. By switching the x- and y-coordinates, we see that $\int_a^b g\,\mathrm{d}f$ can be interpreted as the area of the plane region N bounded by the curve φ and the lines $x=0$, $y=f(a)$, $y=f(b)$, i.e.,

$$N=\{(x,y)\in\mathbb{R}^2 : y=f(t),\ 0\le x\le g(t)\ \text{ for } t\in[a,b]\}.$$

According to Figure 1.3, the total area of M and N is $f(b)\,g(b)-f(a)\,g(a)$,

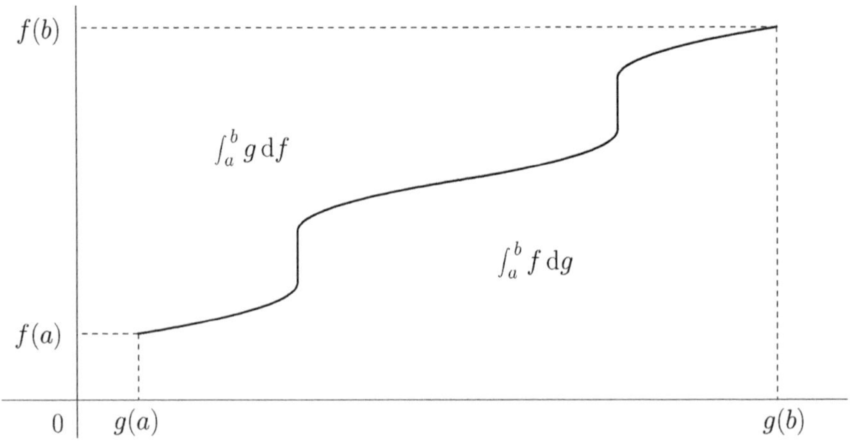

Fig. 1.3 Geometric meaning of the integration by parts formula

which gives

$$\int_a^b f\,\mathrm{d}g + \int_a^b g\,\mathrm{d}f = f(b)\,g(b) - f(a)\,g(a).$$

This relation is known as the integration by parts formula for the Riemann-Stieltjes integral, and it holds under much weaker hypotheses on f and g.

1.2 Center of mass and moments

Consider a system of n particles located on a line, say the x-axis. Let x_j be the coordinate of the j-th particle and m_j its mass. Then it is well known that the center of mass of this system is located at the point

$$\frac{\sum_{j=1}^n x_j m_j}{\sum_{j=1}^n m_j}. \tag{1.2.1}$$

Similarly, if we have a continuous distribution of mass in the interval $[a,b]$ and $\rho(x)$ is the density at x, then the center of mass is given by the formula

$$\frac{\int_a^b x\rho(x)\,\mathrm{d}x}{\int_a^b \rho(x)\,\mathrm{d}x}. \tag{1.2.2}$$

In both (1.2.1) and (1.2.2), the numerator is known as the first moment of the given distribution of mass. More generally, for $k \in \mathbb{N} \cup \{0\}$, the k-th moment is given by $\sum_{j=1}^{n} x_j^k m_j$ in the discrete case, and by $\int_a^b x^k \rho(x)\,\mathrm{d}x$ in the continuous case. The zeroth moment is simply the total mass, which appears in the denominators of (1.2.1) and (1.2.2). As we now show, Stieltjes integrals make it possible to unify the discrete and continuous case; the previous formulas represent particular instances of a more general result.

Suppose that all the mass is contained in an interval $[a,b)$ on the real line, and for each $x \in [a,b]$, let $g(x)$ be the total mass in the interval $[a,x)$. Then, if $a \le x < y \le b$, the difference $g(y) - g(x)$ expresses the total mass in the interval $[x,y)$. Consider a division $\boldsymbol{\alpha} = \{\alpha_0, \alpha_1, \ldots, \alpha_m\}$ of $[a,b]$, and choose $\xi_j \in [\alpha_{j-1}, \alpha_j)$ for each $j \in \{1,\ldots,m\}$. If the division $\boldsymbol{\alpha}$ is sufficiently fine, then the mass in the interval $[\alpha_{j-1}, \alpha_j)$ may be approximated by a particle of mass $g(\alpha_j) - g(\alpha_{j-1})$ located at ξ_j. Consequently, the k-th moment is approximately equal to the sum

$$S(\boldsymbol{\alpha}, \boldsymbol{\xi}) = \sum_{j=1}^{n} \xi_j^k \,(g(\alpha_j) - g(\alpha_{j-1})).$$

In the limit when $\boldsymbol{\alpha}$ becomes finer and finer, the value of $S(\boldsymbol{\alpha}, \boldsymbol{\xi})$ approaches the integral $\int_a^b x^k \,\mathrm{d}g(x)$, which is the precise value of the k-th moment. In particular, the center of mass is located at

$$\frac{\int_a^b x \,\mathrm{d}g(x)}{\int_a^b \mathrm{d}g(x)} = \frac{\int_a^b x \,\mathrm{d}g(x)}{g(b) - g(a)}.$$

Note that in the case of a continuous distribution with density ρ, we have $g(x) = \int_a^x \rho$, and the k-th moment can be expressed in two equivalent ways:

$$\int_a^b x^k \rho(x)\,\mathrm{d}x = \int_a^b x^k \,\mathrm{d}\Big(\int_a^x \rho\Big).$$

This identity is a special case of the substitution theorem, which will be shown to hold in a much more general setting.

1.3 LINE INTEGRALS

LINE INTEGRAL OF THE FIRST KIND

Consider a piece of wire whose shape is described by a space curve $\varphi : [a,b] \to \mathbb{R}^3$ of finite length with no self-intersections. The wire need not be homogeneous; let $f(x)$ be its density at a point x, and suppose that f is continuous. For an arbitrary $t \in [a,b]$, let $v(t)$ be the length of the part of φ between $\varphi(a)$ and $\varphi(t)$. If $[c,d] \subset [a,b]$ is a sufficiently short interval and $\xi \in [c,d]$, then the mass of the part of wire between $\varphi(c)$ and $\varphi(d)$ is approximately $f(\varphi(\xi))\,(v(d)-v(c))$. Hence, if $\boldsymbol{\alpha} = \{\alpha_0, \alpha_1, \dots, \alpha_m\}$ is a division of $[a,b]$ and $\xi_j \in [\alpha_{j-1}, \alpha_j]$ for $j \in \{1, \dots, m\}$, then the sum

$$\sum_{j=1}^{m} f(\varphi(\xi_j))\,(v(\alpha_j) - v(\alpha_{j-1})),$$

approximates the total mass of the wire. In the limit, this sum approaches the value of the Stieltjes integral $\int_a^b f \circ \varphi \, \mathrm{d}v$, which gives the precise total mass of the wire. The integral is known as the *line integral of the first kind* of the function f along the curve φ, and is sometimes denoted by $\int_\varphi f \, \mathrm{d}s$.

LINE INTEGRAL OF THE SECOND KIND

The work performed by a constant force $f \in \mathbb{R}^3$ on a particle moving along a straight line segment given by a vector $v \in \mathbb{R}^3$ is defined as the dot product $f \cdot v$. Consider now a more general situation where the constant force is replaced by a continuous force field $f\colon \mathbb{R}^3 \to \mathbb{R}^3$, and the particle moves along a curve $\varphi\colon [a,b] \to \mathbb{R}^3$ of finite length. If $\boldsymbol{\alpha} = \{\alpha_0, \alpha_1, \dots, \alpha_m\}$ is a division of $[a,b]$ and $\xi_j \in [\alpha_{j-1}, \alpha_j]$ for $j \in \{1, \dots, m\}$, then the dot product $f(\varphi(\xi_j)) \cdot (\varphi(\alpha_j) - \varphi(\alpha_{j-1}))$ approximates the work done by the force field as the particle moves from $\varphi(\alpha_{j-1})$ to $\varphi(\alpha_j)$. Hence, the total work is approximately

$$\sum_{j=1}^{m} f(\varphi(\xi_j)) \cdot (\varphi(\alpha_j) - \varphi(\alpha_{j-1})).$$

The limit of these sums as $\boldsymbol{\alpha}$ becomes finer is the Stieltjes integral

$$\int_a^b (f \circ \varphi)\, \mathrm{d}\varphi = \int_a^b (f_1 \circ \varphi)\, \mathrm{d}\varphi_1 + \int_a^b (f_2 \circ \varphi)\, \mathrm{d}\varphi_2 + \int_a^b (f_3 \circ \varphi)\, \mathrm{d}\varphi_3,$$

which gives the total work done by the force field. The integral is called the *line integral of the second kind* of the vector function f along the curve φ, and is sometimes denoted by $\int_\varphi f \cdot \mathrm{d}s$.

Chapter 2

Functions of bounded variation

The concept of a function of bounded variation was originally discovered in the context of Fourier series, but it turned out to be useful in other areas of mathematics, including integration theory and differential equations. A major part of this chapter deals with the variation of a real function defined on a compact interval. We establish some basic properties of bounded variation functions, such as the decomposition into a difference of two monotone functions, approximation by finite step functions, or decomposition into continuous and jump parts. The final section is concerned with the notion of variation over the so-called elementary sets, i.e., finite unions of bounded (not necessarily closed) intervals.

2.1 Definition and basic properties

A division of an interval $[a,b] \subset \mathbb{R}$ is a finite set $\boldsymbol{\alpha} = \{\alpha_0, \alpha_1, \ldots, \alpha_m\}$ of points from $[a,b]$ such that

$$a = \alpha_0 < \alpha_1 < \cdots < \alpha_m = b. \tag{2.1.1}$$

The symbol $\mathscr{D}[a,b]$ stands for the set of all divisions of $[a,b]$. The elements of a division $\boldsymbol{\alpha} \in \mathscr{D}[a,b]$ are usually denoted by α_j, and we always assume that they are sorted in increasing order, i.e., (2.1.1) holds. Furthermore, we denote by $\nu(\boldsymbol{\alpha})$ the index of the maximal element of $\boldsymbol{\alpha}$ (i.e., $\alpha_{\nu(\boldsymbol{\alpha})} = b$), and

$$|\boldsymbol{\alpha}| = \max_{j \in \{1,\ldots,\nu(\boldsymbol{\alpha})\}} (\alpha_j - \alpha_{j-1}) \quad \text{for } \boldsymbol{\alpha} \in \mathscr{D}[a,b].$$

If $\boldsymbol{\beta} \supset \boldsymbol{\alpha}$, we say that $\boldsymbol{\beta}$ is a *refinement* of $\boldsymbol{\alpha}$.

2.1.1 Definition. For each function $f:[a,b]\to\mathbb{R}$ and a division $\boldsymbol{\alpha}$ of the interval $[a,b]$, we define

$$V(f,\boldsymbol{\alpha})=\sum_{j=1}^{\nu(\boldsymbol{\alpha})}|f(\alpha_j)-f(\alpha_{j-1})| \quad \text{and} \quad \operatorname{var}_a^b f=\sup_{\boldsymbol{\alpha}\in\mathscr{D}[a,b]} V(f,\boldsymbol{\alpha}).$$

If $a=b$, we define $\operatorname{var}_a^a f=0$. The quantity $\operatorname{var}_a^b f$ is called the *variation of the function* f *on the interval* $[a,b]$. If $\operatorname{var}_a^b f<\infty$, we say that f has *bounded variation* on $[a,b]$ (or *is of bounded variation on* $[a,b]$). The set of all functions of bounded variation on $[a,b]$ is denoted by $\mathrm{BV}([a,b])$.

The concept of variation is closely related to the problem of rectifiability of curves. Consider a parametric curve $c:[a,b]\to\mathbb{R}^n$ (c is a continuous mapping). For each division $\boldsymbol{\alpha}$ of the interval $[a,b]$, the sum

$$\lambda(c,\boldsymbol{\alpha})=\sum_{j=1}^{\nu(\boldsymbol{\alpha})}|c(\alpha_j)-c(\alpha_{j-1})|$$

equals the length of the polygonal path connecting the points $c(\alpha_0),\dots,c(\alpha_{\nu(\boldsymbol{\alpha})})$. The length $\Lambda(c,[a,b])$ of the curve c is then

$$\Lambda(c,[a,b])=\sup_{\boldsymbol{\alpha}\in\mathscr{D}[a,b]}\lambda(c,\boldsymbol{\alpha}).$$

The next theorem provides a necessary and sufficient condition for the length of a curve to be finite.

2.1.2 Theorem. *Consider a parametric curve* $c:[a,b]\to\mathbb{R}^n$, *where* $c(t)=(c_1(t),\dots,c_n(t))$ *for each* $t\in[a,b]$. *Then the length of* c *is finite if and only if each of the functions* $c_1,\dots,c_n$ *has bounded variation on* $[a,b]$.

Proof. If $x_1,\dots,x_n$ are arbitrary real numbers, then

$$x_i^2\le x_1^2+\dots+x_n^2\le(|x_1|+\dots+|x_n|)^2,\quad i\in\{1,\dots,n\},$$

and therefore

$$|x_i|\le\sqrt{x_1^2+\dots+x_n^2}\le|x_1|+\dots+|x_n|,\quad i\in\{1,\dots,n\}. \tag{2.1.2}$$

Thus, for an arbitrary division $\boldsymbol{\alpha}$ of $[a,b]$ and each $i\in\{1,\dots,n\}$, we obtain

$$\sum_{j=1}^{\nu(\boldsymbol{\alpha})}|c_i(\alpha_j)-c_i(\alpha_{j-1})|\le\sum_{j=1}^{\nu(\boldsymbol{\alpha})}\sqrt{\sum_{k=1}^{n}(c_k(\alpha_j)-c_k(\alpha_{j-1}))^2}$$

$$\leq \sum_{j=1}^{\nu(\boldsymbol{\alpha})} \sum_{k=1}^{n} |c_k(\alpha_j) - c_k(\alpha_{j-1})| = \sum_{k=1}^{n} \sum_{j=1}^{\nu(\boldsymbol{\alpha})} |c_k(\alpha_j) - c_k(\alpha_{j-1})|.$$

This means that

$$V(c_i, \boldsymbol{\alpha}) \leq \lambda(c, \boldsymbol{\alpha}) \leq \sum_{k=1}^{n} V(c_k, \boldsymbol{\alpha}).$$

Passing to the supremum, we get

$$\operatorname{var}_a^b c_i \leq \Lambda(c, [a,b]) \leq \sum_{k=1}^{n} \operatorname{var}_a^b c_k,$$

wherefrom the statement of the theorem follows immediately. □

In practice, one often deals with planar curves defined by an equation $y = f(x)$, where $f : [a,b] \to \mathbb{R}$ is a continuous function. The corresponding parametric curve $c : [a,b] \to \mathbb{R}^2$ is given by $c(t) = (t, f(t))$ for $t \in [a,b]$, and Theorem 2.1.2 implies the following statement.

2.1.3 Corollary. *The graph of a function $f : [a,b] \to \mathbb{R}$ has finite length if and only if f has bounded variation on $[a,b]$.*

2.1.4 Example. Let $f : [a,b] \to \mathbb{R}$ be continuous and such that $|f'(x)| \leq M$ for all $x \in (a,b)$. Then, by the mean value theorem, the inequality

$$|f(y) - f(x)| \leq M\, |y - x|$$

holds for all $x, y \in [a,b]$. Hence, for each division $\boldsymbol{\alpha}$ of $[a,b]$, we have

$$V(f, \boldsymbol{\alpha}) \leq M \sum_{j=1}^{\nu(\boldsymbol{\alpha})} (\alpha_j - \alpha_{j-1}) = M\,(b-a).$$

We see that *every continuous function on $[a,b]$ which has bounded derivative in (a,b) is of bounded variation.*

If $|f'|$ is in addition Riemann integrable on $[a,b]$ (e.g., if f' is continuous on (a,b)), then the variation of f on $[a,b]$ can be calculated as follows.

2.1.5 Theorem. *If $f : [a,b] \to \mathbb{R}$ is such that $|f'|$ is Riemann integrable, then*

$$\operatorname{var}_a^b f = \int_a^b |f'(x)|\, \mathrm{d}x.$$

Proof. Let $\varepsilon > 0$ be given. The existence of the Riemann integral $\int_a^b |f'(x)| \, dx$ means that there is a $\delta > 0$ such that the inequality

$$\Big| \sum_{j=1}^{\nu(\boldsymbol{\alpha})} |f'(\xi_j)| \, (\alpha_j - \alpha_{j-1}) - \int_a^b |f'(x)| \, dx \Big| < \frac{\varepsilon}{2} \tag{2.1.3}$$

holds for each division $\boldsymbol{\alpha}$ of $[a,b]$ such that $|\boldsymbol{\alpha}| < \delta$ and for every choice of points ξ_j such that

$$\xi_j \in [\alpha_{j-1}, \alpha_j] \quad \text{for} \quad j = 1, \ldots, \nu(\boldsymbol{\alpha}). \tag{2.1.4}$$

On the other hand, by the definition of variation, there exists $\boldsymbol{\alpha} \in \mathscr{D}[a,b]$ such that $|\boldsymbol{\alpha}| < \delta$ and

$$\operatorname{var}_a^b f \geq V(f, \boldsymbol{\alpha}) > \operatorname{var}_a^b f - \frac{\varepsilon}{2}. \tag{2.1.5}$$

By the mean value theorem, there are points ξ_j, $j = 1, \ldots, \nu(\boldsymbol{\alpha})$, satisfying (2.1.4) and such that

$$V(f, \boldsymbol{\alpha}) = \sum_{j=1}^{\nu(\boldsymbol{\alpha})} |f'(\xi_j)| \, (\alpha_j - \alpha_{j-1}).$$

This, together with (2.1.3) and (2.1.5), implies that

$$\begin{aligned} &\Big| \operatorname{var}_a^b f - \int_a^b |f'(x)| \, dx \Big| \\ &\quad \leq |\operatorname{var}_a^b f - V(f, \boldsymbol{\alpha})| + \Big| \sum_{j=1}^{\nu(\boldsymbol{\alpha})} |f'(\xi_j)| \, (\alpha_j - \alpha_{j-1}) - \int_a^b |f'(x)| \, dx \Big| \\ &\quad < \frac{\varepsilon}{2} + \frac{\varepsilon}{2} = \varepsilon. \end{aligned}$$

Since $\varepsilon > 0$ was arbitrary, the proof is complete. □

2.1.6 Exercise. Calculate the variation and estimate the length of the graph of the following functions:

(i) $f(x) = \sin^2 x, \; x \in [0, \pi]$,

(ii) $f(x) = x^3 - 3\,x + 4, \; x \in [0, 2]$,

(iii) $f(x) = \cos x + x \sin x, \; x \in [0, 2\pi]$.

2.1.7 Remark. By Definition 2.1.1, it is obvious that $\operatorname{var}_a^b f$ is nonnegative for every function $f:[a,b]\to\mathbb{R}$. Furthermore,

$$\operatorname{var}_a^b f = \sup_{\boldsymbol{\alpha}\supset\boldsymbol{\beta}} V(f,\boldsymbol{\alpha}) \tag{2.1.6}$$

holds for any division $\boldsymbol{\beta}\in\mathscr{D}[a,b]$. This follows from several elementary observations: First, because

$$\{V(f,\boldsymbol{\alpha}):\boldsymbol{\alpha}\in\mathscr{D}[a,b] \quad\text{and}\quad \boldsymbol{\alpha}\supset\boldsymbol{\beta}\}\subset\{V(f,\boldsymbol{\alpha}):\boldsymbol{\alpha}\in\mathscr{D}[a,b]\},$$

it follows that

$$\sup_{\boldsymbol{\alpha}\supset\boldsymbol{\beta}} V(f,\boldsymbol{\alpha})\le\operatorname{var}_a^b f.$$

Moreover, by the triangle inequality, for arbitrary two divisions $\boldsymbol{\alpha}$, $\boldsymbol{\alpha}'$ of $[a,b]$ such that $\boldsymbol{\alpha}'\supset\boldsymbol{\alpha}$ and for any function $f:[a,b]\to\mathbb{R}$ we have

$$V(f,\boldsymbol{\alpha})\le V(f,\boldsymbol{\alpha}').$$

Finally, if $\boldsymbol{\alpha}\in\mathscr{D}[a,b]$ is given and $\boldsymbol{\alpha}'=\boldsymbol{\alpha}\cup\boldsymbol{\beta}$, then $\boldsymbol{\alpha}'\supset\boldsymbol{\beta}$ and thus

$$V(f,\boldsymbol{\alpha})\le V(f,\boldsymbol{\alpha}').$$

This means that for every $d\in\{V(f,\boldsymbol{\alpha}):\boldsymbol{\alpha}\in\mathscr{D}[a,b]\}$ there exists

$$d'\in\{V(f,\boldsymbol{\alpha}):\boldsymbol{\alpha}\in\mathscr{D}[a,b] \quad\text{and}\quad \boldsymbol{\alpha}\supset\boldsymbol{\beta}\}$$

such that $d\le d'$. Thus

$$\operatorname{var}_a^b f\le\sup_{\boldsymbol{\alpha}\supset\boldsymbol{\beta}} V(f,\boldsymbol{\alpha}),$$

which implies (2.1.6).

2.1.8 Exercise. Prove the following statements:

(i) If $[c,d]\subset[a,b]$, then

$$|f(d)-f(c)|\le\operatorname{var}_c^d f\le\operatorname{var}_a^b f$$

holds for every function $f:[a,b]\to\mathbb{R}$.

(ii) If $f\in\mathrm{BV}([a,b])$, then

$$|f(x)|\le|f(a)|+\operatorname{var}_a^b f \quad \text{for every } x\in[a,b]. \tag{2.1.7}$$

(iii) $\operatorname{var}_a^b f = d \in \mathbb{R}$ if and only if for each $\varepsilon > 0$ there is $\boldsymbol{\alpha}_\varepsilon \in \mathscr{D}[a,b]$ such that

$$d - \varepsilon \le V(f, \boldsymbol{\alpha}) \le d$$

for all refinements $\boldsymbol{\alpha}$ of $\boldsymbol{\alpha}_\varepsilon$.

(iv) $\operatorname{var}_a^b f = \infty$ if and only if for each $K > 0$ there is a division $\boldsymbol{\alpha}_K \in \mathscr{D}[a,b]$ such that $V(f, \boldsymbol{\alpha}_K) \ge K$.

(v) $\operatorname{var}_a^b f = \infty$ if and only if there exists a sequence $\{\boldsymbol{\alpha}^n\}$ of divisions of $[a,b]$ such that

$$\lim_{n\to\infty} V(f, \boldsymbol{\alpha}^n) = \infty.$$

(vi) Suppose there exists an $L \in \mathbb{R}$ such that

$$|f(x) - f(y)| \le L\,|x - y| \quad \text{for all} \quad x, y \in [a,b].$$

(In such a case, we say that f satisfies the *Lipschitz condition* on $[a,b]$, or that f is *Lipschitz continuous* on $[a,b]$.) Show that $\operatorname{var}_a^b f \le L\,(b-a)$.

2.1.9 Remark. The inequality (2.1.7) implies that every function with bounded variation on $[a,b]$ is necessarily bounded on $[a,b]$.

2.1.10 Example. Let

$$f(x) = \begin{cases} 0 & \text{if } x = 0, \\ x\,\sin\left(\dfrac{\pi}{x}\right) & \text{if } x \in (0,2]. \end{cases}$$

Notice that $f(x) = 0$ if and only if $x = 0$ or $x = 1/k$ for some $k \in \mathbb{N}$. Furthermore, for $x \in (0,2]$ we have

$$f(x) = \begin{cases} x & \text{if and only if } \; x = y_k = \dfrac{2}{4k+1} & \text{for some } k \in \mathbb{N} \cup \{0\}, \\ -x & \text{if and only if } \; x = z_k = \dfrac{2}{4k-1} & \text{for some } k \in \mathbb{N}. \end{cases}$$

Thus, for a given $n \in \mathbb{N}$ and for $\boldsymbol{\alpha}^n = \{0, y_n, z_n, \dots, y_1, z_1, 2\}$, we have

$$\begin{aligned} V(f, \boldsymbol{\alpha}^n) &= |f(0) - f(y_n)| + \sum_{k=1}^n |f(y_{k-1}) - f(z_k)| + \sum_{k=1}^n |f(y_k) - f(z_k)| \\ &= y_n + \sum_{k=1}^n (y_{k-1} + z_k) + \sum_{k=1}^n (y_k + z_k) \end{aligned}$$

$$= y_0 + 2\sum_{k=1}^{n}(y_k + z_k) = 2 + 4\sum_{k=1}^{n}\frac{8\,k}{16k^2 - 1} \geq 2\Big(1 + \sum_{k=1}^{n}\frac{1}{k}\Big).$$

Since $\sum_{k=1}^{\infty} 1/k = \infty$, we see that $\lim_{n\to\infty} V(f,\boldsymbol{\alpha}^n) = \infty$, and consequently $\operatorname{var}_0^2 f = \infty$.

We can easily determine the variation of a monotone function.

2.1.11 Theorem. *If f is monotone on $[a,b]$, then*

$$\operatorname{var}_a^b f = |f(b) - f(a)|.$$

Proof. If f is nonincreasing on $[a,b]$ and $\boldsymbol{\alpha} \in \mathscr{D}[a,b]$, then

$$V(f,\boldsymbol{\alpha}) = \sum_{j=1}^{m}\big(f(\alpha_{j-1}) - f(\alpha_j)\big) = f(a) - f(b),$$

i.e., $\operatorname{var}_a^b f = f(a) - f(b) = \big|f(b) - f(a)\big|$.

Similarly, if f is nondecreasing on $[a,b]$, then $\operatorname{var}_a^b f = f(b) - f(a) = \big|f(b) - f(a)\big|$. □

2.1.12 Exercise. (i) Prove that $f\colon [a,b] \to \mathbb{R}$ has bounded variation if and only if there exists a nondecreasing function $h\colon [a,b] \to \mathbb{R}$ such that

$$|f(y) - f(x)| \leq h(y) - h(x) \quad \text{for } x, y \in [a,b], \quad x \leq y.$$

(ii) Prove that the inequality

$$\sum_{j=1}^{\nu(\boldsymbol{\alpha})}\Big(|\Delta^+ f(\alpha_{j-1})| + |f(\alpha_j-) - f(\alpha_{j-1}+)| + |\Delta^- f(\alpha_j)|\Big) \leq \operatorname{var}_a^b f$$

holds for each function f: $[a,b] \to \mathbb{R}$ and each division $\boldsymbol{\alpha} = \{\alpha_0, \ldots, \alpha_{\nu(\boldsymbol{\alpha})}\}$ of $[a,b]$.

Hint: Consider the expressions

$$\sum_{j=1}^{\nu(\boldsymbol{\alpha})}\Big(|f(\alpha_{j-1}+\delta) - f(\alpha_{j-1})| + |f(\alpha_j-\delta) - f(\alpha_{j-1}+\delta)| + |f(\alpha_j) - f(\alpha_j-\delta)|\Big)$$

and let $\delta \to 0$.

2.1.13 Example. (i) A simple example of a bounded variation function which does not have bounded derivative (and hence the statement from Example 2.1.4 (i) is not applicable) is $f(x) = \sqrt{x}$, $x \in [0,1]$. Indeed, since f is increasing on $[0,1]$, by Theorem 2.1.11 we have $\operatorname{var}_0^1 f = 1$.

(ii) An example of a discontinuous function with bounded variation is

$$f(x) = \begin{cases} 0 & \text{if } x = 0, \\ \dfrac{1}{k} & \text{if } x \in (0,1] \text{ and } x \in (\frac{1}{k+1}, \frac{1}{k}] \text{ for some } k \in \mathbb{N}. \end{cases}$$

This function is nondecreasing on the interval $[0,1]$, and therefore $\operatorname{var}_0^1 f = 1$ by Theorem 2.1.11.

2.1.14 Theorem. *For every $c \in [a,b]$ and every function $f : [a,b] \to \mathbb{R}$, we have*

$$\operatorname{var}_a^b f = \operatorname{var}_a^c f + \operatorname{var}_c^b f.$$

P r o o f. Let $f : [a,b] \to \mathbb{R}$ and $c \in [a,b]$ be given. If $c = a$ or $c = b$, the statement of the theorem is trivial. Thus, assume that $c \in (a,b)$.

Let $\widetilde{\boldsymbol{\alpha}} = \{a, c, b\}$ and let $\boldsymbol{\alpha} \in \mathscr{D}[a,b]$ be an arbitrary refinement of $\widetilde{\boldsymbol{\alpha}}$. Then necessarily $c \in \boldsymbol{\alpha}$ and we can split the division $\boldsymbol{\alpha}$ in two parts: the division $\boldsymbol{\alpha}'$ of the interval $[a,c]$ and the division $\boldsymbol{\alpha}''$ of the interval $[c,b]$, i.e., $\boldsymbol{\alpha} = \boldsymbol{\alpha}' \cup \boldsymbol{\alpha}''$, where $\boldsymbol{\alpha}' \in \mathscr{D}[a,c]$ and $\boldsymbol{\alpha}'' \in \mathscr{D}[c,b]$. Obviously, we have

$$V(f, \boldsymbol{\alpha}) = V(f, \boldsymbol{\alpha}') + V(f, \boldsymbol{\alpha}''), \tag{2.1.8}$$

wherefrom, by Remark 2.1.7, we deduce

$$\operatorname{var}_a^b f = \sup_{\boldsymbol{\alpha} \supset \widetilde{\boldsymbol{\alpha}}} V(f, \boldsymbol{\alpha}) \le \operatorname{var}_a^c f + \operatorname{var}_c^b f. \tag{2.1.9}$$

On the other hand, for every two divisions $\boldsymbol{\alpha}' \in \mathscr{D}[a,c]$ and $\boldsymbol{\alpha}'' \in \mathscr{D}[c,b]$, their union $\boldsymbol{\alpha} = \boldsymbol{\alpha}' \cup \boldsymbol{\alpha}''$ is a division of $[a,b]$ and (2.1.8) holds again. This implies

$$\operatorname{var}_a^c f + \operatorname{var}_c^b f = \sup_{\boldsymbol{\alpha}' \in \mathscr{D}[a,c]} V(f, \boldsymbol{\alpha}') + \sup_{\boldsymbol{\alpha}'' \in \mathscr{D}[c,b]} V(f, \boldsymbol{\alpha}'') \le \operatorname{var}_a^b f,$$

which completes the proof. □

2.1.15 Example. Let $n\in\mathbb{N}$. Consider the function $f_n:[0,2]\to\mathbb{R}$ given by

$$f_n(x)=\begin{cases}0 & \text{if } 0\le x\le \frac{1}{n},\\ x\,\sin\left(\dfrac{\pi}{x}\right) & \text{if } \frac{1}{n}<x\le 2.\end{cases}$$

Its derivative

$$f_n'(x)=\begin{cases}0 & \text{if } 0\le x<\frac{1}{n},\\ \sin\left(\dfrac{\pi}{x}\right)-\dfrac{\pi}{x}\,\cos\left(\dfrac{\pi}{x}\right) & \text{if } \frac{1}{n}<x\le 2\end{cases}$$

is bounded on $(0,\frac{1}{n})$ and on $(\frac{1}{n},2)$. Evidently, $\operatorname{var}_0^{1/n} f_n=0$. By Example 2.1.4 (i), $\operatorname{var}_{1/n}^2 f_n<\infty$. Thus, Theorem 2.1.14 yields $\operatorname{var}_0^2 f_n<\infty$ for every $n\in\mathbb{N}$.

The following statement implies that the set $\mathrm{BV}([a,b])$ is closed under pointwise addition of functions and multiplication by scalars.

2.1.16 Lemma. *If $f,g:[a,b]\to\mathbb{R}$ are arbitrary functions and $c\in\mathbb{R}$, then*

$$\operatorname{var}_a^b(f+g)\le\operatorname{var}_a^b f+\operatorname{var}_a^b f, \tag{2.1.10}$$

$$\operatorname{var}_a^b(c\,f)=|c|\,\operatorname{var}_a^b f. \tag{2.1.11}$$

Furthermore, $\operatorname{var}_a^b f=0$ *if and only if f is constant on $[a,b]$.*

Proof. It suffices to notice that

$$V(f+g,\boldsymbol{\alpha})\le V(f,\boldsymbol{\alpha})+V(g,\boldsymbol{\alpha})\quad\text{and}\quad V(c\,f,\boldsymbol{\alpha})=|c|\,V(f,\boldsymbol{\alpha})$$

holds for every division $\boldsymbol{\alpha}$ of $[a,b]$, and that $\operatorname{var}_a^b f=0$ may happen only if $|f(x)-f(a)|=0$ for each $x\in[a,b]$. □

2.1.17 Exercise. Show that the inequality

$$\operatorname{var}_a^b(f+g)\ge|\operatorname{var}_a^b f-\operatorname{var}_a^b g|$$

holds for each pair of functions $f,g\in\mathrm{BV}([a,b])$.

Hint: Notice that $f=(f+g)-g$ and $g=(f+g)-f$, and use (2.1.10).

2.1.18 Remark. The trivial example $f(x)=\cos x$, $g(x)=-\cos x$ with

$$\operatorname{var}_0^{\pi/2} f=\operatorname{var}_0^{\pi/2} g=1\quad\text{and}\quad f+g\equiv 0,$$

shows that in general, the inequality in (2.1.10) cannot be replaced by equality. Nevertheless, the following lemma provides sufficient conditions ensuring that equality holds in (2.1.10).

2.1.19 Lemma. *Let* $f \in \mathrm{BV}([a,b])$ *and* $g \in \mathrm{BV}([a,b])$ *be such that for each* $\varepsilon > 0$ *there are* $n \in \mathbb{N}$ *and* $a_j, b_j \in [a,b]$, $j \in \{1,\dots,n\}$, *satisfying*

$$a \le a_1 \le b_1 \le \cdots \le a_n \le b_n \le b, \tag{2.1.12}$$

$$\sum_{j=1}^{n} \operatorname{var}_{a_j}^{b_j} f > \operatorname{var}_a^b f - \varepsilon, \tag{2.1.13}$$

$$\sum_{j=1}^{n} \operatorname{var}_{a_j}^{b_j} g < \varepsilon. \tag{2.1.14}$$

Then $\operatorname{var}_a^b(f+g) = \operatorname{var}_a^b f + \operatorname{var}_a^b g$.

P r o o f. Let $\varepsilon > 0$ be given and let $\{a_j, b_j\} \subset [a,b]$ with $j \in \{1,\dots,n\}$ be such that relations (2.1.12)–(2.1.14) hold.

Denote $b_0 = a$ and $a_{n+1} = b$. Then

$$\operatorname{var}_a^b f = \sum_{j=1}^{n} \operatorname{var}_{a_j}^{b_j} f + \sum_{j=0}^{n} \operatorname{var}_{b_j}^{a_{j+1}} f.$$

This together with (2.1.13) means that

$$\sum_{j=0}^{n} \operatorname{var}_{b_j}^{a_{j+1}} f < \varepsilon. \tag{2.1.15}$$

Similarly, by (2.1.14),

$$\sum_{j=0}^{n} \operatorname{var}_{b_j}^{a_{j+1}} g > \operatorname{var}_a^b g - \varepsilon. \tag{2.1.16}$$

Now, using (2.1.13)–(2.1.16) and Exercise 2.1.17, we deduce that

$$\begin{aligned}
\operatorname{var}_a^b(f+g) &= \sum_{j=1}^{n} \operatorname{var}_{a_j}^{b_j}(f+g) + \sum_{j=0}^{n} \operatorname{var}_{b_j}^{a_{j+1}}(f+g) \\
&\ge \sum_{j=1}^{n} (\operatorname{var}_{a_j}^{b_j} f - \operatorname{var}_{a_j}^{b_j} g) + \sum_{j=0}^{n} (\operatorname{var}_{b_j}^{a_{j+1}} g - \operatorname{var}_{b_j}^{a_{j+1}} f) \\
&> \operatorname{var}_a^b f - 2\,\varepsilon + \operatorname{var}_a^b g - 2\,\varepsilon = \operatorname{var}_a^b f + \operatorname{var}_a^b g - 4\,\varepsilon.
\end{aligned}$$

Since $\varepsilon > 0$ can be arbitrarily small, we get the inequality

$$\operatorname{var}_a^b(f+g) \geq \operatorname{var}_a^b f + \operatorname{var}_a^b g,$$

which in combination with (2.1.10) completes the proof. □

2.1.20 Remark. Important examples of functions f, g satisfying the assumptions of Lemma 2.1.19 will be provided in Theorems 2.5.6 and 3.3.5.

2.1.21 Theorem (JORDAN). *A function $f : [a,b] \to \mathbb{R}$ has bounded variation if and only if there exist nondecreasing functions $f_1, f_2 : [a,b] \to \mathbb{R}$ such that $f = f_1 - f_2$.*

Proof. If f_1 and f_2 are nondecreasing and $f = f_1 - f_2$, then, by Theorem 2.1.11, both f_1 and f_2 have bounded variation. Hence, by (2.1.10), we have $\operatorname{var}_a^b f < \infty$.

Conversely, assume that $f \in \mathrm{BV}([a,b])$, and define

$$f_1(x) = \operatorname{var}_a^x f \quad \text{and} \quad f_2(x) = f_1(x) - f(x) \quad \text{for } x \in [a,b].$$

Let $x, y \in [a,b]$ and $y \geq x$. Then, by Theorem 2.1.14, $f_1(y) = f_1(x) + \operatorname{var}_x^y f$, and since the variation is always nonnegative, it follows that f_1 is nondecreasing on $[a,b]$. Furthermore, by Theorem 2.1.14 we have

$$f_2(y) = f_1(x) + \operatorname{var}_x^y f - f(y)$$

and

$$f_2(y) - f_2(x) = \operatorname{var}_x^y f - (f(y) - f(x)) \geq 0$$

(see Exercise 2.1.8 (i)). This means that the function f_2 is also nondecreasing and the proof is complete. □

2.1.22 Exercise. Let $f \in \mathrm{BV}([a,b])$. Prove that the functions

$$\mathfrak{p}(x) = \begin{cases} 0 & \text{if } x = a, \\ \sup\limits_{\boldsymbol{\alpha} \in \mathscr{D}[a,x]} \sum\limits_{j=1}^{\nu(\boldsymbol{\alpha})} \big(f(\alpha_j) - f(\alpha_{j-1})\big)^+ & \text{if } x \in (a,b] \end{cases}$$

and

$$\mathfrak{n}(x) = \begin{cases} 0 & \text{if } x = a, \\ \sup\limits_{\boldsymbol{\alpha} \in \mathscr{D}[a,x]} \sum\limits_{j=1}^{\nu(\boldsymbol{\alpha})} \big(f(\alpha_j) - f(\alpha_{j-1})\big)^- & \text{if } x \in (a,b] \end{cases}$$

are nondecreasing and nonnegative on $[a,b]$, and that the relations

$$f(x)=f(a)+\mathfrak{p}(x)-\mathfrak{n}(x) \quad \text{and} \quad \operatorname{var}_a^x f=\mathfrak{p}(x)+\mathfrak{n}(x)$$

hold for all $x\in[a,b]$.

2.1.23 Corollary. *For each function $f\in \mathrm{BV}([a,b])$ there exist finite limits*

$$f(t+)=\lim_{\tau\to t+} f(\tau) \quad \textit{for } t\in[a,b), \quad \textit{and} \quad f(s-)=\lim_{\tau\to s-} f(\tau) \quad \textit{for } s\in(a,b].$$

Proof. By Theorem 2.1.21, we can assume that f is nondecreasing on $[a,b]$. It follows that f is bounded. We claim that

$$f(t+)=\inf_{x\in(t,b]} f(x) \quad \text{if} \quad t\in[a,b).$$

Indeed, let $d=\inf_{x\in(t,b]} f(x)$ and choose an arbitrary $\varepsilon>0$. Then, by the definition of the infimum, there is a $t'\in(t,b]$ such that $d\le f(t')<d+\varepsilon$. By the monotonicity of the function f, it follows that $d\le f(x)<d+\varepsilon$ for every $x\in(t,t']$, which proves the claim.

In a similar way, one can show that

$$f(s-)=\sup_{x\in[a,s)} f(x) \quad \text{if} \quad s\in(a,b],$$

which completes the proof. □

2.2 Space of functions of bounded variation

According to Lemma 2.1.16, the set $\mathrm{BV}([a,b])$ is a linear space with respect to pointwise addition and multiplication by scalars. We will show that, with a suitably chosen norm, $\mathrm{BV}([a,b])$ becomes a normed linear space.

2.2.1 Theorem. $\mathrm{BV}([a,b])$ *is a normed linear space with respect to the norm defined by*

$$\|f\|_{\mathrm{BV}}=|f(a)|+\operatorname{var}_a^b f \quad \textit{for } f\in\mathrm{BV}([a,b]). \tag{2.2.1}$$

Proof. It suffices to verify that (2.2.1) defines a norm. By Lemma 2.1.16, the relations

$$\|f+g\|_{\mathrm{BV}}\le\|f\|_{\mathrm{BV}}+\|g\|_{\mathrm{BV}} \quad \text{and} \quad \|c\,f\|_{\mathrm{BV}}=|c|\,\|f\|_{\mathrm{BV}}$$

hold for all $f, g \in \mathrm{BV}([a,b])$ and every $c \in \mathbb{R}$. Finally, if $\|f\|_{\mathrm{BV}} = 0$, then $f(a) = 0$ and $\mathrm{var}_a^b f = 0$. Hence, by Lemma 2.1.16, $f(x) - f(a) = 0$ on $[a,b]$, i.e., f is the zero function. Consequently, the relation (2.2.1) defines a norm on $\mathrm{BV}([a,b])$. □

We now prove that $\mathrm{BV}([a,b])$ with respect to the above-mentioned norm (2.2.1) is a complete space.

2.2.2 Theorem. $\mathrm{BV}([a,b])$ *is a Banach space.*

Proof. Let $\{f_n\} \subset \mathrm{BV}([a,b])$ be a Cauchy sequence, i.e., assume that

$$\left.\begin{array}{l}\text{for each } \varepsilon > 0 \text{ there is } n_\varepsilon \in \mathbb{N} \text{ such that}\\ |f_n(x) - f_m(x)| \le \|f_n - f_m\|_{\mathrm{BV}} < \varepsilon \text{ for } x \in [a,b] \text{ and } n, m \ge n_\varepsilon.\end{array}\right\} \quad (2.2.2)$$

We will show that $\{f_n\}$ has a limit in $\mathrm{BV}([a,b])$.

By (2.2.2), the sequence $\{f_n(x)\}$ is a Cauchy sequence of real numbers for every $x \in [a,b]$. Hence, for any $x \in [a,b]$ it has a finite limit

$$\lim_{n\to\infty} f_n(x) = f(x).$$

By (2.2.2), there is $n_1 \in \mathbb{N}$ such that

$$\mathrm{var}_a^b f_n \le \|f_n\|_{\mathrm{BV}} \le \|f_{n_1}\|_{\mathrm{BV}} + 1 \quad \text{for } n \ge n_1.$$

Consequently, the real sequence $\{\mathrm{var}_a^b f_n\}$ is bounded from above by a certain $d \in [0, \infty)$. As a result, we have

$$V(f, \boldsymbol{\alpha}) = \lim_{n\to\infty} V(f_n, \boldsymbol{\alpha}) \le d \quad \text{for all } \boldsymbol{\alpha} \in \mathscr{D}[a,b],$$

which implies that $f \in \mathrm{BV}([a,b])$. It remains to show that

$$\lim_{n\to\infty} \|f - f_n\|_{\mathrm{BV}} = 0. \tag{2.2.3}$$

Let an arbitrary $\varepsilon > 0$ be given. By (2.2.2), there exists an $n_\varepsilon \in \mathbb{N}$ such that

$$V(f_n - f_m, \boldsymbol{\alpha}) \le \mathrm{var}_a^b(f_n - f_m) < \varepsilon \text{ for all } \; n, m \ge n_\varepsilon \; \text{ and } \; \boldsymbol{\alpha} \in \mathscr{D}[a,b],$$

wherefrom, by letting $m \to \infty$, we deduce that

$$V(f_n - f, \boldsymbol{\alpha}) = \lim_{m\to\infty} V(f_n - f_m, \boldsymbol{\alpha}) \le \varepsilon \; \text{ for all } n \ge n_\varepsilon \; \text{ and } \boldsymbol{\alpha} \in \mathscr{D}[a,b],$$

and consequently $\lim_{n\to\infty} \mathrm{var}_a^b(f_n - f) = 0$. This fact and the convergence $f_n(a) \to f(a)$ imply that (2.2.3) holds. □

2.3 BOUNDED VARIATION AND CONTINUITY

2.3.1 Definition. Given a function $f:[a,b]\to\mathbb{R}$, we say that $x\in[a,b]$ is a *point of discontinuity of the first kind* if f is discontinuous at x, but the one-sided limits of f at x exist and are finite.

By Corollary 2.1.23, functions of bounded variation can only have discontinuities of the first kind. This section provides a closer look on the properties of bounded variation functions related to continuity.

2.3.2 Theorem. *Every function $f:[a,b]\to\mathbb{R}$ with bounded variation has at most countably many discontinuities.*

Proof. By virtue of Theorem 2.1.21, it is enough to prove the statement for the case when f is a nondecreasing function. Let D be the set of all discontinuity points of f. For each $x\in D$, choose an arbitrary rational $q(x)$ in the interval $(f(x-),f(x+))$. Since f is nondecreasing, it follows that $q(x)\neq q(y)$ whenever $x,y\in D$ and $x\neq y$. Hence, the mapping q provides a one-to-one correspondence between D and a subset of rational numbers. This proves that D is at most countable. □

Given a function $f\in\mathrm{BV}([a,b])$, it is clear that $x\mapsto\operatorname{var}_a^x f$ is a nondecreasing function (cf. the proof of Theorem 2.1.21). We will now show that this function inherits the continuity properties of the function f.

2.3.3 Lemma. *Let $f\in\mathrm{BV}([a,b])$ and $v(x)=\operatorname{var}_a^x f$ for $x\in[a,b]$. Then*

$$\Delta^- v(x)=|\Delta^- f(x)| \quad \textit{for} \quad x\in(a,b], \tag{2.3.1}$$

$$\Delta^+ v(x)=|\Delta^+ f(x)| \quad \textit{for} \quad x\in[a,b). \tag{2.3.2}$$

Proof. a) If $x\in(a,b]$, then

$$v(x)-v(s)=\operatorname{var}_s^x f\geq|f(x)-f(s)| \quad \text{holds for every} \quad s\in[a,x].$$

For $s\to x-$, we get the inequality

$$\Delta^- v(x)\geq|\Delta^- f(x)| \quad \text{for} \quad x\in(a,b]. \tag{2.3.3}$$

Let $\varepsilon>0$ be given. Choose a $\delta>0$ such that

$$|f(x-)-f(s)|<\frac{\varepsilon}{2} \quad \text{for} \quad s\in(x-\delta,x). \tag{2.3.4}$$

Furthermore, choose a division $\boldsymbol{\alpha}=\{\alpha_0,\alpha_1,\dots,\alpha_m\}$ of $[a,x]$ such that

$$\alpha_{m-1}\ge x-\tfrac{\delta}{2} \quad\text{and}\quad v(x)-V(f,\boldsymbol{\alpha})<\frac{\varepsilon}{2}. \tag{2.3.5}$$

Then, by (2.3.5) and (2.3.4), we have

$$\begin{aligned}
v(x)-\sum_{j=1}^{m-1}|f(\alpha_j)-f(\alpha_{j-1})| &< V(f,\boldsymbol{\alpha})+\frac{\varepsilon}{2}-\sum_{j=1}^{m-1}|f(\alpha_j)-f(\alpha_{j-1})|\\
&=|f(x)-f(\alpha_{m-1})|+\frac{\varepsilon}{2}\le|\Delta^-f(x)|+|f(x-)-f(\alpha_{m-1})|+\frac{\varepsilon}{2}\\
&\le|\Delta^-f(x)|+\varepsilon.
\end{aligned}$$

Moreover, since

$$\sum_{j=1}^{m-1}|f(\alpha_j)-f(\alpha_{j-1})|\le v(\alpha_{m-1})$$

and $v(t)\ge v(\alpha_{m-1})$ for every $t\in(\alpha_{m-1},x)$, it follows that

$$\begin{aligned}
v(x)-v(t)\le v(x)-v(\alpha_{m-1})&\le v(x)-\sum_{j=1}^{m-1}|f(\alpha_j)-f(\alpha_{j-1})|\\
&\le|\Delta^-f(x)|+\varepsilon
\end{aligned}$$

for every $t\in(\alpha_{m-1},x)$ and $\varepsilon>0$. Letting $t\to x-$ and recalling that $\varepsilon>0$ can be arbitrary, we get the inequality $\Delta^-v(x)\le|\Delta^-f(x)|$, which together with (2.3.3) proves the equality (2.3.1).

b) The second relation (2.3.2) can be proved by a symmetry argument based on reflecting the graph of f about the vertical axis: Let $\widetilde{f}:[-b,-a]\to\mathbb{R}$ be given by $\widetilde{f}(x)=f(-x)$, $x\in[-b,-a]$. Observe that $\operatorname{var}_c^d f=\operatorname{var}_{-d}^{-c}\widetilde{f}$ whenever $[c,d]\subset[a,b]$, and $\Delta^+f(x)=-\Delta^-\widetilde{f}(-x)$ for each $x\in[a,b)$. Therefore,

$$\begin{aligned}
\Delta^+v(x)&=\lim_{\delta\to0+}\Big(\operatorname{var}_a^{x+\delta}f-\operatorname{var}_a^x f\Big)=\lim_{\delta\to0+}\operatorname{var}_x^{x+\delta}f\\
&=\lim_{\delta\to0+}\operatorname{var}_{-x-\delta}^{-x}\widetilde{f}=\lim_{\delta\to0+}\Big(\operatorname{var}_{-b}^{-x}\widetilde{f}-\operatorname{var}_{-b}^{-x-\delta}\widetilde{f}\Big)\\
&=|\Delta^-\widetilde{f}(-x)|=|\Delta^+f(x)|,
\end{aligned}$$

where the first equality on the last line follows from part a) applied to the function $\widetilde{f}$ at the point $-x$. □

2.3.4 Corollary. *Let $f\in\mathrm{BV}([a,b])$ and $v(x)=\operatorname{var}_a^x f$ for $x\in[a,b]$. Then f is right-continuous at a point $x\in[a,b)$ if and only if v is right-continuous* *at this point. Similarly, f is left-continuous at a point $x \in (a,b]$ if and only if v is left-continuous at this point.*

Our next goal is to show that the sum of absolute values of the jumps of a bounded variation function is always finite. For the proof, we need the following statement.

2.3.5 Lemma. *If $f:[a,b] \to \mathbb{R}$ has bounded variation and the functions $\mathfrak{p}$ and $\mathfrak{n}$ are defined as in Exercise* 2.1.22, *then*

$$\Delta^-\mathfrak{p}(x) = (\Delta^- f(x))^+, \ \Delta^-\mathfrak{n}(x) = (\Delta^- f(x))^- \quad \textit{for} \ \ x \in (a,b],$$
$$\Delta^+\mathfrak{p}(x) = (\Delta^+ f(x))^+, \ \Delta^+\mathfrak{n}(x) = (\Delta^+ f(x))^- \quad \textit{for} \ \ x \in [a,b).$$

The proof can be carried out analogously as the proof of Lemma 2.3.3; it suffices to work with

$$(f(\alpha_j) - f(\alpha_{j-1}))^+ \text{ or } \ (f(\alpha_j) - f(\alpha_{j-1}))^-$$

instead of $|f(\alpha_j) - f(\alpha_{j-1})|$. A detailed proof is left as an exercise for the reader.

2.3.6 Theorem. *Let $f \in \mathrm{BV}([a,b])$ and let $D = \{s_k\}$ be a non-repeating sequence (i.e., $s_k \neq s_\ell$ whenever $k \neq \ell$) of points from the open interval (a,b). Then*

$$|\Delta^+ f(a)| + \sum_{k=1}^{\infty} \Big(|\Delta^+ f(s_k)| + |\Delta^- f(s_k)|\Big) + |\Delta^- f(b)| \leq \operatorname{var}_a^b f. \qquad (2.3.6)$$

Proof. a) First, assume that f is nondecreasing. Then

$$|\Delta^+ f(a)| + \sum_{k=1}^{\infty} \Big(|\Delta^+ f(s_k)| + |\Delta^- f(s_k)|\Big) + |\Delta^- f(b)|$$
$$= \Delta^+ f(a) + \sum_{k=1}^{\infty} \Delta\, f(s_k) + \Delta^- f(b).$$

Choose an arbitrary $n \in \mathbb{N}$. Denote $\alpha_0 = a$, $\alpha_k = s_k$ for $k \in \{1, \dots, n\}$, and

$\alpha_{n+1} = b$. Then

$$\begin{aligned}
\Delta^+ f(a) + \sum_{k=1}^{n} \Delta f(s_k) + \Delta^- f(b) \\
&= \Delta^+ f(\alpha_0) + \sum_{k=1}^{n} \Big(f(\alpha_k+) - f(\alpha_k-) \Big) + \Delta^- f(\alpha_{n+1}) \\
&= -f(\alpha_0) + \sum_{k=0}^{n} \Big(f(\alpha_k+) - f(\alpha_{k+1}-) \Big) + f(\alpha_{n+1}) \\
&\le -f(\alpha_0) + f(\alpha_{n+1}) = f(b) - f(a) = \operatorname{var}_a^b f,
\end{aligned}$$

where the last inequality follows from the fact that f is nondecreasing. Passing to the limit $n \to \infty$, we obtain (2.3.6).

b) Now, let f be an arbitrary function of bounded variation on $[a,b]$ and let the functions $\mathfrak{p}$ and $\mathfrak{n}$ be defined as in Exercise 2.1.22. We know that $f = f(a) + \mathfrak{p} - \mathfrak{n}$ on $[a,b]$. Obviously,

$$\Delta^+ f(t) = \Delta^+ \mathfrak{p}(t) - \Delta^+ \mathfrak{n}(t) \quad \text{and} \quad \Delta^- f(s) = \Delta^- \mathfrak{p}(s) - \Delta^- \mathfrak{n}(s)$$

for $t \in [a,b)$ and $s \in (a,b]$. Finally, using Lemma 2.3.5, we can easily deduce that the relations

$$|\Delta^+ f(t)| = \Delta^+ \mathfrak{p}(t) + \Delta^+ \mathfrak{n}(t) \quad \text{and} \quad |\Delta^- f(s)| = \Delta^- \mathfrak{p}(s) + \Delta^- \mathfrak{n}(s)$$

hold for $t \in [a,b)$, $s \in (a,b]$.

By the first part of the proof we have

$$\Delta^+ \mathfrak{p}(a) + \sum_{k=1}^{\infty} \Big(\Delta^+ \mathfrak{p}(s_k) + \Delta^- \mathfrak{p}(s_k) \Big) + \Delta^- \mathfrak{p}(b) \le \mathfrak{p}(b),$$
$$\Delta^+ \mathfrak{n}(a) + \sum_{k=1}^{\infty} \Big(\Delta^+ \mathfrak{n}(s_k) + \Delta^- \mathfrak{n}(s_k) \Big) + \Delta^- \mathfrak{n}(b) \le \mathfrak{n}(b).$$

By adding these two inequalities, we obtain

$$\begin{aligned}
|\Delta^+ f(a)| + \sum_{k=1}^{\infty} \Big(|\Delta^+ f(s_k)| + |\Delta^- f(s_k)| \Big) + |\Delta^- f(b)| \\
\le \mathfrak{p}(b) + \mathfrak{n}(b) = \operatorname{var}_a^b f,
\end{aligned}$$

and the proof is complete. □

2.3.7 Remark. Let $f:[a,b]\to\mathbb{R}$ have bounded variation and let the set D of its discontinuity points in (a,b) be infinite. By Theorem 2.3.2, the elements of D can be arranged into a sequence $\{s_k\}$. (Naturally, there is an infinite number of such sequences.) By Theorem 2.3.6, the series

$$\sum_{k=1}^{\infty}\Big(|\Delta^+ f(s_k)|+|\Delta^- f(s_k)|\Big)$$

is absolutely convergent and its sum does not depend on the ordering of points in $\{s_k\}$. Since for $x\in(a,b)$ the expression $|\Delta^+ f(x)|+|\Delta^- f(x)|$ is nonzero if and only if $x\in D$, it makes sense to define

$$\sum_{a<x<b}\Big(|\Delta^+ f(x)|+|\Delta^- f(x)|\Big)=\sum_{x\in D}\Big(|\Delta^+ f(x)|+|\Delta^- f(x)|\Big)$$
$$=\sum_{k=1}^{\infty}\Big(|\Delta^+ f(s_k)|+|\Delta^- f(s_k)|\Big),$$

where $\{s_k\}$ is an arbitrary non-repeating sequence of points from (a,b) consisting of all points of D. The symbols

$$\sum_{a\le x<b},\ \sum_{a<x\le b},\ \sum_{a\le x\le b},\quad \text{or}\quad \sum_{x\in[a,b)},\ \sum_{x\in(a,b]},\ \sum_{x\in[a,b]}\quad \text{etc.}$$

should be understood in an analogous way.

2.3.8 Corollary. *Each function* $f\in\mathrm{BV}([a,b])$ *satisfies the inequality*

$$\sum_{x\in[a,b)}|\Delta^+ f(x)|+\sum_{x\in(a,b]}|\Delta^- f(x)|\le \operatorname{var}_a^b f.$$

2.4 DERIVATIVES OF BOUNDED VARIATION FUNCTIONS

In this section we consider differentiability of functions with bounded variation. To this aim, let us recall the concept of the Lebesgue outer measure.

For an arbitrary set $M\subset\mathbb{R}$, we define its *Lebesgue outer measure*

$$\mu^*(M)=\inf\sum_k (b_k-a_k),$$

where the infimum is taken over all at most countable collections $\{(a_k,b_k)\}$

of open intervals in $\mathbb{R}$ such that

$$M \subset \bigcup_k (a_k, b_k).$$

The Lebesgue outer measure is either a nonnegative real number, or ∞. Furthermore, $\mu^*(M_1) \le \mu^*(M_2)$ whenever $M_1 \subset M_2$.

Obviously, for any finite collection $\{I_k\}$ of disjoint open intervals in $\mathbb{R}$, we have

$$\mu^*\Big(\bigcup_k I_k\Big) = \sum_k |I_k|.$$

We say that a certain property holds *almost everywhere* (a.e.) in the interval $[a,b]$ or for almost all $x \in [a,b]$ if it holds for all $x \in [a,b] \setminus N$, where $N \subset [a,b]$ is a set whose Lebesgue outer measure is zero.

If not stated otherwise, in what follows by *outer measure* we always mean the Lebesgue outer measure.

2.4.1 Exercise. Prove the following assertions:

(i) Every countable set $S \subset \mathbb{R}$ has outer measure equal to zero.

(ii) If I is an interval, then $\mu^*(I)$ equals its length $|I|$.

(iii) If $\{M_k\}$ is a countable collection of subsets of $\mathbb{R}$, then

$$\mu^*\Big(\bigcup_k M_k\Big) \le \sum_k \mu^*(M_k).$$

(iv) The union of countably many sets whose outer measure is zero has outer measure equal to zero.

The next theorem is well known and we present it without proof; see the bibliographic remarks in the end of the chapter.

2.4.2 Theorem (LEBESGUE DIFFERENTIATION THEOREM). *Every monotone function $f:[a,b] \to \mathbb{R}$ has a finite derivative $f'(x)$ for almost all $x \in [a,b]$.*

2.4.3 Remark. In particular, by Theorem 2.1.21, each function of bounded variation has bounded derivative almost everywhere on $[a,b]$. Moreover, it is known (see Theorem 3.2.5) that the derivative is Lebesgue integrable. However, the seemingly natural identity

$$f(b) - f(a) = \int_a^b f'(t)\,\mathrm{d}t$$

does not necessarily hold. For example, there exist functions $f \in \mathrm{BV}([a,b])$ which are non-constant and satisfy $f'=0$ a.e. on $[a,b]$. A well-known example of such a function is the Cantor function, which is continuous and nondecreasing.

2.4.4 Definition. A function $f \in \mathrm{BV}([a,b])$ is called *singular* if $f'(x)=0$ for almost all $x \in [a,b]$.

2.5 Step functions

This section is concerned with *finite step functions* and the more general *step functions*. Both types of functions represent simple examples of singular functions.

2.5.1 Definition. A function $f:[a,b]\to\mathbb{R}$ is *a finite step function* if there exists a division $\boldsymbol{\alpha}=\{\alpha_0,\alpha_1,\dots,\alpha_m\}$ of $[a,b]$ such that f is constant on each of the open intervals (α_{j-1},α_j), $j=1,\dots,m$. The set of all finite step functions on the interval $[a,b]$ is denoted by $\mathrm{S}([a,b])$.

By definition, a function $f:[a,b]\to\mathbb{R}$ is a finite step function if and only if there exists an $m\in\mathbb{N}$, numbers $\widetilde{c}_0,\dots,\widetilde{c}_m\in\mathbb{R}$ and $\widetilde{d}_1,\dots,\widetilde{d}_m\in\mathbb{R}$, and a division $a=s_0<s_1<\dots<s_m=b$ of $[a,b]$ such that

$$\begin{aligned} f(s_j)&=\widetilde{c}_j \quad \text{for } j=0,1,\dots,m,\\ f(x)&=\widetilde{d}_j \quad \text{for } x\in(s_{j-1},s_j) \text{ and } j=1,\dots,m. \end{aligned}$$

Thus, we have the representation

$$\begin{aligned} f(x)&=\sum_{j=0}^{m}\widetilde{c}_j\,\chi_{[s_j]}(x)+\sum_{j=1}^{m}\widetilde{d}_j\,\chi_{(s_{j-1},s_j)}(x)\\ &=\widetilde{c}_0\,(\chi_{[a,b]}(x)-\chi_{(a,b]}(x))+\sum_{j=1}^{m-1}\widetilde{c}_j\,(\chi_{[s_j,b]}-\chi_{(s_j,b]}(x))\\ &\qquad+\widetilde{c}_m\,\chi_{[b]}(x)+\sum_{j=1}^{m-1}\widetilde{d}_j\,(\chi_{(s_{j-1},b]}(x)-\chi_{[s_j,b]}(x)) \end{aligned}$$

$$
\begin{aligned}
&+\widetilde{d}_m\left(\chi_{(s_{m-1},b]}(x)-\chi_{[b]}(x)\right)\\
&=\widetilde{c}_0+\sum_{j=0}^{m-1}\left(\widetilde{d}_{j+1}-\widetilde{c}_j\right)\chi_{(s_j,b]}(x)\\
&\quad+\sum_{j=1}^{m-1}\left(\widetilde{c}_j-\widetilde{d}_j\right)\chi_{[s_j,b]}(x)+\left(\widetilde{c}_m-\widetilde{d}_m\right)\chi_{[b]}(x).
\end{aligned}
$$

Equivalently,

$$
f(x)=c+\sum_{j=0}^{m-1}c_j\,\chi_{(s_j,b]}(x)+\sum_{j=1}^{m-1}d_j\,\chi_{[s_j,b]}(x)+d\,\chi_{[b]}(x), \tag{2.5.1}
$$

where $c=\widetilde{c}_0$, $d=\widetilde{c}_m-\widetilde{d}_m=\Delta^- f(b)$,

$$
\begin{aligned}
c_j&=\widetilde{d}_{j+1}-\widetilde{c}_j=\Delta^+ f(s_j) \quad \text{for } j=0,1,\dots,m-1,\\
d_j&=\widetilde{c}_j-\widetilde{d}_j=\Delta^- f(s_j) \qquad \text{for } j=1,\dots,m-1.
\end{aligned}
$$

Obviously, f is continuous at all points $[a,b]\setminus\{s_0,\dots,s_m\}$.

A more general concept is that of a *step function*, also known as a *jump function.*

2.5.2 Definition. A function $f:[a,b]\to\mathbb{R}$ is *a step function* if either f is a finite step function or there exist $c,c_0,d\in\mathbb{R}$, a non-repeating sequence $\{s_k\}\subset(a,b)$, and sequences $\{c_k\}\subset\mathbb{R}$ and $\{d_k\}\subset\mathbb{R}$ such that

$$
\sum_{k=1}^{\infty}\left(|c_k|+|d_k|\right)<\infty \tag{2.5.2}
$$

and

$$
f(x)=c+c_0\,\chi_{(a,b]}(x)+\sum_{k=1}^{\infty}\Big(c_k\,\chi_{(s_k,b]}(x)+d_k\,\chi_{[s_k,b]}(x)\Big)+d\,\chi_{[b]}(x) \tag{2.5.3}
$$

for all $x\in[a,b]$. The set of all step functions on the interval $[a,b]$ is denoted by $\mathrm{B}([a,b])$.

If $f\in\mathrm{B}([a,b])$ is not a finite step function, then the sequence $\{s_k\}$ from Definition 2.5.2 is infinite and, in general, it is not possible to reorder it into an increasing sequence. However, thanks to condition (2.5.2) we have

$$
\sum_{k=1}^{\infty}\left|c_k\,\chi_{(s_k,b]}(x)+d_k\,\chi_{[s_k,b]}(x)\right|\le\sum_{k=1}^{\infty}\left(|c_k|+|d_k|\right)<\infty. \tag{2.5.4}
$$

This means that the series on the right-hand side of (2.5.3) is absolutely convergent for each $x \in [a,b]$. Hence, the values $f(x)$ do not depend on the particular ordering of the sequence $\{s_k\}$. Consequently, for each $x \in [a,b]$ the relation (2.5.3) can be equivalently rewritten as

$$f(x)=\begin{cases} c & \text{if } x=a, \\ c+c_0+\sum\limits_{a<s_k<x} c_k+\sum\limits_{a<s_k\le x} d_k & \text{if } x\in(a,b), \\ c+c_0+\sum\limits_{a<s_k<b} c_k+\sum\limits_{a<s_k<b} d_k+d & \text{if } x=b, \end{cases} \tag{2.5.5}$$

where the sum $\sum_{a<s_k<x} c_k$ runs over all indices $k \in \mathbb{N}$ such that $s_k \in (a,x)$, the sum $\sum_{a<s_k\le x} c_k$ runs over all indices $k\in\mathbb{N}$ such that $s_k \in (a,x]$, and analogously in the case of $\sum_{a<s_k<b} d_k$.

Clearly, the function $f \in \mathrm{B}([a,b])$ defined by (2.5.3) satisfies $f(a)=c$. Since

$$\lim_{t\to a+}\Big(\sum_{a<s_k<t} c_k+\sum_{a<s_k\le t} d_k\Big)$$

is the limit of the remainder of an absolutely convergent series, it must be zero. Thus,

$$f(a+)=c+c_0.$$

A similar argument can be used to prove the following formulas:

$$\left.\begin{aligned} f(x-)&=c+c_0+\sum_{a<s_k<x} c_k+\sum_{a<s_k<x} d_k \quad \text{if} \quad x\in(a,b], \\ f(x+)&=c+c_0+\sum_{a<s_k\le x} c_k+\sum_{a<s_k\le x} d_k \quad \text{if} \quad x\in[a,b). \end{aligned}\right\} \tag{2.5.6}$$

Subtracting (2.5.5) from (2.5.6) leads to

$$f(x-)=f(x)=f(x+) \quad \text{for } x\in(a,b)\setminus\{s_k\}$$

and

$$\left.\begin{aligned} \Delta^+ f(s_k)&=c_k \quad \text{for } k\in\mathbb{N}, \quad \Delta^+ f(a)=c_0, \\ \Delta^- f(s_k)&=d_k \quad \text{for } k\in\mathbb{N}, \quad \Delta^- f(b)=d. \end{aligned}\right\} \tag{2.5.7}$$

Thus, the relation (2.5.3) from Definition 2.5.2 can be reformulated in further two equivalent ways:

$$f(x)=\begin{cases} c & \text{if } x=a,\\ c+\Delta^+ f(a)+\sum\limits_{a<s_k<x}\Delta^+ f(s_k)+\sum\limits_{a<s_k\le x}\Delta^- f(s_k) & \text{if } x\in(a,b),\\ c+\Delta^+ f(a)+\sum\limits_{a<s_k<b}\Delta^+ f(s_k)+\sum\limits_{a<s_k<b}\Delta^- f(s_k)+\Delta^- f(b) & \text{if } x=b,\end{cases}$$

or

$$\begin{aligned} f(x)=f(a)+\sum_{d\in D}\big(\Delta^+ f(d)\,\chi_{(d,b]}(x)+\Delta^- f(d)\,\chi_{[d,b]}(x)\big)&\\ +\Delta^+ f(a)\,\chi_{(a,b]}(x)+\Delta^- f(b)\,\chi_{[b]}(x)\quad \text{for } x\in[a,b],& \end{aligned}$$

where $D=\{s_k : k\in\mathbb{N}\}$. Recall that we assume that $D\subset(a,b)$. Hence, D is the set of all discontinuity points of f in (a,b). (The points a, b might also be discontinuity points of f.)

2.5.3 Theorem. $\mathrm{S}([a,b])\subset\mathrm{B}([a,b])\subset\mathrm{BV}([a,b])$ *and the inequality*

$$\operatorname{var}_a^b f=|\Delta^+ f(a)|+\sum_{x\in(a,b)}\Big(|\Delta^+ f(x)|+|\Delta^- f(x)|\Big)+|\Delta^- f(b)|<\infty \tag{2.5.8}$$

holds for each step function $f\in\mathrm{B}([a,b])$.

P r o o f. By Definition 2.5.1, we have

$$\mathrm{S}([a,b])\subset\mathrm{B}([a,b])\quad\text{and}\quad \mathrm{S}([a,b])\subset\mathrm{BV}([a,b]).$$

Obviously, (2.5.8) holds for all finite step functions $f\in\mathrm{S}([a,b])$. Thus, we can restrict ourselves to the case that $f\in\mathrm{B}([a,b])\setminus\mathrm{S}([a,b])$, i.e., we may assume that f is given by (2.5.5), and (2.5.2) holds. Using (2.5.7) we get

$$\left.\begin{aligned} &|\Delta^+ f(a)|+\sum_{a<x<b}\Big(|\Delta^+ f(x)|+|\Delta^- f(x)|\Big)+|\Delta^- f(b)|\\ &\qquad=|c_0|+\sum_{k=1}^{\infty}\Big(|c_k|+|d_k|\Big)+|d|<\infty. \end{aligned}\right\} \tag{2.5.9}$$

Notice that

$$|f(y)-f(x)|\le\sum_{x\le s_k<y}|c_k|+\sum_{x<s_k\le y}|d_k|$$

holds for arbitrary $x,\, y \in (a,b)$ such that $x<y$. Furthermore,

$$|f(y)-f(a)| \le |c_0| + \sum_{a<s_k<y} |c_k| + \sum_{a<s_k\le y} |d_k| \quad \text{if } a<y<b,$$
$$|f(b)-f(x)| \le \sum_{x\le s_k<b} |c_k| + \sum_{x<s_k<b} |d_k| + |d| \quad \text{if } a<x<b.$$

Hence, for any division $\boldsymbol{\alpha}$ of $[a,b]$ with $\nu(\boldsymbol{\alpha})\ge 3$ we obtain

$$\begin{aligned} V(f,\boldsymbol{\alpha}) &= \sum_{j=1}^{\nu(\boldsymbol{\alpha})} |f(\alpha_j)-f(\alpha_{j-1})| \\ &\le |c_0| + \sum_{a<s_k<\alpha_1} |c_k| + \sum_{a<s_k\le\alpha_1} |d_k| + \sum_{j=2}^{\nu(\boldsymbol{\alpha})-1} \Big(\sum_{\alpha_{j-1}\le s_k<\alpha_j} |c_k| + \sum_{\alpha_{j-1}<s_k\le\alpha_j} |d_k| \Big) \\ &\quad + \sum_{\alpha_{\nu(\boldsymbol{\alpha})-1}\le s_k<b} |c_k| + \sum_{\alpha_{\nu(\boldsymbol{\alpha})-1}<s_k<b} |d_k| + |d| \le |c_0| + \sum_{k=1}^{\infty} \big(|c_k|+|d_k|\big) + |d|. \end{aligned}$$

Consequently, by (2.5.2) we have

$$\operatorname{var}_a^b f \le |c_0| + \sum_{k=1}^{\infty} \big(|c_k|+|d_k|\big) + |d| < \infty, \tag{2.5.10}$$

i.e., $f\in \mathrm{BV}([a,b])$ and $\mathrm{B}([a,b]) \subset \mathrm{BV}([a,b])$. Finally, Theorem 2.3.6 yields

$$|\Delta^+ f(a)| + \sum_{x\in(a,b)} \big(|\Delta^+ f(x)| + |\Delta^- f(x)|\big) + |\Delta^- f(b)| \le \operatorname{var}_a^b f. \tag{2.5.11}$$

By combining (2.5.9), (2.5.10) and (2.5.11), we obtain (2.5.8). □

Obviously, if f is a finite step function on $[a,b]$, then $f'=0$ on $[a,b]\setminus M$, where $M\subset[a,b]$ is a finite (possibly empty) set. Thus, finite step functions are singular. We will show that every step function is also singular. For this purpose, we need the following statement known as the *little Fubini theorem.*

2.5.4 Theorem. *Let $\{f_k\}$ be a sequence of nondecreasing functions on $[a,b]$ such that the series $f(x)=\sum_{k=1}^{\infty} f_k(x)$ converges for every $x\in[a,b]$. Then $f'(x)=\sum_{k=1}^{\infty} f_k'(x)\in\mathbb{R}$ for almost all $x\in[a,b]$.*

Proof. Denote

$$g(x)=f(x)-f(a), \quad g_k(x)=f_k(x)-f_k(a) \quad \text{for } k\in\mathbb{N},\ x\in[a,b],$$

$$s_n(x)=\sum_{k=1}^{n} g_k(x) \quad \text{for } n\in\mathbb{N},\ x\in[a,b].$$

The functions g, g_k, $k\in\mathbb{N}$, are nonnegative and nondecreasing on $[a,b]$, and

$$g(x)=\sum_{k=1}^{\infty} g_k(x)=\lim_{n\to\infty} s_n(x) \quad \text{for} \quad x\in[a,b].$$

By Theorem 2.4.2, for every $k\in\mathbb{N}$ there exists a set $D_k\subset[a,b]$ of zero outer measure such that the function g_k has a finite derivative $g_k'(x)$ for every $x\in[a,b]\setminus D_k$. Similarly, there exists a finite derivative $g'(x)$ for every $x\in[a,b]\setminus D$, where $D\subset[a,b]$ has zero outer measure. Thus, if we let $\widetilde{D}=D\cup\bigcup_{k=1}^{\infty} D_k$, we conclude that there exist finite derivatives $g'(x)$, $g_k'(x)$ for all $k\in\mathbb{N}$ and $x\in[a,b]\setminus\widetilde{D}$. By Exercise 2.4.1 (iv) the set $\widetilde{D}$ also has zero outer measure.

For any $x\in[a,b]\setminus\widetilde{D}$ and $\xi\in[a,b]$ such that $\xi\neq x$, we have

$$\sum_{k=1}^{\infty}\frac{g_k(\xi)-g_k(x)}{\xi-x}=\frac{g(\xi)-g(x)}{\xi-x}.$$

Since every term in the sum on the left-hand side is nonnegative (because g_k is nondecreasing), it follows that

$$\frac{s_n(\xi)-s_n(x)}{\xi-x}=\sum_{k=1}^{n}\frac{g_k(\xi)-g_k(x)}{\xi-x}\le\frac{g(\xi)-g(x)}{\xi-x}$$

holds for any $x\in[a,b]$, $\xi\in[a,b]\setminus\{x\}$ and $n\in\mathbb{N}$. Letting $\xi\to x$ we get

$$s_n'(x)=\sum_{k=1}^{n} g_k'(x)\le g'(x) \quad \text{for} \quad x\in[a,b]\setminus\widetilde{D} \quad \text{and} \quad n\in\mathbb{N}.$$

Since $g_k'(x)\ge 0$ for $x\in[a,b]\setminus\widetilde{D}$ and $k\in\mathbb{N}$, the sequence $\{s_n'(x)\}$ is bounded and nondecreasing for each $x\in[a,b]\setminus\widetilde{D}$. Thus, for every $x\in[a,b]\setminus\widetilde{D}$ there exists a finite limit

$$\lim_{n\to\infty} s_n'(x)=\sum_{k=1}^{\infty} g_k'(x)\le g'(x), \tag{2.5.12}$$

i.e., the series $\sum_{k=1}^{\infty} g_k'(x)$ converges for almost all $x \in [a,b]$.

On the other hand, for every $\ell \in \mathbb{N}$ there exists n_ℓ such that

$$0 \le g(b) - s_{n_\ell}(b) < \frac{1}{2^\ell}.$$

Since both g and s_{n_ℓ} are nondecreasing on $[a,b]$, it follows that

$$0 \le g(x) - s_{n_\ell}(x) = \sum_{k=n_\ell+1}^{\infty} g_k(x) \le \sum_{k=n_\ell+1}^{\infty} g_k(b) = g(b) - s_{n_\ell}(b) < \frac{1}{2^\ell}$$

and hence also

$$0 \le \sum_{\ell=1}^{\infty} \big(g(x) - s_{n_\ell}(x)\big) \le \sum_{\ell=1}^{\infty} \frac{1}{2^\ell} = 1 \quad \text{for } x \in [a,b].$$

Repeating the considerations above with $\{g_k\}$ replaced by $\{g(x) - s_{n_\ell}(x)\}$ we deduce that the series $\sum_{\ell=1}^{\infty} \big(g'(x) - s_{n_\ell}'(x)\big)$ is convergent for almost all $x \in [a,b]$. In particular,

$$\lim_{\ell\to\infty} \big(g'(x) - s_{n_\ell}'(x))\big) = 0 \quad \text{for almost all } x \in [a,b].$$

Naturally, this could not be true if the inequality in (2.5.12) was strict. Consequently,

$$\begin{aligned} f'(x) &= (f(x) - f(a))' = g'(x) = \sum_{k=1}^{\infty} g_k'(x) \\ &= \sum_{k=1}^{\infty} \big(f_k(x) - f(a)\big)' = \sum_{k=1}^{\infty} f_k'(x) \quad \text{for almost all } x \in [a,b]. \end{aligned}$$

This completes the proof. □

2.5.5 Theorem. *Every step function on* $[a,b]$ *is singular.*

Proof. Let f be a step function on $[a,b]$. If f is a finite step function, the statement of the theorem is obvious. Otherwise, f has the form (2.5.3) and (2.5.2) holds. Define

$$v_a(x) = \begin{cases} 0 & \text{if } x = a, \\ |c_0| & \text{if } x > a, \end{cases} \qquad v_b(x) = \begin{cases} 0 & \text{if } x < b, \\ |d| & \text{if } x = b, \end{cases}$$

and

$$v_k(x) = \begin{cases} 0 & \text{if } \ x < s_k, \\ |d_k| & \text{if } \ x = s_k, \\ |c_k| + |d_k| & \text{if } \ x > s_k \end{cases}$$

for all $k \in \mathbb{N}$. All these functions are nondecreasing on $[a,b]$,

$$v_a'(x) = 0 \ \text{ for } \ x \neq a, \quad v_b'(x) = 0 \ \text{ for } \ x \neq b,$$

$$v_k'(x) = 0 \ \text{ for } \ k \in \mathbb{N} \ \text{ and } \ x \neq s_k.$$

Moreover, by (2.5.2) we have

$$\sum_{k=1}^{\infty} |v_k(x)| \le \sum_{k=1}^{\infty} \big(|c_k| + |d_k|\big) < \infty \quad \text{for } x \in [a,b].$$

Thus, the series $\sum_{k=1}^{\infty} v_k(x)$ is absolutely convergent for each $x \in [a,b]$ and the function

$$v(x) = v_a(x) + \sum_{k=1}^{\infty} v_k(x) + v_b(x)$$

is well defined for each $x \in [a,b]$. In view of Theorem 2.5.4, we have

$$v'(x) = v_a'(x) + \sum_{k=1}^{\infty} v_k'(x) + v_b'(x) = 0 \quad \text{for almost all } x \in [a,b].$$

Now, since

$$\left| \frac{f(x) - f(y)}{x - y} \right| \le \left| \frac{v(x) - v(y)}{x - y} \right|$$

holds for all $x, y \in [a,b]$ such that $x \neq y$, it follows easily that $f'(x) = 0$ for almost all $x \in [a,b]$. □

The next assertion is interesting in the context of Remark 2.1.18 and Lemma 2.1.19.

2.5.6 Theorem. *If f is a step function on $[a,b]$ and $g \in \mathrm{BV}([a,b])$ is continuous, then* $\operatorname{var}_a^b(f+g) = \operatorname{var}_a^b f + \operatorname{var}_a^b g$.

Proof. We will verify that the assumptions of Lemma 2.1.19 are satisfied.

Let f be given as in Definition 2.5.2, where $K = \mathbb{N}$ and $D = \{s_k\}$ is the set of discontinuity points of f in (a,b). By Theorem 2.5.3, we have

$$|\Delta^+ f(a)| + \sum_{k=1}^{\infty} \Big(|\Delta^- f(s_k)| + |\Delta^+ f(s_k)| \Big) + |\Delta^- f(b)| = \operatorname{var}_a^b f < \infty.$$

Let $\varepsilon > 0$ be given and let $n \in \mathbb{N}$ be such that

$$|\Delta^+ f(a)| + \sum_{k=1}^{n} \Big(|\Delta^- f(s_k)| + |\Delta^+ f(s_k)| \Big) + |\Delta^- f(b)| > \operatorname{var}_a^b f - \frac{\varepsilon}{2}.$$

Let $x_0, x_1, \ldots, x_n, x_{n+1}$ be such that

$$\{s_k\}_{k=1}^n = \{x_k\}_{k=1}^n \quad \text{and} \quad a = x_0 < x_1 < \cdots < x_n < x_{n+1} = b.$$

Further, choose points a_k, b_k, $k \in \{0, 1, \ldots, n, n+1\}$, in such a way that

$$a = x_0 = a_0 < b_0 < a_1 < x_1 < b_1 < \cdots < a_n < x_n < b_n < a_{n+1} < x_{n+1} = b_{n+1} = b,$$
$$|f(b_k) - f(x_k)| > |\Delta^+ f(x_k)| - \frac{\varepsilon}{4(n+2)} \quad \text{for } k \in \{0, 1, \ldots, n\},$$
$$|f(x_k) - f(a_k)| > |\Delta^- f(x_k)| - \frac{\varepsilon}{4(n+2)} \quad \text{for } k \in \{1, \ldots, n, n+1\}.$$

Furthermore, as $v_g(x) = \operatorname{var}_a^x g$ is continuous on $[a, b]$ by Corollary 2.3.4, we can also assume that

$$v_g(b_k) - v_g(a_k) < \frac{\varepsilon}{n+2} \quad \text{for all } k \in \{0, \ldots, n+1\}.$$

To summarize, we have

$$\sum_{k=0}^{n+1} \operatorname{var}_{a_k}^{b_k} g < \varepsilon,$$
$$\sum_{k=0}^{n+1} \operatorname{var}_{a_k}^{b_k} f \geq \sum_{k=0}^{n+1} \Big(|f(b_k) - f(x_k)| + |f(x_k) - f(a_k)| \Big)$$
$$> \sum_{k=0}^{n+1} \Big(|\Delta^+ f(x_k)| + |\Delta^- f(x_k)| - \frac{\varepsilon}{2(n+2)} \Big) > \operatorname{var}_a^b f - \varepsilon.$$

Thus, the assumptions of Lemma 2.1.19 hold and the proof is complete. □

2.6 DECOMPOSITION INTO CONTINUOUS AND JUMP PARTS

Besides Jordan's decomposition of a bounded variation function into the difference of two nondecreasing functions, there exists another useful type of decomposition, which is described in the next theorem.

2.6.1 Theorem. *Every function $f \in \mathrm{BV}([a,b])$ is the sum of a step function and a continuous function with bounded variation.*

If $f = f_1 + f_2 = \widetilde{f}_1 + \widetilde{f}_2$ for two couples $f_1, \widetilde{f}_1 \in \mathrm{C}([a,b]) \cap \mathrm{BV}([a,b])$ and $f_2, \widetilde{f}_2 \in \mathrm{B}([a,b])$, then there is a $c \in \mathbb{R}$ such that

$$f_1(x) - \widetilde{f}_1(x) = \widetilde{f}_2(x) - f_2(x) = c \quad \textit{for all} \quad x \in [a,b].$$

Proof. a) Let D be the set of all discontinuity points of f in (a,b). The set D has at most countably many points, i.e.,

$$D = \{s_k \in (a,b) \colon k \in K\}, \text{ where } K = \{1, \dots, m\} \text{ for some } m \in \mathbb{N}, \text{ or } K = \mathbb{N}.$$

Define

$$\left.\begin{aligned} f_2(x) = {} & f(a) + \Delta^+ f(a)\, \chi_{(a,b]}(x) \\ & + \sum_{k \in K} \big(\Delta^- f(s_k)\, \chi_{[s_k,b]}(x) + \Delta^+ f(s_k)\, \chi_{(s_k,b]}(x)\big) \\ & + \Delta^- f(b)\, \chi_{[b]}(x). \end{aligned}\right\} \tag{2.6.1}$$

By Corollary 2.3.8 we have

$$|\Delta^+ f(a)| + \sum_{k \in K} \Big(|\Delta^+ f(s_k)| + |\Delta^- f(s_k)|\Big) + |\Delta^- f(b)| \le \mathrm{var}_a^b f < \infty$$

and thus $f_2 \in \mathrm{B}([a,b])$. Further, using (2.5.7) we get

$$\left.\begin{aligned} & \Delta^+ f_2(t) = \Delta^+ f(t) \quad \text{and} \quad \Delta^- f_2 = \Delta^- f(s) \\ & \qquad\qquad\qquad\qquad \text{for} \quad t \in [a,b) \quad \text{and} \quad s \in (a,b]. \end{aligned}\right\} \tag{2.6.2}$$

Consequently,

$$\begin{aligned} \Delta^+(f - f_2)(t) &= \Delta^+ f(t) - \Delta^+ f_2(t) = 0 \quad \text{for } t \in [a,b), \\ \Delta^-(f - f_2)(s) &= \Delta^- f(s) - \Delta^- f_2(s) = 0 \quad \text{for } s \in (a,b]. \end{aligned}$$

This means that the function $f_1 = f - f_2$ is continuous on $[a,b]$.

b) Let $f = \widetilde{f}_1 + \widetilde{f}_2$, where $\widetilde{f}_1 \in \mathrm{C}([a,b]) \cap \mathrm{BV}([a,b])$ and $\widetilde{f}_2 \in \mathrm{B}([a,b])$. Then

$$\begin{aligned} 0 = \Delta^+ \widetilde{f}_1(t) = \Delta^+(f - \widetilde{f}_2)(t) &= \Delta^+ f(t) - \Delta^+ \widetilde{f}_2(t) \quad \text{for } t \in [a,b), \\ 0 = \Delta^- \widetilde{f}_1(s) = \Delta^-(f - \widetilde{f}_2)(s) &= \Delta^- f(s) - \Delta^- \widetilde{f}_2(s) \quad \text{for } s \in (a,b]. \end{aligned}$$

Taking into account (2.6.2), we deduce that

$$\left.\begin{aligned} &\Delta^+\widetilde{f}_2(t)=\Delta^+ f_2(t)=\Delta^+ f(t) &&\text{for } t\in[a,b),\\ &\Delta^-\widetilde{f}_2(s)=\Delta^- f_2(s)=\Delta^- f(s) &&\text{for } s\in(a,b]. \end{aligned}\right\} \tag{2.6.3}$$

Since $\widetilde{f}_2$ is a step function whose discontinuities are contained in D, there exist real numbers $\widetilde{c}$, $\widetilde{c}_0$, $\widetilde{d}$ and sequences $\{\widetilde{c}_k\}$, $\{\widetilde{d}_k\}$ such that

$$\widetilde{f}_2(x)=\widetilde{c}+\widetilde{c}_0\,\chi_{(a,b]}(x)+\sum_{k\in K}\Big(\widetilde{c}_k\,\chi_{(s_k,b]}(x)+\widetilde{d}_k\,\chi_{[s_k,b]}(x)\Big)+\widetilde{d}\,\chi_{[b]}(x)$$

for all $x\in[a,b]$, where

$$\sum_{k\in K}(|\widetilde{c}_k|+|\widetilde{d}_k|)<\infty,$$

$$\Delta^+\widetilde{f}_2(s_k)=\widetilde{c}_k \ \text{ for } k\in\mathbb{N}, \quad \Delta^+\widetilde{f}_2(a)=\widetilde{c}_0,$$

$$\Delta^-\widetilde{f}_2(s_k)=\widetilde{d}_k \ \text{ for } k\in\mathbb{N}, \quad \Delta^-\widetilde{f}_2(b)=d.$$

Using (2.6.3), we obtain

$$\begin{aligned} \widetilde{f}_2(x)&=\widetilde{c}+\Delta^+ f(a)\,\chi_{(a,b]}(x)+\Delta^- f(b)\,\chi_{[b]}(x)\\ &\qquad+\sum_{k\in K}\Big(\Delta^+ f(s_k)\,\chi_{(s_k,b]}(x)+\Delta^- f(s_k)\,\chi_{[s_k,b]}(x)\Big)\\ &=(\widetilde{c}-f(a))+f_2(x) \quad \text{for } \ x\in[a,b]. \end{aligned}$$

It follows that on the whole interval $[a,b]$, $\widetilde{f}_2-f_2$ is equal to the constant $c=\widetilde{c}-f(a)$. Hence also

$$f_1(x)-\widetilde{f}_1(x)=(f(x)-f_2(x))-(f(x)-\widetilde{f}_2(x))=\widetilde{f}_2(x)-f_2(x)=c$$

for all $x\in[a,b]$, which completes the proof. □

2.6.2 Definition. Every function f_2 assigned to f by Theorem 2.6.1 is called the *jump part* (or the *saltus part*) of the function f, and the corresponding function $f_1=f-f_2$ is then called the *continuous part* of the function f. The jump part and the continuous part of the function f are usually denoted by $f^{\,\mathrm{B}}$ and $f^{\,\mathrm{C}}$, respectively.

2.6.3 Remark. According to the proof of Theorem 2.6.1, if f has bounded variation on $[a,b]$, then the unique decomposition $f=f^{\,\mathrm{B}}+f^{\,\mathrm{C}}$ in the sense of Definition 2.6.2 such that $f^{\rm B}(a)=f(a)$ and $f^{\rm C}(a)=0$ is given by the formulas

$$\begin{aligned} f^{\rm B}(x) &= f(a)+\Delta^+ f(a)\,\chi_{(a,b]}(x) \\ &\quad +\sum_{k\in K}\big(\Delta^- f(s_k)\,\chi_{[s_k,b]}(x)+\Delta^+ f(s_k)\,\chi_{(s_k,b]}(x)\big)+\Delta^- f(b)\,\chi_{[b]}(x), \\ f^{\rm C}(x) &= f(x)-f^{\rm B}(x), \end{aligned}$$

where $\{s_k\}_{k\in K}$ is the set of all discontinuity points of f in (a,b).

2.6.4 Exercise. Let $\{f_n\}$ be a sequence of functions with bounded variations on $[a,b]$ and let $\{f_n^{\rm C}\}$ and $\{f_n^{\rm B}\}$ be the sequences of continuous and jump parts of $\{f_n\}$, respectively.

(i) Show that $\operatorname{var}_a^b f_n = \operatorname{var}_a^b f_n^{\rm C} + \operatorname{var}_a^b f_n^{\rm B}$ for each $n\in\mathbb{N}$.

(ii) Prove that $\lim_{n\to\infty} \operatorname{var}_a^b f_n = 0$ if and only if

$$\lim_{n\to\infty}(\operatorname{var}_a^b f_n^{\rm C}) = \lim_{n\to\infty}(\operatorname{var}_a^b f_n^{\rm B}) = 0.$$

For dealing with step functions, it is useful to know that every step function may be approximated in the norm of the space $\mathrm{BV}([a,b])$ by finite step functions. This is the content of the following lemma, which will be particularly useful in Chapter 6.

2.6.5 Lemma. *For each step function $f:[a,b]\to\mathbb{R}$ there is a sequence of finite step functions $\{f_n\}\subset \mathrm{S}([a,b])$ such that*

$$\lim_{n\to\infty}\|f-f_n\|_{\rm BV}=0.$$

If f has the form

$$\left.\begin{aligned} f(x) &= f(a)+\Delta^+ f(a)\,\chi_{(a,b]}(x) \\ &\quad +\sum_{k=1}^{\infty}\Big(\Delta^+ f(s_k)\,\chi_{(s_k,b]}(x)+\Delta^- f(s_k)\,\chi_{[s_k,b]}(x)\Big) \\ &\quad +\Delta^- f(b)\,\chi_{[b]}(x) \quad \text{for } x\in[a,b], \end{aligned}\right\} \tag{2.6.4}$$

with $\{s_k\}$ being the set of all discontinuity points of f in (a,b), then $\{f_n\}$

might be chosen as

$$\left.\begin{aligned} f_n(x) = f(a) + \Delta^+ f(a)\, \chi_{(a,b]}(x) \\ + \sum_{k=1}^{n} \Big(\Delta^+ f(s_k)\, \chi_{(s_k,b]}(x) + \Delta^- f(s_k)\, \chi_{[s_k,b]}(x) \Big) \\ + \Delta^- f(b)\, \chi_{[b]}(x) \quad \textit{for } x \in [a,b] \quad \textit{and} \quad n \in \mathbb{N}. \end{aligned}\right\} \tag{2.6.5}$$

Proof. Let $f \in \mathrm{B}([a,b])$ and let D be the set of its discontinuity points in (a,b). If D is finite, the statement is obvious. Therefore, we can assume that $D = \{s_k\}$ is a non-repeating infinite sequence. By Theorem 2.3.6 and Corollary 2.3.8, the series

$$\sum_{k=1}^{\infty} \Big(\Delta^+ f(s_k)\, \chi_{(s_k,b]}(x) + \Delta^- f(s_k)\, \chi_{[s_k,b]}(x) \Big)$$

is absolutely convergent for $x \in [a,b]$. Therefore (cf. (2.5.3) and (2.5.7)), f can be expressed in the form (2.6.4).

Consider the finite step functions f_n, $n \in \mathbb{N}$, given by (2.6.5). Then

$$f(x) - f_n(x) = \sum_{k=n+1}^{\infty} \Big(\Delta^+ f(s_k)\, \chi_{(s_k,b]}(x) + \Delta^- f(s_k)\, \chi_{[s_k,b]}(x) \Big).$$

Consequently, by Theorem 2.5.3 we have

$$\mathrm{var}_a^b (f - f_n) = \sum_{k=n+1}^{\infty} \big(|\Delta^- f(s_k)| + |\Delta^+ f(s_k)| \big).$$

The right-hand side is the remainder of an absolutely convergent series, and therefore tends to 0 as $n \to \infty$. This means that

$$\lim_{n \to \infty} \mathrm{var}_a^b (f - f_n) = 0.$$

Since $f_n(a) = f(a)$ for each $n \in \mathbb{N}$, we get $\lim_{n \to \infty} \|f - f_n\|_{\mathrm{BV}} = 0$. □

2.7 Pointwise convergence

2.7.1 Example. Consider again the functions

$$f_n(x) = \begin{cases} 0 & \text{if } 0 \le x \le \frac{1}{n}, \\ x \sin\left(\frac{\pi}{x}\right) & \text{if } \frac{1}{n} \le x \le 2 \end{cases} \qquad \text{for } n \in \mathbb{N}$$

and

$$f(x) = \begin{cases} 0 & \text{if } x = 0, \\ x \sin\left(\frac{\pi}{x}\right) & \text{if } x > 0. \end{cases}$$

From Example 2.1.15 we know that $\operatorname{var}_0^2 f_n < \infty$ for each $n \in \mathbb{N}$. It can be easily verified that $\{f_n\}$ converges to f uniformly on $[0,2]$, while by Example 2.1.10, f has unbounded variation on $[0,2]$.

The previous example shows that the limit of bounded variation functions need not have bounded variation even if the convergence is uniform. On the other hand, the following theorem shows that uniform boundedness of variations of f_n together with the pointwise convergence $f_n \to f$ already guarantee that the limit function f has bounded variation. Notice that, using the argument from Example 2.1.10, it is possible to show that the sequence $\{f_n\}$ from Examples 2.1.15 and 2.7.1 satisfies $\lim_{n\to\infty} \operatorname{var}_0^2 f_n = \infty$, and therefore $\sup_{n\in\mathbb{N}} \operatorname{var}_0^2 f_n = \infty$.

2.7.2 Theorem. *Let $f : [a,b] \to \mathbb{R}$ be given and let $\{f_n\}$ be a sequence of functions such that*

$$\operatorname{var}_a^b f_n \le \varkappa < \infty \quad \textit{for } n \in \mathbb{N}, \quad \textit{and} \quad \lim_{n\to\infty} f_n(x) = f(x) \quad \textit{for } x \in [a,b].$$

Then $\operatorname{var}_a^b f \le \varkappa$.

P r o o f. Given an arbitrary $\boldsymbol{\alpha} \in \mathscr{D}[a,b]$, we have

$$V(f, \boldsymbol{\alpha}) = \lim_{n\to\infty} V(f_n, \boldsymbol{\alpha}) \le \varkappa.$$

Consequently, $\operatorname{var}_a^b f \le \varkappa$. □

2.7.3 Exercise. Let

$$f(x) = \begin{cases} 2^{-k} & \text{if } x = \frac{1}{k+1} \text{ for some } k \in \mathbb{N}, \\ 0 & \text{otherwise.} \end{cases}$$

Prove that $f \in \mathrm{BV}([0,1])$.

We now formulate and prove Helly's selection theorem, which states that every sequence of functions with uniformly bounded variations has a subsequence which is pointwise convergent to a function of bounded variation. In other words, the theorem provides a sufficient condition for the relative compactness of a subset of $\mathrm{BV}([a,b])$.

2.7.4 Theorem (HELLY'S SELECTION THEOREM). *Let $\{f_n\}\subset\mathrm{BV}([a,b])$, $\varkappa\in\mathbb{R}$, $|f_n(a)|\le\varkappa$ and $\mathrm{var}_a^b f_n\le\varkappa$ for all $n\in\mathbb{N}$. Then there exist a function $f\in\mathrm{BV}([a,b])$ and a subsequence $\{f_{n_k}\}$ such that*

$$|f(a)|\le\varkappa,\quad \mathrm{var}_a^b f\le\varkappa \quad and \quad \lim_{k\to\infty} f_{n_k}(x)=f(x) \quad for\ \ x\in[a,b].$$

To prove Theorem 2.7.4 we need the following two assertions.

Assertion 1. *Let $|f_n(x)|\le M<\infty$ for all $x\in[a,b]$ and all $n\in\mathbb{N}$. Then for any countable subset P of $[a,b]$, there is a subsequence $\{f_{n_k}\}$ such that $\lim_{k\to\infty} f_{n_k}(p)$ exists and is finite for all $p\in P$.*

Proof. Let $P=\{p_k\}$. We have $|f_n(p_k)|\le M<\infty$ for all $n,k\in\mathbb{N}$. Hence, by the Bolzano-Weierstrass theorem, there is a sequence $\{n_{k,1}:k\in\mathbb{N}\}$ and a number $q_1\in\mathbb{R}$ such that

$$\lim_{k\to\infty} f_{n_{k,1}}(p_1)=q_1.$$

Similarly, there are $\{f_{n_{k,2}}:k\in\mathbb{N}\}\subset\{f_{n_{k,1}}:k\in\mathbb{N}\}$ and $q_2\in\mathbb{R}$ such that

$$\lim_{k\to\infty} f_{n_{k,2}}(p_2)=q_2\in\mathbb{R},\ \text{ and } \lim_{k\to\infty} f_{n_{k,2}}(p_1)=q_1\in\mathbb{R}.$$

In this way, for each $j\in\{2,3,\dots\}$ we can find a sequence

$$\{f_{n_{k,j}}:k\in\mathbb{N}\}\subset\{f_{n_{k,j-1}}:k\in\mathbb{N}\}$$

and a number $q_j\in\mathbb{R}$ such that

$$\lim_{k\to\infty} f_{n_{k,\ell}}(p_\ell)=q_\ell\in\mathbb{R}\quad\text{for all } \ell\in\{1,\dots,j\}.$$

Put $f_{n_k}=f_{n_{k,k}}$ for $k\in\mathbb{N}$. Then

$$\lim_{k\to\infty} f_{n_k}(p_j)=q_j\in\mathbb{R}\quad\text{for } j\in\mathbb{N},$$

and the assertion is proved. □

Assertion 2. *Let $\{f_n\}$ be a sequence of nondecreasing functions on $[a,b]$ and let $\|f_n\|\le M<\infty$ for all $n\in\mathbb{N}$. Then there is a subsequence $\{f_{n_k}\}$ and* *a nondecreasing function* $f:[a,b]\to\mathbb{R}$ *such that*

$$\lim_{k\to\infty} f_{n_k}(x)=f(x) \quad \textit{for} \quad x\in[a,b].$$

Proof. Let $P=(\mathbb{Q}\cap(a,b))\cup\{a\}\cup\{b\}$ be the set of all rational numbers from the open interval (a,b) together with the points a, b. Then P is countable and $[a,b]\setminus P\subset(a,b)$. By Assertion 1 there is a subsequence $\{f_{n_k}\}$ and a function $\varphi:P\to\mathbb{R}$ such that

$$\lim_{k\to\infty} f_{n_k}(p)=\varphi(p) \quad \text{for} \quad p\in P.$$

Obviously, $\varphi(p')\le\varphi(p'')$ if p', $p''\in P$ and $p'\le p''$. Furthermore, define

$$\varphi(x)=\sup_{p\in P\cap[a,x)}\varphi(p) \quad \text{for } x\in(a,b)\setminus P.$$

The function φ is nondecreasing on $[a,b]$ and

$$\varphi(x)=\lim_{\substack{p\to x-\\ p\in P}}\varphi(p) \quad \text{for } x\in(a,b)\setminus P.$$

We will show that

$$\lim_{k\to\infty} f_{n_k}(x_0)=\varphi(x_0) \tag{2.7.1}$$

whenever $x_0\in(a,b)$ and φ is continuous at x_0. Indeed, assume that $x_0\in(a,b)$ is a continuity point of φ and let an arbitrary $\varepsilon>0$ be given. Then there is a $\delta_\varepsilon>0$ such that

$$\varphi(x_0)-\varepsilon<\varphi(x)<\varphi(x_0)+\varepsilon \quad \text{for all } x\in(x_0-\delta_\varepsilon,x_0+\delta_\varepsilon).$$

Further, let us choose $r'\in P\cap(x_0-\delta_\varepsilon,x_0)$ and $r''\in P\cap(x_0,x_0+\delta_\varepsilon)$ arbitrarily. Then

$$\varphi(x_0)-\varepsilon<\varphi(r')\le\varphi(x_0)\le\varphi(r'')<\varphi(x_0)+\varepsilon.$$

Moreover, there is a k_ε such that

$$\varphi(r')-\varepsilon<f_{n_k}(r')<\varphi(r')+\varepsilon$$

and

$$\varphi(r'')-\varepsilon<f_{n_k}(r'')<\varphi(r'')+\varepsilon$$

for all $k \geq k_\varepsilon$. Hence, for each $k \geq k_\varepsilon$ we have

$$\begin{aligned}\varphi(x_0) - 2\,\varepsilon < \varphi(r') - \varepsilon < f_{n_k}(r') \leq f_{n_k}(x_0) \\ \leq f_{n_k}(r'') < \varphi(r'') + \varepsilon < \varphi(x_0) + 2\,\varepsilon.\end{aligned}$$

Thus, (2.7.1) is true.

To summarize, we have proved that if D is the set of all discontinuity points of the function φ in (a,b), then

$$\lim_{k\to\infty} f_{n_k}(x) = \varphi(x) \quad \text{for} \quad x \in [a,b] \setminus D.$$

By Theorem 2.3.2, the set D is countable. Thus, we can use once more Assertion 1 to prove that there is a subsequence

$$\{f_{n_{k_\ell}} : \ell \in \mathbb{N}\} \subset \{f_{n_k} : k \in \mathbb{N}\}$$

of $\{f_{n_k}\}$ which has a limit $\psi(x) \in \mathbb{R}$ for each $x \in D$. Now, define

$$f(x) = \begin{cases} \varphi(x), & \text{if} \quad x \in [a,b] \setminus D, \\ \psi(x), & \text{if} \quad x \in D. \end{cases}$$

Then

$$\lim_{\ell\to\infty} f_{n_{k_\ell}}(x) = f(x) \quad \text{for } x \in [a,b],$$

and f is nondecreasing on $[a,b]$ because it is the pointwise limit of a sequence of nondecreasing functions. The proof of Assertion 2 is complete. □

Proof of Theorem 2.7.4

For each $n \in \mathbb{N}$ and $x \in [a,b]$, let

$$g_n(x) = \operatorname{var}_a^x f_n \quad \text{and} \quad h_n(x) = g_n(x) - f_n(x).$$

Then $f_n = g_n - h_n$ and the functions g_n, h_n with $n \in \mathbb{N}$ are nondecreasing on $[a,b]$ (see Exercise 2.1.22). Furthermore,

$$\|g_n\| \leq \operatorname{var}_a^b f_n \leq \varkappa \quad \text{and} \quad \|h_n\| \leq \|f_n\| + \|g_n\| \leq 3\,\varkappa \quad \text{for } n \in \mathbb{N}.$$

By Assertion 2, there exist functions $g, h \in \mathrm{BV}([a,b])$ and a sequence $\{n_k\} \subset \mathbb{N}$ such that

$$\|g\| \leq \varkappa, \ \|h\| \leq 2\,\varkappa, \ \operatorname{var}_a^b g \leq \varkappa, \ \operatorname{var}_a^b h \leq 2\,\varkappa,$$

$$\lim_{k\to\infty} g_{n_k}(x) = g(x) \quad \text{and} \quad \lim_{k\to\infty} h_{n_k}(x) = h(x) \quad \text{for all } x \in [a,b].$$

Denote $f = g - h$. Then

$$\lim_{k\to\infty} f_{n_k}(x) = \lim_{k\to\infty} \big(g_{n_k}(x) - h_{n_k}(x)\big) = g(x) - h(x) = f(x)$$

for all $x \in [a,b]$. Obviously, $|f(a)| \le \varkappa$. Finally, Theorem 2.7.2 implies that $\operatorname{var}_a^b f \le \varkappa$. This completes the proof. □

2.8 Variation on elementary sets

The aim of this section is to extend the notion of variation from closed intervals to arbitrary intervals, and later proceed to elementary sets, i.e., finite unions of intervals.

2.8.1 Definition. Let J be a bounded interval in $\mathbb{R}$. We say that a finite set

$$\boldsymbol{\alpha} = \{\alpha_0, \alpha_1, \ldots, \alpha_{\nu(\boldsymbol{\alpha})}\} \subset J$$

is a *generalized division of* J if $\alpha_0 < \alpha_1 < \cdots < \alpha_{\nu(\boldsymbol{\alpha})}$.

The set of all generalized divisions of the interval J is denoted by $\mathscr{D}^*(J)$.

Let $f : [a,b] \to \mathbb{R}$ and let J be an arbitrary subinterval of $[a,b]$. Then we define the variation of f on J by

$$\operatorname{var}(f, J) = \sup\Big\{ \sum_{j=1}^{\nu(\boldsymbol{\alpha})} |f(\alpha_j) - f(\alpha_{j-1})| : \boldsymbol{\alpha} \in \mathscr{D}^*(J) \Big\}.$$

We say that f is of bounded variation on J if $\operatorname{var}(f, J) < \infty$. In such a case, we write $f \in \mathrm{BV}(J)$. We also set $\operatorname{var}(f, \emptyset) = 0$ and $\operatorname{var}(f, [c]) = 0$ for $c \in [a,b]$.

2.8.2 Remark. It is easy to see that Definition 2.8.1 coincides with the definition of the variation in the sense of Definition 2.1.1 if J is a compact interval; that is, for $f : [a,b] \to \mathbb{R}$ and $J = [c,d] \subset [a,b]$, we have

$$\operatorname{var}(f, J) = \operatorname{var}_c^d f.$$

For this reason, in the case of a compact interval J, we may always restrict ourselves to divisions containing the endpoints of J.

Moreover, it is easy to see that if J is a bounded interval and $f \in \mathrm{BV}(J)$, then f is bounded on J.

The next assertion follows immediately from Definition 2.8.1.

2.8.3 Theorem. *Let $f:[a,b]\to\mathbb{R}$ and let J_1 and J_2 be subintervals of $[a,b]$ such that $J_2\subset J_1$. Then $\operatorname{var}(f,J_2)\le\operatorname{var}(f,J_1)$.*

In particular, if J is a subinterval of $[a,b]$ and $f\in\mathrm{BV}(J)$, then $f\in\mathrm{BV}(I)$ for every interval $I\subset J$.

2.8.4 Exercise. Prove Theorem 2.8.3.

The next theorem presents formulas for the variation over half-open and open intervals.

2.8.5 Theorem. *Let $f:[a,b]\to\mathbb{R}$ and $c,\ d\in[a,b]$, with $c<d$.*

(i) *If $f\in\mathrm{BV}([c,d))$, then*

$$\operatorname{var}(f,[c,d))=\lim_{\delta\to0+}\operatorname{var}_c^{d-\delta}f=\sup_{t\in[c,d)}\operatorname{var}_c^t f.$$

(ii) *If $f\in\mathrm{BV}((c,d])$, then*

$$\operatorname{var}(f,(c,d])=\lim_{\delta\to0+}\operatorname{var}_{c+\delta}^{d}f=\sup_{t\in(c,d]}\operatorname{var}_t^d f.$$

(iii) *If $f\in\mathrm{BV}((c,d))$, then*

$$\operatorname{var}(f,(c,d))=\lim_{\delta\to0+}\operatorname{var}_{c+\delta}^{d-\delta}f.$$

Proof. We prove only the first assertion; the proof of the remaining two is similar and is left to the reader.

For a fixed $\delta>0$, consider a division $\boldsymbol{\alpha}=\{\alpha_0,\alpha_1,\dots,\alpha_{\nu(\boldsymbol{\alpha})}\}$ of $[c,d-\delta]$. Of course, $\boldsymbol{\alpha}$ is also a generalized division of $[c,d)$ and hence

$$\sum_{j=1}^{\nu(\boldsymbol{\alpha})}|f(\alpha_j)-f(\alpha_{j-1})|\le\operatorname{var}(f,[c,d)).$$

Thus, taking the supremum over all divisions of $[c,d-\delta]$, we get

$$\operatorname{var}_c^{d-\delta}f\le\operatorname{var}(f,[c,d)).$$

Since this inequality holds for every $\delta>0$, it follows that

$$M=\sup_{t\in[c,d)}\operatorname{var}_c^t f=\lim_{\delta\to0+}\operatorname{var}_c^{d-\delta}f\le\operatorname{var}(f,[c,d)).$$

Now, assume that $M < \operatorname{var}(f,[c,d))$. Then, for $\varepsilon = \operatorname{var}(f,[c,d)) - M$, there exists a generalized division $\boldsymbol{\alpha} = \{\alpha_0, \alpha_1, \ldots, \alpha_{\nu(\boldsymbol{\alpha})}\}$ of $[c,d)$ such that

$$M = \operatorname{var}(f,[c,d)) - \varepsilon < \sum_{j=1}^{\nu(\boldsymbol{\alpha})} |f(\alpha_j) - f(\alpha_{j-1})| \le \operatorname{var}_c^{\alpha_{\nu(\boldsymbol{\alpha})}} f \le M,$$

a contradiction. This completes the proof of (i). □

The next theorem connects the variation over arbitrary intervals and the usual variation over a compact interval. To this aim, the next lemma will help.

2.8.6 Lemma. *Let $f:[a,b] \to \mathbb{R}$, $a \le c < d \le b$ and $f \in \mathrm{BV}((c,d))$. Then both the limits $f(c+)$ and $f(d-)$ exist.*

Proof. Let $\varepsilon > 0$ and an increasing sequence $\{t_n\} \subset (c,d)$ tending to d be given. By Theorem 2.8.5 (iii) there is $\delta > 0$ such that

$$0 < \operatorname{var}(f,(c,d)) - \operatorname{var}_{c+\delta}^{d-\delta} f < \varepsilon.$$

Choose $n_0 \in \mathbb{N}$ in such a way that $t_n > d - \delta$ for every $n \ge n_0$. Hence, for $n > m \ge n_0$, we have

$$\begin{aligned} |f(t_n) - f(t_m)| &\le \operatorname{var}_{t_m}^{t_n} f = \operatorname{var}_{c+\delta}^{t_n} f - \operatorname{var}_{c+\delta}^{t_m} f \\ &\le \operatorname{var}(f,(c,d)) - \operatorname{var}_{c+\delta}^{d-\delta} f < \varepsilon, \end{aligned}$$

wherefrom the existence of the limit $f(d-)$ follows immediately.

The existence of the limit $f(c+)$ can be proved analogously. □

2.8.7 Theorem. *Let $f:[a,b] \to \mathbb{R}$ and $a \le c < d \le b$.*

(i) *If $f \in \mathrm{BV}([c,d))$, then $f(d-)$ exists and*

$$\operatorname{var}_c^d f = \operatorname{var}(f,[c,d)) + |\Delta^- f(d)|.$$

(ii) *If $f \in \mathrm{BV}((c,d])$, then $f(c+)$ exists and*

$$\operatorname{var}_c^d f = \operatorname{var}(f,(c,d]) + |\Delta^+ f(c)|.$$

(iii) *If $f \in \mathrm{BV}((c,d))$, then both the limits $f(c+)$ and $f(d-)$ exist and*

$$\operatorname{var}_c^d f = \operatorname{var}(f,(c,d)) + |\Delta^+ f(c)| + |\Delta^- f(d)|.$$

Proof. The existence of all the limits follows from Lemma 2.8.6.

Assume that $f \in \mathrm{BV}([c,d))$. Let $\varepsilon > 0$ and

$$\boldsymbol{\alpha} = \{\alpha_0, \alpha_1, \dots, \alpha_{m+1}\} \in \mathcal{D}[c,d]$$

be given. We can choose $\xi \in [c,d]$ in such a way that

$$\alpha_m < \xi < d \quad \text{and} \quad |f(d-) - f(\xi)| < \varepsilon.$$

Consequently,

$$\begin{aligned}
&\sum_{j=1}^{m+1} |f(\alpha_j) - f(\alpha_{j-1})| \\
&\quad \le \sum_{j=1}^{m} |f(\alpha_j) - f(\alpha_{j-1})| + |f(\xi) - f(\alpha_m)| \\
&\qquad + |f(d-) - f(\xi)| + |\Delta^- f(d)| \\
&\quad < \operatorname{var}_c^{\xi} f + \varepsilon + |\Delta^- f(d)| \le \operatorname{var}(f,[c,d)) + \varepsilon + |\Delta^- f(d)|.
\end{aligned}$$

Since $\boldsymbol{\alpha} \in \mathcal{D}[c,d]$ and $\varepsilon > 0$ were chosen arbitrarily, we conclude that

$$\operatorname{var}_c^d f \le \operatorname{var}(f,[c,d)) + |\Delta^- f(d)|.$$

On the other hand, for any $\delta > 0$ we have

$$|f(d) - f(d-\delta)| \le \operatorname{var}_{d-\delta}^d f = \operatorname{var}_c^d f - \operatorname{var}_c^{d-\delta} f.$$

Hence, letting $\delta \to 0+$ we get

$$|\Delta^- f(d)| \le \operatorname{var}_c^d f - \lim_{\delta \to 0+} \operatorname{var}_c^{d-\delta} f = \operatorname{var}_c^d f - \operatorname{var}(f,[c,d)),$$

wherefrom we conclude that

$$\operatorname{var}_c^d f \ge \operatorname{var}(f,[c,d)) + |\Delta^- f(d)|.$$

This completes the proof of (i).

Similarly, we can prove the assertions (ii) and (iii). □

2.8.8 Corollary. *Let $f : [a,b] \to \mathbb{R}$ and $c,\ d \in [a,b]$, with $c < d$. Then the following assertions are equivalent:*

(i) $f \in \mathrm{BV}([c,d])$,

(ii) $f \in \mathrm{BV}((c,d])$,

(iii) $f \in \mathrm{BV}([c,d))$,

(iv) $f \in \mathrm{BV}((c,d))$.

2.8.9 Remark. In view of Theorem 2.8.7, we can also observe that for every $f:[a,b] \to \mathbb{R}$ and $c \in [a,b]$ we have

$$\lim_{\delta \to 0+} \operatorname{var}_{c-\delta}^{c+\delta} f = |\Delta^- f(c)| + |\Delta^+ f(c)|$$

provided the one-sided limits exist at the point c. Furthermore, by Theorem 2.8.7, the relations

$$\operatorname{var}(f,[c,d)) = \operatorname{var}(f,(c,d)) = \operatorname{var}(f,(c,d]) = \operatorname{var}_c^d f$$

hold for c, $d \in [a,b]$ such that $c<d$, whenever $f \in \mathrm{BV}([a,b]) \cap \mathrm{C}([a,b])$.

The relations established in Theorem 2.8.7 allow us to estimate the variation of the jump part of an arbitrary function with bounded variation; such an estimate will be useful later.

2.8.10 Theorem. *Let $f \in \mathrm{BV}([a,b])$, D be the set of discontinuity points of f in $[a,b]$ and let f^{B} be the jump part of f. Then*

$$\operatorname{var}(f^{\mathrm{B}}, J) \le \sum_{s \in J \cap D} \big(|\Delta^+ f(s)| + |\Delta^- f(s)|\big)$$

for every subinterval J of $[a,b]$.

Proof. If J is closed, we know from Theorem 2.5.3 that the equality holds. Next, assume that $J=[c,d)$, for some $c<d$ in $[a,b]$. Note that, by Theorems 2.8.7 and 2.5.3 we have

$$\begin{aligned}\operatorname{var}(f^{\mathrm{B}}, J) &= \operatorname{var}_c^d f^{\mathrm{B}} - |\Delta^- f^{\mathrm{B}}(d)| \\ &= |\Delta^+ f^{\mathrm{B}}(c)| + \sum_{s \in (c,d)} \big(|\Delta^+ f^{\mathrm{B}}(s)| + |\Delta^- f^{\mathrm{B}}(s)|\big) \\ &\le \sum_{s \in J \cap D} \big(|\Delta^+ f^{\mathrm{B}}(s)| + |\Delta^- f^{\mathrm{B}}(s)|\big).\end{aligned}$$

Recalling that $\Delta^- f^{\mathrm{B}} = \Delta^- f$ and $\Delta^+ f^{\mathrm{B}} = \Delta^+ f$, the result follows.

The cases $J=(c,d]$ and $J=(c,d)$ can be proved in similar way. □

The next result provides some information on the variation over the union of intervals.

2.8.11 Theorem. *Let* $f \in \mathrm{BV}([a,b]) \cap \mathrm{C}([a,b])$ *and let* J_1 *and* J_2 *be subintervals of* $[a,b]$ *such that* $J_1 \cup J_2$ *is an interval. Then*

$$\mathrm{var}(f, J_1 \cup J_2) \le \mathrm{var}(f, J_1) + \mathrm{var}(f, J_2). \tag{2.8.1}$$

In particular, if J_1 *and* J_2 *are nonoverlapping subintervals of* $[a,b]$, *then*

$$\mathrm{var}(f, J_1 \cup J_2) = \mathrm{var}(f, J_1) + \mathrm{var}(f, J_2).$$

Proof. The inequality (2.8.1) is obvious in the cases when $J_1 \subset J_2$ or $J_2 \subset J_1$. Otherwise, denote $a_i = \inf J_i$ and $b_i = \sup J_i$, $i = 1, 2$. We distinguish three cases:

The first case is when either $J_1 \cup J_2 = [a_1, b_1]$ or $J_1 \cup J_2 = [a_2, b_2]$. Without loss of generality, assume that the first equality holds. Thus, using Theorem 2.8.7 and the continuity of f we obtain

$$\mathrm{var}(f, J_1 \cup J_2) = \mathrm{var}_{a_1}^{b_1} f = \mathrm{var}(f, J_1) \le \mathrm{var}(f, J_1) + \mathrm{var}(f, J_2).$$

The second case is when $J_1 \cap J_2 \ne \emptyset$. Without loss of generality, assume $a_1 < a_2 < b_1 < b_2$. Choosing an arbitrary $c \in (a_2, b_1)$, from Theorems 2.8.7 and 2.8.3 it follows that

$$\mathrm{var}(f, J_1 \cup J_2) = \mathrm{var}_{a_1}^{b_2} f = \mathrm{var}_{a_1}^{c} f + \mathrm{var}_{c}^{b_2} f \le \mathrm{var}(f, J_1) + \mathrm{var}(f, J_2).$$

The third case is when J_1 and J_2 are nonoverlapping. Without loss of generality, assume $a_1 < b_1 = a_2 < b_2$. Therefore, using Theorem 2.8.7 and the continuity of f we obtain

$$\mathrm{var}(f, J_1 \cup J_2) = \mathrm{var}_{a_1}^{b_2} f = \mathrm{var}_{a_1}^{b_1} f + \mathrm{var}_{a_2}^{b_2} f = \mathrm{var}(f, J_1) + \mathrm{var}(f, J_2),$$

and the proof is complete. □

We now extend the notion of the variation over intervals to variation over elementary sets.

2.8.12 Definition. Let $E \subset \mathbb{R}$ be bounded. We say that E is an *elementary set* if it is a finite union of intervals.

A collection of intervals $\{J_k : k = 1, \dots, N\}$ is called a *minimal decomposition* of E if $E = \bigcup_{k=1}^{N} J_k$ and the union $J_k \cup J_\ell$ is not an interval whenever $k \ne \ell$.

If $S \subset \mathbb{R}$, then $\mathcal{E}(S)$ stands for the set of all elementary subsets of S.

Note that the minimal decomposition of an elementary set is uniquely determined. Moreover, the intervals of such decomposition are pairwise disjoint. Having this in mind, we define the variation over elementary sets as follows.

2.8.13 Definition. Given a function $f:[a,b]\to\mathbb{R}$ and an elementary subset E of $[a,b]$, the variation of f over E is

$$\operatorname{var}(f,E)=\sum_{k=1}^{N}\operatorname{var}(f,J_k),$$

where $\{J_k\colon k=1,\dots,N\}$ is the minimal decomposition of E.

The remaining theorems of this section show that the variation in the sense of Definition 2.8.13 has some properties of a measure. We start by proving the monotonicity of the variation over elementary sets.

2.8.14 Theorem. *Let $f:[a,b]\to\mathbb{R}$ and let E_1 and E_2 be elementary subsets of $[a,b]$ such that $E_2\subset E_1$. Then*

$$\operatorname{var}(f,E_2)\le\operatorname{var}(f,E_1).$$

In particular, $\operatorname{var}(f,E)\le\operatorname{var}_a^b f$ *for any* $E\in\mathcal{E}([a,b])$.

Proof. Assume $\{I_k^{(i)}:k=1,\dots,m_i\}$ is the minimal decomposition of E_i for $i=1,2$. Since $E_2\subset E_1$, it is clear that each $I_k^{(2)}$, $k=1,\dots,m_2$, is contained in $I_\ell^{(1)}$ for some $\ell\in\{1,\dots,m_1\}$. Let Λ be the set of indexes $\ell\in\{1,\dots,m_1\}$ such that $I_\ell^{(1)}$ contains at least one interval from the minimal decomposition of E_2. Thus, applying Theorems 2.8.3 and 2.8.11, we obtain

$$\begin{aligned}\operatorname{var}(f,E_2)&=\sum_{k=1}^{m_2}\operatorname{var}_{I_k^{(2)}} f\le\sum_{\ell\in\Lambda}\operatorname{var}_{I_\ell^{(1)}} f\\&\le\sum_{k=1}^{m_1}\operatorname{var}_{I_k^{(1)}} f=\operatorname{var}(f,E_1),\end{aligned}$$

and the proof is complete. □

Next we discuss additivity properties. Note that the notion of nonoverlapping intervals can be naturally extended to elementary sets as follows: two elementary sets are nonoverlapping if the union of their miminal decompositions forms a system of nonoverlapping intervals.

2.8.15 Theorem. *Let $f \in \mathrm{BV}([a,b]) \cap \mathrm{C}([a,b])$ and let E_1 and E_2 be elementary subsets of $[a,b]$. Then*

$$\operatorname{var}(f, E_1 \cup E_2) \leq \operatorname{var}(f, E_1) + \operatorname{var}(f, E_2). \tag{2.8.2}$$

In particular, if E_1 and E_2 are nonoverlapping elementary subsets of $[a,b]$, then

$$\operatorname{var}(f, E_1 \cup E_2) = \operatorname{var}(f, E_1) + \operatorname{var}(f, E_2).$$

Proof. Let $\{J_k : k = 1, \dots, N\}$ be the minimal decomposition of $E_1 \cup E_2$ and put

$$\Lambda_0 = \big\{k \in \{1, \dots, N\} : J_k \cap E_1 \cap E_2 \neq \emptyset\big\},$$
$$\Lambda_1 = \big\{k \in \{1, \dots, N\} : J_k \subset (E_1 \setminus E_2)\big\},$$
$$\Lambda_2 = \big\{k \in \{1, \dots, N\} : J_k \subset (E_2 \setminus E_1)\big\}.$$

Note that each index $k \in \{1, \dots, N\}$ belongs to only one of the sets above. Moreover, for each $k \in \Lambda_0$, we can write $J_k = J_k^{(1)} \cup J_k^{(2)}$, where $J_k^{(i)} = J_k \cap E_i$ for $i = 1, 2$. It is not difficult to see that

$$E_i = \Big(\bigcup_{k \in \Lambda_i} J_k\Big) \cup \Big(\bigcup_{k \in \Lambda_0} J_k^{(i)}\Big) \quad \text{for } i = 1, 2,$$

and

$$\operatorname{var}(f, E_i) = \sum_{k \in \Lambda_i} \operatorname{var}(f, J_k) + \sum_{k \in \Lambda_0} \operatorname{var}(f, J_k^{(i)}).$$

This together with Theorems 2.8.11 and 2.8.14 yields

$$\begin{aligned}\operatorname{var}(f, E_1 \cup E_2) &= \sum_{k \in \Lambda_1} \operatorname{var}(f, J_k) + \sum_{k \in \Lambda_2} \operatorname{var}(f, J_k) + \sum_{k \in \Lambda_0} \operatorname{var}(f, J_k^{(1)} \cup J_k^{(2)}) \\ &\leq \operatorname{var}(f, E_1) + \operatorname{var}(f, E_2),\end{aligned}$$

proving (2.8.2).

Now assume that E_1 and E_2 are nonoverlapping elementary subsets of $[a,b]$, and let $\{I_k^{(i)} : k = 1, \dots, m_i\}$ denote the minimal decomposition of E_i for $i = 1, 2$. Each J_k is a finite union of nonoverlapping intervals from the minimal decomposition of E_1 or E_2, that is,

$$J_k = \Big(\bigcup_{\ell \in \Lambda_k^{(1)}} I_\ell^{(1)}\Big) \cup \Big(\bigcup_{\ell \in \Lambda_k^{(2)}} I_\ell^{(2)}\Big),$$

where $\Lambda_k^{(i)} \subset \{1,\ldots,m_i\}$ for $i=1,2$. By Theorem 2.8.11, we have

$$\operatorname{var}(f,J_k)=\sum_{\ell\in\Lambda_k^{(1)}}\operatorname{var}(f,I_\ell^{(1)})+\sum_{\ell\in\Lambda_k^{(2)}}\operatorname{var}(f,I_\ell^{(2)}).$$

On the other hand, since each interval from the minimal decomposition of E_i is contained in some J_k, we have $\bigcup_{k=1}^{N}\Lambda_k^{(i)}=\{1,\ldots,m_i\}$ for $i=1,2$. Therefore,

$$\begin{aligned}\operatorname{var}(f,E_1)+\operatorname{var}(f,E_2)&=\sum_{k=1}^{m_1}\operatorname{var}(f,I_k^{(1)})+\sum_{k=1}^{m_2}\operatorname{var}(f,I_k^{(2)})\\&=\sum_{k=1}^{N}\Big(\sum_{\ell\in\Lambda_k^{(1)}}\operatorname{var}(f,I_\ell^{(1)})+\sum_{\ell\in\Lambda_k^{(2)}}\operatorname{var}(f,I_\ell^{(2)})\Big)\\&=\sum_{k=1}^{N}\operatorname{var}(f,J_k)=\operatorname{var}(f,E_1\cup E_2),\end{aligned}$$

and the proof is complete. □

We now show that the variation of an elementary set can be approximated by the variation over a closed elementary set.

2.8.16 Theorem. *Let $f\in\mathrm{BV}([a,b])$. If E is an elementary subset of $[a,b]$, then for every $\varepsilon>0$, there exists $H_\varepsilon\in\mathcal{E}(E)$ such that H_ε is closed in $[a,b]$ and*

$$\operatorname{var}(f,E)-\varepsilon<\operatorname{var}(f,H_\varepsilon).$$

Proof. Let $\{J_k\colon k=1,\ldots,N\}$ be the minimal decomposition of E. Given $\varepsilon>0$, for each $k=1,\ldots,N$, we can choose a generalized division $\boldsymbol{\alpha}^k=\{\alpha_0^k,\ldots,\alpha_{\nu(\boldsymbol{\alpha}^k)}^k\}$ of J_k such that

$$\operatorname{var}(f,J_k)-\frac{\varepsilon}{N}<\sum_{j=1}^{\nu(\boldsymbol{\alpha}^k)}|f(\alpha_j^k)-f(\alpha_{j-1}^k)|\le\operatorname{var}(f,I_k),$$

where $I_k=[\alpha_0^k,\alpha_{\nu(\boldsymbol{\alpha}^k)}^k]$. Define $H_\varepsilon=\bigcup_{k=1}^{N}I_k$. Since $I_k\subset J_k$ for each $k=1,\ldots,N$, we clearly have $H_\varepsilon\in\mathcal{E}(E)$, with H_ε closed in $[a,b]$. Moreover the inequality above yields

$$\operatorname{var}(f,E)-\varepsilon<\sum_{k=1}^{N}\operatorname{var}(f,I_k)=\operatorname{var}(f,H_\varepsilon),$$

and the result follows. □

Consider a function $f \in \mathrm{BV}([a,b])$. For each set $A \subset [a,b]$, define

$$v(A) = \sup\{\mathrm{var}(f,E) : E \in \mathcal{E}(A)\}.$$

Our next goal is to prove, in Theorem 2.8.18, an assertion which resembles the property of continuity from above for measures. We begin by establishing this property in the case when f is a continuous function with bounded variation.

Recall that a sequence $\{A_n\}$ of subsets of $\mathbb{R}$ is said to be *decreasing* if $A_{n+1} \subset A_n$ holds for $n \in \mathbb{N}$.

2.8.17 Lemma. *Let $f \in \mathrm{BV}([a,b]) \cap \mathrm{C}([a,b])$. If $\{A_n\}$ is a decreasing sequence of subsets of $[a,b]$ with empty intersection, then*

$$\lim_{n\to\infty} v(A_n) = 0.$$

Proof. Note that if $A_n = \emptyset$ for some $n \in \mathbb{N}$, then $v(A_n) = 0$. In this case the result is obvious as the sequence $\{v(A_n)\}$ is nonincreasing. For contradiction, assume that $v(A_n)$ does not tend to zero. Thus, there exists $\varepsilon > 0$ such that

$$v(A_n) > \varepsilon \quad \text{for all} \quad n \in \mathbb{N}. \tag{2.8.3}$$

Consequently, for each $n \in \mathbb{N}$, there is an $E_n \in \mathcal{E}(A_n)$ such that

$$\mathrm{var}(f, E_n) > v(A_n) - \frac{\varepsilon}{2^n}\,. \tag{2.8.4}$$

Due to Theorem 2.8.16, we can assume that E_n is closed. Let

$$H_n = \bigcap_{j=1}^{n} E_j \quad \text{for} \quad n \in \mathbb{N}.$$

Clearly, H_n is closed and $H_n \subset A_n$, $n \in \mathbb{N}$. We claim that

$$\mathrm{var}(f,E) < \varepsilon \quad \text{for every} \quad E \in \mathcal{E}(A_n \setminus H_n). \tag{2.8.5}$$

Given $n \in \mathbb{N}$, consider $E \in \mathcal{E}(A_n \setminus H_n)$ and let $F_j = (E \setminus E_j)$ for $j = 1, \ldots, n$. It is not difficult to see that $F_j \in \mathcal{E}(A_n \setminus E_j) \subset \mathcal{E}(A_j \setminus E_j)$, and using Theorem 2.8.15 we get

$$\mathrm{var}(f, F_j) + \mathrm{var}(f, E_j) = \mathrm{var}(f, F_j \cup E_j) \le v(A_j) \quad \text{for} \quad j = 1, \ldots, n.$$

Therefore, it follows from (2.8.4) that $\mathrm{var}(f, F_j) < \varepsilon/2^j$. Since $E \subset A_n \setminus H_n$, each element $t \in E$ is such that $t \notin H_n = \bigcap_{j=1}^n E_j$, that is, $t \notin E_j$ for some $j \in \{1, \dots, n\}$. In other words, $E = \bigcup_{j=1}^n F_j$, and Theorem 2.8.15 implies

$$\mathrm{var}(f, E) \leq \sum_{j=1}^{n} \mathrm{var}(f, F_j) < \sum_{j=1}^{n} \frac{\varepsilon}{2^j} < \varepsilon,$$

that is, inequality (2.8.5) holds. In view of this, we must have $H_n \neq \emptyset$ for every $n \in \mathbb{N}$. Otherwise, if $H_n = \emptyset$ for some $n \in \mathbb{N}$, by (2.8.5) we would have

$$v(A_n) = \sup\{\, \mathrm{var}(f, E) \colon E \in \mathcal{E}(A_n \setminus H_n)\,\} \leq \varepsilon,$$

contradicting (2.8.3). Therefore, $H_n \neq \emptyset$ for every $n \in \mathbb{N}$.

Since $\{H_n\}$ is a decreasing sequence of nonempty, closed and bounded sets, using Cantor's intersection theorem we get $\bigcap_n H_n \neq \emptyset$. However, this means that

$$\bigcap_n A_n \neq \emptyset,$$

a contradiction. Therefore, $\lim_{n\to\infty} v(A_n) = 0$ must hold. □

2.8.18 Theorem. *Let $f \in \mathrm{BV}([a,b])$. If $\{A_n\}$ is a decreasing sequence of subsets of $[a,b]$ with empty intersection, then*

$$\lim_{n\to\infty} v(A_n) = 0.$$

Proof. By Theorem 2.6.1 we can write $f = f^{\mathrm{C}} + f^{\mathrm{B}}$, where f^{C} and f^{B} are the continuous part and the jump part of the function f, respectively. For $A \subset [a,b]$ let

$$v^{\mathrm{C}}(A) = \sup\{\, \mathrm{var}(f^{\mathrm{C}}, E) \colon E \in \mathcal{E}(A)\}, \;\; v^{\mathrm{B}}(A) = \sup\{\, \mathrm{var}(f^{\mathrm{B}}, E) \colon E \in \mathcal{E}(A)\}.$$

Clearly $v(A) \leq v^{\mathrm{C}}(A) + v^{\mathrm{B}}(A)$ for every $A \subset [a,b]$. In view of Lemma 2.8.17 it suffices to prove that $\lim_{n\to\infty} v^{\mathrm{B}}(A_n) = 0$.

According to Remark 2.6.3 we can write

$$f^{\mathrm{B}}(t) = f(a) + \sum_{k=1}^{\infty} \big(\Delta^+ f(s_k)\, \chi_{(s_k,b]}(t) + \Delta^- f(s_k)\, \chi_{[s_k,b]}(t)\big) \quad \text{for} \quad t \in [a,b],$$

where $\{s_k\}$ is the set of discontinuity points of f in $[a,b]$. Given $\varepsilon > 0$, there exists $n_0 \in \mathbb{N}$ such that

$$\sum_{k=n_0+1}^{\infty} \bigl(|\Delta^+ f(s_k)| + |\Delta^- f(s_k)|\bigr) < \frac{\varepsilon}{2} \tag{2.8.6}$$

(due to the absolute convergence of the series, Corollary 2.3.8). For each $n \in \mathbb{N}$, by the definition of $v^{\mathrm{B}}(A_n)$, we can choose $E_n \in \mathcal{E}(A_n)$ such that

$$v^{\mathrm{B}}(A_n) - \frac{\varepsilon}{2} < \operatorname{var}(f^{\mathrm{B}}, E_n). \tag{2.8.7}$$

Note that by Theorem 2.8.10 we have

$$\begin{aligned} \operatorname{var}(f^{\mathrm{B}}, E_n) &\le \sum_{s_k \in E_n} \bigl(|\Delta^+ f(s_k)| + |\Delta^- f(s_k)|\bigr) \\ &\le \sum_{s_k \in A_n} \bigl(|\Delta^+ f(s_k)| + |\Delta^- f(s_k)|\bigr). \end{aligned}$$

Denote $T = \{s_1, \dots, s_{n_0}\}$. Since $\bigcap_n A_n = \emptyset$, there exists a number $N \in \mathbb{N}$ such that $T \cap A_n = \emptyset$ for all $n \ge N$, and as a consequence of (2.8.6) we obtain

$$\operatorname{var}(f^{\mathrm{B}}, E_n) \le \sum_{s_k \in A_n} \bigl(|\Delta^+ f(s_k)| + |\Delta^- f(s_k)|\bigr) < \frac{\varepsilon}{2} \quad \text{for } n \ge N.$$

This together with (2.8.7) yields $v^{\mathrm{B}}(A_n) < \varepsilon$ for $n \ge \max\{n_0, N\}$, showing that $\lim_{n\to\infty} v^{\mathrm{B}}(A_n) = 0$ and concluding the proof. □

Bibliographic remarks

Functions of bounded variation ("fonctions à oscillation limitée") were introduced in C. Jordan's paper on Fourier series [Jordan (1881)], where he also presented the decomposition into a difference of two monotone functions as described in Exercise 2.1.22.

Basic properties of functions of bounded variation are presented in many monographs and textbooks. In particular, we highlight [Hildebrandt (1963)]. Further, we can mention e.g. [Apostol (1974)], [Appell, Banás, Merentes (2014)], [Garg (1992)], [Kolmogorov, Fomin (2012)], [Saks (1937)].

The proof of Theorem 2.4.2 (Lebesgue differentiation theorem) can be found in many real analysis textbooks, see e.g. Appendix E in [Bartle (2001)], Theorem 7.9 in [Bressoud (2008)], Theorem 4.9 in [Gordon (1994)] or Theorem 6.2.1 in [Rana (2002)].

The definition of the Cantor function and the proof that it is continuous, nondecreasing and singular (as stated in Remark 2.4.3) may be found e.g. on pages 14–15 of [Gordon (1994)].

Theorem 2.7.4 (Helly's selection theorem) was originally published in [Helly (1912)] (see Section 6 there).

Definition 2.8.1 is motivated by the notion of variation over arbitrary subsets E of $[a,b]$ presented in Definition 6.1 from [Gordon (1994)], see also [Saks (1937)]. Unfortunately, the extension of Definition 2.8.1 to arbitrary sets $E \subset [a,b]$ has a major drawback: the resulting notion of variation need not be additive. Indeed, let $a < c < d < b$ and $E = [a,c] \cup [d,b]$. According to the extended definition, we would have

$$\operatorname{var}(f,E) \geq \operatorname{var}_a^c f + \operatorname{var}_d^b f + |f(d) - f(c)| > \operatorname{var}(f,[a,c]) + \operatorname{var}(f,[d,b])$$

whenever $f(d) \neq f(c)$. On the other hand, Definition 2.8.13 provides us with a notion of variation over elementary sets which is compatible with the variation over intervals.

Theorem 2.8.7 is a particular case of Corollary 4.7 in [Chistyakov (1998)]; a paper devoted to an extensive study of the properties of the variation over arbitrary sets in the frame of metric space-valued functions.

The observation mentioned in Remark 2.8.9 is already known from Proposition I.2.8 in [Hönig (1973)].

Theorem 2.8.18 was inspired by a result from [Lewin (1986)], which is concerned with the Lebesgue measure. The proof of Lemma 2.8.17 has been delivered in [Hanung, Monteiro, Tvrdý (2016)].

Chapter 3

Absolutely continuous functions

Absolutely continuous functions represent an important special case of continuous functions with bounded variation. Here we summarize their basic properties and recall their importance in integration theory.

3.1 Definition and basic properties

We say that $\{[\alpha_j,\beta_j]: j\in J\}$ is a system of nonoverlapping intervals in $[a,b]$ if $[\alpha_j,\beta_j]\subset[a,b]$ for each $j\in J$, and $(\alpha_j,\beta_j)\cap(\alpha_k,\beta_k)=\emptyset$ for $j\neq k$.

3.1.1 Definition. A function $f:[a,b]\to\mathbb{R}$ is *absolutely continuous* on the interval $[a,b]$ if for every $\varepsilon>0$ there exists a $\delta>0$ such that

$$\sum_{j=1}^{m}|f(\beta_j)-f(\alpha_j)|<\varepsilon$$

holds for every finite system $\{[\alpha_j,\beta_j]: j=1,2,\dots,m\}$ of nonoverlapping intervals in $[a,b]$ satisfying $\sum_{j=1}^{m}(\beta_j-\alpha_j)<\delta$.

The set of all absolutely continuous functions on $[a,b]$ will be denoted by $\mathrm{AC}([a,b])$.

Clearly, each absolutely continuous function is continuous.

3.1.2 Exercise. Prove that every Lipschitz-continuous function on $[a,b]$ (see Exercise 2.1.8 (vi)) is absolutely continuous on $[a,b]$. Furthermore, show that each differentiable function whose derivative is bounded on $[a,b]$ is Lipschitz-continuous and therefore absolutely continuous on $[a,b]$.

3.1.3 Theorem. *If f is absolutely continuous on $[a,b]$ and $[c,d] \subset [a,b]$, then f is absolutely continuous on $[c,d]$.*

If $a<c<b$ and f is absolutely continuous on $[a,c]$ and $[c,b]$, then it is absolutely continuous on $[a,b]$.

Proof. The first statement is obvious.

Suppose that $f:[a,b]\to\mathbb{R}$ is absolutely continuous on $[a,c]$ and $[c,b]$, where $c\in(a,b)$. Let $\varepsilon>0$ be given. By definition, there is a $\delta>0$ such that

$$\sum_{j=1}^{m}|f(\beta_j)-f(\alpha_j)|<\frac{\varepsilon}{2} \tag{3.1.1}$$

holds for each system $\{[\alpha_j,\beta_j]:j=1,2,\dots,m\}$ of nonoverlapping intervals in $[a,c]$ satisfying $\sum_{j=1}^{m}(\beta_j-\alpha_j)<\delta$, and

$$\sum_{j=1}^{n}|f(\delta_j)-f(\gamma_j)|<\frac{\varepsilon}{2} \tag{3.1.2}$$

holds for each system $\{[\gamma_j,\delta_j]:j=1,\dots,n\}$ of nonoverlapping intervals in $[c,b]$ satisfying $\sum_{j=1}^{n}(\delta_j-\gamma_j)<\delta$.

Now, consider a system $\{[a_j,b_j]:j=1,2,\dots,p\}$ of nonoverlapping intervals in $[a,b]$ such that $\sum_{j=1}^{p}(b_j-a_j)<\delta$. Suppose first that each interval $[a_j,b_j]$ is a subset of either $[a,c]$ or $[c,b]$. Using (3.1.1) and (3.1.2), we get

$$\begin{aligned}&\sum_{j=1}^{p}|f(b_j)-f(a_j)|\\ &\quad=\sum_{j:[a_j,b_j]\subset[a,c]}|f(b_j)-f(a_j)|+\sum_{j:[a_j,b_j]\subset[c,b]}|f(b_j)-f(a_j)|<\frac{\varepsilon}{2}+\frac{\varepsilon}{2}=\varepsilon.\end{aligned}$$

Otherwise, there exists a unique $k\in\{1,\dots,p\}$ such that $c\in(a_k,b_k)$. Using again (3.1.1) and (3.1.2), we obtain

$$\begin{aligned}\sum_{j=1}^{p}|f(b_j)-f(a_j)|\le&\sum_{j:[a_j,b_j]\subset[a,c]}|f(b_j)-f(a_j)|+|f(c)-(a_k)|\\ &+\sum_{j:[a_j,b_j]\subset[c,b]}|f(b_j)-f(a_j)|+|f(b_k)-f(c)|<\frac{\varepsilon}{2}+\frac{\varepsilon}{2}=\varepsilon.\end{aligned}$$

This shows that f is absolutely continuous on $[a,b]$. □

3.1.4 Example. A simple example of an absolutely continuous function which is not continuously differentiable on $[a,b]$ is

$$f(x)=\begin{cases} x-a & \text{for } x\in[a,\frac{a+b}{2}],\\ b-x & \text{for } x\in[\frac{a+b}{2},b]. \end{cases}$$

This function is absolutely continuous on the intervals $[a,\frac{a+b}{2}]$ and $[\frac{a+b}{2},b]$ (because it is continuously differentiable there). By Theorem 3.1.3, f is absolutely continuous on $[a,b]$.

The following exercise shows that the property described in the definition of absolutely continuous functions remains valid also for countable systems of nonoverlapping intervals.

3.1.5 Exercise. Prove the following statement: *If f is absolutely continuous on $[a,b]$, then for every $\varepsilon>0$ there exists a $\delta>0$ such that*

$$\sum_{j=1}^{\infty}|f(\beta_j)-f(\alpha_j)|<\varepsilon$$

holds for every system $\{[\alpha_j,\beta_j]:j\in\mathbb{N}\}$ of nonoverlapping intervals in $[a,b]$ such that $\sum_{j=1}^{\infty}(\beta_j-\alpha_j)<\delta$.

3.1.6 Theorem. *If $f,g:[a,b]\to\mathbb{R}$ are absolutely continuous on $[a,b]$, then*

$$|f|,\ f+g,\ f\,g,\ \max\{f,g\},\ \min\{f,g\}$$

are absolutely continuous on $[a,b]$. Moreover, if $f(x)\neq 0$ on $[a,b]$, then $1/f$ is absolutely continuous on $[a,b]$.

Proof. Let $f,\ g\in \mathrm{AC}([a,b])$.

a) By the reverse triangle inequality $||x|-|y||\le|x-y|$, we get

$$\sum_{j=1}^{m}\big||f(\beta_j)|-|f(\alpha_j)|\big|\le\sum_{j=1}^{m}|f(\beta_j)-f(\alpha_j)|$$

for each system $\{[\alpha_j,\beta_j]:j=1,2,\ldots,m\}$, which shows that $|f|\in\mathrm{AC}([a,b])$.

b) The statements $f+g\in\mathrm{AC}([a,b])$ and $f\,g\in\mathrm{AC}([a,b])$ follow from the inequalities

$$|(f(x)+g(x))-(f(y)+g(y))|\le|f(x)-f(y)|+|g(x)-g(y)|$$

and

$$|f(x)\,g(x) - f(y)\,g(y)| \le \|f\|\,|g(x) - g(y)| + \|g\|\,|f(x) - f(y)|$$

(note that $\|f\|$ and $\|g\|$ are finite, because f and g are continuous).

c) For any $x \in [a,b]$ we have

$$\max\{f(x), g(x)\} = \frac{1}{2}\Big(f(x) + g(x) + |f(x) - g(x)|\Big),$$
$$\min\{f(x), g(x)\} = \frac{1}{2}\Big(f(x) + g(x) - |f(x) - g(x)|\Big).$$

Therefore $\max\{f,g\} \in \mathrm{AC}([a,b])$ and $\min\{f,g\} \in \mathrm{AC}([a,b])$ as a consequence of a) and b).

d) If $f(x) \ne 0$ on $[a,b]$, then there exists a $\mu > 0$ such that $|f(x)| \ge \mu$ for all $x \in [a,b]$. Hence

$$\left|\frac{1}{f(x)} - \frac{1}{f(y)}\right| \le \frac{|f(x) - f(y)|}{\mu^2} \quad \text{for all } x, y \in [a,b],$$

which easily implies that $1/f \in \mathrm{AC}([a,b])$. □

The following two results provide a link between absolutely continuous functions and functions of bounded variation.

3.1.7 Theorem. *If $f : [a,b] \to \mathbb{R}$ is absolutely continuous, then it has bounded variation on $[a,b]$.*

Proof. Choose a $\delta > 0$ such that

$$\sum_{j=1}^{m} |f(b_j) - f(a_j)| < 1$$

holds for every finite system $\{[a_j, b_j] : j = 1, 2, , \dots, m\}$ of nonoverlapping intervals in $[a,b]$ satisfying $\sum_{j=1}^{n}(b_j - a_j) < \delta$. Next, choose a division $a = x_0 < x_1 < \dots < x_k = b$ such that

$$0 < x_i - x_{i-1} < \delta \quad \text{for every } i \in \{1, \dots, k\}.$$

Thus, if $\boldsymbol{\alpha}^i$ is an arbitrary division of $[x_{i-1}, x_i]$, we have $V(f, \boldsymbol{\alpha}^i) \le 1$. Consequently (cf. Theorem 2.1.14),

$$\operatorname{var}_a^b f = \sum_{i-1}^{k} \operatorname{var}_{x_{i-1}}^{x_i} f = \sum_{i=1}^{k} \sup_{\boldsymbol{\alpha}^i \in \mathscr{D}[x_{i-1}, x_i]} V(f, \boldsymbol{\alpha}^i) \le k < \infty,$$

i.e., the variation of f is finite. □

3.1.8 Lemma. *Let $f:[a,b]\to\mathbb{R}$ be absolutely continuous and $v(x)=\operatorname{var}_a^x f$ for $x\in[a,b]$. Then v is also absolutely continuous on $[a,b]$.*

Proof. Consider an $\varepsilon>0$ and let $\delta>0$ be such that

$$\sum_{j=1}^{m}|f(b_j)-f(a_j)|<\frac{\varepsilon}{2}$$

holds for every finite system $\{[a_j,b_j]:j=1,2,,\dots,m\}$ of nonoverlapping intervals in $[a,b]$ satisfying $\sum_{j=1}^{m}(b_j-a_j)<\delta$. Let such a system be given. If $\boldsymbol{\alpha}^j$ is an arbitrary division of $[a_j,b_j]$ for each $j\in\{1,\dots,m\}$, then

$$\sum_{j=1}^{n}V(f,\boldsymbol{\alpha}^j)<\frac{\varepsilon}{2}.$$

Hence,

$$\sum_{j=1}^{n}\big(v(b_j)-v(a_j)\big)=\sum_{j=1}^{n}\operatorname{var}_{a_j}^{b_j} f=\sum_{j=1}^{n}\Big(\sup_{\boldsymbol{\alpha}^j\in\mathscr{D}[a_j,b_j]}V(f,\boldsymbol{\alpha}^j)\Big)\le\frac{\varepsilon}{2}<\varepsilon,$$

which completes the proof. □

3.1.9 Corollary. *A function $f:[a,b]\to\mathbb{R}$ is absolutely continuous if and only if there exist nondecreasing and absolutely continuous functions $f_1,f_2:[a,b]\to\mathbb{R}$ such that $f=f_1-f_2$.*

Proof. Let $f:[a,b]\to\mathbb{R}$ be absolutely continuous. Then f has bounded variation and, by Theorem 2.1.21, there exist nondecreasing functions $f_1,f_2:[a,b]\to\mathbb{R}$ such that $f=f_1-f_2$. According to the proof of this theorem, we can take

$$f_1(x)=\operatorname{var}_a^x f \quad\text{and}\quad f_2(x)=f_1(x)-f(x) \quad\text{for } x\in[a,b].$$

Then f_1 and f_2 are absolutely continuous by Lemma 3.1.8 and Theorem 3.1.6, respectively.

The converse implication follows immediately from Theorem 3.1.6. □

3.2 ABSOLUTELY CONTINUOUS FUNCTIONS AND THE LEBESGUE INTEGRAL

It is useful to keep in mind (see Theorem 2.4.2) that every function of bounded variation on $[a,b]$ has a finite derivative almost everywhere on $[a,b]$. Thus, by Theorem 3.1.7, absolutely continuous functions have the same property. Nevertheless, there is an important difference: Consider a continuous function $f:[a,b]\to\mathbb{R}$ with bounded variation. If we know that $f'=0$ almost everywhere in $[a,b]$, it is not necessarily true that f is constant. However, the conclusion becomes true if we know that f is absolutely continuous.

The proof relies on the Vitali covering lemma, which is based on the notion of a *Vitali cover.*

3.2.1 Definition. Let $E\subset\mathbb{R}$. A collection $\mathcal{V}$ of nondegenerate intervals is called a Vitali cover of E if for each $\varepsilon>0$ and $x\in E$, there exists an interval $I\in\mathcal{V}$ such that $x\in I$ and $|I|<\varepsilon$.

3.2.2 Lemma (VITALI COVERING LEMMA). *Let $E\subset\mathbb{R}$ with $\mu^*(E)<\infty$. If $\mathcal{V}$ is a Vitali cover of E, then for every $\varepsilon>0$, there is a finite collection $\{I_1,\dots,I_N\}$ of disjoint intervals from $\mathcal{V}$ such that*

$$\mu^*(E\setminus\bigcup_{j=1}^{N} I_j)<\varepsilon.$$

As in Section 2.4, the symbol μ^* stands for the outer Lebesgue measure. The proof of the Vitali covering lemma is available in many sources; see the bibliographic remarks in the end of the chapter.

3.2.3 Lemma. *If $f:[a,b]\to\mathbb{R}$ is absolutely continuous and $f'=0$ almost everywhere in $[a,b]$, then f is constant on $[a,b]$.*

Proof. Choose an arbitrary $c\in(a,b]$; we will prove that $f(c)=f(a)$. Let

$$E=\{x\in(a,c):f'(x)=0\}$$

and note that $\mu^*((a,c)\setminus E)=0$. Consider an arbitrary $\varepsilon>0$. Since f is absolutely continuous, there exists a $\delta>0$ such

$$\sum_{j=1}^{m}|f(\beta_j)-f(\alpha_j)|<\varepsilon$$

holds for every finite system $\{[\alpha_j,\beta_j]:j=1,2,\dots,m\}$ of nonoverlapping intervals in $[a,b]$ satisfying $\sum_{j=1}^{m}(\beta_j-\alpha_j)<\delta$.

The system

$$\mathcal{V} = \left\{ [x,y] : x \in E,\ x < y < c \ \text{ and } \ \left| \frac{f(y)-f(x)}{y-x} \right| < \varepsilon \right\}$$

is a Vitali cover of E. Thus, by Lemma 3.2.2 (Vitali covering lemma), there exists a system $\{[x_j, y_j] : j = 1, \ldots, n\}$ of disjoint intervals from $\mathcal{V}$ such that

$$\mu^*(E \setminus \bigcup_{j=1}^{n} [x_j, y_j]) < \delta.$$

Since

$$(a,c) \setminus \bigcup_{j=1}^{n} [x_j, y_j] \subset \Big((a,c) \setminus E \Big) \cup \Big(E \setminus \bigcup_{j=1}^{n} [x_j, y_j] \Big)$$

and $\mu^*((a,c) \setminus E) = 0$, we obtain

$$\mu^*((a,c) \setminus \bigcup_{j=1}^{n} [x_j, y_j]) < \delta.$$

Without loss of generality, we assume that $a < x_1 < y_1 < \cdots < x_n < y_n < c$, and therefore

$$\mu^*((a,c) \setminus \bigcup_{j=1}^{n} [x_j, y_j]) = (x_1 - a) + \sum_{j=2}^{n} (x_j - y_{j-1}) + (c - y_n).$$

Using the previous two relations and the definitions of δ and $\mathcal{V}$, we get

$$\begin{aligned} |f(c) - f(a)| &\le |f(x_1) - f(a)| + \sum_{j=1}^{n} |f(y_j) - f(x_j)| \\ &\quad + \sum_{j=2}^{n} |f(x_j) - f(y_{j-1})| + |f(c) - f(y_n)| \\ &\le \varepsilon + \sum_{j=1}^{n} |f(y_j) - f(x_j)| < \varepsilon + \varepsilon \sum_{j=1}^{n} (y_j - x_j) \le \varepsilon + \varepsilon(c - a). \end{aligned}$$

Since ε can be arbitrarily small, we conclude that $f(c) = f(a)$. □

The previous result plays an important role in the context of the Lebesgue integration theory. We now summarize some well known results that provide a link between absolute continuity and Lebesgue integrals. Readers who are not familiar with Lebesgue integration theory may skip this part or consult Section 6.12. All integrals in this chapter are understood in the sense of Lebesgue, and the phrase "$f:[a,b]\to\mathbb{R}$ is Lebesgue integrable" means that the Lebesgue integral $\int_a^b f(x)\,\mathrm{d}x$ exists and is finite.

The next theorem, which can be can interpreted as one part of the fundamental theorem of calculus, shows that the indefinite Lebesgue integral is an absolutely continuous function.

3.2.4 Theorem. *Let $f:[a,b]\to\mathbb{R}$ be a Lebesgue integrable function, and*

$$F(x)=\int_a^x f(t)\,\mathrm{d}t \quad \textit{for} \quad x\in[a,b].$$

Then F is absolutely continuous and $F'=f$ almost everywhere in $[a,b]$.

To establish the second part of the fundamental theorem of calculus, suppose that $F:[a,b]\to\mathbb{R}$ satisfies $F'=f$ almost everywhere in $[a,b]$. We have to deal with two questions: 1) Is f Lebesgue integrable? 2) It is true that $\int_a^b f(x)\,\mathrm{d}x=F(b)-F(a)$? In general, the answer is negative (a classical counterexample is the Cantor function), but we have the following result.

3.2.5 Theorem. *If $F:[a,b]\to\mathbb{R}$ has bounded variation, then its derivative $F'=f$ is Lebesgue integrable on $[a,b]$.*

Moreover, if F is nondecreasing, then

$$0\le\int_a^b f(x)\,\mathrm{d}x\le F(b)-F(a).$$

The proofs of Theorems 3.2.4 and 3.2.5 are based on some auxiliary results from the Lebesgue integration theory, and are beyond the scope of this book; see the bibliographic remarks in the end of the chapter.

The conclusion of Theorem 3.2.5 can be strengthened if we assume that F is absolutely continuous.

3.2.6 Theorem. *If $F:[a,b]\to\mathbb{R}$ is absolutely continuous, then its derivative $F'=f$ is Lebesgue integrable on $[a,b]$, and*

$$\int_a^b f(x)\,\mathrm{d}x=F(b)-F(a).$$

Proof. By Theorems 3.1.7 and 3.2.5, the derivative $F'=f$ is Lebesgue integrable on $[a,b]$. Let

$$H(x)=\int_a^x f(t)\,\mathrm{d}t \quad \text{and} \quad G(x)=F(x)-H(x) \quad \text{for } x\in[a,b].$$

By Theorem 3.2.4, the function H is absolutely continuous and $H'=f$ almost everywhere on $[a,b]$. Hence, $G'=(F-H)'=0$ almost everywhere on $[a,b]$. Therefore, by Lemma 3.2.3, G is a constant function. We get

$$\int_a^b f(t)\,\mathrm{d}t=H(b)=F(b)-G(b)=F(b)-G(a)=F(b)-F(a),$$

and the proof is complete. □

Theorems 3.2.4 and 3.2.6 lead to the following characterization of Lebesgue integrable or absolutely continuous functions.

3.2.7 Corollary. *The following statements hold:*

(i) *A function $f:[a,b]\to\mathbb{R}$ is Lebesgue integrable if and only if there exists an absolutely continuous function $F:[a,b]\to\mathbb{R}$ such that $F'=f$ almost everywhere on $[a,b]$. In this case,*

$$\int_a^b f(t)\,\mathrm{d}t=F(b)-F(a).$$

(ii) *A function $f:[a,b]\to\mathbb{R}$ is absolutely continuous if and only if there exists a Lebesgue integrable function $g:[a,b]\to\mathbb{R}$ such that*

$$f(x)-f(a)=\int_a^x g(t)\,\mathrm{d}t \qquad \text{for } x\in[a,b].$$

In this case, $f'=g$ almost everywhere on $[a,b]$.

The next result is an extension of Theorem 2.1.5:

3.2.8 Theorem. *If $f:[a,b]\to\mathbb{R}$ is absolutely continuous, then*

$$\operatorname{var}_a^b f=\int_a^b |f'(t)|\,\mathrm{d}t.$$

Proof. Let $v(x)=\operatorname{var}_a^x f$ for all $x\in[a,b]$. By Lemma 3.1.8, v is absolutely continuous, and Theorem 3.2.6 yields

$$v(x)=\int_a^x v'(t)\,\mathrm{d}t \quad \text{for } x\in[a,b].$$

If $t \in [a,b)$ is such that both $f'(t)$ and $v'(t)$ exist and are finite, then

$$|f'(t)| = \lim_{h\to 0+} \left|\frac{f(t+h)-f(t)}{h}\right| \leq \lim_{h\to 0+} \frac{\operatorname{var}_t^{t+h} f}{h} = v'(t).$$

Hence, $|f'(t)| \leq v'(t)$ for almost all $t \in [a,b]$. Consequently,

$$\operatorname{var}_a^b f = v(b) = \int_a^b v'(t)\,\mathrm{d}t \geq \int_a^b |f'(t)|\,\mathrm{d}t. \tag{3.2.1}$$

On the other hand, for an arbitrary division $\boldsymbol{\alpha}$ of $[a,b]$ we have

$$V(f,\boldsymbol{\alpha}) = \sum_{j=1}^{\nu(\boldsymbol{\alpha})} \Big|\int_{\alpha_{j-1}}^{\alpha_j} f'(t)\,\mathrm{d}t\Big| \leq \sum_{j=1}^{\nu(\boldsymbol{\alpha})} \int_{\alpha_{j-1}}^{\alpha_j} |f'(t)|\,\mathrm{d}t = \int_a^b |f'(t)|\,\mathrm{d}t,$$

and therefore

$$\operatorname{var}_a^b f \leq \int_a^b |f'(t)|\,\mathrm{d}t. \tag{3.2.2}$$

By combining (3.2.1) and (3.2.2), we obtain $\operatorname{var}_a^b f = \int_a^b |f'(t)|\,\mathrm{d}t$. □

3.2.9 Exercise. Prove the following assertion:

If $f:[a,b]\to\mathbb{R}$ is absolutely continuous and $v(x) = \operatorname{var}_a^x f$ for all $x\in[a,b]$, then $v'(t) = |f'(t)|$ for almost all $t \in [a,b]$.

3.3 LEBESGUE DECOMPOSITION OF FUNCTIONS OF BOUNDED VARIATION

According to Section 2.6, each function of bounded variation can be decomposed into the sum of a continuous function with bounded variation (referred to as the continuous part) and a step function (referred to as the jump part). We now show that the continous part can be further decomposed into the sum of an absolutely continuous function and a continuous singular function (Definition 2.4.4); in this way, we obtain the so-called Lebesgue decomposition of a function with bounded variation.

3.3.1 Theorem (LEBESGUE DECOMPOSITION). *If $f:[a,b]\to\mathbb{R}$ has bounded variation, it can be written in the form*

$$f = f^{\,\mathrm{AC}} + f^{\,\mathrm{SC}} + f^{\,\mathrm{B}},$$

where $f^{\,\mathrm{AC}}$ *is an absolutely continuous function,* $f^{\,\mathrm{SC}}$ *is a singular continuous function, and* $f^{\,\mathrm{B}}$ *is a step function. Moreover, these functions are unique up to constants.*

Proof. By Theorem 2.6.1, there exists a step function $f^{\,\mathrm{B}}$ such that the function $f^{\,\mathrm{C}} = f - f^{\,\mathrm{B}}$ is continuous on $[a,b]$. Furthermore, f' is Lebesgue integrable due to Theorem 3.2.5. Let

$$f^{\,\mathrm{AC}}(x) = \int_a^x f'(t)\,\mathrm{d}t \quad \text{and} \quad f^{\,\mathrm{SC}}(x) = f^{\,\mathrm{C}}(x) - f^{\,\mathrm{AC}}(x) \quad \text{for } x \in [a,b].$$

According to Theorem 3.2.4, $f^{\,\mathrm{AC}}$ is absolutely continuous. By Theorems 2.5.5 and 3.2.4 we have $(f^{\,\mathrm{B}})' = 0$ and $(f^{\,\mathrm{AC}})' = f'$ almost everywhere on $[a,b]$. This means that

$$(f^{\,\mathrm{SC}})' = f' - (f^{\,\mathrm{B}})' - (f^{\,\mathrm{AC}})' = 0 \quad \text{almost everywhere on} \quad [a,b],$$

i.e., $f^{\,\mathrm{SC}}$ is a singular continuous function.

By Theorem 2.6.1, the function $f^{\,\mathrm{B}}$ is unique up to a constant. Every admissible function $f^{\,\mathrm{AC}}$ is absolutely continuous and satisfies

$$(f^{\,\mathrm{AC}})' = (f - f^{\,\mathrm{B}} - f^{\,\mathrm{SC}}) = f' \quad \text{almost everywhere on} \quad [a,b],$$

because $(f^{\,\mathrm{B}})' = 0$ and $(f^{\,\mathrm{SC}})' = 0$. Hence, by Lemma 3.2.3, the difference of each two admissible functions $f^{\,\mathrm{AC}}$ is constant. Consequently, $f^{\,\mathrm{SC}}$ is also unique up to a constant. □

3.3.2 Definition. If $f : [a,b] \to \mathbb{R}$ has bounded variation, then each three functions $f^{\,\mathrm{AC}}$, $f^{\,\mathrm{SC}}$ and $f^{\,\mathrm{B}}$ assigned to f by Theorem 3.3.1 are referred to as the *absolutely continuous part*, *continuous singular part*, and the *jump part* of the function f, respectively. Moreover, the sum $f^{\,\mathrm{SC}} + f^{\,\mathrm{B}}$ is called the *singular part* of f and denoted by $f^{\,\mathrm{SING}}$.

3.3.3 Remark. According to Remark 2.6.3 and the proof of Theorem 3.3.1, if $f : [a,b] \to \mathbb{R}$ has bounded variation, then the unique decomposition $f = f^{\,\mathrm{AC}} + f^{\,\mathrm{SC}} + f^{\,\mathrm{B}}$ in the sense of Definition 3.3.2 such that $f^{\,\mathrm{B}}(a) = f(a)$ and $f^{\,\mathrm{AC}}(a) = f^{\,\mathrm{SC}}(a) = 0$ is given by the formulas

$$\begin{aligned} f^{\,\mathrm{B}}(x) = {} & f(a) + \Delta^+ f(a)\,\chi_{(a,b]}(x) + \Delta^- f(b)\,\chi_{[b]}(x) \\ & + \sum_{k \in K} \big(\Delta^- f(s_k)\,\chi_{[s_k,b]}(x) + \Delta^+ f(s_k)\,\chi_{(s_k,b]}(x)\big), \end{aligned}$$

$$f^{\mathrm{AC}}(x) = \int_a^x f'(t)\,\mathrm{d}t,$$
$$f^{\mathrm{SC}}(x) = f(x) - f^{\mathrm{B}}(x) - f^{\mathrm{AC}}(x),$$

where $\{s_k\}_{k\in K}$ is the set of all discontinuity points of f in (a,b).

The next assertion provides more information in the special case when f is nondecreasing.

3.3.4 Theorem. *If $f:[a,b]\to\mathbb{R}$ is nondecreasing, then the functions $f^{\mathrm{AC}}, f^{\mathrm{SC}}$, and f^{B} from Theorem 3.3.1 are also nondecreasing.*

Proof. Without loss of generality, we can assume that $f^{\mathrm{AC}}, f^{\mathrm{SC}}$, and f^{B} are given by the formulas from Remark 3.3.3. Since f is nondecreasing, we have $\Delta^+ f(t) \ge 0$ for all $t\in[a,b)$, and $\Delta^- f(s)\ge 0$ for all $s\in(a,b]$. Thus, if $a<x<y<b$, then

$$f^{\mathrm{B}}(y) - f^{\mathrm{B}}(x) = \sum_{x<s_k\le y} \Delta^- f(s_k) + \sum_{x\le s_k<y} \Delta^+ f(s_k) \ge 0.$$

One can easily check that the same inequality holds also if $x=a$ or $y=b$. This shows that f^{B} is nondecreasing.

The continuous part $f^{\mathrm{C}} = f - f^{\mathrm{B}}$ is also nondecreasing, because Corollary 2.3.8 implies

$$f^{\mathrm{B}}(y) - f^{\mathrm{B}}(x) \le \operatorname{var}_x^y f = f(y) - f(x),$$

and hence

$$f^{\mathrm{C}}(y) - f^{\mathrm{C}}(x) = \big(f(y) - f(x)\big) - \big(f^{\mathrm{B}}(y) - f^{\mathrm{B}}(x)\big) \ge 0.$$

The derivative $f'(t)$ exists and is finite for almost all $t\in[a,b]$. Since f is nondecreasing, we have $f'\ge 0$ almost everywhere on $[a,b]$. Thus,

$$f^{\mathrm{AC}}(y) - f^{\mathrm{AC}}(x) = \int_x^y f'(t)\,\mathrm{d}t \ge 0 \quad \text{whenever } a\le x<y\le b.$$

This means that f^{AC} is nondecreasing on $[a,b]$.

By Theorem 2.5.5, $(f^{\mathrm{B}})' = 0$ almost everywhere on $[a,b]$, and hence $(f^{\mathrm{C}})' = f' - (f^{\mathrm{B}})' = f'$ almost everywhere on $[a,b]$. By Theorem 3.2.5, we deduce that

$$f^{\mathrm{C}}(y) - f^{\mathrm{C}}(x) \ge \int_x^y (f^{\mathrm{C}})'(t)\,\mathrm{d}t = \int_x^y f'(t)\,\mathrm{d}t = f^{\mathrm{AC}}(y) - f^{\mathrm{AC}}(x)$$

whenever $a \le x < y \le b$. Consequently,

$$f^{\mathrm{SC}}(y) - f^{\mathrm{SC}}(x) = \big(f^{\mathrm{C}}(y) - f^{\mathrm{AC}}(y)\big) - \big(f^{\mathrm{C}}(x) - f^{\mathrm{AC}}(x)\big) \ge 0,$$

which shows that f^{SC} is nondecreasing. □

The next result is in some sense complementary to Theorem 2.5.6.

3.3.5 Theorem. *If $f : [a,b] \to \mathbb{R}$ is singular and $g : [a,b] \to \mathbb{R}$ is absolutely continuous, then* $\operatorname{var}_a^b (f+g) = \operatorname{var}_a^b f + \operatorname{var}_a^b g$.

Proof. We will verify that the assumptions of Lemma 2.1.19 are satisfied.

Let $\varepsilon > 0$ be given. By Lemma 3.1.8, the function $x \mapsto \operatorname{var}_a^x g$ is absolutely continuous on $[a,b]$. Hence, there is a $\delta > 0$ such that

$$\sum_{j=1}^{n} \operatorname{var}_{a_j}^{b_j} g < \varepsilon \tag{3.3.1}$$

for every finite system $\{[a_j, b_j] : j = 1, 2, , \dots, n\}$ of nonoverlapping intervals in $[a,b]$ satisfying $\sum_{j=1}^{n} (b_j - a_j) < \delta$.

Further, choose a division $\boldsymbol{\sigma}$ of $[a,b]$ such that

$$V(f, \boldsymbol{\sigma}) > \operatorname{var}_a^b f - \frac{\varepsilon}{2}.$$

This implies that for any refinement $\boldsymbol{\beta}$ of $\boldsymbol{\sigma}$ we have

$$V(f, \boldsymbol{\beta}) \ge V(f, \boldsymbol{\sigma}) > \operatorname{var}_a^b f - \frac{\varepsilon}{2}. \tag{3.3.2}$$

As f is singular, there is a set $N \subset [a,b]$ such that

$$f'(x) = 0 \quad \text{for all} \quad x \in [a,b] \setminus N \quad \text{and} \quad \mu(N) = 0.$$

The system

$$\mathcal{V} = \Big\{ [\xi, \eta] \subset [a,b] \setminus \boldsymbol{\sigma} : \xi < \eta \ \text{ and } \ \Big| \frac{f(\eta) - f(\xi)}{\eta - \xi} \Big| < \frac{\varepsilon}{2(b-a)} \Big\}$$

is a Vitali cover of the set $E = [a,b] \setminus (N \cup \boldsymbol{\sigma})$. Thus, by Lemma 3.2.2 (Vitali covering lemma), there is a finite system $\big\{[\xi_j, \eta_j] : j = 1, \dots, r\big\}$ of disjoint intervals in $\mathcal{V}$ such that

$$\mu^*\Big(E \setminus \bigcup_{j=1}^{r} [\xi_j, \eta_j]\Big) < \delta.$$

We can now write the interval $[a,b]$ as a union of nonoverlapping intervals

$$[a,b]=\Big(\bigcup_{k=1}^{r}[\xi_j,\eta_j]\Big)\cup\Big(\bigcup_{j=1}^{n}[a_j,b_j]\Big),$$

where a_j, b_j can be chosen in such a way that $\boldsymbol{\sigma}\subset\{a_j,b_j: j=1,\dots,n\}$. Since

$$\bigcup_{j=1}^{n}(a_j,b_j)\subset[a,b]\setminus\bigcup_{j=1}^{r}[\xi_j,\eta_j]\subset\Big([a,b]\setminus E\Big)\cup\Big(E\setminus\bigcup_{j=1}^{r}[\xi_j,\eta_j]\Big)$$

and $\mu^*([a,b]\setminus E)=0$, we obtain

$$\sum_{j=1}^{n}(b_j-a_j)=\mu^*\Big(\bigcup_{j=1}^{n}(a_j,b_j)\Big)\le\mu^*\Big(E\setminus\bigcup_{j=1}^{r}[\xi_j,\eta_j]\Big)<\delta.$$

Consequently, we can apply (3.3.1) to get

$$\sum_{j=1}^{n}\operatorname{var}_{a_j}^{b_j} g<\varepsilon.$$

On the other hand, since $\boldsymbol{\sigma}\subset\{a_j,b_j: j=1,\dots,n\}$, the division $\boldsymbol{\beta}$ defined as the union $\{a_j,b_j: j=1,\dots,n\}\cup\{\xi_j,\eta_j: j=1,\dots,r\}$ is a refinement of $\boldsymbol{\sigma}$. Thus, from (3.3.2) and the definition of $\mathcal{V}$ it follows that

$$\begin{aligned}\operatorname{var}_a^b f-\frac{\varepsilon}{2}&<\sum_{j=1}^{r}|f(\eta_j)-f(\xi_j)|+\sum_{j=1}^{n}|f(b_j)-f(a_j)|\\&\le\sum_{j=1}^{r}\frac{\varepsilon(\eta_j-\xi_j)}{2(b-a)}+\sum_{j=1}^{n}\operatorname{var}_{a_j}^{b_j} f\le\frac{\varepsilon}{2}+\sum_{j=1}^{n}\operatorname{var}_{a_j}^{b_j} f,\end{aligned}$$

and consequently

$$\sum_{j=1}^{n}\operatorname{var}_{a_j}^{b_j} f>\operatorname{var}_a^b f-\varepsilon.$$

The assumptions of Lemma 2.1.19 are now verified, and the proof is complete. □

3.3.6 Exercise. Use Theorem 3.3.5 to prove the following statements:

(i) *Let $\{f_n\}$ be a sequence of functions with bounded variations on $[a,b]$, and let $\{f_n^{\mathrm{AC}}\}$ and $\{f_n^{\mathrm{SING}}\}$ be sequences of their absolutely continuous* *and singular parts. Then*

$$\lim_{n\to\infty}(\operatorname{var}_a^b f_n) = 0 \text{ if and only if } \lim_{n\to\infty}(\operatorname{var}_a^b f_n^{\mathrm{AC}}) = \lim_{n\to\infty}(\operatorname{var}_a^b f_n^{\mathrm{SING}}) = 0.$$

(ii) *If* $f : [a,b] \to \mathbb{R}$ *has bounded variation, then* $\operatorname{var}_a^b f \ge \int_a^b |f'(t)|\,dt$.

By Theorems 3.1.6 and 3.1.7, the space $\mathrm{AC}([a,b])$ is a linear subspace of $\mathrm{BV}([a,b])$. Using Theorem 3.3.5, we can show that it is a closed subspace. Consequently, $\mathrm{AC}([a,b])$ is a Banach space with respect to the BV norm.

3.3.7 Theorem. $\mathrm{AC}([a,b])$ *is a closed subspace of* $\mathrm{BV}([a,b])$.

Proof. Assume that $\{f_n\}$ is a sequence of absolutely continuous functions which is convergent in the BV norm to a function $f : [a,b] \to \mathbb{R}$, i.e.,

$$\lim_{n\to\infty} \|f_n - f\|_{\mathrm{BV}} = 0.$$

Clearly, $f \in \mathrm{BV}([a,b])$. Thus, we can write $f = f^{\mathrm{AC}} + f^{\mathrm{SING}}$, where f^{AC} is absolutely continuous and f^{SING} is singular. Note also that $\lim_{n\to\infty} f_n(a) = f(a)$. Using Theorem 3.3.5, we obtain

$$\begin{aligned} 0 &= \lim_{n\to\infty} \|f - f_n\|_{\mathrm{BV}} = \lim_{n\to\infty} |f(a) - f_n(a)| + \lim_{n\to\infty} \operatorname{var}_a^b (f - f_n) \\ &= \lim_{n\to\infty} \operatorname{var}_a^b (f^{\mathrm{AC}} + f^{\mathrm{SING}} - f_n) = \lim_{n\to\infty} \operatorname{var}_a^b (f^{\mathrm{AC}} - f_n) + \operatorname{var}_a^b f^{\mathrm{SING}}, \end{aligned}$$

which is possible only if $\operatorname{var}_a^b f^{\mathrm{SING}} = 0$, i.e., if f^{SING} is identically zero. It follows that $f = f^{\mathrm{AC}} \in \mathrm{AC}([a,b])$, which completes the proof. □

On the other hand, note that the space $\mathrm{AC}([a,b])$ is not a closed subspace of $\mathrm{C}([a,b])$ with the supremum norm.

3.3.8 Exercise. Let $\{f_n\}$ and f be as in Example 2.7.1. Show that f_n are absolutely continuous and uniformly convergent to f, which is not absolutely continuous.

BIBLIOGRAPHIC REMARKS

The proofs of Lemma 3.2.2 (Vitali covering lemma), Theorems 3.2.4 and 3.2.5 can be found e.g. in [Gordon (1994)] (see Lemma 4.6, Theorems 4.10 and Theorem 4.12 there) or in [Rana (2002)] (see Theorems 4.4.5, 6.2.9 and 6.3.1 there). An interesting alternative proof of Lemma 3.2.3, which relies on Lemma 6.2.3 (Cousin's lemma) instead of the Vitali covering lemma, is given in [Gordon (1998)].

Most results presented in this chapter are standard; for more information about absolutely continuous functions, see e.g. [Appell, Banás, Merentes (2014)], [Gordon (1994)], [Kolmogorov, Fomin (2012)], or [Lukeš, Malý (2017)]. Theorem 3.3.5 is taken from Chapter IX of the Czech textbook [Jarník (1976)].

Chapter 4

Regulated functions

Functions of bounded variation play a crucial role in the Stieltjes integration theory. Of similar importance is the class of regulated functions, which are functions whose one-sided limits at all points exist and are finite. Thus, regulated functions represent a very natural generalization of both continuous functions and functions of bounded variation. The present chapter provides an overview of the most important facts about regulated functions.

Throughout the rest of the book, we use the following notation: Given a function $f:[a,b]\to\mathbb{R}$, we denote

$$\|f\| = \sup_{t\in[a,b]} |f(t)|.$$

$\|f\|$ is a nonnegative real number if f is bounded, and $\|f\|=\infty$ otherwise.

4.1 Definition and basic properties

4.1.1 Definition. A function $f:[a,b]\to\mathbb{R}$ is said to be regulated if the left limit $f(t+)$ exists and is finite for every $t\in[a,b)$, and the right limit $f(t-)$ exists and is finite for every $t\in(a,b]$. The set of all regulated functions on $[a,b]$ will be denoted by $\mathrm{G}([a,b])$. Recall that

$$\Delta^+ f(t)=f(t+)-f(t) \quad\text{and}\quad \Delta^- f(s)=f(s)-f(s-) \quad\text{for } t\in[a,b),\, s\in(a,b].$$

4.1.2 Remark. Evidently, the following relations hold:

$$\mathrm{BV}([a,b])\cup\mathrm{C}([a,b])\subset\mathrm{G}([a,b]),$$

$$\mathrm{G}([a,b])\setminus\mathrm{C}([a,b])\neq\emptyset \quad\text{and}\quad \mathrm{G}([a,b])\setminus\mathrm{BV}([a,b])\neq\emptyset\,.$$

For an example of a regulated function which does not have bounded variation, see Example 2.1.10.

4.1.3 Theorem. *If a sequence $\{f_n\}$ of regulated functions converges uniformly on $[a,b]$ to a function $f:[a,b]\to\mathbb{R}$, then f is regulated.*

Proof. Let $t\in[a,b)$ and let $\{t_k\}\subset(t,b]$ be an arbitrary decreasing sequence such that $t_k\to t$ for $k\to\infty$. Given an arbitrary $\varepsilon>0$, choose $n_0\in\mathbb{N}$ and $k_0\in\mathbb{N}$ such that

$$\|f-f_{n_0}\|<\frac{\varepsilon}{3} \quad\text{and}\quad |f_{n_0}(t_k)-f_{n_0}(t_\ell)|<\frac{\varepsilon}{3} \quad \text{for all} \quad k,\ell\geq k_0.$$

Then

$$\begin{aligned}|f(t_k)-f(t_\ell)| &\leq |f(t_k)-f_{n_0}(t_k)|+|f_{n_0}(t_k)-f_{n_0}(t_\ell)|+|f(t_\ell)-f_{n_0}(t_\ell)|\\ &\leq 2\,\|f-f_{n_0}\|+|f_{n_0}(t_k)-f_{n_0}(t_\ell)|<\varepsilon \quad \text{for} \quad k,\ell\geq k_0.\end{aligned}$$

Consequently, there exists a finite limit

$$f(t+)=\lim_{k\to\infty} f(t_k).$$

Similarly, one can show that for every $t\in(a,b]$ there exists a finite limit $f(t-)$. □

4.1.4 Exercise. (i) In the context of Theorem 4.1.3, prove that

$$\begin{aligned}f(t+)&=\lim_{n\to\infty} f_n(t+) \quad \text{for every } t\in[a,b),\\ f(t-)&=\lim_{n\to\infty} f_n(t-) \quad \text{for every } t\in(a,b].\end{aligned}$$

(This statement represents a special case of the Moore-Osgood theorem; an even stronger result will be obtained in Lemma 4.2.3.)

(ii) Let $f(x)=x$ if $x=1/k$ for a certain $k\in\mathbb{N}$, and $f(x)=0$ otherwise. Show that f is regulated on $[0,1]$.

(iii) Let $f_D(x)=1$ if x is a rational number, and $f_D(x)=0$ otherwise (f_D is the *Dirichlet function*). Show that f_D is not regulated on $[0,1]$.

Let us now formulate a crucial result of this chapter.

4.1.5 Theorem (HÖNIG). *The following three statements are equivalent:*

(i) $f\in\mathrm{G}([a,b])$.

(ii) *There exists a sequence* $\{f_n\} \subset \mathrm{S}([a,b])$ *which converges uniformly to* f *on* $[a,b]$.

(iii) *For every* $\varepsilon > 0$ *there exists a division* $\boldsymbol{\alpha}$ *of* $[a,b]$ *such that*

$$|f(t) - f(s)| < \varepsilon$$

holds for every $j \in \{1,\dots,\nu(\boldsymbol{\alpha})\}$ *and each pair* $t,s \in (\alpha_{j-1},\alpha_j)$.

Proof. a) The implication (ii) $\Longrightarrow$ (i) is proved by Theorem 4.1.3.

b) Assume (i) holds and let an arbitrary $\varepsilon > 0$ be given. Denote by B the set of all points $\tau \in (a,b]$ with the following property:

$$\left.\begin{array}{l}\text{There is a division } \boldsymbol{\alpha} \text{ of } [a,\tau] \text{ such that } |f(t)-f(s)|<\varepsilon \\ \text{for each pair } t,s \in (\alpha_{j-1},\alpha_j), \text{ where } j \in \{1,\dots,\nu(\boldsymbol{\alpha})\}.\end{array}\right\} \quad (4.1.1)$$

Our goal is to prove that $b \in B$. First, we show that B is nonempty. By Definition 4.1.1, there is a $\delta_a \in (0, b-a)$ such that

$$|f(t) - f(a+)| < \frac{\varepsilon}{2} \quad \text{holds for all} \quad t \in (a, a+\delta_a).$$

Thus, for arbitrary $t,s \in (a, a+\delta_a)$, we get

$$|f(t) - f(s)| \le |f(t) - f(a+)| + |f(s) - f(a+)| < \varepsilon.$$

Denote $\tau = a + \delta_a$. Then $\{a,\tau\}$ is a division of $[a,\tau]$ satisfying (4.1.1). This means that the set B is nonempty and $\tau^* := \sup B \in (a,b]$.

Next, we will show that $\tau^* \in B$. Indeed, by Definition 4.1.1 we can choose a $\delta_1 \in (0, \tau^* - a)$ in such a way that

$$|f(t) - f(\tau^*-)| < \frac{\varepsilon}{2} \quad \text{holds for all} \quad t \in (\tau^* - \delta_1, \tau^*).$$

Hence, for arbitrary $t,s \in (\tau^* - \delta_1, \tau^*)$, we have

$$|f(t) - f(s)| \le |f(t) - f(\tau^*-)| + |f(s) - f(\tau^*-)| < \varepsilon. \quad (4.1.2)$$

Further, by the definition of the supremum, there is a $\tau \in B \cap (\tau^* - \delta_1, \tau^*)$. Let $\boldsymbol{\alpha}$ be a division of $[a,\tau]$ such that (4.1.1) is true and let $\widetilde{\boldsymbol{\alpha}} = \boldsymbol{\alpha} \cup \{\tau^*\}$. Then $\widetilde{\boldsymbol{\alpha}} = \{\alpha_0, \alpha_1, \dots, \tau, \tau^*\}$ is a division of $[a,\tau^*]$ with $\nu(\widetilde{\boldsymbol{\alpha}}) = \nu(\boldsymbol{\alpha}) + 1$, whose division points are

$$\widetilde{\alpha}_j = \begin{cases} \alpha_j & \text{if} \quad j \in \{1,\dots,\nu(\boldsymbol{\alpha})\}, \\ \tau^* & \text{if} \quad j = \nu(\widetilde{\boldsymbol{\alpha}}). \end{cases}$$

Using (4.1.1) and (4.1.2), we get $|f(t)-f(s)|<\varepsilon$ for all $t,s\in(\widetilde{\alpha}_{j-1},\widetilde{\alpha}_j)$, where $j\in\{1,\dots,\nu(\widetilde{\boldsymbol{\alpha}})\}$. This means that $\tau^*\in B$.

Finally, we prove that $\tau^*=b$. Assume, on the contrary, that $\tau^*<b$. By Definition 4.1.1, we can choose a $\delta_2\in(0,b-\tau^*)$ in such a way that

$$|f(t)-f(\tau^*+)|<\frac{\varepsilon}{2} \quad \text{holds for all } \ t\in(\tau^*,\tau^*+\delta_2).$$

Similarly as before in this proof, we can deduce that the inequality

$$|f(t)-f(s)|\le|f(t)-f(\tau^*+)|+|f(s)-f(\tau^*+)|<\varepsilon \tag{4.1.3}$$

holds for arbitrary $t,s\in(\tau^*,\tau^*+\delta_2)$. Let $\boldsymbol{\alpha}$ be a division of the interval $[a,\tau^*]$ such that (4.1.1) holds. Set $\tau=\tau^*+\delta_2$ and $\widetilde{\boldsymbol{\alpha}}=\boldsymbol{\alpha}\cup\{\tau\}$. Then

$$\widetilde{\boldsymbol{\alpha}}=\{\alpha_0,\alpha_1,\dots,\tau^*,\tau\}$$

is a division of $[a,\tau]$ with $\nu(\widetilde{\boldsymbol{\alpha}})=\nu(\boldsymbol{\alpha})+1$, whose division points are

$$\widetilde{\alpha}_j=\begin{cases}\alpha_j & \text{if } \ j\in\{1,\dots,\nu(\boldsymbol{\alpha})\},\\ \tau & \text{if } \ j=\nu(\widetilde{\boldsymbol{\alpha}}).\end{cases}$$

Using (4.1.1) and (4.1.3), we have

$$|f(t)-f(s)|<\varepsilon \quad \text{for all } t,s\in(\widetilde{\alpha}_{j-1},\widetilde{\alpha}_j) \ \text{ and } \ j\in\{1,\dots,\nu(\widetilde{\boldsymbol{\alpha}})\}.$$

It follows that $\tau\in B$. However, since we have $\tau>\tau^*$, this contradicts the definition of $\tau^*=\sup B$. Hence, $\tau^*=b$, and the proof of the implication (i) $\Longrightarrow$ (iii) is complete.

c) Assume that (iii) holds. Let $n\in\mathbb{N}$ be given and let $\boldsymbol{\alpha}$ be a division of $[a,b]$ such that $|f(t)-f(s)|<1/n$ for all $t,s\in(\alpha_{j-1},\alpha_j)$ and $j\in\{1,\dots,\nu(\boldsymbol{\alpha})\}$.

For every $j\in\{1,\dots,\nu(\boldsymbol{\alpha})\}$ choose an arbitrary $\tau_j\in(\alpha_{j-1},\alpha_j)$ and put

$$f_n(t)=\begin{cases}f(t) & \text{if } \ t\in\boldsymbol{\alpha},\\ f(\tau_j) & \text{if } \ t\in(\alpha_{j-1},\alpha_j).\end{cases}$$

Obviously, $f_n\in\mathrm{S}([a,b])$ and $\|f-f_n\|<1/n$ for every $n\in\mathbb{N}$, i.e. $f_n\rightrightarrows f$ on $[a,b]$ when $n\to\infty$. This proves the implication (iii) $\Longrightarrow$ (ii). □

4.1.6 Corollary. *Every regulated function $f:[a,b]\to\mathbb{R}$ is bounded.*

Proof. By statement (iii) of Theorem 4.1.5, there is a division $\boldsymbol{\alpha}$ of $[a,b]$ such that

$$|f(t)-f(s)|\leq 1 \quad \text{whenever} \quad t,s\in(\alpha_{j-1},\alpha_j) \quad \text{and} \quad j\in\{1,\ldots,\nu(\boldsymbol{\alpha})\}.$$

For every $j\in\{1,\ldots,\nu(\boldsymbol{\alpha})\}$, choose an arbitrary $\tau_j\in(\alpha_{j-1},\alpha_j)$. Then

$$|f(t)|\leq|f(\tau_j)|+1 \quad \text{for} \quad t\in(\alpha_{j-1},\alpha_j) \quad \text{and} \quad j\in\{1,\ldots,\nu(\boldsymbol{\alpha})\}.$$

Hence, if we denote

$$M=\max\big\{|f(\alpha_0)|,\ldots,|f(\alpha_{\nu(\boldsymbol{\alpha})})|,|f(\tau_1)|+1,\ldots,|f(\tau_{\nu(\boldsymbol{\alpha})})|+1\big\},$$

then $|f(t)|\leq M$ for all $t\in[a,b]$. □

4.1.7 Corollary. *For every regulated function $f:[a,b]\to\mathbb{R}$ and every $\varepsilon>0$, there are at most finitely many points $t\in[a,b]$ such that*

$$t\in[a,b) \quad \textit{and} \quad |\Delta^+f(t)|>\varepsilon \quad \textit{or} \quad t\in(a,b] \quad \textit{and} \quad |\Delta^-f(x)|>\varepsilon.$$

Proof. Let $\varepsilon>0$ be given. By the statement (iii) from Theorem 4.1.5, we can find a division $\boldsymbol{\alpha}$ of $[a,b]$ such that

$$|f(t)-f(s)|<\varepsilon \quad \text{for} \quad t,s\in(\alpha_{j-1},\alpha_j) \quad \text{and} \quad j\in\{1,\ldots,\nu(\boldsymbol{\alpha})\}.$$

This implies that $|\Delta^+f(t)|\leq\varepsilon$ and $|\Delta^-f(t)|\leq\varepsilon$ for $t\in[a,b]\setminus\boldsymbol{\alpha}$, and the proof is complete. □

4.1.8 Theorem. *Every regulated function $f:[a,b]\to\mathbb{R}$ has at most countably many discontinuities.*

Proof. For each $k\in\mathbb{N}$, denote

$$D_k^+=\big\{t\in[a,b):|\Delta^+f(t)|>\tfrac{1}{k}\big\} \quad \text{and} \quad D_k^-=\big\{t\in(a,b]:|\Delta^-f(t)|>\tfrac{1}{k}\big\}.$$

Then

$$D^+=\bigcup_{k\in\mathbb{N}}D_k^+=\big\{t\in[a,b):|\Delta^+f(t)|>0\big\}$$

is the set of all points where f is discontinuous from the right, and

$$D^-=\bigcup_{k\in\mathbb{N}}D_k^-=\{t\in(a,b]:|\Delta^-f(t)|>0\}$$

is the set of all points where the function f is discontinuous from the left. Obviously, $D=D^+\cup D^-$ is the set of all discontinuity points of f on $[a,b]$.

By Corollary 4.1.7 every set D_k^+, D_k^-, $k \in \mathbb{N}$, is finite. As a result, D is at most countable. □

4.1.9 Corollary. *Let* $f \in \mathrm{G}([a,b])$ *and*

$$\widetilde{f}(t) = \begin{cases} f(t+) & \text{if } t \in [a,b), \\ f(b) & \text{if } t = b, \end{cases} \tag{4.1.4}$$

$$\widehat{f}(t) = \begin{cases} f(a) & \text{if } x = a, \\ f(t-) & \text{if } t \in (a,b]. \end{cases} \tag{4.1.5}$$

Then both $\widetilde{f}$ *and* $\widehat{f}$ *are regulated on* $[a,b]$ *and*

$$\widetilde{f}(t+) = f(t+) \text{ if } t \in [a,b), \quad \widetilde{f}(t-) = f(t-) \text{ if } t \in (a,b], \tag{4.1.6}$$

$$\widehat{f}(t+) = f(t+) \text{ if } t \in [a,b), \quad \widehat{f}(t-) = f(t-) \text{ if } t \in (a,b]. \tag{4.1.7}$$

Proof. a) Let $\varepsilon > 0$ be given. By Theorem 4.1.5 (iii), there exists a division $\boldsymbol{\alpha}$ of $[a,b]$ such that

$$|f(t) - f(s)| < \frac{\varepsilon}{2}$$

whenever $t, s \in (\alpha_{j-1}, \alpha_j)$ for some $j \in \{1, \dots, \nu(\boldsymbol{\alpha})\}$. In particular,

$$|f(t+\delta) - f(s+\delta)| < \frac{\varepsilon}{2}$$

holds for every pair $t, s \in [\alpha_{j-1}, \alpha_j)$ with $j \in \{1, \dots, \nu(\boldsymbol{\alpha})\}$ and every $\delta > 0$ such that $t+\delta, s+\delta \in (\alpha_{j-1}, \alpha_j)$. Therefore

$$|f(t+) - f(s+)| = \lim_{\delta \to 0+} |f(t+\delta) - f(s+\delta)| \le \frac{\varepsilon}{2} < \varepsilon$$

holds for each $j \in \{1, \dots, \nu(\boldsymbol{\alpha})\}$ and each pair $t, s \in [\alpha_{j-1}, \alpha_j)$, as well. In other words,

$$|\widetilde{f}(t) - \widetilde{f}(s)| < \varepsilon \text{ for every } j \in \{1, \dots, \nu(\boldsymbol{\alpha})\} \text{ and } t, s \in [\alpha_{j-1}, \alpha_j). \tag{4.1.8}$$

Similarly, it can be shown that

$$|\widehat{f}(t) - \widehat{f}(s)| < \varepsilon \text{ for every } j \in \{1, \dots, \nu(\boldsymbol{\alpha})\} \text{ and } t, s \in (\alpha_{j-1}, \alpha_j]. \tag{4.1.9}$$

By Theorem 4.1.5, it follows that both $\widetilde{f}$ and $\widehat{f}$ are regulated on $[a,b]$.

b) Let $x \in [a,b)$ and $\varepsilon > 0$ be given, and let $\boldsymbol{\alpha}$ be a division of $[a,b]$ such that $|f(t) - f(s)| < \varepsilon/2$ for all $t, s \in (\alpha_{j-1}, \alpha_j)$ and $j \in \{1, \dots, \nu(\boldsymbol{\alpha})\}$. There is a unique index $i \in \{1, \dots, \nu(\boldsymbol{\alpha})\}$ such that $x \in [\alpha_{i-1}, \alpha_i)$. By (4.1.8), we have

$$|\widetilde{f}(t) - f(x+)| = |\widetilde{f}(t) - \widetilde{f}(x)| < \varepsilon \quad \text{for} \quad t \in (x, \alpha_i).$$

In other words, $\widetilde{f}(x+) = f(x+)$. This proves the first statement from (4.1.6).

c) Analogously, let $x \in (a, b]$, $\varepsilon > 0$, and let $\boldsymbol{\alpha}$ be a division of $[a, b]$ such that $|f(t) - f(s)| < \varepsilon/2$ for all $t, s \in (\alpha_{j-1}, \alpha_j)$ and $j \in \{1, \dots, \nu(\boldsymbol{\alpha})\}$. There is a unique $i \in \{1, \dots, \nu(\boldsymbol{\alpha})\}$ such that $x \in (\alpha_{i-1}, \alpha_i]$. If $t \in (\alpha_{i-1}, x)$ and $0 < \delta < \min\{x - t, x - \alpha_{i-1}\}$, then

$$\alpha_{i-1} < x - \delta < \alpha_i \quad \text{and} \quad \alpha_{i-1} < t + \delta < x.$$

Therefore, by the definition of the division $\boldsymbol{\alpha}$, we have

$$|f(x-) - \widetilde{f}(t)| = \lim_{\delta \to 0+} |f(x - \delta) - f(t + \delta)| \le \frac{\varepsilon}{2} < \varepsilon \quad \text{for} \quad t \in (\alpha_{i-1}, x).$$

In other words $\widetilde{f}(x-) = f(x-)$ and this completes the proof of the second statement from (4.1.6).

d) The relations (4.1.7) can be proved similarly to (4.1.6). □

4.2 SPACE OF REGULATED FUNCTIONS AND ITS SUBSPACES

Clearly, the set $\mathrm{G}([a, b])$ is a linear space with respect to pointwise addition and multiplication by scalars. It is also easy to verify that

$$\|f\| = \sup_{x \in [a,b]} |f(x)|$$

defines a norm on $\mathrm{G}([a, b])$. We now prove that $\mathrm{G}([a, b])$ with respect to this norm is a complete space.

4.2.1 Theorem. $\mathrm{G}([a, b])$ *is a Banach space.*

P r o o f. Let $\{f_n\} \subset \mathrm{G}([a, b])$ be a Cauchy sequence, i.e., assume that

$$\left.\begin{array}{l} \text{for each } \varepsilon > 0 \text{ there is } n_\varepsilon \in \mathbb{N} \text{ such that} \\ |f_n(x) - f_m(x)| \le \|f_n - f_m\| < \varepsilon \text{ for } x \in [a, b] \text{ and } n, m \ge n_\varepsilon. \end{array}\right\} \tag{4.2.1}$$

We will show that $\{f_n\}$ has a limit in $\mathrm{G}([a, b])$.

By (4.2.1), the sequence $\{f_n(x)\}$ is a Cauchy sequence of real numbers for every $x\in[a,b]$. Hence, for each $x\in[a,b]$ it has a finite limit

$$\lim_{n\to\infty} f_n(x)=f(x).$$

Let an arbitrary $\varepsilon>0$ be given and let $n_\varepsilon\in\mathbb{N}$ be determined by (4.2.1). Then for every $x\in[a,b]$ we have

$$|f(x)-f_{n_\varepsilon}(x)|=\lim_{n\to\infty}|f_n(x)-f_{n_\varepsilon}(x)|\le\varepsilon,$$

and hence the inequalities

$$|f(x)-f_n(x)|\le|f(x)-f_{n_\varepsilon}(x)|+|f_{n_\varepsilon}(x)-f_n(x)|<2\,\varepsilon \tag{4.2.2}$$

hold for all $n\ge n_\varepsilon$ and $x\in[a,b]$. This means that $\{f_n\}$ is uniformly convergent to f on $[a,b]$. Hence, Theorem 4.1.3 implies that $f\in\mathrm{G}([a,b])$. On the other hand, (4.2.2) implies

$$\|f-f_n\|\le 2\,\varepsilon \quad \text{for } n\ge n_\varepsilon,$$

and it follows that $\{f_n\}$ is convergent to f in the norm of $\mathrm{G}([a,b])$; this completes the proof. □

4.2.2 Remark. (i) By Definition 2.5.1, $f\in\mathrm{S}([a,b])$ if and only if there exists a division $\boldsymbol{\alpha}$ of the interval $[a,b]$ such that f is constant on every subinterval (α_{j-1},α_j). Every function from $\mathrm{S}([a,b])$ is a finite linear combination of functions of the form $\chi_{(\alpha,\beta)}$ and $\chi_{[\tau]}$, where (α,β) is an arbitrary subinterval in $[a,b]$ and τ is any point in $[a,b]$. Note that

$$\chi_{(\alpha,\beta)}=\chi_{(\alpha,b]}-\chi_{[\beta,b]} \quad \text{for all } \alpha,\beta\in[a,b],\ \alpha<\beta$$

and

$$\chi_{[\tau]}=\chi_{[\tau,b]}-\chi_{(\tau,b]} \quad \text{for all } \tau\in[a,b).$$

Hence $f\in\mathrm{S}([a,b])$ if and only if f is a finite linear combination of functions of the forms $\chi_{[b]}$, $\chi_{[\tau,b]}$, $\chi_{(\tau,b]}$, where τ can be an arbitrary point in $[a,b)$, i.e.,

$$\mathrm{S}([a,b])=\mathrm{Lin}\Big(\{\chi_{[\tau,b]},\,\chi_{(\tau,b]},\,\chi_{[b]}:\tau\in[a,b)\}\Big), \tag{4.2.3}$$

where $\mathrm{Lin}\,(M)$ denotes the linear span of the set M.

(ii) Similarly, we can show that also

$$\mathrm{S}([a,b])=\mathrm{Lin}\Big(\{\chi_{[a,\tau]},\,\chi_{[a,\tau)},\,\chi_{[a]}:\tau\in(a,b]\}\Big). \tag{4.2.4}$$

(iii) By Theorem 4.1.5, the set $\mathrm{S}([a,b])$ is dense in $\mathrm{G}([a,b])$, i.e., the closure of $\mathrm{S}([a,b])$ is $\mathrm{G}([a,b])$.

4.2.3 Lemma. *Let* $\{f_n\} \subset \mathrm{G}([a,b])$ *and* $f_n \rightrightarrows f$ *on* $[a,b]$. *For* $n \in \mathbb{N}$, *set*

$$\widetilde{f}_n(t) = \begin{cases} f_n(t+), & \text{if } t \in [a,b), \\ f_n(b), & \text{if } t = b, \end{cases} \qquad \widehat{f}_n(t) = \begin{cases} f_n(a), & \text{if } t = a, \\ f_n(t-), & \text{if } t \in (a,b], \end{cases}$$

and

$$\widetilde{f}(t) = \begin{cases} f(t+), & \text{if } t \in [a,b), \\ f(b), & \text{if } t = b, \end{cases} \qquad \widehat{f}(t) = \begin{cases} f(a), & \text{if } t = a, \\ f(t-), & \text{if } t \in (a,b]. \end{cases}$$

Then $\widetilde{f}_n \rightrightarrows \widetilde{f}$ *and* $\widehat{f}_n \rightrightarrows \widehat{f}$ *on* $[a,b]$.

P r o o f. By Corollary 4.1.9, the functions $\widetilde{f}$, $\widetilde{f}_n$, $\widehat{f}$, $\widehat{f}_n$, $n \in \mathbb{N}$, are regulated on $[a,b]$. Let $\varepsilon > 0$ be given. Choose $n_\varepsilon \in \mathbb{N}$ such that $|f_n(s) - f(s)| < \varepsilon/2$ for every $n \ge n_\varepsilon$ and every $s \in [a,b]$. Letting $s \to t$ from the right we get that

$$|\widetilde{f}_n(t) - \widetilde{f}(t)| = \lim_{s \to t+} |f_n(s) - f(s)| \le \tfrac{\varepsilon}{2} < \varepsilon$$

holds for every $t \in [a,b)$ and every $n \ge n_\varepsilon$. Consequently,

$$\lim_{n \to \infty} \|\widetilde{f}_n - \widetilde{f}\| = 0, \ \text{ i.e., } \ \widetilde{f}_n \rightrightarrows \widetilde{f} \ \text{ on } [a,b].$$

Similarly, we would show that $\widehat{f}_n \rightrightarrows \widehat{f}$ on $[a,b]$. □

Note that if the assumptions of Lemma 4.2.3 are satisfied, then

$$f(t+) = \lim_{n \to \infty} f_n(t+) \ \text{ for each } t \in [a,b),$$
$$f(t-) = \lim_{n \to \infty} f_n(t-) \ \text{ for each } t \in (a,b];$$

this observation leads to the following corollary.

4.2.4 Corollary. *The sets*

$$\mathrm{G_L}([a,b]) = \{f \in \mathrm{G}([a,b]) : f(t-) = f(t) \text{ for } t \in (a,b]\},$$

$$\mathrm{G_R}([a,b]) = \{f \in \mathrm{G}([a,b]) : f(t+) = f(t) \text{ for } t \in [a,b)\},$$

$$\mathrm{G_{reg}}([a,b]) = \{f \in \mathrm{G}([a,b]) : f(t-) + f(t+) = 2\,f(t) \text{ for } t \in (a,b), \\ f(a+) = f(a),\ f(b-) = f(b)\}$$

are closed in $\mathrm{G}([a,b])$.

Functions from the space $G_{reg}([a,b])$ are said to be *regular* on $[a,b]$.

4.2.5 Exercise. Prove the following assertions:

(i) Left-continuous step functions are dense in $G_L([a,b])$.

(ii) Right-continuous step functions are dense in $G_R([a,b])$.

(iii) Regular step functions are dense in $G_{reg}([a,b])$.

Hint: Make use of Theorem 4.1.5.

4.2.6 Exercise. Prove the following relations:

(i) $G_L([a,b]) \cap S([a,b]) = \text{Lin}\Big(\{\chi_{[a,\tau]} : \tau \in [a,b]\}\Big)$

(ii) $G_R([a,b]) \cap S([a,b]) = \text{Lin}\Big(\{\chi_{[\tau,b]} : \tau \in [a,b]\}\Big)$

(iii) $G_{reg}([a,b]) \cap S([a,b]) = \text{Lin}\Big(\{\chi_{[a,b]}, \tfrac{1}{2}\,\chi_{[\tau]} + \chi_{(\tau,b]} : \tau \in (a,b)\}\Big)$

Hint: See Remark 4.2.2.

4.3 Relatively compact subsets of $\mathbf{G}([a,b])$

Recall that a subset M of a Banach space X is *relatively compact* if any sequence of its elements contains a convergent subsequence. It is known (see e.g. Theorem I.6.15 in [Dunford, Schwartz (1958,1963)], Theorem 4.17 in [Schechter M. (2001)], or page 93 of [Shilov (1996)]) that M is relatively compact if and only if it is *totally bounded*, i.e., if for each $\varepsilon > 0$ there is a finite set $D_\varepsilon \subset X$ such that for every $x \in M$ there exists a $d \in D_\varepsilon$ satisfying $\|x-d\|_X < \varepsilon$. Such a set D_ε is called an *ε-net* for M in X.

The following assertion is not surprising.

4.3.1 Lemma. *Each totally bounded set is bounded.*

Proof. Let $M \subset X$ be totally bounded and let $D = \{d_1, d_2, \dots, d_m\} \subset X$ be such that for each $x \in M$, there is $\widetilde{d}_x \in D$ satisfying $\|x - \widetilde{d}_x\|_X < 1$. Thus, for an arbitrary $x \in M$ we have

$$\|x\|_X \le \|x - \widetilde{d}_x\|_X + \|\widetilde{d}_x\|_X \le K,$$

where $K = 1 + \max\{\|d_1\|_X, \dots, \|d_n\|_X\}$ does not depend on $x \in M$. □

In the space $C([a,b])$ of continuous functions we have the following criterion for relative compactness known as the Arzelà-Ascoli theorem. Its proof can be found in many functional analysis textbooks, see e.g. [Kolmogorov, Fomin (2012)].

4.3.2 Theorem (ARZELÀ-ASCOLI). *A subset M of the space* $\mathrm{C}([a,b])$ *is relatively compact if and only if the following conditions are satisfied:*

(i) *There is a constant $c^* \geq 0$ such that $\|f\| \leq c^*$ for each $f \in M$.*

(ii) *For each $\varepsilon > 0$ there is a $\delta > 0$ such that $|f(t) - f(s)| < \varepsilon$ holds for each $f \in M$ and each $t, s \in [a,b]$ satisfying $|t - s| < \delta$.*

If the condition (i) from the Arzelà-Ascoli theorem is satisfied, we say that the set M is *uniformly bounded,* while if the condition (ii) is satisfied, we say that the set M is *equicontinuous.* Thus, Theorem 4.3.2 can be reformulated as follows:

A subset of $\mathrm{C}([a,b])$ *is relatively compact if and only if it is uniformly bounded and equicontinuous.*

We will derive an analogous criterion for subsets of the space $\mathrm{G}([a,b])$ with the notion of equicontinuity replaced by the related notion of *equiregulatedness.* The definition of this notion resembles the definition of equicontinuity with ordinary limits replaced by the one-sided ones.

4.3.3 Definition. A subset M of $\mathrm{G}([a,b])$ is called *equiregulated* if the following conditions hold:

- For each $\varepsilon > 0$ and $\tau \in (a,b]$ there is a $\delta_1(\tau) \in (0, \tau - a)$ such that

$$|f(\tau-) - f(t)| < \varepsilon \quad \text{for all } t \in (\tau - \delta_1(\tau), \tau) \quad \text{and} \quad f \in M.$$

- For each $\varepsilon > 0$ and $\tau \in [a,b)$ there is a $\delta_2(\tau) \in (0, b - \tau)$ such that

$$|f(\tau+) - f(t)| < \varepsilon \quad \text{for all } t \in (\tau, \tau + \delta_2(\tau)) \quad \text{and} \quad f \in M.$$

The next lemma provides a useful characterization of equiregulated sets of functions.

4.3.4 Lemma. *The following statements are equivalent:*

(i) *$M \subset \mathrm{G}([a,b])$ is equiregulated.*

(ii) *For every $\varepsilon > 0$ there exists a division $\boldsymbol{\alpha}$ of $[a,b]$ such that for every $f \in M$, $j \in \{1, \dots, \nu(\boldsymbol{\alpha})\}$ and $s, t \in (\alpha_{j-1}, \alpha_j)$, we have $|f(s) - f(t)| < \varepsilon$.*

Proof. a) The proof of the implication (i) $\Longrightarrow$ (ii) is almost identical with the proof of the implication (i) $\Longrightarrow$ (iii) in Theorem 4.1.5; we leave it as an exercise for the reader.

b) Let an arbitrary $\varepsilon > 0$ be given, and let $\boldsymbol{\alpha}$ be the corresponding division from condition (ii).

Choose an arbitrary $\tau \in (a, b]$. There is a unique $j \in \{1, \dots, \nu(\boldsymbol{\alpha})\}$ such that $\tau \in (\alpha_{j-1}, \alpha_j]$. For all $t, s \in (\alpha_{j-1}, \tau)$ and $f \in M$ we have $|f(s) - f(t)| < \varepsilon$. Letting $s \to \tau-$ we get

$$|f(\tau-) - f(t)| \le \varepsilon \quad \text{for all } t \in (\tau - \delta_1, \tau) \text{ and } f \in M,$$

where $\delta_1 = \tau - \alpha_{j-1}$.

Analogously, if $\tau \in [a, b)$, there exists a unique $j \in \{1, \dots, \nu(\boldsymbol{\alpha})\}$ such that $\tau \in [\alpha_{j-1}, \alpha_j)$. Hence, $|f(s) - f(t)| < \varepsilon$ holds for all $t, s \in (\tau, \alpha_j)$ and $f \in M$. Letting $s \to \tau+$ yields

$$|f(\tau+) - f(t)| \le \varepsilon \quad \text{for all } \ t \in (\tau, \tau + \delta_2), \quad f \in M,$$

where $\delta_2 = \alpha_j - \tau$.

This shows that $M \subset \mathrm{G}([a, b])$ is equiregulated. □

4.3.5 Exercise. Prove the implication (i) $\Longrightarrow$ (ii) from Lemma 4.3.4.

We now proceed to the analogue of the Arzelà-Ascoli theorem in the world of regulated functions, which reads as follows.

4.3.6 Theorem (FRAŇKOVÁ). *A set $M \subset \mathrm{G}([a, b])$ is relatively compact if and only if it is uniformly bounded and equiregulated.*

Proof. a) Let $M \subset \mathrm{G}([a, b])$ be relatively compact. We will show that M is uniformly bounded and equiregulated. The uniform boundedness of functions from M follows from Lemma 4.3.1. It remains to show that M is equiregulated.

Let $\varepsilon > 0$ and $\tau \in [a, b]$ be given and let $F = \{f_1, f_2, \dots, f_m\}$ be an $\varepsilon/3-$net for the set M in $\mathrm{G}([a, b])$. This means that

$$\text{for any } \ f \in M \ \text{ there is an } \widetilde{f} \in F \ \text{ such that } \ \|f - \widetilde{f}\| < \frac{\varepsilon}{3}. \tag{4.3.1}$$

Consequently, the inequalities

$$\left.\begin{aligned} &|f(t-)-\widetilde{f}(t-)| \le \varepsilon/3 \quad \text{for} \quad t \in (a,b], \\ &|f(t+)-\widetilde{f}(t+)| \le \varepsilon/3 \quad \text{for} \quad t \in [a,b) \end{aligned}\right\} \tag{4.3.2}$$

hold for any $f \in M$ and any $\widetilde{f}$ satisfying (4.3.1). All the functions $f_k \in F$ are regulated on $[a,b]$. Hence, for a given $\tau \in (a,b]$ and every $k \in \{1,2,\ldots,m\}$, there is a $\delta_k^1 \in (0, \tau - a)$ such that

$$|f_k(t) - f_k(\tau-)| < \varepsilon \quad \text{for} \quad t \in (\tau - \delta_k^1, \tau). \tag{4.3.3}$$

Similarly, for a given $\tau \in [a,b)$ and every $k \in \{1,2,\ldots,m\}$, there is a $\delta_k^2 \in (0, b-\tau)$ such that

$$|f_k(t) - f_k(\tau+)| < \varepsilon \quad \text{for} \quad t \in (\tau, \tau + \delta_k^2). \tag{4.3.4}$$

Set

$$\delta = \begin{cases} \min\{\delta_k^i : i=1,2;\ k=1,\ldots,m\} & \text{if} \quad \tau \in (a,b), \\ \min\{\delta_k^1 : k=1,\ldots,m\} & \text{if} \quad \tau = b, \\ \min\{\delta_k^2 : k=1,\ldots,m\} & \text{if} \quad \tau = a. \end{cases}$$

If $a \le \tau - \delta < t < \tau \le b$, then by (4.3.1)–(4.3.4) the inequalities

$$|f(t) - f(\tau-)| \le |f(t) - \widetilde{f}(t)| + |\widetilde{f}(t) - \widetilde{f}(\tau-)| + |\widetilde{f}(\tau-) - f(\tau-)| < \varepsilon$$

hold for any $f \in M$ and any $\widetilde{f}$ corresponding to f by (4.3.1). Similarly, we can prove that

$$|f(t) - f(\tau+)| < \varepsilon \quad \text{for all} \quad f \in M$$

whenever $a \le \tau - \delta < t < \tau \le b$. Consequently, the set M is equiregulated.

b) Now, assume that M is uniformly bounded and equiregulated. We will show that M is relatively compact in $\mathrm{G}([a,b])$. It suffices to show that M is totally bounded, i.e., that for every $\varepsilon > 0$ the set M has a finite ε-net in $\mathrm{G}([a,b])$.

Let an arbitrary $\varepsilon > 0$ be given, and let $\boldsymbol{\alpha}$ be the corresponding division of $[a,b]$ from part (ii) of Lemma 4.3.4.

Since M is uniformly bounded, there is a $c^* > 0$ such that $\|f\| \le c^*$ for all $f \in M$. Let $\boldsymbol{z} = \{z_0, z_1, \ldots, z_n\}$ be a division of $[-c^*, c^*]$ such that

$$|\boldsymbol{z}| = \max_{1 \le j \le n} (z_j - z_{j-1}) < \frac{\varepsilon}{2}.$$

Let F be the set of all functions $\widetilde{f}:[a,b]\to\mathbb{R}$ which are constant on each of the intervals (α_{j-1},α_j), $j=1,\dots,\nu(\boldsymbol{\alpha})$, and whose values belong to the set $\boldsymbol{z}$. The number of elements of F is obviously finite.

We will show that F is an ε-net for M in $\mathrm{G}([a,b])$. To this aim, consider an arbitrary function $f\in M$. By the definition of $\boldsymbol{z}$, we know that

- for each $j\in\{0,1,\dots,\nu(\boldsymbol{\alpha})\}$ there is a $k_j\in\{0,1,\dots,n\}$ such that
$$\left|f(\alpha_j)-z_{k_j}\right|<\frac{\varepsilon}{2},$$
- for each $j\in\{1,\dots,\nu(\boldsymbol{\alpha})\}$ there is an $\ell_j\in\{0,1,\dots,n\}$ such that
$$\left|f\Big(\frac{\alpha_{j-1}+\alpha_j}{2}\Big)-z_{\ell_j}\right|<\frac{\varepsilon}{2}.$$

Furthermore, by the definition of $\boldsymbol{\alpha}$, we have

$$|f(t)-z_{\ell_j}|\le\left|f(t)-f\Big(\frac{\alpha_{j-1}+\alpha_j}{2}\Big)\right|+\left|f\Big(\frac{\alpha_{j-1}+\alpha_j}{2}\Big)-z_{\ell_j}\right|<\varepsilon$$

for all $j\in\{1,\dots,\nu(\boldsymbol{\alpha})\}$ and $t\in(\alpha_{j-1},\alpha_j)$, as well. Let us define

$$\widetilde{f}(t)=\begin{cases} z_{k_j} & \text{if } \ t=\alpha_j \quad \text{for some } \ j\in\{0,1,\dots,m\},\\ z_{\ell_j} & \text{if } \ t\in(\alpha_{j-1},\alpha_j) \quad \text{for some } \ j\in\{1,\dots,m\}.\end{cases}$$

Obviously, $\widetilde{f}\in F$ and $\|f-\widetilde{f}\|<\varepsilon$. Thus F is an ε-net for M in $\mathrm{G}([a,b])$, and the proof is complete. □

The next assertion shows that the condition of uniform boundedness can be weakened.

4.3.7 Corollary. *A subset M of the space $\mathrm{G}([a,b])$ is relatively compact if and only if it is equiregulated and*

$$\textit{the set } \{f(t):f\in M\} \ \textit{ is bounded for each } t\in[a,b]. \tag{4.3.5}$$

Proof. If $M\subset\mathrm{G}([a,b])$ is uniformly bounded, then it obviously satisfies condition (4.3.5). Hence, by Theorem 4.3.6, any relatively compact subset M of $\mathrm{G}([a,b])$ is equiregulated and satisfies (4.3.5). It remains to prove the reverse implication. To this aim, assume that M is equiregulated and satisfies condition (4.3.5). We will show that M is uniformly bounded. By Lemma 4.3.4, we can choose a division $\boldsymbol{\alpha}$ of $[a,b]$ such that

$$\left.\begin{aligned}&|f(t)-f(s)|<1\\ &\qquad\text{for all } t,s\in(\alpha_{j-1},\alpha_j),\ j\in\{1,\dots,m\} \ \text{ and } \ f\in M,\end{aligned}\right\}\tag{4.3.6}$$

where $m = \nu(\boldsymbol{\alpha})$. By our assumption (4.3.5), there exist constants

$$\gamma_j,\ j=0,1,\dots,m, \quad \text{and} \quad \widetilde{\gamma}_j,\ j=1,\dots,m,$$

such that the estimates

$$\left.\begin{aligned} |f(\alpha_j)| &\le \gamma_j && \text{for } j=0,1,\dots,m,\\ \big|f(\tfrac{1}{2}(\alpha_{j-1}+\alpha_j))\big| &\le \widetilde{\gamma}_j && \text{for } j=1,\dots,m \end{aligned}\right\} \tag{4.3.7}$$

hold for all $f \in M$. This, together with (4.3.6), implies that the estimate

$$\left.\begin{aligned} |f(t)| < \big|f(\tfrac{1}{2}(\alpha_{j-1}+\alpha_j))\big| + 1 \le \widetilde{\gamma}_j + 1 \\ \text{if } t \in (\alpha_{j-1},\alpha_j) \text{ and } j \in \{1,\dots,m\} \end{aligned}\right\} \tag{4.3.8}$$

holds for each $f \in M$. According to (4.3.7) and (4.3.8) we have $\|f\| < c^*$ for any $f \in M$, where

$$\gamma^* = \max\{\gamma_j : j=0,1,\dots,m\}, \quad \widetilde{\gamma}^* = \max\{\widetilde{\gamma}_j : j=1,\dots,m\},$$
$$c^* = \max\{\gamma^*, \widetilde{\gamma}^*\} + 1.$$

Hence, the set M is uniformly bounded and the proof is complete. □

We conclude this section by another useful criterion for the relative compactness in $\mathrm{G}([a,b])$.

4.3.8 Corollary. *Let $M \subset \mathrm{G}([a,b])$. Assume that the set $\{f(a) : f \in M\}$ is bounded and there exists a nondecreasing function $h : [a,b] \to \mathbb{R}$ such that*

$$|f(t) - f(s)| \le |h(t) - h(s)| \quad \textit{for all } t,s \in [a,b] \textit{ and } f \in M. \tag{4.3.9}$$

Then M is relatively compact in $\mathrm{G}([a,b])$.

Proof. By assumption, there is $K \in [0,\infty)$ such that $|f(a)| \le K$ for any $f \in M$. Consequently,

$$|f(x)| \le |f(a)| + |f(x) - f(a)| \le K + h(b) - h(a)$$

for all $x \in [a,b]$ and $f \in M$. Thus, the set M is uniformly bounded.

Now, let an arbitrary $\varepsilon > 0$ be given. Obviously, $h \in \mathrm{G}([a,b])$. Hence, by Theorem 4.1.5, there is a division $\boldsymbol{\alpha}$ of $[a,b]$ such that

$$|h(t) - h(s)| < \varepsilon \text{ for all } t,s \in (\alpha_{j-1},\alpha_j) \quad \text{and} \quad j \in \{1,\dots,\nu(\boldsymbol{\alpha})\}.$$

It follows that

$$|f(t) - f(s)| \le |h(t) - h(s)| < \varepsilon$$

for arbitrary $t, s \in (\alpha_{j-1}, \alpha_j)$, $j \in \{1, \dots, \nu(\boldsymbol{\alpha})\}$, and $f \in M$. By Lemma 4.3.4, the set M is equiregulated. Finally, by Theorem 4.3.6, M is relatively compact in $\mathrm{G}([a,b])$. □

4.3.9 Remark. Corollary 4.3.8 provides a sufficient condition for the relative compactness of sets in $\mathrm{G}([a,b])$. This condition is not necessary: If (4.3.9) holds, it is easy to verify that $\operatorname{var}_a^b f \le h(b) - h(a)$ for each $f \in M$. However, in general, regulated functions need not have bounded variation.

The next assertion shows that if $M \subset \mathrm{G}([a,b])$ is a pointwise convergent sequence of functions which satisfy (4.3.9), then necessarily this sequence converges uniformly.

4.3.10 Corollary. *Assume that* $\{f_n\} \subset \mathrm{G}([a,b])$ *is a sequence which is pointwise convergent to* $f\colon [a,b] \to \mathbb{R}$. *Moreover, suppose there exists a nondecreasing function* $h\colon [a,b] \to \mathbb{R}$ *such that*

$$|f_n(t) - f_n(s)| \le |h(t) - h(s)| \quad \textit{for all } t, s \in [a,b] \textit{ and } n \in \mathbb{N}.$$

Then $\{f_n\}$ *converges uniformly to* f *on* $[a,b]$.

P r o o f. Corollary 4.3.8 implies that each subsequence of $\{f_n\}$ has a subsequence which is uniformly convergent. Obviously, the uniform limit of this subsequence is necessarily f.

Now, suppose that $\{f_n\}$ is not uniformly convergent to f. Then there exists an $\varepsilon > 0$ such that for each $k \in \mathbb{N}$, there is an index $n_k \in \mathbb{N}$ with the property that $\|f_{n_k} - f\| \ge \varepsilon$. It follows that $\{f_{n_k}\}_k$ has no subsequence which is uniformly convergent to f, and this is a contradiction. □

BIBLIOGRAPHIC REMARKS

Regulated functions ("Regelfunktionen") were introduced by G. Aumann in Section 7.3 of [Aumann (1954)]. Later, Ch. S. Hönig in Section I.3 of [Hönig (1975)] presented some of their basic properties needed for the study of a class of generalized integral equations. Other useful results (e.g., a characterization of compact sets in $\mathrm{G}([a,b])$ or an analogue of Helly's selection theorem) are included in [Fraňková (1991)].

Chapter 5

Riemann-Stieltjes integral

This chapter focuses on the simplest Stieltjes-type integral – the Riemann-Stieltjes integral, which makes it possible to deal with the problems mentioned in Chapter 1. It is well known that there are three equivalent definitions of the classical Riemann integral: the first is based on partitions whose norm approaches zero (Riemann's original definition), the second involves refinements of partitions, and the third rests on upper and lower sums (Darboux's definition). The situation becomes more complicated when we consider Stieltjes integrals. The analogues of the first two definitions are no longer equivalent, and the third makes sense only if the integrator is non-decreasing. The goal of this chapter is to clarify the relations between the three definitions, to obtain necessary and sufficient conditions for the existence of the integral $\int_a^b f\,\mathrm{d}g$, and to establish some basic results including integration by parts, substitution, and various convergence theorems.

5.1 Definition and basic properties

Recall that a set $\boldsymbol{\alpha}=\{\alpha_0,\alpha_1,\dots,\alpha_m\}$ of points from an interval $[a,b]$ is called a *division of the interval* $[a,b]$ if

$$a=\alpha_0<\alpha_1<\dots<\alpha_m=b.$$

The set of all divisions of the interval $[a,b]$ is denoted by $\mathscr{D}[a,b]$. The elements of a division $\boldsymbol{\alpha}$ of $[a,b]$ are usually denoted by α_j, $\nu(\boldsymbol{\alpha})$ is the index of the maximal element of $\boldsymbol{\alpha}$ (i.e., $\alpha_{\nu(\boldsymbol{\alpha})}=b$) and

$$|\boldsymbol{\alpha}|=\max_{j=1,\dots,m}(\alpha_j-\alpha_{j-1}).$$

We say that a division $\boldsymbol{\alpha}'$ of $[a,b]$ is a *refinement* of $\boldsymbol{\alpha}$ if $\boldsymbol{\alpha}' \supset \boldsymbol{\alpha}$.

5.1.1 Definition. A *partition* of the interval $[a,b]$ is a pair $P=(\boldsymbol{\alpha},\boldsymbol{\xi})$, where $\boldsymbol{\alpha}=\{\alpha_0,\ldots,\alpha_{\nu(\boldsymbol{\alpha})}\}$ is a division of $[a,b]$, and $\boldsymbol{\xi}=(\xi_1,\ldots,\xi_{\nu(\boldsymbol{\alpha})})$ is a vector such that

$$\xi_j \in [\alpha_{j-1},\alpha_j] \quad \text{for} \quad j \in \{1,\ldots,\nu(\boldsymbol{\alpha})\}.$$

Each point ξ_j is referred to as the *tag* of the subinterval $[\alpha_{j-1},\alpha_j]$, and $\boldsymbol{\xi}$ will be called the *vector of tags* for the division $\boldsymbol{\alpha}$. For each partition $P=(\boldsymbol{\alpha},\boldsymbol{\xi})$, we denote $\nu(P)=\nu(\boldsymbol{\alpha})$. Thus, $\nu(P)$ is the number of subintervals or tags determined by P.

5.1.2 Remark. Many textbooks define a partition as a finite collection of tagged intervals. In other words, a partition $P=(\boldsymbol{\alpha},\boldsymbol{\xi})$ in the sense of Definition 5.1.1 can be regarded as a collection of interval-point pairs $([\alpha_{j-1},\alpha_j],\xi_j)$, where ξ_j is the corresponding tag of the subinterval $[\alpha_{j-1},\alpha_j]$, $j=1,\ldots,\nu(\boldsymbol{\alpha})$. In certain situations, this alternative interpretation of partitions might be more convenient.

Sequences of divisions or partitions will be denoted by $\{\boldsymbol{\alpha}^n\}$ or $\{(\boldsymbol{\beta}^n,\boldsymbol{\eta}^n)\}$, respectively; we use upper indices to avoid confusion with the elements of $\boldsymbol{\alpha}$, $\boldsymbol{\beta}$ or coordinates of $\boldsymbol{\eta}$, etc.

Let us introduce the following convention: Given $a<c<b$, assume that $R=(\boldsymbol{\rho},\boldsymbol{\eta})$ is a partition of $[a,c]$, and $S=(\boldsymbol{\sigma},\boldsymbol{\zeta})$ is a partition of $[c,b]$. Then $P=R\cup S$ denotes the partition of $[a,b]$ having the form $P=(\boldsymbol{\alpha},\boldsymbol{\xi})$, with $\boldsymbol{\alpha}=\boldsymbol{\rho}\cup\boldsymbol{\sigma}$ and $\boldsymbol{\xi}=(\eta_1,\ldots,\eta_{\nu(R)},\zeta_1,\ldots,\zeta_{\nu(S)})$.

5.1.3 Definition. Given a pair of functions $f,g\colon[a,b]\to\mathbb{R}$ and a partition $P=(\boldsymbol{\alpha},\boldsymbol{\xi})$ of the interval $[a,b]$, we define

$$S(f,\mathrm{d}g,P)=\sum_{j=1}^{\nu(\boldsymbol{\alpha})} f(\xi_j)\,(g(\alpha_j)-g(\alpha_{j-1})).$$

If f,g are fixed and no misunderstanding can arise, we write $S(P)$ instead of $S(f,\mathrm{d}g,P)$.

5.1.4 Definition. Let $f,g\colon[a,b]\to\mathbb{R}$.

(i) We say that the *Riemann-Stieltjes* (δ)*-integral* of f with respect to g

$$(\delta)\int_a^b f(x)\,\mathrm{d}g(x) \quad \text{(we also write } (\delta)\int_a^b f\,\mathrm{d}g)$$

exists and equals $I \in \mathbb{R}$ if

$$\left.\begin{array}{l}\text{for every } \varepsilon > 0 \text{ there is a } \delta_\varepsilon > 0 \text{ such that} \\ \quad |S(P) - I| < \varepsilon \\ \text{for all partitions } P = (\boldsymbol{\alpha}, \boldsymbol{\xi}) \text{ of } [a,b] \text{ such that } |\boldsymbol{\alpha}| < \delta_\varepsilon.\end{array}\right\} \tag{5.1.1}$$

(ii) We say that the *Riemann-Stieltjes* (σ)*-integral* of f with respect to g

$$(\sigma)\int_a^b f(x)\,\mathrm{d}g(x) \quad \Big(\text{we also write } (\sigma)\int_a^b f\,\mathrm{d}g\Big)$$

exists and equals $I \in \mathbb{R}$ if

$$\left.\begin{array}{l}\text{for every } \varepsilon > 0 \text{ there is a division } \boldsymbol{\alpha}_\varepsilon \text{ of } [a,b] \text{ such that} \\ \quad |S(P) - I| < \varepsilon \\ \text{for all partitions } P = (\boldsymbol{\alpha}, \boldsymbol{\xi}) \text{ of } [a,b] \text{ such that } \boldsymbol{\alpha} \supset \boldsymbol{\alpha}_\varepsilon.\end{array}\right\} \tag{5.1.2}$$

(iii) For any $c \in [a,b]$ we set

$$(\delta)\int_c^c f\,\mathrm{d}g = (\sigma)\int_c^c f\,\mathrm{d}g = 0.$$

If the integral $(\delta)\int_a^b f\,\mathrm{d}g$ or $(\sigma)\int_a^b f\,\mathrm{d}g$ exists, we define

$$(\delta)\int_b^a f\,\mathrm{d}g = -(\delta)\int_a^b f\,\mathrm{d}g \quad \text{or} \quad (\sigma)\int_b^a f\,\mathrm{d}g = -(\sigma)\int_a^b f\,\mathrm{d}g,$$

respectively.

5.1.5 Remark. The classical Riemann integral is a special case of the Riemann-Stieltjes (δ)-integral corresponding to the choice $g(x) = x$ for all $x \in [a,b]$. It is denoted by $\int_a^b f(x)\,\mathrm{d}x$, or simply $\int_a^b f$.

If we speak about the Riemann-Stieltjes integral without distinguishing between the (δ) or (σ) variant, we mean that the given statement holds for both integrals. In such and other cases when no misunderstanding can occur we do not include the symbol (δ) or (σ) before the integral sign.

The function f in the integral $\int_a^b f\,\mathrm{d}g$ is called the *integrand*, while the function g is called the *integrator*.

5.1.6 Exercise. Prove that the value of the integral in Definition 5.1.4 is determined uniquely, i.e., if $I_1 \in \mathbb{R}$ and $I_2 \in \mathbb{R}$ satisfy (5.1.1) with I replaced by I_1 or I_2, then $I_1 = I_2$ (and similarly for (5.1.2)).

5.1.7 Exercise. Prove the following assertions for both kinds of the Riemann-Stieltjes integral:

(i) If the function g is constant, then $\int_a^b f\,\mathrm{d}g = 0$ for any function $f\colon[a,b]\to\mathbb{R}$.

(ii) If the function f is constant, then $\int_a^b f\,\mathrm{d}g = f(a)\,(g(b)-g(a))$ for any function $g\colon[a,b]\to\mathbb{R}$.

From Definition 5.1.4 we can easily conclude that the Riemann-Stieltjes (δ)-integral is a special case of the Riemann-Stieltjes (σ)-integral in the following sense.

5.1.8 Theorem. *If* $(\delta)\int_a^b f\,\mathrm{d}g$ *exists, then* $(\sigma)\int_a^b f\,\mathrm{d}g$ *exists as well and has the same value.*

Proof. The statement follows from the fact that the inequality $|\boldsymbol{\alpha}''|\le|\boldsymbol{\alpha}'|$ holds for all divisions $\boldsymbol{\alpha}',\boldsymbol{\alpha}''$ of $[a,b]$ such that $\boldsymbol{\alpha}''\supset\boldsymbol{\alpha}'$. □

5.1.9 Remark. Let an arbitrary $\delta_0>0$ be given. Then in Definition 5.1.1 (i) the condition (5.1.1) can be replaced by the following weaker condition:

$$\left.\begin{array}{l}\text{For every } \varepsilon>0 \text{ there is a } \delta_\varepsilon\in(0,\delta_0) \text{ such that}\\[4pt] \qquad |S(P)-I|<\varepsilon\\[4pt] \text{for all partitions } P=(\boldsymbol{\alpha},\boldsymbol{\xi}) \text{ of } [a,b] \text{ such that } |\boldsymbol{\alpha}|<\delta_\varepsilon.\end{array}\right\}\qquad(5.1.1')$$

Similarly, if a division $\boldsymbol{\alpha}_0$ of $[a,b]$ is given, then in Definition 5.1.4 (ii) the condition (5.1.2) can be replaced by the following weaker condition:

$$\left.\begin{array}{l}\text{For every } \varepsilon>0 \text{ there is an } \boldsymbol{\alpha}_\varepsilon\in\mathscr{D}[a,b] \text{ such that } \boldsymbol{\alpha}_\varepsilon\supset\boldsymbol{\alpha}_0\\ \text{and}\\[4pt] \qquad |S(P)-I|<\varepsilon\\[4pt] \text{for all partitions } P=(\boldsymbol{\alpha},\boldsymbol{\xi}) \text{ of } [a,b] \text{ such that } \boldsymbol{\alpha}\supset\boldsymbol{\alpha}_\varepsilon.\end{array}\right\}\qquad(5.1.2')$$

5.1.10 Exercise. Verify the statements mentioned in Remark 5.1.9.

5.1.11 Example. Let $a=-1$, $b=1$ and

$$f(x)=\begin{cases}0 & \text{if } x\le 0,\\ 1 & \text{if } x>0,\end{cases}\qquad\text{and}\qquad g(x)=\begin{cases}1 & \text{if } x<0,\\ 0 & \text{if } x\ge 0.\end{cases}$$

Set $\boldsymbol{\alpha}_0=\{-1,0,1\}$. Then for every division $\boldsymbol{\alpha}$ of $[-1,1]$ which is a refinement of $\boldsymbol{\alpha}_0$ and for every partition $P=(\boldsymbol{\alpha},\boldsymbol{\xi})$ of $[a,b]$, there is a $k\in\{1,\dots,\nu(\boldsymbol{\alpha})\}$ such that $\alpha_k=0$. Then

$$S(P)=f(\xi_k)\,(g(0)-g(\alpha_{k-1}))+f(\xi_{k+1})\,(g(\alpha_{k+1})-g(0))=0,$$

because

$$f(\xi_k)=0 \quad \text{and} \quad g(\alpha_{k+1})-g(0)=0.$$

By the second part of Remark 5.1.9, we see that $(\sigma)\int_{-1}^{1} f\,\mathrm{d}g=0$.

On the other hand, for every partition $P=(\boldsymbol{\alpha},\boldsymbol{\xi})$ of the interval $[-1,1]$ such that $0\notin\boldsymbol{\alpha}$ there is a $k\in\{1,\dots,\nu(\boldsymbol{\alpha})\}$ such that $\alpha_{k-1}<0<\alpha_k$, and consequently

$$S(P)=f(\xi_k)\,(g(\alpha_k)-g(\alpha_{k-1}))=-f(\xi_k)=-\begin{cases}0 & \text{if } \xi_k\le 0,\\ 1 & \text{if } \xi_k>0.\end{cases}$$

Now it is clear that the integral $(\delta)\int_{-1}^{1} f\,\mathrm{d}g$ does not exist.

The following two lemmas hold for both kinds of the Riemann-Stieltjes integral and are direct corollaries of Definition 5.1.4.

5.1.12 Lemma. (i) *If the integral $\int_a^b f\,\mathrm{d}g$ exists, then*

$$\Big|\int_a^b f\,\mathrm{d}g\Big|\le\|f\|\,\mathrm{var}_a^b\,g.$$

(ii) *If, in addition, $g\in\mathrm{BV}([a,b])$ and the integral $\int_a^b |f(x)|\,\mathrm{d}(\mathrm{var}_a^x\,g)$ exists, then*

$$\Big|\int_a^b f\,\mathrm{d}g\Big|\le\int_a^b |f(x)|\,\mathrm{d}(\mathrm{var}_a^x\,g)\le\|f\|\,\mathrm{var}_a^b\,g.$$

5.1.13 Remark. We will show later (cf. Corollary 5.3.9) that if g has bounded variation, then for both kinds of the Riemann-Stieltjes integral, the existence of the integral $\int_a^b |f(x)|\,\mathrm{d}(\mathrm{var}_a^x\,g)$ already follows from the existence of the integral $\int_a^b f\,\mathrm{d}g$.

5.1.14 Lemma. *Let $c_1,c_2\in\mathbb{R}$ be arbitrary numbers.*

(i) *If $f_1,f_2,g\colon[a,b]\to\mathbb{R}$ are such that the integrals $\int_a^b f_1\,\mathrm{d}g$ and $\int_a^b f_2\,\mathrm{d}g$ exist, then*

$$\int_a^b (c_1\,f_1+c_2\,f_2)\,\mathrm{d}g=c_1\int_a^b f_1\,\mathrm{d}g+c_2\int_a^b f_2\,\mathrm{d}g.$$

(ii) *If* $f, g_1, g_2 : [a,b] \to \mathbb{R}$ *are such that the integrals* $\int_a^b f\,\mathrm{d}g_1$ *and* $\int_a^b f\,\mathrm{d}g_2$ *exist, then*

$$\int_a^b f\,\mathrm{d}(c_1\,g_1 + c_2\,g_2) = c_1 \int_a^b f\,\mathrm{d}g_1 + c_2 \int_a^b f\,\mathrm{d}g_2.$$

5.1.15 Exercise. (i) Prove Lemmas 5.1.12 and 5.1.14.

(ii) Prove the following statement for both kinds of the Riemann-Stieltjes integral: *If* $f, g : [a,b] \to \mathbb{R}$ *are such that* g *is nondecreasing and the integral* $\int_a^b f\,\mathrm{d}g$ *exists, then*

$$\Big(\inf_{x\in[a,b]} f(x)\Big)\,(g(b)-g(a)) \le \int_a^b f\,\mathrm{d}g \le \Big(\sup_{x\in[a,b]} f(x)\Big)\,(g(b)-g(a)).$$

Both kinds of the Riemann-Stieltjes integral are, in a sense, generalized limits of integral sums with respect to partitions. It is thus not surprising that the following statement, which is an analogue of the classical Bolzano-Cauchy condition, holds.

5.1.16 Theorem (BOLZANO-CAUCHY CONDITION).
Given a pair of functions $f, g : [a,b] \to \mathbb{R}$, *the integral* $(\delta)\int_a^b f\,\mathrm{d}g$ *exists if and only if*

$$\left.\begin{aligned} &\text{for every } \varepsilon>0 \text{ there is a } \delta_\varepsilon>0 \text{ such that}\\ &\qquad |S(P)-S(Q)|<\varepsilon\\ &\text{for all partitions } P=(\boldsymbol{\alpha},\boldsymbol{\xi}),\ Q=(\boldsymbol{\beta},\boldsymbol{\eta}) \text{ of } [a,b]\\ &\text{such that } |\boldsymbol{\alpha}|<\delta_\varepsilon \text{ and } |\boldsymbol{\beta}|<\delta_\varepsilon. \end{aligned}\right\} \tag{5.1.3}$$

Similarly, the integral $(\sigma)\int_a^b f\,\mathrm{d}g$ *exists if and only if*

$$\left.\begin{aligned} &\text{for every } \varepsilon>0 \text{ there is a division } \boldsymbol{\alpha}_\varepsilon \text{ of } [a,b] \text{ such that}\\ &\qquad |S(P)-S(Q)|<\varepsilon\\ &\text{for all partitions } P=(\boldsymbol{\alpha},\boldsymbol{\xi}), Q=(\boldsymbol{\beta},\boldsymbol{\eta}) \text{ of } [a,b]\\ &\text{such that } \boldsymbol{\alpha}\cap\boldsymbol{\beta}\supset\boldsymbol{\alpha}_\varepsilon. \end{aligned}\right\} \tag{5.1.4}$$

Proof. The necessity of the conditions (5.1.3) and (5.1.4) for the existence of the corresponding integrals is obvious from Definition 5.1.4.

We will prove that condition (5.1.4) guarantees the existence of the integral $(\sigma)\int_a^b f\,\mathrm{d}g$. If (5.1.4) is satisfied, there is a sequence $\{P_k\} = \{(\boldsymbol{\alpha}^k, \boldsymbol{\xi}^k)\}$

of partitions of $[a, b]$ such that $\boldsymbol{\alpha}^k \subset \boldsymbol{\alpha}^\ell$ for $\ell \geq k$ and

$$\left. \begin{array}{l} |S(P) - S(P_k)| < \frac{1}{k} \\ \qquad \text{for all partitions } P = (\boldsymbol{\alpha}, \boldsymbol{\xi}) \text{ of } [a,b] \text{ such that } \boldsymbol{\alpha} \supset \boldsymbol{\alpha}^k. \end{array} \right\} \tag{5.1.5}$$

In particular,

$$|S(P_k) - S(P_\ell)| < \tfrac{1}{k} \quad \text{whenever } k, \ell \in \mathbb{N} \text{ and } \ell \geq k. \tag{5.1.6}$$

The sequence $\{S(P_k)\}$ is a Cauchy sequence of real numbers. Hence there exists an $I \in \mathbb{R}$ such that $\lim\limits_{k \to \infty} S(P_k) = I$. Now, let $\varepsilon > 0$ be given. Choose k_ε in such a way that

$$\frac{1}{k_\varepsilon} < \frac{\varepsilon}{2} \quad \text{and} \quad |S(P_{k_\varepsilon}) - I| < \frac{\varepsilon}{2}. \tag{5.1.7}$$

Then, combining (5.1.5) and (5.1.7), we deduce that

$$|S(P) - I| \leq |S(P) - S(P_{k_\varepsilon})| + |S(P_{k_\varepsilon}) - I| < \varepsilon$$

for every partition $P = (\boldsymbol{\alpha}, \boldsymbol{\xi})$ of $[a, b]$ such that $\boldsymbol{\alpha} \supset \boldsymbol{\alpha}^{k_\varepsilon}$. Therefore $I = (\sigma) \int_a^b f \, dg$.

In a similar way we can prove that condition (5.1.3) implies the existence of the integral $(\delta) \int_a^b f \, dg$. □

5.1.17 Exercise. (i) Prove the assertion of Theorem 5.1.16 for Riemann-Stieltjes (δ)-integrals.

(ii) Prove that conditions (5.1.3) or (5.1.4) are respectively equivalent to the following ones:

$$\left. \begin{array}{l} \text{For every } \varepsilon > 0 \text{ there is a } \delta_\varepsilon > 0 \text{ such that} \\ \qquad |S(P) - S(Q)| < \varepsilon \\ \text{for all partitions } P = (\boldsymbol{\alpha}, \boldsymbol{\xi}),\ Q = (\boldsymbol{\beta}, \boldsymbol{\eta}) \text{ of } [a, b] \\ \text{such that } |\boldsymbol{\alpha}| < \delta_\varepsilon,\ \boldsymbol{\beta} \supset \boldsymbol{\alpha}. \end{array} \right\} \tag{5.1.3'}$$

$$\left. \begin{array}{l} \text{For every } \varepsilon > 0 \text{ there is a division } \boldsymbol{\alpha}_\varepsilon \text{ of } [a, b] \text{ such that} \\ \qquad |S(P) - S(Q)| < \varepsilon \\ \text{for all partitions } P = (\boldsymbol{\alpha}, \boldsymbol{\xi}),\ Q = (\boldsymbol{\beta}, \boldsymbol{\eta}) \text{ of } [a, b] \\ \text{such that } \boldsymbol{\beta} \supset \boldsymbol{\alpha} \supset \boldsymbol{\alpha}_\varepsilon. \end{array} \right\} \tag{5.1.4'}$$

The following theorem is a direct corollary of Theorem 5.1.16. It is valid for both kinds of the Riemann-Stieltjes integral.

5.1.18 Theorem. *If the integral* $\int_a^b f\,\mathrm{d}g$ *exists and* $[c,d]\subset[a,b]$, *then the integral* $\int_c^d f\,\mathrm{d}g$ *exists, as well.*

Proof. Assume that the integral $(\sigma)\int_a^b f\,\mathrm{d}g$ exists and $a<c<d<b$; the cases when $a=c$ or $d=b$ are simpler and their proof is left to the reader.

Let $\varepsilon>0$ be given. By Theorem 5.1.16 there is a division $\boldsymbol{\alpha}_\varepsilon$ of $[a,b]$ such that

$$|S(P)-S(P')|<\varepsilon \tag{5.1.8}$$

holds for all partitions $P=(\boldsymbol{\alpha},\boldsymbol{\xi})$, $P'=(\boldsymbol{\alpha}',\boldsymbol{\xi}')$ of $[a,b]$ such that $\boldsymbol{\alpha}\cap\boldsymbol{\alpha}'\supset\boldsymbol{\alpha}_\varepsilon$. By Remark 5.1.9, we can assume that $\{c,d\}\subset\boldsymbol{\alpha}_\varepsilon$ and thus we can decompose $\boldsymbol{\alpha}_\varepsilon$ in such a way that

$$\boldsymbol{\alpha}_\varepsilon=\boldsymbol{\beta}_\varepsilon^-\cup\boldsymbol{\beta}_\varepsilon\cup\boldsymbol{\beta}_\varepsilon^+,$$

where $\boldsymbol{\beta}_\varepsilon^-$ is a division of $[a,c]$, $\boldsymbol{\beta}_\varepsilon$ is a division of $[c,d]$, and $\boldsymbol{\beta}_\varepsilon^+$ is a division of $[d,b]$.

Now, let us fix partitions $Q^-=(\boldsymbol{\beta}_\varepsilon^-,\boldsymbol{\eta}^-)$ of $[a,c]$, and $Q^+=(\boldsymbol{\beta}_\varepsilon^+,\boldsymbol{\eta}^+)$ of $[c,d]$ and let $Q=(\boldsymbol{\beta},\boldsymbol{\eta})$ and $Q'=(\boldsymbol{\beta}',\boldsymbol{\eta}')$ be partitions of $[c,d]$ such that $\boldsymbol{\beta}\cap\boldsymbol{\beta}'\supset\boldsymbol{\beta}_\varepsilon$. Consider the partitions $P=(\boldsymbol{\alpha},\boldsymbol{\xi})=Q^-\cup Q\cup Q^+$ and $P'=(\boldsymbol{\alpha}',\boldsymbol{\xi}')=Q^-\cup Q'\cup Q^+$. Since $\boldsymbol{\alpha}\cap\boldsymbol{\alpha}'\supset\boldsymbol{\alpha}_\varepsilon$ and

$$S(P)=S(Q^-)+S(Q)+S(Q^+),\quad S(P')=S(Q^-)+S(Q')+S(Q^+),$$

using (5.1.8) we get

$$|S(Q)-S(Q')|=|S(P)-S(P')|<\varepsilon.$$

The existence of the integral $\int_c^d f\,\mathrm{d}g$ now follows by Theorem 5.1.16.

The statement of the theorem for the Riemann-Stieltjes (δ)-integral can be proved analogously; we leave it as an exercise to the reader. □

5.1.19 Exercise. Prove Theorem 5.1.18 for the Riemann-Stieltjes (δ)-integral.

The following statement holds for both kinds of the Riemann-Stieltjes integral.

5.1.20 Theorem. *If the integral* $\int_a^b f\,\mathrm{d}g$ *exists and* $c\in[a,b]$, *then also both the integrals* $\int_a^c f\,\mathrm{d}g$ *and* $\int_c^b f\,\mathrm{d}g$ *exist and satisfy*

$$\int_a^b f\,\mathrm{d}g=\int_a^c f\,\mathrm{d}g+\int_c^b f\,\mathrm{d}g.$$

Proof. If $c=a$ or $c=b$, the statement of the theorem is trivial. Thus, let $c\in(a,b)$ and let the integral $\int_a^b f\,\mathrm{d}g$ exist. Then the existence of the integrals $\int_a^c f\,\mathrm{d}g$ and $\int_c^b f\,\mathrm{d}g$ is guaranteed by Theorem 5.1.18.

Let $\varepsilon>0$. Choose partitions P' of $[a,c]$ and P'' of $[c,b]$ in such a way that

$$\left.\begin{aligned}&\Big|S(P')-\int_a^c f\,\mathrm{d}g\Big|+\Big|S(P'')-\int_c^b f\,\mathrm{d}g\Big|+\Big|S(P)-\int_a^b f\,\mathrm{d}g\Big|<\varepsilon,\\ &\text{where } P=P'\cup P''.\end{aligned}\right\} \tag{5.1.9}$$

Obviously, $S(P)=S(P')+S(P'')$. Hence,

$$\begin{aligned}&\Big|\int_a^b f\,\mathrm{d}g-\int_a^c f\,\mathrm{d}g-\int_c^b f\,\mathrm{d}g\Big|\\ &\quad\le\Big|\int_a^b f\,\mathrm{d}g-S(P)\Big|+\Big|S(P)-S(P')-S(P'')\Big|\\ &\qquad+\Big|S(P')-\int_a^c f\,\mathrm{d}g\Big|+\Big|S(P'')-\int_c^b f\,\mathrm{d}g\Big|<\varepsilon.\end{aligned}$$

As $\varepsilon>0$ was arbitrary, this completes the proof. □

5.1.21 Exercise. Why does the existence of the integrals

$$\int_a^b f\,\mathrm{d}g,\quad \int_a^c f\,\mathrm{d}g,\quad \int_c^b f\,\mathrm{d}g$$

imply the existence of partitions P' of $[a,c]$ and P'' of $[c,b]$ such that (5.1.9) holds?

The converse of Theorem 5.1.20 is easily shown to be valid for the (σ) Riemann-Stieltjes integral.

5.1.22 Theorem. *Let $f,g:[a,b]\to\mathbb{R}$ and $c\in(a,b)$. If both integrals $(\sigma)\int_a^c f\,\mathrm{d}g$ and $(\sigma)\int_c^b f\,\mathrm{d}g$ exist, then the integral $(\sigma)\int_a^b f\,\mathrm{d}g$ exist as well, and*

$$(\sigma)\int_a^b f\,\mathrm{d}g=(\sigma)\int_a^c f\,\mathrm{d}g+(\sigma)\int_c^b f\,\mathrm{d}g.$$

Proof. Denote $I_1=(\sigma)\int_a^c f\,\mathrm{d}g$ and $I_2=(\sigma)\int_c^b f\,\mathrm{d}g$. Let $\varepsilon>0$ be given.

Choose divisions $\boldsymbol{\alpha}'_\varepsilon$ of $[a,c]$ and $\boldsymbol{\alpha}''_\varepsilon$ of $[c,b]$ in such a way that

$$\begin{aligned}&\big|S(P')-I_1\big|<\varepsilon\\ &\qquad\text{for all partitions } P'=(\boldsymbol{\alpha}',\boldsymbol{\xi}') \text{ of } [a,c] \text{ such that } \boldsymbol{\alpha}'\supset\boldsymbol{\alpha}'_\varepsilon,\\ &\big|S(P'')-I_2\big|<\varepsilon\\ &\qquad\text{for all partitions } P''=(\boldsymbol{\alpha}'',\boldsymbol{\xi}'') \text{ of } [c,b] \text{ such that } \boldsymbol{\alpha}''\supset\boldsymbol{\alpha}''_\varepsilon.\end{aligned}$$

Now, let $\boldsymbol{\alpha}_\varepsilon=\boldsymbol{\alpha}'_\varepsilon\cup\boldsymbol{\alpha}''_\varepsilon$. Since $c\in\boldsymbol{\alpha}_\varepsilon$, every partition $P=(\boldsymbol{\alpha},\boldsymbol{\xi})$ of $[a,b]$ satisfying $\boldsymbol{\alpha}\supset\boldsymbol{\alpha}_\varepsilon$ can be decomposed to partitions $P'=(\boldsymbol{\alpha}',\boldsymbol{\xi}')$ of $[a,c]$ and $P''=(\boldsymbol{\alpha}'',\boldsymbol{\xi}'')$ of $[c,b]$ in such a way that

$$P=P'\cup P'', \quad \boldsymbol{\alpha}'\supset\boldsymbol{\alpha}'_\varepsilon \text{ and } \boldsymbol{\alpha}''\supset\boldsymbol{\alpha}''_\varepsilon.$$

Moreover, $S(P)=S(P')+S(P'')$, and the definitions of $\boldsymbol{\alpha}'_\varepsilon$ and $\boldsymbol{\alpha}''_\varepsilon$ imply

$$\big|S(P)-(I_1+I_2)\big|\le\big|S(P')-I_1\big|+\big|S(P'')-I_2\big|<2\,\varepsilon$$

for every partition $P=(\boldsymbol{\alpha},\boldsymbol{\xi})$ of $[a,b]$ such that $\boldsymbol{\alpha}\supset\boldsymbol{\alpha}_\varepsilon$. This completes the proof of the theorem. □

5.1.23 Remark. In general, an analogue of Theorem 5.1.22 does not hold for the Riemann-Stieltjes (δ)-integral. For instance, consider the functions $f,g\colon[-1,1]\to\mathbb{R}$ introduced in Example 5.1.11. It is clear that $(\delta)\int_{-1}^0 f\,\mathrm{d}g=(\delta)\int_0^1 f\,\mathrm{d}g=0$, however, as shown in Example 5.1.11, the integral $(\delta)\int_{-1}^1 f\,\mathrm{d}g$ does not exist.

This problem will be investigated in the next section, where we show that an analogue of Theorem 5.1.22 for the (δ)-integral holds if f, g satisfy the condition of pseudo-additivity at the point c; see Definition 5.2.1 and Exercise 5.2.8.

For the existence of the Riemann-Stieltjes (δ)-integral, we have the following necessary and sufficient condition.

5.1.24 Theorem. *For every pair $f,g\colon[a,b]\to\mathbb{R}$, the integral $(\delta)\int_a^b f\,\mathrm{d}g$ exists if and only if*

$$\left.\begin{aligned}&\lim_{n\to\infty} S(P_n)\in\mathbb{R} \textit{ for every sequence } \{P_n\}=\{(\boldsymbol{\alpha}^n,\boldsymbol{\xi}^n)\}\\ &\qquad\qquad \textit{of partitions of } [a,b] \textit{ such that } \lim_{n\to\infty}|\boldsymbol{\alpha}^n|=0.\end{aligned}\right\}\quad(5.1.10)$$

Proof. The necessity of the condition (5.1.10) for the existence of the integral $(\delta)\int_a^b f\,dg$ is obvious; it remains to prove its sufficiency.

Thus, assume that (5.1.10) holds and let $\{P_n\}=\{(\boldsymbol{\alpha}^n,\boldsymbol{\xi}^n)\}$ and $\{\widetilde{P}_n\}=\{(\widetilde{\boldsymbol{\alpha}}^n,\widetilde{\boldsymbol{\xi}}^n)\}$ be the sequences of partitions of $[a,b]$ such that

$$\lim_{n\to\infty}|\boldsymbol{\alpha}^n| = \lim_{n\to\infty}|\widetilde{\boldsymbol{\alpha}}^n|=0,$$

$$\lim_{n\to\infty} S(P_n)=I\in\mathbb{R} \quad\text{and}\quad \lim_{n\to\infty} S(\widetilde{P}_n)=\widetilde{I}\in\mathbb{R}.$$

Now, consider the sequence $\{Q_n\}$ of partitions of $[a,b]$ given by

$$Q_{2k-1}=P_k,\quad Q_{2k}=\widetilde{P}_k \quad\text{for}\quad k\in\mathbb{N}.$$

By our assumption, the sequence $\{S(Q_n)\}$ has a finite limit $J\in\mathbb{R}$, and since it contains both the sequences $\{S(P_n)\}$ and $\{S(\widetilde{P}_n)\}$, we necessarily have $I=\widetilde{I}=J$. This means that the value of the limit

$$I=\lim_{n\to\infty} S(P_n)$$

does not depend on the choice of the sequence $\{P_n\}=\{(\boldsymbol{\alpha}^n,\boldsymbol{\xi}^n)\}$ of partitions of $[a,b]$ for which $\lim_{n\to\infty}|\boldsymbol{\alpha}^n|=0$.

Now, assume that $(\delta)\int_a^b f\,dg\neq I$. Then there exists an $\widetilde{\varepsilon}>0$ such that for every $k\in\mathbb{N}$ there is a partition $P_k=(\boldsymbol{\alpha}^k,\boldsymbol{\xi}^k)$ of $[a,b]$ such that

$$|\boldsymbol{\alpha}^k|<1/k \quad\text{and}\quad |S(P_k)-I|>\widetilde{\varepsilon}.$$

In other words, we have a sequence $\{P_k\}=\{(\boldsymbol{\alpha}^k,\boldsymbol{\xi}^k)\}$ of partitions of $[a,b]$ such that

$$\lim_{k\to\infty}|\boldsymbol{\alpha}^k|=0 \quad\text{and}\quad \lim_{k\to\infty} S(P_k)\neq I,$$

which is a contradiction. Therefore $(\delta)\int_a^b f\,dg=I$. □

5.1.25 Theorem. *Consider functions $f,g\colon[a,b]\to\mathbb{R}$.*

(i) *Suppose that $(\delta)\int_a^b f\,dg$ exists. Then for each $x\in[a,b]$, at least one of f and g is continuous at x.*

(ii) *Suppose that $(\sigma)\int_a^b f\,dg$ exists. Then for each $x\in(a,b]$, at least one of f and g is left-continuous at x, and for each $x\in[a,b)$, at least one of f and g is right-continuous at x.*

Proof. (i) Assume that $(\delta)\int_a^b f\,dg$ exists and f is discontinuous from the right at a point $x\in[a,b)$; we will show that g is continuous at x.

There exists an $\varepsilon_0 > 0$ such that for each $\delta > 0$, there is a $y \in (x, x+\delta)$ satisfying $|f(y) - f(x)| \geq \varepsilon_0$.

Choose an arbitrary $\varepsilon > 0$. According to Theorem 5.1.16, there is a $\delta_\varepsilon > 0$ such that $|S(P) - S(Q)| < \varepsilon \cdot \varepsilon_0$ for all partitions $P = (\boldsymbol{\alpha}, \boldsymbol{\xi})$, $Q = (\boldsymbol{\beta}, \boldsymbol{\eta})$ of $[a,b]$ such that $|\boldsymbol{\alpha}| < \delta_\varepsilon$ and $|\boldsymbol{\beta}| < \delta_\varepsilon$.

Now, let $\tau' \in (x - \delta_\varepsilon/2, x] \cap [a,b]$, $\tau'' \in [x, x + \delta_\varepsilon/2) \cap [a,b]$ and let $y \in (x, \tau'']$ be such that $|f(y) - f(x)| \geq \varepsilon_0$.

Choose partitions $P = (\boldsymbol{\alpha}, \boldsymbol{\xi})$, $Q = (\boldsymbol{\beta}, \boldsymbol{\eta})$ of $[a,b]$ such that $\boldsymbol{\alpha} = \boldsymbol{\beta}$, $|\boldsymbol{\alpha}| < \delta_\varepsilon$, the points τ', τ'' are two adjacent division points of $\boldsymbol{\alpha}$, and the tags of P and Q coincide on all subintervals except $[\tau', \tau'']$, where we take tags x and y, respectively. Then $|S(P) - S(Q)| = |f(x) - f(y)|\,|g(\tau'') - g(\tau')|$, and therefore

$$|f(x) - f(y)|\,|g(\tau'') - g(\tau')| < \varepsilon \cdot \varepsilon_0.$$

Recalling that $|f(x) - f(y)| \geq \varepsilon_0$, we necessarily have

$$|g(\tau'') - g(\tau')| < \varepsilon$$

whenever $\tau' \in (x - \delta_\varepsilon/2, x] \cap [a,b]$, $\tau'' \in [x, x + \delta_\varepsilon/2) \cap [a,b]$. This proves that g is continuous at x.

Similarly, if f is discontinuous from the left at a point $x \in (a,b]$, one can show that g is continuous at x.

(ii) Assume that $(\sigma)\int_a^b f\,\mathrm{d}g$ exists and f is discontinuous from the right at a point $x \in [a,b)$; we will show that g is right-continuous at x.

There exists an $\varepsilon_0 > 0$ such that for each $\delta > 0$, there is a $y \in (x, x+\delta)$ satisfying $|f(y) - f(x)| \geq \varepsilon_0$.

Choose an arbitrary $\varepsilon > 0$. According to Theorem 5.1.16, there is a division $\boldsymbol{\alpha}_\varepsilon$ of $[a,b]$ such that $|S(P) - S(Q)| < \varepsilon \cdot \varepsilon_0$ for all partitions $P = (\boldsymbol{\alpha}, \boldsymbol{\xi})$, $Q = (\boldsymbol{\beta}, \boldsymbol{\eta})$ of $[a,b]$ such that $\boldsymbol{\alpha}_\varepsilon \subset \boldsymbol{\alpha}$ and $\boldsymbol{\alpha}_\varepsilon \subset \boldsymbol{\beta}$.

Now, find a $\delta_\varepsilon > 0$ such that $(x, x+\delta_\varepsilon] \cap \boldsymbol{\alpha}_\varepsilon = \emptyset$, and consider an arbitrary $\tau \in (x, x+\delta_\varepsilon] \cap [a,b]$. Find a $y \in (x, \tau]$ satisfying $|f(y) - f(x)| \geq \varepsilon_0$.

Choose partitions $P = (\boldsymbol{\alpha}, \boldsymbol{\xi})$, $Q = (\boldsymbol{\beta}, \boldsymbol{\eta})$ of $[a,b]$ such that $\boldsymbol{\alpha} = \boldsymbol{\beta}$, $\boldsymbol{\alpha}_\varepsilon \subset \boldsymbol{\alpha}$, the points x, τ are two adjacent division points of $\boldsymbol{\alpha}$, and the tags of P and Q coincide on all subintervals except $[x, \tau]$, where we take tags x and y, respectively. Then $|S(P) - S(Q)| = |f(x) - f(y)|\,|g(\tau) - g(x)|$, and therefore

$$|f(x) - f(y)|\,|g(\tau) - g(x)| < \varepsilon \cdot \varepsilon_0.$$

Recalling that $|f(x) - f(y)| \geq \varepsilon_0$, we necessarily have

$$|g(\tau) - g(x)| < \varepsilon$$

whenever $\tau \in (x, x + \delta_\varepsilon] \cap [a, b]$. This proves that g is right-continuous at x.

Similarly, if f is discontinuous from the left at a point $x \in (a, b]$, one can show that g is left-continuous at x. $\square$

We conclude the present section with a brief overview of the Darboux-Stieltjes integral.

5.1.26 Definition. Let $g : [a, b] \to \mathbb{R}$ be nondecreasing. Given a bounded function $f : [a, b] \to \mathbb{R}$ and a division $\boldsymbol{\alpha}$ of the interval $[a, b]$, denote

$$\overline{S}(f, \mathrm{d}g, \boldsymbol{\alpha}) = \sum_{j=1}^{\nu(\boldsymbol{\alpha})} \Big(\sup_{x \in [\alpha_{j-1}, \alpha_j]} f(x) \Big) (g(\alpha_j) - g(\alpha_{j-1})),$$

$$\underline{S}(f, \mathrm{d}g, \boldsymbol{\alpha}) = \sum_{j=1}^{\nu(\boldsymbol{\alpha})} \Big(\inf_{x \in [\alpha_{j-1}, \alpha_j]} f(x) \Big) (g(\alpha_j) - g(\alpha_{j-1})).$$

Furthermore, define the *upper integral* of f with respect to g as

$$\overline{\int_a^b} f \, \mathrm{d}g = \inf \left\{ \overline{S}(f, \mathrm{d}g, \boldsymbol{\alpha}) : \boldsymbol{\alpha} \text{ is a division of } [a, b] \right\},$$

and the *lower integral* of f with respect to g as

$$\underline{\int_a^b} f \, \mathrm{d}g = \sup \left\{ \underline{S}(f, \mathrm{d}g, \boldsymbol{\alpha}) : \boldsymbol{\alpha} \text{ is a division of } [a, b] \right\}.$$

If the functions f and g are fixed, we write simply $\overline{S}(\boldsymbol{\alpha})$ and $\underline{S}(\boldsymbol{\alpha})$ instead of $\overline{S}(f, \mathrm{d}g, \boldsymbol{\alpha})$ and $\underline{S}(f, \mathrm{d}g, \boldsymbol{\alpha})$. Finally, if

$$\overline{\int_a^b} f \, \mathrm{d}g = \underline{\int_a^b} f \, \mathrm{d}g \in \mathbb{R},$$

then the common value of both integrals is called the *Darboux-Stieltjes integral* of f with respect to g.

5.1.27 Exercise. Consider the functions $f, g : [-1, 1] \to \mathbb{R}$ given by

$$f(x) = \begin{cases} 0 & \text{if } x \leq 0, \\ 1 & \text{if } x > 0, \end{cases} \qquad g(x) = \begin{cases} 0 & \text{if } x < 0, \\ 1 & \text{if } x \geq 0. \end{cases}$$

Determine the values of the upper and lower integral of f with respect to g and decide whether the Darboux-Stieltjes integral exists.

The next result implies that if g is nondecreasing, then the Darboux-Stieltjes integral is equivalent to the Riemann-Stieltjes (σ)-integral.

5.1.28 Theorem. *Let $f:[a,b]\to\mathbb{R}$ be bounded and let $g:[a,b]\to\mathbb{R}$ be nondecreasing. Then*

$$\overline{\int_a^b} f\,\mathrm{d}g = \underline{\int_a^b} f\,\mathrm{d}g = I\in\mathbb{R} \tag{5.1.11}$$

if and only if $(\sigma)\int_a^b f\,\mathrm{d}g = I$.

Proof. a) Assume that (5.1.11) holds. Since g is nondecreasing, it follows directly from Definition 5.1.26 that

$$\underline{S}(\boldsymbol{\alpha})\le S(\boldsymbol{\alpha},\boldsymbol{\xi})\le \overline{S}(\boldsymbol{\alpha}) \quad \text{for all partitions } (\boldsymbol{\alpha},\boldsymbol{\xi}) \text{ of } [a,b],$$

and

$$\underline{S}(\widetilde{\boldsymbol{\alpha}})\ge \underline{S}(\boldsymbol{\alpha}) \quad \text{and} \quad \overline{S}(\widetilde{\boldsymbol{\alpha}})\le \overline{S}(\boldsymbol{\alpha}) \quad \text{if} \quad \widetilde{\boldsymbol{\alpha}}\supset\boldsymbol{\alpha}.$$

Using the first fact, it is not difficult to verify that for each $k\in\mathbb{N}$ there is a division $\boldsymbol{\alpha}^k$ of $[a,b]$ such that the inequalities

$$I-\frac{1}{k}<\underline{S}(\boldsymbol{\alpha}^k)\le S(\boldsymbol{\alpha}^k,\boldsymbol{\xi}^k)\le\overline{S}(\boldsymbol{\alpha}^k)<I+\frac{1}{k}$$

hold whenever $(\boldsymbol{\alpha}^k,\boldsymbol{\xi}^k)$ is a partition of $[a,b]$. For a given $\varepsilon>0$, choose a $k_\varepsilon>1/\varepsilon$ and let $\boldsymbol{\alpha}_\varepsilon=\boldsymbol{\alpha}^{k_\varepsilon}$. Then

$$I-\varepsilon<\underline{S}(\boldsymbol{\alpha}^{k_\varepsilon})\le\underline{S}(\boldsymbol{\alpha})\le S(\boldsymbol{\alpha},\boldsymbol{\xi})\le\overline{S}(\boldsymbol{\alpha})\le\overline{S}(\boldsymbol{\alpha}^{k_\varepsilon})<I+\varepsilon$$

whenever $\boldsymbol{\alpha}\supset\boldsymbol{\alpha}_\varepsilon$ and $(\boldsymbol{\alpha},\boldsymbol{\xi})$ is a partition of $[a,b]$. It follows that $(\sigma)\int_a^b f\,\mathrm{d}g=I$.

b) Assume that $(\sigma)\int_a^b f\,\mathrm{d}g$ exists. Let $\varepsilon>0$ be given. By Theorem 5.1.16 there exists a division $\boldsymbol{\alpha}$ such that the inequality $\left|S(\boldsymbol{\alpha},\boldsymbol{\xi})-S(\boldsymbol{\alpha},\boldsymbol{\eta})\right|<\varepsilon/2$, or equivalently

$$\Big|\sum_{j=1}^{\nu(\boldsymbol{\alpha})}\big(f(\xi_j)-f(\eta_j)\big)\big(g(\alpha_j)-g(\alpha_{j-1})\big)\Big|<\frac{\varepsilon}{2},$$

holds for all vectors of tags $\boldsymbol{\xi},\boldsymbol{\eta}$ for the division $\boldsymbol{\alpha}$. Passing to the supremum and infimum on every interval $[\alpha_{j-1},\alpha_j]$, we get

$$\begin{aligned}0 &\le \overline{S}(\boldsymbol{\alpha})-\underline{S}(\boldsymbol{\alpha})\\ &=\sum_{j=1}^{\nu(\boldsymbol{\alpha})}\Big(\sup_{x\in[\alpha_{j-1},\alpha_j]} f(x)-\inf_{x\in[\alpha_{j-1},\alpha_j]} f(x)\Big)(g(\alpha_j)-g(\alpha_{j-1}))\le\frac{\varepsilon}{2}<\varepsilon.\end{aligned}$$

Consequently,

$$\overline{\int_a^b} f\,\mathrm{d}g\le\overline{S}(\boldsymbol{\alpha})<\underline{S}(\boldsymbol{\alpha})+\varepsilon\le\underline{\int_a^b} f\,\mathrm{d}g+\varepsilon$$

and finally also

$$0\le\overline{\int_a^b} f\,\mathrm{d}g-\underline{\int_a^b} f\,\mathrm{d}g<\varepsilon.$$

Since $\varepsilon>0$ is arbitrary, we conclude that

$$\overline{\int_a^b} f\,\mathrm{d}g=\underline{\int_a^b} f\,\mathrm{d}g.$$

By the first part of the proof, both integrals are equal to $(\sigma)\int_a^b f\mathrm{d}g$. □

5.2 PSEUDO-ADDITIVITY

We know that the existence of the Riemann-Stieltjes (δ)-integral implies the existence of the Riemann-Stieltjes (σ)-integral (see Theorem 5.1.8). In this section, we clarify the relationship between these integrals in the opposite direction. We introduce the concept of *pseudo-additivity* and show that $(\delta)\int_a^b f\,\mathrm{d}g$ exists if and only if $(\sigma)\int_a^b f\,\mathrm{d}g$ exists and f,g satisfy the condition of pseudo-additivity at every point $x\in(a,b)$.

5.2.1 Definition. We say that functions $f,g\colon[a,b]\to\mathbb{R}$ satisfy the condition of pseudo-additivity at a point $x\in(a,b)$, if for every $\varepsilon>0$ there is a $\delta_\varepsilon>0$ such that

$$\big|f(\xi)\,(g(x'')-g(x'))-f(\xi')\,(g(x)-g(x'))-f(\xi'')\,(g(x'')-g(x))\big|<\varepsilon$$

for all $x'\in(x-\delta_\varepsilon,x)$, $x''\in(x,x+\delta_\varepsilon)$, $\xi\in[x',x'']$, $\xi'\in[x',x]$, $\xi''\in[x,x'']$.

5.2.2 Lemma. *If $f,g:[a,b]\to\mathbb{R}$ are such that the integral $(\delta)\int_a^b f\,\mathrm{d}g$ exists, then they satisfy the condition of pseudo-additivity at every point $x\in(a,b)$.*

Proof. Assume that the integral $(\delta)\int_a^b f\,\mathrm{d}g$ exists and at the same time, the condition of pseudo-additivity does not hold at some point $x\in(a,b)$. Then there exists an $\widetilde{\varepsilon}>0$ such that for every $\delta>0$ it is possible to find points

$$x'\in(x-\delta,x),\ x''\in(x,x+\delta),\ \eta\in[x',x''],\ \eta'\in[x',x] \ \text{ and } \ \eta''\in[x,x'']$$

such that

$$\big|f(\eta)\,(g(x'')-g(x'))-f(\eta')\,(g(x)-g(x'))-f(\eta'')\,(g(x'')-g(x))\big|\geq\widetilde{\varepsilon}.$$

Let $\delta>0$ be given and let $P=(\boldsymbol{\alpha},\boldsymbol{\xi})$ with $\nu(P)=m$ be a partition of $[a,b]$ such that

$$|\boldsymbol{\alpha}|<\delta,\ \alpha_{k-1}=x'<x<x''=\alpha_k \ \text{ and } \ \xi_k=\eta \quad \text{for some } k\in\{1,\dots,m\}.$$

Put $\widetilde{\boldsymbol{\alpha}}=\boldsymbol{\alpha}\cup\{x\}$, $\widetilde{\boldsymbol{\xi}}=(\xi_1,\dots,\xi_{k-1},\eta',\eta'',\xi_{k+1},\dots,\xi_m)$ and $\widetilde{P}=(\widetilde{\boldsymbol{\alpha}},\widetilde{\boldsymbol{\xi}})$. Note that $|\widetilde{\boldsymbol{\alpha}}|<\delta$ and

$$\begin{aligned}|S(P)-S(\widetilde{P})|&=\big|f(\xi_k)\,(g(\alpha_k)-g(\alpha_{k-1}))\\&\quad-f(\eta')\,(g(x)-g(\alpha_{k-1}))-f(\eta'')\,(g(\alpha_k)-g(x))\big|\\&=\big|f(\eta)\,(g(x'')-g(x'))-f(\eta')\,(g(x)-g(x'))-f(\eta'')\,(g(x'')-g(x))\big|\geq\widetilde{\varepsilon}.\end{aligned}$$

This means that the Bolzano-Cauchy condition (5.1.3) is not satisfied and hence, in view of Theorem 5.1.16, the integral $(\delta)\int_a^b f\,\mathrm{d}g$ does not exist. □

5.2.3 Theorem. *Let $f,g:[a,b]\to\mathbb{R}$. Then the integral $(\delta)\int_a^b f\,\mathrm{d}g$ exists if and only if the integral $(\sigma)\int_a^b f\,\mathrm{d}g$ exists and the pair f,g satisfies the condition of pseudo-additivity at every point $x\in(a,b)$.*

Proof. If $(\delta)\int_a^b f\,\mathrm{d}g$ exists, then, by Theorem 5.1.8, $(\sigma)\int_a^b f\,\mathrm{d}g$ exists as well. By Lemma 5.2.2, the functions f,g satisfy the condition of pseudo-additivity at every point $x\in(a,b)$.

Conversely, assume that the integral $(\sigma)\int_a^b f\,\mathrm{d}g=I$ exists and the functions f,g satisfy the condition of pseudo-additivity at every point $x\in(a,b)$.

Let $\varepsilon > 0$ be given and let the division $\boldsymbol{\alpha}_\varepsilon = \{s_0, s_1, \dots, s_r\}$ of $[a,b]$ be such that $r \geq 2$ and

$$\left.\begin{aligned} &|S(Q) - I| < \varepsilon \\ &\qquad \text{for all partitions } Q = (\boldsymbol{\beta}, \boldsymbol{\eta}) \text{ of } [a,b] \text{ such that } \boldsymbol{\beta} \supset \boldsymbol{\alpha}_\varepsilon. \end{aligned}\right\} \tag{5.2.1}$$

Set

$$\delta_* = \min\{s_i - s_{i-1} : i \in \{1, \dots, r\}\}. \tag{5.2.2}$$

Since the functions f, g satisfy the condition of pseudo-additivity on (a,b), there exists a $\delta_\varepsilon \in (0, \delta_*)$ such that for every $i \in \{1, \dots, r-1\}$, the inequality

$$\left.\begin{aligned} &\big|f(\xi)\,(g(s_i'') - g(s_i')) \\ &\qquad -f(\xi')\,(g(s_i) - g(s_i')) - f(\xi'')\,(g(s_i'') - g(s_i))\big| < \frac{\varepsilon}{r-1} \\ &\text{holds whenever} \\ &\qquad s_i' \in (s_i - \delta_\varepsilon, s_i),\ \ s_i'' \in (s_i, s_i + \delta_\varepsilon), \\ &\qquad \xi \in [s_i', s_i''],\ \ \xi' \in [s_i', s_i],\ \ \xi'' \in [s_i, s_i'']. \end{aligned}\right\} \tag{5.2.3}$$

Let $P = (\boldsymbol{\alpha}, \boldsymbol{\xi})$ be a partition of $[a,b]$, $m = \nu(P)$ and $|\boldsymbol{\alpha}| < \delta_\varepsilon$.

By (5.2.2), for any $j \in \{1, \dots, m\}$ the set $(\alpha_{j-1}, \alpha_j) \cap \boldsymbol{\alpha}_\varepsilon$ is either a singleton or empty. Let

$$U_1 = \{j \in \{1, \dots, m\} : (\alpha_{j-1}, \alpha_j) \cap \boldsymbol{\alpha}_\varepsilon = \emptyset\},$$
$$U_2 = \{1, \dots, m\} \setminus U_1.$$

Then for every $j \in U_2$ there is a unique $i(j) \in \{1, \dots, r-1\}$ such that $s_{i(j)} \in (\alpha_{j-1}, \alpha_j)$. Thus, the cardinality of U_2 does not exceed $r - 1$.

Now, set $\boldsymbol{\beta} = \boldsymbol{\alpha} \cup \boldsymbol{\alpha}_\varepsilon$. Then

$$|\boldsymbol{\beta}| < \delta_\varepsilon < \delta_* \tag{5.2.4}$$

and for every $j \in U_1$ there exists a unique $k(j) \in \{1, \dots, \nu(\boldsymbol{\beta})\}$ such that

$$[\beta_{k(j)-1}, \beta_{k(j)}] = [\alpha_{j-1}, \alpha_j]. \tag{5.2.5}$$

If $j \in U_2$, then there exists exactly one $\ell(j) \in \{1, \dots, \nu(\boldsymbol{\beta}) - 1\}$ such that

$$\beta_{\ell(j)-1} = \alpha_{j-1},\ \ \beta_{\ell(j)} = s_{i(j)},\ \ \beta_{\ell(j)+1} = \alpha_j. \tag{5.2.6}$$

Choose a partition $Q = (\boldsymbol{\beta}, \boldsymbol{\eta})$ of $[a,b]$ such that

$$\eta_{k(j)} = \xi_j, \quad \text{if } j \in U_1. \tag{5.2.7}$$

Now, we compare the integral sums $S(P)$ and $S(Q)$. We have

$$S(P) = \sum_{j \in U_1} f(\xi_j)\,(g(\alpha_j) - g(\alpha_{j-1})) + \sum_{j \in U_2} f(\xi_j)\,(g(\alpha_j) - g(\alpha_{j-1})).$$

Let $V_1 = \{k(j) : j \in U_1\}$, $V_2 = \{1, \dots, \nu(Q)\} \setminus V_1$. Then by (5.2.5)–(5.2.7),

$$\begin{aligned}
S(Q) &= \sum_{k \in V_1} f(\eta_k)\,(g(\beta_k) - g(\beta_{k-1})) + \sum_{k \in V_2} f(\eta_k)\,(g(\beta_k) - g(\beta_{k-1}))\\
&= \sum_{j \in U_1} f(\eta_{k(j)})\,(g(\beta_{k(j)}) - g(\beta_{k(j)-1})) + \sum_{k \in V_2} f(\eta_k)\,(g(\beta_k) - g(\beta_{k-1}))\\
&= \sum_{j \in U_1} f(\xi_j)\,(g(\alpha_j) - g(\alpha_{j-1}))\\
&\quad + \sum_{j \in U_2} \big(f(\eta_{\ell(j)})\,(g(\beta_{\ell(j)}) - g(\beta_{\ell(j)-1})) + f(\eta_{\ell(j)+1})\,(g(\beta_{\ell(j)+1}) - g(\beta_{\ell(j)}))\big)\\
&= \sum_{j \in U_1} f(\xi_j)\,(g(\alpha_j) - g(\alpha_{j-1}))\\
&\quad + \sum_{j \in U_2} \big(f(\eta_{\ell(j)})\,(g(s_{i(j)}) - g(\alpha_{j-1})) + f(\eta_{\ell(j)+1})\,(g(\alpha_j) - g(s_{i(j)}))\big).
\end{aligned}$$

Hence

$$\begin{aligned}
S(P) - S(Q) &= \sum_{j \in U_2} f(\xi_j)\,(g(\alpha_j) - g(\alpha_{j-1}))\\
&\quad - \sum_{j \in U_2} \big(f(\eta_{\ell(j)})\,(g(s_{i(j)}) - g(\alpha_{j-1})) + f(\eta_{\ell(j)+1})\,(g(\alpha_j) - g(s_{i(j)}))\big),
\end{aligned}$$

i.e., $|S(P) - S(Q)| \le \sum_{j \in U_2} |W_j|$, where

$$\begin{aligned}
W_j &= f(\xi_j)\,(g(\alpha_j) - g(\alpha_{j-1}))\\
&\quad - f(\eta_{\ell(j)})\,(g(s_{i(j)}) - g(\alpha_{j-1})) - f(\eta_{\ell(j)+1})\,(g(\alpha_j) - g(s_{i(j)})).
\end{aligned}$$

Let us recall that by (5.2.4) and (5.2.6) we have

$$[\alpha_{j-1}, \alpha_j] \subset (s_{i(j)} - \delta_\varepsilon, s_{i(j)} + \delta_\varepsilon), \ \xi_j \in [\alpha_{j-1}, \alpha_j],$$
$$\eta_{\ell(j)} \in [\alpha_{j-1}, s_{i(j)}], \ \eta_{\ell(j)+1} \in [s_{i(j)}, \alpha_j] \quad \text{for } j \in U_2.$$

By (5.2.3), $|W_j| < \varepsilon/(r-1)$ for every $j \in U_2$, and therefore (using the fact that the cardinality of U_2 does not exceed $r-1$) we have

$$|S(P) - S(Q)| \le \sum_{j \in U_2} |W_j| < \varepsilon.$$

Finally, by (5.2.1) and in view of the definition of β we get

$$|S(P)-I| \leq |S(P)-S(Q)| + |S(Q)-I| < 2\,\varepsilon.$$

This means that $(\delta)\int_a^b f\,\mathrm{d}g = I$. □

5.2.4 Example. Consider again the functions $f, g\colon[-1,1]\to\mathbb{R}$ from Example 5.1.11:

$$f(x)=\begin{cases}0 & \text{if } x\leq 0,\\ 1 & \text{if } x>0,\end{cases} \qquad g(x)=\begin{cases}1 & \text{if } x\leq 0,\\ 0 & \text{if } x>0.\end{cases}$$

We know that $(\sigma)\int_{-1}^1 f\,\mathrm{d}g$ exists, but $(\delta)\int_{-1}^1 f\,\mathrm{d}g$ does not. Thus, by Theorem 5.2.3, the functions f, g fail to satisfy the condition of pseudo-additivity at a certain point $x\in(-1,1)$. Let us show that this is indeed the case for $x=0$: If $x'<0<x''$, $\xi\in[x',x'']$, $\xi'\in[x',0]$, $\xi''\in[0,x'']$, then

$$\begin{aligned}&\big|f(\xi)\,(g(x'')-g(x'))-f(\xi')\,(g(0)-g(x'))-f(\xi'')\,(g(x'')-g(0))\big|\\ &\quad=\big|-f(\xi)+f(\xi'')\big|=1\end{aligned}$$

whenever $\xi\leq 0$ and $\xi''>0$. Thus, the functions f, g do not satisfy the condition of pseudo-additivity at the point 0.

The next lemma provides some necessary and sufficient conditions for f and g to satisfy the condition of pseudo-additivity.

5.2.5 Lemma. *If $f, g\colon[a,b]\to\mathbb{R}$ satisfy the condition of pseudo-additivity at $x\in(a,b)$, then at least one of the functions f, g is continuous at x.*

Conversely, if one of the functions f, g is continuous at $x\in(a,b)$ and the other one is bounded on a neighborhood of x, then the functions f, g satisfy the condition of pseudo-additivity at the point x.

Proof. a) Let $x\in(a,b)$ and let f, g satisfy the condition of pseudo-additivity at x. Using Definition 5.2.1 with $\xi=\xi'$, we get that for each $\varepsilon>0$ there is a $\delta_\varepsilon>0$ such that

$$|f(\xi')-f(\xi'')|\,|g(x'')-g(x)|<\varepsilon$$

holds whenever

$$x'\in(x-\delta_\varepsilon,x),\ x''\in(x,x+\delta_\varepsilon)\ \text{ and }\ \xi'\in[x',x],\ \xi''\in[x,x''].$$

It follows that if g is not continuous from the right at x, then f has to be continuous at x. Similarly, using Definition 5.2.1 with $\xi = \xi''$, we can prove that if g is not continuous from the left at x, then f has to be continuous at x.

b) Let $x \in (a,b)$, $x' \in [a,x)$, $x'' \in (x,b]$, $\xi \in [x',x'']$, $\xi' \in [x',x]$, $\xi'' \in [x,x'']$. Then

$$\begin{aligned}
&|f(\xi)\,(g(x'')-g(x'))-f(\xi')\,(g(x)-g(x'))-f(\xi'')\,(g(x'')-g(x))| \\
&\quad= \big|(f(\xi)-f(\xi'))\,(g(x)-g(x'))-(f(\xi'')-f(\xi))\,(g(x'')-g(x))\big| \\
&\quad\le |f(\xi)-f(\xi')|\,|g(x)-g(x')|+|f(\xi'')-f(\xi)|\,|g(x'')-g(x)| \\
&\quad\le \big(|f(\xi)-f(x)|+|f(x)-f(\xi')|\big)\,|g(x)-g(x')| \\
&\qquad\quad+\big(|f(\xi'')-f(x)|+|f(x)-f(\xi)|\big)\,|g(x'')-g(x)|.
\end{aligned}$$

It follows easily that functions f,g satisfy the condition of pseudo-additivity at $x \in (a,b)$ whenever one of them is continuous at x and the other is bounded on some neighborhood of x. □

5.2.6 Corollary. *Assume that $(\sigma)\int_a^b f\,\mathrm{d}g$ exists and for each $x \in (a,b)$, at least one of the functions $f,g\colon [a,b] \to \mathbb{R}$ is continuous at x, while the other is bounded on a neighborhood of x. Then $(\delta)\int_a^b f\,\mathrm{d}g$ exists and equals $(\sigma)\int_a^b f\,\mathrm{d}g$.*

P r o o f. By Lemma 5.2.5, the functions f,g satisfy the condition of pseudo-additivity at every point $x \in (a,b)$. Hence, the integral $(\delta)\int_a^b f\,\mathrm{d}g$ exists by Theorem 5.2.3, and is equal to $(\sigma)\int_a^b f\,\mathrm{d}g$ by Theorem 5.1.8. □

5.2.7 Remark. In particular, if $g(x) = x$ and f is bounded on $[a,b]$, then the definitions of the integrals $(\delta)\int_a^b f(x)\,\mathrm{d}x$ and $(\sigma)\int_a^b f(x)\,\mathrm{d}x$ coincide.

5.2.8 Exercise. Prove the following statement:

If $c \in [a,b]$, the integrals $(\delta)\int_a^c f\,\mathrm{d}g$ and $(\delta)\int_c^b f\,\mathrm{d}g$ exist, and f,g satisfy the condition of pseudo-additivity at c, then the integral $(\delta)\int_a^b f\,\mathrm{d}g$ exists and

$$(\delta)\int_a^b f\,\mathrm{d}g = (\delta)\int_a^c f\,\mathrm{d}g + (\delta)\int_c^b f\,\mathrm{d}g.$$

Hint: Make use of Theorems 5.1.22 and 5.2.3.

5.3 Absolute integrability

The goal of this section is to clarify the relations between the existence of the integrals $\int_a^b f(x)\,\mathrm{d}g(x)$, $\int_a^b |f(x)|\,\mathrm{d}g(x)$, and $\int_a^b f(x)\,\mathrm{d}(\mathrm{var}_a^x\, g)$, provided that g has bounded variation.

Along the way, we discover another useful result: If $\int_a^b f\,\mathrm{d}g$ exists, then f has to be bounded on $[a,b]$ with the exception of finitely many intervals where g is constant.

We begin by introducing the following notation.

5.3.1 Definition. Let $f,g\colon[a,b]\to\mathbb{R}$. If $a\le c<d\le b$, we define

$$\omega(S_{f\Delta g};[c,d])=\sup\Big\{|S(P)-S(P')|:P,P' \text{ are partitions of } [c,d]\Big\}.$$

We can now reformulate the Bolzano-Cauchy condition as follows.

5.3.2 Theorem. *Let $f,g\colon[a,b]\to\mathbb{R}$. Then the following assertions hold:*

(i) *The integral $(\delta)\int_a^b f\,\mathrm{d}g$ exists if and only if*

$$\left.\begin{array}{l} \text{for every } \varepsilon>0 \text{ there is a } \delta_\varepsilon>0 \text{ such that} \\ \displaystyle\sum_{j=1}^{\nu(\boldsymbol{\alpha})}\omega(S_{f\Delta g};[\alpha_{j-1},\alpha_j])<\varepsilon \\ \text{holds for all divisions } \boldsymbol{\alpha} \text{ of } [a,b] \text{ such that } |\boldsymbol{\alpha}|<\delta_\varepsilon. \end{array}\right\} \tag{5.3.1}$$

(ii) *The integral $(\sigma)\int_a^b f\,\mathrm{d}g$ exists if and only if*

$$\left.\begin{array}{l} \text{for every } \varepsilon>0 \text{ there is a division } \boldsymbol{\alpha}_\varepsilon \text{ of } [a,b] \text{ such that} \\ \displaystyle\sum_{j=1}^{\nu(\boldsymbol{\alpha})}\omega(S_{f\Delta g};[\alpha_{j-1},\alpha_j])<\varepsilon \\ \text{holds for all divisions } \boldsymbol{\alpha} \text{ of } [a,b] \text{ such that } \boldsymbol{\alpha}\supset\boldsymbol{\alpha}_\varepsilon. \end{array}\right\} \tag{5.3.2}$$

Proof. We will show that (5.3.1) is a necessary and sufficient condition for the existence of the Riemann-Stieltjes (δ)-integral.

a) Assume that (5.1.3) holds. Let $\widetilde{\varepsilon}>0$, $\varepsilon=\widetilde{\varepsilon}/2$, and let δ_ε be defined by the condition (5.1.3). Let a division $\boldsymbol{\alpha}$ of $[a,b]$ be such that $|\boldsymbol{\alpha}|<\delta_\varepsilon$. Denote $m=\nu(\boldsymbol{\alpha})$ and, for every $j\in\{1,\dots,m\}$, choose partitions $P_j=(\boldsymbol{\alpha}^j,\boldsymbol{\xi}^j)$, $\widetilde{P}_j=(\widetilde{\boldsymbol{\alpha}}^j,\widetilde{\boldsymbol{\xi}}^j)$ of $[\alpha_{j-1},\alpha_j]$ such that

$$\omega(S_{f\Delta g},[\alpha_{j-1},\alpha_j])<S(P_j)-S(\widetilde{P}_j)+\frac{\varepsilon}{m}. \tag{5.3.3}$$

Further, define $Q = P_1 \cup \cdots \cup P_m$ and $\widetilde{Q} = \widetilde{P}_1 \cup \cdots \cup \widetilde{P}_m$. Then $Q = (\boldsymbol{\beta}, \boldsymbol{\eta})$ and $\widetilde{Q} = (\widetilde{\boldsymbol{\beta}}, \widetilde{\boldsymbol{\eta}})$ are partitions of $[a, b]$ such that

$$|\boldsymbol{\beta}| < \delta_\varepsilon \quad \text{and} \quad |\widetilde{\boldsymbol{\beta}}| < \delta_\varepsilon.$$

Hence, using (5.1.3) and (5.3.3) we get

$$\begin{aligned}\sum_{j=1}^m \omega\left(S_{f\Delta g}; [\alpha_{j-1}, \alpha_j]\right) &< \sum_{j=1}^m \left(S(P_j) - S(\widetilde{P}_j) + \frac{\varepsilon}{m}\right) \\ &= S(Q) - S(\widetilde{Q}) + \varepsilon < 2\,\varepsilon = \widetilde{\varepsilon}.\end{aligned}$$

Since $\widetilde{\varepsilon} > 0$ was arbitrary, it follows that condition (5.3.1) is satisfied.

b) For the proof of the reverse implication, assume that (5.3.1) holds. We will prove that condition (5.1.3') is satisfied.

Choose an arbitrary $\varepsilon > 0$. Let δ_ε be defined by (5.3.1) and let $P = (\boldsymbol{\alpha}, \boldsymbol{\xi})$ and $\widetilde{P} = (\widetilde{\boldsymbol{\alpha}}, \widetilde{\boldsymbol{\xi}})$ be partitions of $[a, b]$ such that $|\boldsymbol{\alpha}| < \delta_\varepsilon$ and $\widetilde{\boldsymbol{\alpha}} \supset \boldsymbol{\alpha}$. Set $m = \nu(P)$. Then for each $j \in \{1, \ldots, m\}$ there is a partition $P_j = (\boldsymbol{\alpha}^j, \boldsymbol{\xi}^j)$ of $[\alpha_{j-1}, \alpha_j]$ such that $\widetilde{P} = P_1 \cup \cdots \cup P_m$. In view of the assumption (5.3.1), we get

$$\begin{aligned}|S(P) - S(\widetilde{P})| &\le \sum_{j=1}^m \left|f(\xi_j)\,(g(\alpha_j) - g(\alpha_{j-1})) - S(P_j)\right| \\ &\le \sum_{j=1}^m \omega(S_{f\Delta g}; [\alpha_{j-1}, \alpha_j]) < \varepsilon.\end{aligned}$$

Therefore, (5.1.3') holds.

The equivalence of the condition (5.3.2) with the Bolzano-Cauchy condition for the existence of the Riemann-Stieltjes (σ)-integral can be proved analogously; the detailed proof is left as an exercise for the reader. □

5.3.3 Exercise. Prove the statement of Theorem 5.3.2 for Riemann-Stieltjes (σ)-integrals.

The next lemma provides a lower and an upper bound for $\omega(S_{f\Delta g}; [c, d])$; the latter estimate is useful if g has bounded variation.

5.3.4 Lemma. *Let $f, g : [a, b] \to \mathbb{R}$ and $[c, d] \subset [a, b]$. Then*

$$\omega_{[c,d]}(f)\,|g(d) - g(c)| \le \omega(S_{f\Delta g}; [c, d]) \le \omega_{[c,d]}(f)\ \mathrm{var}_c^d\, g, \qquad (5.3.4)$$

where

$$\omega_I(f) = \sup_{s,t \in I} |f(t) - f(s)|$$

denotes the modulus of oscillation of f over an interval I.

Proof. Let $\boldsymbol{\alpha} = \boldsymbol{\beta} = \{c, d\}$, and $\boldsymbol{\xi} = (\xi)$, $\boldsymbol{\eta} = (\eta)$, for $\xi, \eta \in [c, d]$. Then $P = (\boldsymbol{\alpha}, \boldsymbol{\xi})$ and $\widetilde{P} = (\boldsymbol{\beta}, \boldsymbol{\eta})$ are partitions of $[c, d]$, and thus

$$|f(\xi) - f(\eta)|\,|g(d) - g(c)| = |S(P) - S(\widetilde{P})| \leq \omega(S_{f\Delta g}; [c, d]).$$

Passing to the supremum with respect to ξ and η, we get the inequality $\omega_{[c,d]}(f)\,|g(d) - g(c)| \leq \omega(S_{f\Delta g}; [c, d])$.

To prove the second inequality, assume that $Q = (\boldsymbol{\beta}, \boldsymbol{\eta})$ and $\widetilde{Q} = (\widetilde{\boldsymbol{\beta}}, \widetilde{\boldsymbol{\eta}})$ are arbitrary partitions of $[c, d]$. Then $\boldsymbol{\alpha} = \boldsymbol{\beta} \cup \widetilde{\boldsymbol{\beta}}$ is a division of $[c, d]$ and

$$\big|S(Q) - S(\widetilde{Q})\big| = \Big|\sum_{j=1}^{\nu(\boldsymbol{\alpha})} \big(f(\eta'_j) - f(\widetilde{\eta}'_j)\big)\,\big(g(\alpha_j) - g(\alpha_{j-1})\big)\Big|,$$

where $\eta'_j = \eta_k$ if $[\alpha_{j-1}, \alpha_j] \subset [\beta_{k-1}, \beta_k]$ and $\widetilde{\eta}'_j = \widetilde{\eta}_k$ if $[\alpha_{j-1}, \alpha_j] \subset [\widetilde{\beta}_{k-1}, \widetilde{\beta}_k]$. Consequently,

$$\begin{aligned}\big|S(Q) - S(\widetilde{Q})\big| &\leq \sum_{j=1}^{\nu(\boldsymbol{\alpha})} \big|f(\eta'_j) - f(\widetilde{\eta}'_j)\big|\,\big|g(\alpha_j) - g(\alpha_{j-1})\big| \\ &\leq \omega_{[c,d]}(f)\, V(g, \boldsymbol{\alpha}) \leq \omega_{[c,d]}(f)\ \mathrm{var}_c^d\, g.\end{aligned}$$

Passing to the supremum with respect to Q and $\widetilde{Q}$, we get the inequality $\omega(S_{f\Delta g}; [c, d]) \leq \omega_{[c,d]}(f)\ \mathrm{var}_c^d\, g$. □

The following result indicates the role of bounded functions in the Riemann-Stieltjes integration theory. It implies that if $\int_a^b f\,\mathrm{d}g$ exists, then f is bounded on the complement of a finite collection of intervals where g is constant.

5.3.5 Theorem. *Let $f, g : [a, b] \to \mathbb{R}$ be such that $\int_a^b f\,\mathrm{d}g$ exists. Then there exists a division $\boldsymbol{\alpha}$ of $[a, b]$ such that for each $j \in \{1, \dots, \nu(\boldsymbol{\alpha})\}$, f is bounded on $[\alpha_{j-1}, \alpha_j]$, or g is constant on $[\alpha_{j-1}, \alpha_j]$.*

Proof. Without loss of generality, assume that the integral $(\sigma)\int_a^b f\,\mathrm{d}g$ exists. Choose an arbitrary $\varepsilon > 0$. By Theorem 5.3.2, there is a division $\boldsymbol{\alpha}$

of $[a,b]$ such that

$$\sum_{j=1}^{\nu(\boldsymbol{\beta})} \omega(S_{f\Delta g};[\beta_{j-1},\beta_j]) < \varepsilon$$

holds for all divisions $\boldsymbol{\beta}$ of $[a,b]$ satisfying $\boldsymbol{\beta} \supset \boldsymbol{\alpha}$. In particular, we have

$$\omega(S_{f\Delta g};[c,d]) < \varepsilon \quad \text{for each } [c,d] \subset [\alpha_{j-1},\alpha_j] \text{ with } j \in \{1,\ldots,\nu(\boldsymbol{\alpha})\}.$$

For each $[c,d] \subset [\alpha_{j-1},\alpha_j]$, we also have (cf. the proof of Lemma 5.3.4)

$$|f(\xi)-f(\eta)|\,|g(d)-g(c)| \le \omega(S_{f\Delta g};[c,d]) < \varepsilon \quad \text{whenever} \quad \xi,\eta \in [c,d].$$

Observe that if $\omega_{[c,d]}(f) = \infty$, then the previous inequality necessarily implies that $|g(d)-g(c)| = 0$, i.e., $g(c) = g(d)$.

Let us prove that for each $j \in \{1,\ldots,\nu(\boldsymbol{\alpha})\}$, f is bounded on $[\alpha_{j-1},\alpha_j]$, or g is constant on $[\alpha_{j-1},\alpha_j]$.

If f is unbounded on $[\alpha_{j-1},\alpha_j]$ for a certain $j \in \{1,\ldots,\nu(\boldsymbol{\alpha})\}$, then

$$\omega_{[\alpha_{j-1},\alpha_j]}(f) = \infty.$$

Our previous reasoning leads to the conclusion that

$$g(\alpha_j) = g(\alpha_{j-1}) = \gamma \quad \text{for a certain } \gamma \in \mathbb{R}.$$

It remains to show that $g(t) = \gamma$ for each $t \in (\alpha_{j-1},\alpha_j)$. Note that we necessarily have

$$\omega_{[\alpha_{j-1},t]}(f) = \infty \quad \text{or} \quad \omega_{[t,\alpha_j]}(f) = \infty.$$

The former possibility implies that $g(t) = g(\alpha_{j-1}) = \gamma$, while the latter implies $g(t) = g(\alpha_j) = \gamma$. This completes the proof. □

5.3.6 Remark. Assume that $\int_a^b f\,\mathrm{d}g$ exists. Since the value of the integral does not change if we change arbitrarily the values of f on the intervals where g is constant, we see from Theorem 5.3.5 that it is always possible to find a bounded function $\widetilde{f}:[a,b]\to\mathbb{R}$ such that $\int_c^d f\,\mathrm{d}g = \int_c^d \widetilde{f}\,\mathrm{d}g$ whenever $[c,d] \subset [a,b]$.

We can now describe the relations between the existence of the integrals $\int_a^b f(x)\,\mathrm{d}g(x)$, $\int_a^b |f(x)|\,\mathrm{d}g(x)$, and $\int_a^b f(x)\,\mathrm{d}(\mathrm{var}_a^x\, g)$.

5.3.7 Theorem. *Let $f:[a,b]\to\mathbb{R}$, $g \in \mathrm{BV}([a,b])$ and $v(x) = \mathrm{var}_a^x\, g$ for $x \in [a,b]$. Then the integral $\int_a^b f\,\mathrm{d}g$ exists if and only if the integral $\int_a^b f\,\mathrm{d}v$ exists.*

Proof. a) For every interval $[c,d]\subset[a,b]$ we have $\operatorname{var}_c^d v = v(d)-v(c)$. Hence, by Lemma 5.3.4 (with g replaced by v),

$$\sum_{j=1}^{\nu(\boldsymbol{\alpha})} \omega_{[\alpha_{j-1},\alpha_j]}(f)\,\big(v(\alpha_j)-v(\alpha_{j-1})\big) = \sum_{j=1}^{\nu(\boldsymbol{\alpha})} \omega(S_{f\,\Delta\,v};[\alpha_{j-1},\alpha_j])$$

holds for any division $\boldsymbol{\alpha}$ of $[a,b]$. Consequently, using Lemma 5.3.4 for an arbitrary division $\boldsymbol{\alpha}$ of $[a,b]$ we deduce

$$\sum_{j=1}^{\nu(\boldsymbol{\alpha})} \omega(S_{f\,\Delta\,g};[\alpha_{j-1},\alpha_j]) \le \sum_{j=1}^{\nu(\boldsymbol{\alpha})} \omega_{[\alpha_{j-1},\alpha_j]}(f)\ \operatorname{var}_{\alpha_{j-1}}^{\alpha_j} g$$
$$= \sum_{j=1}^{\nu(\boldsymbol{\alpha})} \omega_{[\alpha_{j-1},\alpha_j]}(f)\,\big(v(\alpha_j)-v(\alpha_{j-1})\big) = \sum_{j=1}^{\nu(\boldsymbol{\alpha})} \omega(S_{f\,\Delta\,v};[\alpha_{j-1},\alpha_j]).$$

Using Theorem 5.3.2, we can now easily prove that for both kinds of the Riemann-Stieltjes integral, the existence of $\int_a^b f\,\mathrm{d}v$ implies the existence of $\int_a^b f\,\mathrm{d}g$.

b) Assume that the integral $(\sigma)\int_a^b f\,\mathrm{d}g$ exists. We will prove that then the integral $(\sigma)\int_a^b f\,\mathrm{d}v$ exists as well.

Let $\varepsilon>0$ be given. By Theorem 5.3.2, there exists a division $\boldsymbol{\beta}$ of $[a,b]$ such that

$$\sum_{j=1}^{\nu(\boldsymbol{\alpha})} \omega\left(S_{f\Delta\,g};[\alpha_{j-1},\alpha_j]\right)<\varepsilon \tag{5.3.5}$$

holds for each its refinement $\boldsymbol{\alpha}\supset\boldsymbol{\beta}$. We can also assume that

$$0\le \operatorname{var}_a^b g - V(g,\boldsymbol{\alpha})<\varepsilon \tag{5.3.6}$$

holds for every division $\boldsymbol{\alpha}$ of $[a,b]$ such that $\boldsymbol{\alpha}\supset\boldsymbol{\beta}$. Finally, according to Theorem 5.3.5, we can suppose that for each $\boldsymbol{\alpha}\supset\boldsymbol{\beta}$ and $j\in\{1,\dots,\nu(\boldsymbol{\alpha})\}$, f is bounded on $[\alpha_{j-1},\alpha_j]$, or g is constant on $[\alpha_{j-1},\alpha_j]$. For each such division, denote by $J_{\boldsymbol{\alpha}}$ the set of all $j\in\{1,\dots,\nu(\boldsymbol{\alpha})\}$ such that f is bounded on $[\alpha_{j-1},\alpha_j]$.

Let $\boldsymbol{\alpha}$ be a division of $[a,b]$ such that $\boldsymbol{\alpha}\supset\boldsymbol{\beta}$. Using Lemma 5.3.4 we obtain

$$\begin{aligned}
\sum_{j=1}^{\nu(\boldsymbol{\alpha})} \omega(S_{f\,\Delta\,v};[\alpha_{j-1},\alpha_j]) &= \sum_{j\in J_{\boldsymbol{\alpha}}} \omega(S_{f\,\Delta\,v};[\alpha_{j-1},\alpha_j]) \\
&\le \sum_{j\in J_{\boldsymbol{\alpha}}} \omega_{[\alpha_{j-1},\alpha_j]}(f)\ \operatorname{var}_{\alpha_{j-1}}^{\alpha_j} g \\
&\le \sum_{j\in J_{\boldsymbol{\alpha}}} \omega_{[\alpha_{j-1},\alpha_j]}(f)\,|g(\alpha_j)-g(\alpha_{j-1})| \\
&\quad + \sum_{j\in J_{\boldsymbol{\alpha}}} \omega_{[\alpha_{j-1},\alpha_j]}(f)\left(\operatorname{var}_{\alpha_{j-1}}^{\alpha_j} g - |g(\alpha_j)-g(\alpha_{j-1})|\right) \\
&\le \sum_{j\in J_{\boldsymbol{\alpha}}} \omega(S_{f\,\Delta\,g};[\alpha_{j-1},\alpha_j]) \\
&\quad + \Big(\max_{j\in J_{\boldsymbol{\alpha}}} \omega_{[\alpha_{j-1},\alpha_j]}(f)\Big) \sum_{j\in J_{\boldsymbol{\alpha}}} \left(\operatorname{var}_{\alpha_{j-1}}^{\alpha_j} g - |g(\alpha_j)-g(\alpha_{j-1})|\right) \\
&< \varepsilon + \Big(\max_{j\in J_{\boldsymbol{\alpha}}} \omega_{[\alpha_{j-1},\alpha_j]}(f)\Big)\left(\operatorname{var}_a^b g - V(g,\boldsymbol{\alpha})\right) < \varepsilon + \Big(\max_{j\in J_{\boldsymbol{\beta}}} \omega_{[\beta_{j-1},\beta_j]}(f)\Big)\varepsilon.
\end{aligned}$$

By Theorem 5.3.2 we conclude that the integral $(\sigma)\int_a^b f\,\mathrm{d}v$ exists.

c) It remains to prove that the existence of the integral $(\delta)\int_a^b f\,\mathrm{d}g$ implies the existence of the integral $(\delta)\int_a^b f\,\mathrm{d}v$. By Theorem 5.3.5, there exists a division $\boldsymbol{\alpha}$ of $[a,b]$ such that for each $j\in\{1,\dots,\nu(\boldsymbol{\alpha})\}$, f is bounded on $[\alpha_{j-1},\alpha_j]$, or g is constant on $[\alpha_{j-1},\alpha_j]$. Consider the function $\widetilde{f}:[a,b]\to\mathbb{R}$ which equals f on each interval $[\alpha_{j-1},\alpha_j]$ where f is bounded, and vanishes elsewhere. Then $\widetilde{f}$ is bounded, $(\delta)\int_a^b \widetilde{f}\,\mathrm{d}g$ exists and equals $(\delta)\int_a^b f\,\mathrm{d}g$.

Since Theorem 5.1.8 implies that the integral $(\sigma)\int_a^b \widetilde{f}\,\mathrm{d}g$ exists, then the integral $(\sigma)\int_a^b \widetilde{f}\,\mathrm{d}v$ exists by part b) of this proof. On the other hand, it follows from the existence of $(\delta)\int_a^b \widetilde{f}\,\mathrm{d}g$ and Theorem 5.1.25 that $\widetilde{f}, g$ have no common discontinuity points in (a,b). As consequence of Lemma 2.3.3 the functions $\widetilde{f}$, v have no common discontinuity points in (a,b). Having this in mind and applying Corollary 5.2.6 we conclude that the integral $(\delta)\int_a^b \widetilde{f}\,\mathrm{d}v$ exists. (The function v is bounded because g has bounded variation on $[a,b]$.)

Note that v is constant on each interval $[\alpha_{j-1},\alpha_j]$ where g is constant, which means that $\int_{\alpha_{j-1}}^{\alpha_j} \widetilde{f}\,\mathrm{d}v = \int_{\alpha_{j-1}}^{\alpha_j} f\,\mathrm{d}v = 0$. By the definition of $\widetilde{f}$, we observe that $(\delta)\int_a^b f\,\mathrm{d}v$ exists and equals $(\delta)\int_a^b \widetilde{f}\,\mathrm{d}v$. □

5.3.8 Theorem. *Assume that* $f:[a,b]\to\mathbb{R}$, $g\in\mathrm{BV}([a,b])$, *and the integral* $\int_a^b f\,\mathrm{d}g$ *exists. Then the integral* $\int_a^b |f|\,\mathrm{d}g$ *exists as well.*

Proof. By Theorem 2.1.21, we have the Jordan decomposition $g=g_1-g_2$, where g_1, g_2 are nondecreasing, $g_1(x)=\mathrm{var}_a^x\, g$, and $g_2(x)=g_1(x)-g(x)$ for all $x\in[a,b]$. By Theorem 5.3.7 and Lemma 5.1.14, the integrals $\int_a^b f\,\mathrm{d}g_1$ and $\int_a^b f\,\mathrm{d}g_2$ exist. Thus, it is enough to prove the theorem for the special case when g is nondecreasing on $[a,b]$.

If g is nondecreasing, then $\mathrm{var}_c^d\, g=g(d)-g(c)$ for each $[c,d]\subset[a,b]$. Thus, due to Lemma 5.3.4, the relations

$$\sum_{j=1}^{\nu(\boldsymbol{\alpha})}\omega(S_{f\Delta g};[\alpha_{j-1},\alpha_j])=\sum_{j=1}^{\nu(\boldsymbol{\alpha})}\omega_{[\alpha_{j-1},\alpha_j]}(f)\,\big(g(\alpha_j)-g(\alpha_{j-1})\big)$$

and

$$\sum_{j=1}^{\nu(\boldsymbol{\alpha})}\omega(S_{|f|\Delta g};[\alpha_{j-1},\alpha_j])=\sum_{j=1}^{\nu(\boldsymbol{\alpha})}\omega_{[\alpha_{j-1},\alpha_j]}(|f|)\,\big(g(\alpha_j)-g(\alpha_{j-1})\big)$$

hold for all divisions $\boldsymbol{\alpha}$ of $[a,b]$. On the other hand,

$$\big||f(x)|-|f(y)|\big|\le|f(x)-f(y)|\quad\text{for all}\quad x,y\in[a,b].$$

Thus we have $\omega_{[c,d]}(|f|)\le\omega_{[c,d]}(f)$ for any $[c,d]\subset[a,b]$ and, therefore,

$$\begin{aligned}\sum_{j=1}^{\nu(\boldsymbol{\alpha})}\omega(S_{|f|\Delta g};[\alpha_{j-1},\alpha_j])&=\sum_{j=1}^{\nu(\boldsymbol{\alpha})}\omega_{[\alpha_{j-1},\alpha_j]}(|f|)\,\big(g(\alpha_j)-g(\alpha_{j-1})\big)\\&\le\sum_{j=1}^{\nu(\boldsymbol{\alpha})}\omega_{[\alpha_{j-1},\alpha_j]}(f)\,\big(g(\alpha_j)-g(\alpha_{j-1})\big)=\sum_{j=1}^{\nu(\boldsymbol{\alpha})}\omega(S_{f\Delta g};[\alpha_{j-1},\alpha_j]).\end{aligned}$$

The statement of the theorem now follows from Theorem 5.3.2. □

The next assertion is a direct corollary of Lemma 5.1.12 and Theorems 5.3.7 and 5.3.8.

5.3.9 Corollary. *Let* $f:[a,b]\to\mathbb{R}$, $g\in\mathrm{BV}([a,b])$, *and* $v(x)=\mathrm{var}_a^x\, g$ *for* $x\in[a,b]$. *If the integral* $\int_a^b f\,\mathrm{d}g$ *exists, then the integral* $\int_a^b |f|\,\mathrm{d}v$ *exists as well, and*

$$\Big|\int_a^b f\,\mathrm{d}g\Big|\le\int_a^b|f|\,\mathrm{d}v\le\|f\|\,\mathrm{var}_a^b\, g.$$

5.4 SUBSTITUTION

We now derive several substitution theorems, which hold in the same form for both kinds of the Riemann-Stieltjes integral. Throughout this section, we use the symbol $\omega(S_{f\Delta g};[c,d])$ introduced in Definition 5.3.1.

5.4.1 Lemma. *If the integral $\int_a^b f\,\mathrm{d}g$ exists, then the inequality*

$$\Big|\int_a^b f\,\mathrm{d}g - S(P)\Big| \le \sum_{j=1}^{\nu(P)} \omega(S_{f\Delta g};[\alpha_{j-1},\alpha_j])$$

holds for any partition $P=(\boldsymbol{\alpha},\boldsymbol{\xi})$ of $[a,b]$.

Proof. Let $P=(\boldsymbol{\alpha},\boldsymbol{\xi})$ be an arbitrary partition of $[a,b]$ and put $m=\nu(P)$. Given $\varepsilon>0$, for both kinds of the Riemann-Stieltjes integral, we can choose a partition $\widetilde{P}=(\widetilde{\boldsymbol{\alpha}},\widetilde{\boldsymbol{\xi}})$ of $[a,b]$ such that $\widetilde{\boldsymbol{\alpha}}\supset\boldsymbol{\alpha}$ and

$$\Big|\int_a^b f\,\mathrm{d}g - S(\widetilde{P})\Big| < \varepsilon.$$

Since $\widetilde{\boldsymbol{\alpha}}$ is a refinement of $\boldsymbol{\alpha}$, we can split it so that

$$\widetilde{\boldsymbol{\alpha}} = \bigcup_{j=1}^{m} \widetilde{\boldsymbol{\alpha}}^j, \quad \text{where } \widetilde{\boldsymbol{\alpha}}^j \text{ is a division of } [\alpha_{j-1},\alpha_j] \text{ for } j\in\{1,\dots,m\}.$$

Consequently, we can write $\widetilde{P}=\widetilde{P}_1\cup\dots\cup\widetilde{P}_m$, where $\widetilde{P}_j$ is a partition of $[\alpha_{j-1},\alpha_j]$. Therefore

$$\begin{aligned}\Big|\int_a^b f\,\mathrm{d}g - S(P)\Big| &\le \Big|\int_a^b f\,\mathrm{d}g - S(\widetilde{P})\Big| + \big|S(\widetilde{P}) - S(P)\big| \\ &< \varepsilon + \sum_{j=1}^{m} \big|f(\xi_j)\,(g(\alpha_j)-g(\alpha_{j-1})) - S(\widetilde{P}_j)\big| \\ &\le \varepsilon + \sum_{j=1}^{m} \omega(S_{f\Delta g};[\alpha_{j-1},\alpha_j]).\end{aligned}$$

Since $\varepsilon>0$ can be arbitrarily small, the proof is complete. □

5.4.2 Corollary. *If the integral $\int_a^b f\,\mathrm{d}g$ exists and $[c,d]\subset[a,b]$, then the inequality*

$$\Big|\int_c^d f\,\mathrm{d}g - f(\xi)\,(g(d)-g(c))\Big| \le \omega(S_{f\Delta g};[c,d])$$

holds for every $\xi \in [c,d]$.

The following substitution theorem is another corollary of Lemma 5.4.1.

5.4.3 Theorem (SUBSTITUTION THEOREM). *Let* $f, g, h : [a,b] \to \mathbb{R}$ *be such that* f *is bounded and* $\int_a^b g\,\mathrm{d}h$ *exists. If one of the integrals*

$$\int_a^b f(x)\,\mathrm{d}\Big(\int_a^x g\,\mathrm{d}h\Big) \quad \textit{and} \quad \int_a^b fg\,\mathrm{d}h$$

exists, then the other exists as well, and

$$\int_a^b f(x)\,\mathrm{d}\Big(\int_a^x g\,\mathrm{d}h\Big) = \int_a^b fg\,\mathrm{d}h.$$

Proof. Consider the function $w : [a,b] \to \mathbb{R}$ given by

$$w(x) = \int_a^x g\,\mathrm{d}h, \quad x \in [a,b].$$

For any partition $P = (\boldsymbol{\alpha}, \boldsymbol{\xi})$ of $[a,b]$ we have

$$\begin{aligned}
&|S(fg, \mathrm{d}h, P) - S(f, \mathrm{d}w, P)| \\
&\quad = \Big| \sum_{j=1}^{\nu(P)} f(\xi_j)\, g(\xi_j)\, (h(\alpha_j) - h(\alpha_{j-1})) - \sum_{j=1}^{\nu(P)} f(\xi_j)\, (w(\alpha_j) - w(\alpha_{j-1}))\Big| \\
&\quad \le \sum_{j=1}^{\nu(P)} |f(\xi_j)| \, \Big| g(\xi_j)\, (h(\alpha_j) - h(\alpha_{j-1})) - \int_{\alpha_{j-1}}^{\alpha_j} g\,\mathrm{d}h \Big| \\
&\quad \le \|f\| \Big(\sum_{j=1}^{\nu(P)} \Big| g(\xi_j)\, (h(\alpha_j) - h(\alpha_{j-1})) - \int_{\alpha_{j-1}}^{\alpha_j} g\,\mathrm{d}h \Big| \Big).
\end{aligned}$$

Now, Corollary 5.4.2 yields

$$\big|S(fg, \mathrm{d}h, P) - S(f, \mathrm{d}w, P)\big| \le \|f\| \sum_{j=1}^{\nu(P)} \omega(S_{g\Delta h}; [\alpha_{j-1}, \alpha_j]),$$

and the proof is completed by using Theorem 5.3.2. □

Setting $h(t) = t$ in Theorem 5.4.3, we get the following statement.

5.4.4 Corollary. *Assume that* $f:[a,b]\to\mathbb{R}$ *is bounded,* $g:[a,b]\to\mathbb{R}$ *is Riemann integrable, and* $p(x)=\int_a^x g(t)\,\mathrm{d}t$. *If one of the integrals*

$$\int_a^b f(x)\,\mathrm{d}p(x) \quad \textit{and} \quad \int_a^b f(x)\,g(x)\,\mathrm{d}x$$

exists, then the other exists as well, and

$$\int_a^b f(x)\,\mathrm{d}p(x)=\int_a^b f(x)\,g(x)\,\mathrm{d}x.$$

5.4.5 Theorem (SECOND SUBSTITUTION THEOREM). *Suppose that* $\phi:[c,d]\to\mathbb{R}$ *is continuous, strictly monotone and maps the interval* $[c,d]$ *onto* $[a,b]$. *If* $f,g:[a,b]\to\mathbb{R}$ *are such that* $\int_a^b f(x)\,\mathrm{d}g(x)$ *exists, then* $\int_c^d f(\phi(x))\,\mathrm{d}g(\phi(x))$ *exists as well, and*

$$\pm\int_c^d f(\phi(x))\,\mathrm{d}g(\phi(x))=\int_a^b f(x)\,\mathrm{d}g(x)$$

holds with the plus sign if ϕ *is increasing, and with the minus sign if* ϕ *is decreasing.*

Proof. Assume that ϕ is decreasing. Then $b=\phi(c)$ and $a=\phi(d)$. For a given partition $P=(\boldsymbol{\alpha},\boldsymbol{\xi})$ of the interval $[c,d]$, set

$$\beta_{\nu(P)-j}=\phi(\alpha_j) \quad \text{and} \quad \eta_{\nu(P)-j}=\phi(\xi_j) \quad \text{for } j\in\{1,\dots,\nu(P)\}.$$

Then $Q=(\boldsymbol{\beta},\boldsymbol{\eta})$, where $\boldsymbol{\beta}=\{\beta_0,\beta_1,\dots,\beta_{\nu(P)}\}$, $\boldsymbol{\eta}=(\eta_1,\eta_2,\dots,\eta_{\nu(P)})$, is a partition of $[a,b]$. We write $\boldsymbol{\beta}=\phi(\boldsymbol{\alpha})$, $\boldsymbol{\eta}=\phi(\boldsymbol{\xi})$, and $Q=\phi(P)$. Obviously, if $\boldsymbol{\alpha}\supset\boldsymbol{\alpha}'$, then also $\phi(\boldsymbol{\alpha})\supset\phi(\boldsymbol{\alpha}')$. Since ϕ is uniformly continuous on $[c,d]$, we have

$$\lim_{|\boldsymbol{\alpha}|\to 0}|\phi(\boldsymbol{\alpha})|=0.$$

Moreover,

$$\sum_{j=1}^{\nu(P)} f(\phi(\xi_j))\,\big(g(\phi(\alpha_j))-g(\phi(\alpha_{j-1}))\big)=-\sum_{i=1}^{\nu(Q)} f(\eta_j)\,\big(g(\beta_j)-g(\beta_{j-1})\big)$$

holds for every partition $P=(\boldsymbol{\alpha},\boldsymbol{\xi})$ of $[c,d]$. This fact easily implies the statement of the theorem; we leave the details to the reader. The case when ϕ is increasing can be handled in a similar way. □

The following theorem is yet another variant of the substitution theorem. Its proof is left as an exercise for the reader.

5.4.6 Theorem. *Let $\phi:[a,b]\to[\phi(a),\phi(b)]$ be increasing and continuous, and let $\psi:[\phi(a),\phi(b)]\to[a,b]$ be the inverse of ϕ. Moreover, let an arbitrary function $f:[a,b]\to\mathbb{R}$ be given. Then, if one of the integrals*

$$\int_a^b f(x)\,\mathrm{d}x, \qquad \int_{\phi(a)}^{\phi(b)} f(\psi(x))\,\mathrm{d}\psi(x)$$

exists, the other exists as well, and

$$\int_a^b f(x)\,\mathrm{d}x = \int_{\phi(a)}^{\phi(b)} f(\psi(x))\,\mathrm{d}\psi(x).$$

5.4.7 Exercise. Prove Theorem 5.4.6 for both kinds of the Riemann-Stieltjes integral.

5.5 INTEGRATION BY PARTS

The following statement is a generalization of the classical integration by parts formula for the Riemann integral (see Exercise 5.5.4).

5.5.1 Theorem (INTEGRATION BY PARTS). *If one of the integrals $\int_a^b f\,\mathrm{d}g$ and $\int_a^b g\,\mathrm{d}f$ exists, then the other exists as well, and we have*

$$\int_a^b f\,\mathrm{d}g + \int_a^b g\,\mathrm{d}f = f(b)g(b) - f(a)g(a). \tag{5.5.1}$$

P r o o f. a) Let an arbitrary partition $P=(\boldsymbol{\alpha},\boldsymbol{\xi})$ of $[a,b]$ be given. Denote $m=\nu(P)$. Rearranging the terms in the sum $S(f,\mathrm{d}g,P)$, we get

$$\begin{aligned} S(f,\mathrm{d}g,P) &= f(\xi_1)\,(g(\alpha_1)-g(a)) + f(\xi_2)\,(g(\alpha_2)-g(\alpha_1)) \\ &\quad + \cdots + f(\xi_m)\,(g(b)-g(\alpha_{m-1})) \\ &= -f(a)\,g(a) - (f(\xi_1)-f(a))\,g(a) - (f(\xi_2)-f(\alpha_1))\,g(\alpha_1) \\ &\quad - (f(\alpha_1)-f(\xi_1))\,g(\alpha_1) - \cdots - (f(\xi_m)-f(\alpha_{m-1}))\,g(\alpha_{m-1}) \\ &\quad - (f(\alpha_{m-1})-f(\xi_{m-1}))\,g(\alpha_{m-1}) - (f(b)-f(\xi_m))\,g(b) + f(b)g(b) \\ &= f(b)\,g(b) - f(a)\,g(a) - S(g,\mathrm{d}f,Q), \end{aligned}$$

where the partition $Q=(\boldsymbol{\beta},\boldsymbol{\eta})$ of $[a,b]$ is such that

$$\boldsymbol{\beta}=\{a,\xi_1,\alpha_1,\xi_2,\alpha_2,\dots,\alpha_{m-1},\xi_m,b\},$$
$$\boldsymbol{\eta}=(a,\alpha_1,\alpha_1,\alpha_2,\alpha_2,\dots,\alpha_{m-1},\alpha_{m-1},b).$$

(If $\xi_j=\alpha_{j-1}$ or $\xi_j=\alpha_j$ for some j, we omit ξ_j from $\boldsymbol{\beta}$ and the corresponding tag from $\boldsymbol{\eta}$.) Clearly, $\boldsymbol{\beta}$ is a refinement of $\boldsymbol{\alpha}$.

b) Assume that the integral $(\sigma)\int_a^b g\,\mathrm{d}f$ exists and let $\varepsilon>0$ be given. Choose a division $\boldsymbol{\beta}_\varepsilon$ of $[a,b]$ such that

$$\left|S(g,\mathrm{d}f,Q)-(\sigma)\int_a^b g\,\mathrm{d}f\right|<\varepsilon$$

holds for all partitions $Q=(\boldsymbol{\beta},\boldsymbol{\eta})$ of $[a,b]$ such that $\boldsymbol{\beta}\supset\boldsymbol{\beta}_\varepsilon$.

Let $P=(\boldsymbol{\alpha},\boldsymbol{\xi})$ be an arbitrary partition of $[a,b]$ such that $\boldsymbol{\alpha}\supset\boldsymbol{\beta}_\varepsilon$. By the first part of the proof there is a partition $Q=(\boldsymbol{\beta},\boldsymbol{\eta})$ of $[a,b]$ such that

$$\begin{aligned}S(f,\mathrm{d}g,P)-f(b)\,g(b)+f(a)\,g(a)+(\sigma)\int_a^b g\,\mathrm{d}f\\ =(\sigma)\int_a^b g\,\mathrm{d}f-S(g,\mathrm{d}f,Q),\end{aligned}$$

while $\boldsymbol{\beta}\supset\boldsymbol{\alpha}$ and hence also $\boldsymbol{\beta}\supset\boldsymbol{\beta}_\varepsilon$. Consequently,

$$\begin{aligned}\left|S(f,\mathrm{d}g,P)-f(b)\,g(b)+f(a)\,g(a)+(\sigma)\int_a^b g\,\mathrm{d}f\right|\\ =\left|(\sigma)\int_a^b g\,\mathrm{d}f-S(g,\mathrm{d}f,Q)\right|<\varepsilon.\end{aligned}$$

This means that the integral $(\sigma)\int_a^b f\,\mathrm{d}g$ exists and (5.5.1) holds.

By interchanging the roles of f and g, we immediately see that if $(\sigma)\int_a^b f\,\mathrm{d}g$ exists, then $(\sigma)\int_a^b g\,\mathrm{d}f$ exists as well and (5.5.1) holds.

c) The statement of the theorem for Riemann-Stieltjes (δ)-integrals also follows easily from the relation (5.5.1); the details are left to the reader. □

5.5.2 Exercise. Prove Theorem 5.5.1 for the Riemann-Stieltjes (δ)-integral.

5.5.3 Exercise. Determine the value of the integral $(\delta)\int_0^1 x^2\,\mathrm{d}g(x)$, where

$$g(x)=\begin{cases}x & \text{if } x\in[0,\frac{1}{2}],\\ 1/x & \text{if } x\in(\frac{1}{2},1].\end{cases}$$

5.5.4 Exercise. The classical integration by parts theorem for the Riemann integral reads as follows: *Assume that* $f, g : [a,b] \to \mathbb{R}$ *are Riemann integrable and let* $F, G : [a,b] \to \mathbb{R}$ *be given by*

$$F(x) = \int_a^x f(t)\,\mathrm{d}t, \quad G(x) = \int_a^x g(t)\,\mathrm{d}t \quad \text{for } x \in [a,b].$$

If one of the integrals $\int_a^b f\,G$ *and* $\int_a^b F\,g$ *exists, then the other exists as well, and we have*

$$\int_a^b f(x)\,G(x)\,\mathrm{d}x + \int_a^b F\,g = F(b)\,G(b) - F(a)\,G(a).$$

Show that this result is a consequence of Theorem 5.5.1.

Hint: Use Corollary 5.4.4.

5.6 Existence of the integral

We know (see Theorem 5.1.25) that the integral $(\delta)\int_a^b f\,\mathrm{d}g$ can exist only if f, g have no common discontinuities in $[a,b]$, i.e., at least one of them is continuous at each point $x \in [a,b]$. Similarly, the integral $(\sigma)\int_a^b f\,\mathrm{d}g$ can exist only if f, g have no common one-sided discontinuities in $[a,b]$, i.e., at least one of them is left-continuous at each point $x \in (a,b]$, and at least one is right-continuous at each point $x \in [a,b)$.

These conditions alone are not sufficient to guarantee the existence of $\int_a^b f\,\mathrm{d}g$; this is easily seen by taking $g(x) = x$ and letting f to be the Dirichlet function (see Exercise 4.1.4 (iii) and Remark 6.2.15). The next theorem provides simple sufficient conditions for the existence of both Riemann-Stieltjes integrals.

Throughout the proof we use the symbol $\omega(S_{f\Delta g};[c,d])$ introduced in Definition 5.3.1, and $\omega_I(f)$ denotes the modulus of oscillation of a function f over an interval I (see Lemma 5.3.4).

5.6.1 Theorem. *Suppose that* $f, g : [a,b] \to \mathbb{R}$ *are regulated and one of them has bounded variation.*

(i) *If* f, g *have no common one-sided discontinuities in* $[a,b]$, *then the integral* $(\sigma)\int_a^b f\,\mathrm{d}g$ *exists.*

(ii) *If f, g have no common discontinuities in $[a,b]$, then the integral $(\delta)\int_a^b f\,\mathrm{d}g$ exists.*

Proof. Without loss of generality (cf. Theorem 5.5.1), we can assume that f is regulated and g has bounded variation.

(i) Assume that f, g have no common one-sided discontinuities in $[a,b]$, and consider an arbitrary $\varepsilon > 0$. By Theorem 4.1.5, there exists a division $\boldsymbol{\tau}$ of $[a,b]$ such that

$$|f(t)-f(s)| < \varepsilon \quad \text{for all } t,s \in (\tau_{j-1},\tau_j), \quad j \in \{1,\dots,\nu(\boldsymbol{\tau})\}. \tag{5.6.1}$$

For each $j \in \{1,\dots,\nu(\boldsymbol{\tau})\}$, since at least one of f,g is right-continuous at τ_{j-1}, there exists a $u_j \in (\tau_{j-1},\tau_j)$ such that

$$\min\{\omega_{[\tau_{j-1},u_j]}(f), \omega_{[\tau_{j-1},u_j]}(g)\} < \frac{\varepsilon}{\nu(\boldsymbol{\tau})}. \tag{5.6.2}$$

Similarly, since at least one of f, g is left-continuous at τ_j, there exists a $v_j \in (u_j,\tau_j)$ such that

$$\min\{\omega_{[v_j,\tau_j]}(f), \omega_{[v_j,\tau_j]}(g)\} < \frac{\varepsilon}{\nu(\boldsymbol{\tau})}. \tag{5.6.3}$$

Thus, for each $j \in \{1,\dots,\nu(\boldsymbol{\tau})\}$, we have $\tau_{j-1} < u_j < v_j < \tau_j$. Now, let

$$\boldsymbol{\alpha}_\varepsilon = \boldsymbol{\tau} \cup \{u_j, v_j : j = 1,\dots,\nu(\boldsymbol{\tau})\}.$$

We consider an arbitrary division $\boldsymbol{\alpha}$ of $[a,b]$ such that $\boldsymbol{\alpha}_\varepsilon \subset \boldsymbol{\alpha}$, and will try to obtain an estimate for the sum $\sum_{\ell=1}^{\nu(\boldsymbol{\alpha})} \omega(S_{f\Delta g};[\alpha_{\ell-1},\alpha_\ell])$. Since $\boldsymbol{\alpha}$ is a refinement of $\boldsymbol{\alpha}_\varepsilon$, each interval $[\alpha_{\ell-1},\alpha_\ell]$ falls into exactly one of the following three categories:

The first case is when $\alpha_{\ell-1} = \tau_{j-1}$ for a certain $j \in \{1,\dots,\nu(\boldsymbol{\tau})\}$, and consequently $\alpha_\ell \le u_j$. To obtain an estimate for $\omega(S_{f\Delta g};[\alpha_{\ell-1},\alpha_\ell])$, consider arbitrary two partitions $Q = (\boldsymbol{\beta},\boldsymbol{\eta})$ and $\widetilde{Q} = (\widetilde{\boldsymbol{\beta}},\widetilde{\boldsymbol{\eta}})$ of $[\alpha_{\ell-1},\alpha_\ell]$. As in the proof of Lemma 5.3.4, we see that $\boldsymbol{\gamma} = \boldsymbol{\beta} \cup \widetilde{\boldsymbol{\beta}}$ is a division of $[\alpha_{\ell-1},\alpha_\ell]$, and

$$|S(Q) - S(\widetilde{Q})| = \Big|\sum_{i=1}^{\nu(\boldsymbol{\gamma})} \big(f(\eta_i') - f(\widetilde{\eta}_i')\big)\big(g(\gamma_i) - g(\gamma_{i-1})\big)\Big|,$$

where $\eta_i' = \eta_k$ if $[\gamma_{i-1},\gamma_i] \subset [\beta_{k-1},\beta_k]$ and $\widetilde{\eta}_i' = \widetilde{\eta}_k$ if $[\gamma_{i-1},\gamma_i] \subset [\widetilde{\beta}_{k-1},\widetilde{\beta}_k]$.

Consequently,

$$
\begin{aligned}
\big|S(Q)-S(\widetilde{Q})\big| &\le \sum_{i=1}^{\nu(\gamma)} \big|f(\eta_i')-f(\widetilde{\eta}_i')\big|\,\big|g(\gamma_i)-g(\gamma_{i-1})\big| \\
&= |f(\eta_1')-f(\widetilde{\eta}_1')|\,|g(\gamma_1)-g(\gamma_0)| + \sum_{i=2}^{\nu(\gamma)} |f(\eta_i')-f(\widetilde{\eta}_i')|\,|g(\gamma_i)-g(\gamma_{i-1})| \\
&\le \min\{\omega_{[\gamma_0,\gamma_1]}(f),\omega_{[\gamma_0,\gamma_1]}(g)\}2\max\{\|f\|,\|g\|\} + \sum_{i=2}^{\nu(\gamma)} \omega_{[\gamma_{i-1},\gamma_i]}(f)\operatorname{var}_{\gamma_{i-1}}^{\gamma_i} g.
\end{aligned}
$$

Since $[\gamma_0,\gamma_1]\subset[\tau_{j-1},u_j]$ and $[\gamma_{i-1},\gamma_i]\subset(\tau_{j-1},\tau_j)$ for all $i=2,\dots,\nu(\gamma)$, the estimates (5.6.2) and (5.6.1) lead to the inequality

$$
\big|S(Q)-S(\widetilde{Q})\big| \le \frac{\varepsilon}{\nu(\boldsymbol{\tau})}\,2\max\{\|f\|,\|g\|\} + \varepsilon\sum_{i=2}^{\nu(\gamma)} \operatorname{var}_{\gamma_{i-1}}^{\gamma_i} g,
$$

which in turn implies

$$
\omega(S_{f\Delta g};[\alpha_{\ell-1},\alpha_\ell]) \le \frac{\varepsilon}{\nu(\boldsymbol{\tau})}\,2\max\{\|f\|,\|g\|\} + \varepsilon\operatorname{var}_{\alpha_{\ell-1}}^{\alpha_\ell} g. \tag{5.6.4}
$$

The second case is when $\alpha_\ell=\tau_j$ for a certain $j\in\{1,\dots,\nu(\boldsymbol{\tau})\}$, and consequently $\alpha_{\ell-1}\ge v_j$. This case is symmetric to the previous one, and similar reasoning based on (5.6.3) leads again to the inequality (5.6.4).

The third case is when $[\alpha_{\ell-1},\alpha_\ell]\subset(\tau_{j-1},\tau_j)$ for some $j\in\{1,\dots,\nu(\boldsymbol{\tau})\}$. In this case, we apply Lemma 5.3.4 and (5.6.1) to get the estimate

$$
\omega(S_{f\Delta g};[\alpha_{\ell-1},\alpha_\ell]) \le \omega_{[\alpha_{\ell-1},\alpha_\ell]}(f)\ \operatorname{var}_{\alpha_{\ell-1}}^{\alpha_\ell} g \le \varepsilon\ \operatorname{var}_{\alpha_{\ell-1}}^{\alpha_\ell} g.
$$

If we consider all the intervals $[\alpha_{\ell-1},\alpha_\ell]$ with $\ell\in\{1,\dots,\nu(\boldsymbol{\alpha})\}$, then $\nu(\boldsymbol{\tau})$ of them are of the first type, $\nu(\boldsymbol{\tau})$ are of the second type, and the remaining ones are of the third type. By combining the estimates for $\omega(S_{f\Delta g};[\alpha_{\ell-1},\alpha_\ell])$ obtained above, we get

$$
\sum_{\ell=1}^{\nu(\boldsymbol{\alpha})} \omega(S_{f\Delta g};[\alpha_{\ell-1},\alpha_\ell]) \le 4\,\varepsilon\ \max\{\|f\|,\|g\|\} + \varepsilon\ \operatorname{var}_a^b g.
$$

The existence of the integral $(\sigma)\int_a^b f\,\mathrm{d}g$ now follows from Theorem 5.3.2.

(ii) Assume that f,g have no common discontinuities in $[a,b]$. By the previous part of the proof, the integral $(\sigma)\int_a^b f\,\mathrm{d}g$ exists. According to the second part of Lemma 5.2.5, the functions f, g satisfy the condition of pseudo-additivity at each point $x \in (a,b)$. Hence, by Theorem 5.2.3, the integral $(\delta)\int_a^b f\,\mathrm{d}g$ exists as well. □

5.6.2 Corollary. *If one of the functions $f, g: [a,b] \to \mathbb{R}$ has bounded variation and the other is continuous, then the integral $\int_a^b f\,\mathrm{d}g$ exists.*

5.7 Convergence theorems

This section focuses on convergence theorems for the Riemann-Stieltjes integral. The simplest result is the following uniform convergence theorem.

5.7.1 Theorem. *Assume that $g \in \mathrm{BV}([a,b])$, $f_n : [a,b] \to \mathbb{R}$, $n \in \mathbb{N}$, are such that the integral $\int_a^b f_n\,\mathrm{d}g$ exists for every $n \in \mathbb{N}$, and $\{f_n\}$ is uniformly convergent to a bounded function $f: [a,b] \to \mathbb{R}$. Then the integral $\int_a^b f\,\mathrm{d}g$ exists as well and*

$$\lim_{n\to\infty} \int_a^b f_n\,\mathrm{d}g = \int_a^b f\,\mathrm{d}g. \tag{5.7.1}$$

Proof. We deal with (σ)-integrals and leave the case of (δ)-integrals to the reader. If $\mathrm{var}_a^b\, g = 0$, then, by Lemma 2.1.16, g is constant on $[a,b]$ and the statement of the theorem is obvious. Therefore, assume that $\mathrm{var}_a^b\, g > 0$.

Let $\varepsilon > 0$ be given. Since $\{f_n\}$ is uniformly convergent to f, we can choose $n_\varepsilon \in \mathbb{N}$ such that

$$\|f_n - f\| < \frac{\varepsilon}{\mathrm{var}_a^b\, g} \quad \text{and} \quad \|f_n\| < \|f\| + 1 \quad \text{for all } n \geq n_\varepsilon. \tag{5.7.2}$$

Furthermore, Lemma 5.1.12 (i) implies that

$$\left|\int_a^b f_n\,\mathrm{d}g\right| \leq \|f_n\|\,\mathrm{var}_a^b\, g \leq (\|f\|+1)\,\mathrm{var}_a^b\, g \quad \text{for all } \ n \geq n_\varepsilon.$$

Hence, there is a subsequence $\{f_{n_k}\}$ and $I \in \mathbb{R}$ such that

$$\lim_{k\to\infty} \int_a^b f_{n_k}\,\mathrm{d}g = I.$$

In particular, there exists a $k_\varepsilon \in \mathbb{N}$ such that

$$n_{k_\varepsilon} \geq n_\varepsilon \quad \text{and} \quad \left|\int_a^b f_{n_k}\,\mathrm{d}g - I\right| < \varepsilon.$$

Now, let $\boldsymbol{\alpha}_\varepsilon$ be a division of $[a,b]$ such that

$$\left|S(f_{n_{k_\varepsilon}}, \mathrm{d}g, P) - \int_a^b f_{n_{k_\varepsilon}} \mathrm{d}g\right| < \varepsilon$$

whenever $P = (\boldsymbol{\alpha}, \boldsymbol{\xi})$ is a partition of $[a,b]$ such that $\boldsymbol{\alpha} \supset \boldsymbol{\alpha}_\varepsilon$.

Since $n_{k_\varepsilon} \geq n_\varepsilon$, it follows from (5.7.2) that

$$\left|S(f, \mathrm{d}g, P) - S(f_{n_{k_\varepsilon}}, \mathrm{d}g, P)\right| \leq \|f - f_{n_{k_\varepsilon}}\| \operatorname{var}_a^b g < \varepsilon$$

for every partition $P = (\boldsymbol{\alpha}, \boldsymbol{\xi})$ of $[a,b]$. Using the previous estimates, we deduce that

$$\begin{aligned}\left|S(f, \mathrm{d}g, P) - I\right| &\leq \left|S(f, \mathrm{d}g, P) - S(f_{n_{k_\varepsilon}}, \mathrm{d}g, P))\right| \\ &\quad + \left|S(f_{n_{k_\varepsilon}}, \mathrm{d}g, P) - \int_a^b f_{n_{k_\varepsilon}} \mathrm{d}g\right| + \left|\int_a^b f_{n_{k_\varepsilon}} \mathrm{d}g - I\right| < 3\,\varepsilon\end{aligned}$$

for each partition $P = (\boldsymbol{\alpha}, \boldsymbol{\xi})$ of $[a,b]$ such that $\boldsymbol{\alpha} \supset \boldsymbol{\alpha}_\varepsilon$. Thus,

$$\int_a^b f \,\mathrm{d}g = I.$$

Finally, since by Lemmas 5.1.12 and 5.1.14 we have

$$\left|\int_a^b f_n \,\mathrm{d}g - \int_a^b f \,\mathrm{d}g\right| \leq \|f_n - f\| \operatorname{var}_a^b g,$$

equality (5.7.1) follows from the uniform convergence of $\{f_n\}$ to f. □

5.7.2 Exercise. Prove the statement of Theorem 5.7.1 for Riemann-Stieltjes (δ)-integrals.

The next convergence theorem deals with a more general situation where not only the integrands but also the integrators depend on $n \in \mathbb{N}$. On the other hand, the integrands are required to be continuous.

5.7.3 Theorem. *Assume that the sequence $\{g_n\} \subset \mathrm{BV}([a,b])$ satisfies $\sup_{n\in\mathbb{N}} \operatorname{var}_a^b g_n < \infty$ and there exists a function $g \in \mathrm{BV}([a,b])$ such that $\lim_{n\to\infty} g_n(x) = g(x)$ for all $x \in D$, where D is a dense subset of $[a,b]$ containing the points a and b. Moreover, let $\{f_n\} \subset \mathrm{C}([a,b])$ be uniformly convergent to a function $f : [a,b] \to \mathbb{R}$. Then*

$$\lim_{n\to\infty} \int_a^b f_n \,\mathrm{d}g_n = \int_a^b f \,\mathrm{d}g.$$

Proof. By Corollary 5.6.2, all the integrals $\int_a^b f_n \,\mathrm{d}g_n$, $n \in \mathbb{N}$, and $\int_a^b f \,\mathrm{d}g$ exist. The hypotheses ensure that there is a $\gamma > 0$ such that $\operatorname{var}_a^b g \leq \gamma$ and $\operatorname{var}_a^b g_n \leq \gamma$ for all $n \in \mathbb{N}$.

Let $\varepsilon > 0$ be given. Continuity of f implies the existence of a $\delta > 0$ such that

$$|f(x) - f(y)| < \frac{\varepsilon}{4\gamma} \quad \text{for all } \; x, y \in [a,b] \text{ with } |x - y| < \delta. \tag{5.7.3}$$

Since D is a dense subset of $[a,b]$ containing a and b, there exists a partition $P = (\boldsymbol{\alpha}, \boldsymbol{\xi})$ of $[a,b]$ such that $|\boldsymbol{\alpha}| < \delta$ and $\boldsymbol{\alpha} \subset D$. Then for each $n \in \mathbb{N}$ we have

$$\begin{aligned}\int_a^b f \,\mathrm{d}g_n - S(f, \mathrm{d}g_n, P) &= \sum_{j=1}^{\nu(\boldsymbol{\alpha})} \Big(\int_{\alpha_{j-1}}^{\alpha_j} f(x) \,\mathrm{d}g_n(x) - f(\xi_j) \int_{\alpha_{j-1}}^{\alpha_j} \mathrm{d}g_n(x) \Big) \\ &= \sum_{j=1}^{\nu(\boldsymbol{\alpha})} \int_{\alpha_{j-1}}^{\alpha_j} \big(f(x) - f(\xi_j)\big) \,\mathrm{d}g_n(x).\end{aligned}$$

Using (5.7.3) and Lemma 5.1.12, we get

$$\Big| \int_a^b f \,\mathrm{d}g_n - S(f, \mathrm{d}g_n, P) \Big| \leq \frac{\varepsilon}{4\gamma} \sum_{j=1}^{\nu(\boldsymbol{\alpha})} \operatorname{var}_{\alpha_{j-1}}^{\alpha_j} g_n \leq \frac{\varepsilon}{4\gamma}\, \gamma = \frac{\varepsilon}{4}.$$

Similarly, we can derive the following inequality with the function g_n replaced by g:

$$\Big| \int_a^b f \,\mathrm{d}g - S(f, \mathrm{d}g, P) \Big| < \frac{\varepsilon}{4}.$$

Since $g_n(x) \to g(x)$ for every $x \in D$ and $\boldsymbol{\alpha} \subset D$, it is clear that

$$\lim_{n \to \infty} \big| S(f, \mathrm{d}g_n, P) - S(f, \mathrm{d}g, P) \big| = 0.$$

Thus, there exists an $n_0 \in \mathbb{N}$ such that

$$\big| S(f, \mathrm{d}g_n, P) - S(f, \mathrm{d}g, P) \big| < \frac{\varepsilon}{4} \quad \text{for } n \geq n_0.$$

Since $\{f_n\}$ is uniformly convergent to f, there exists an $n_1 \in \mathbb{N}$ such that

$$\|f_n - f\| < \frac{\varepsilon}{4\gamma} \quad \text{for } n \geq n_1.$$

Using the previous estimates and Lemma 5.1.12, we finally get

$$\begin{aligned}
&\left|\int_a^b f_n \,\mathrm{d}g_n - \int_a^b f \,\mathrm{d}g\right| \\
&\quad \le \left|\int_a^b (f_n - f)\,\mathrm{d}g_n\right| + \left|\int_a^b f \,\mathrm{d}(g_n - g)\right| \\
&\quad \le \|f_n - f\| \operatorname{var}_a^b g_n + \left|\int_a^b f \,\mathrm{d}g_n - S(f, \mathrm{d}g_n, P)\right| \\
&\qquad + \left|S(f, \mathrm{d}g_n, P) - S(f, \mathrm{d}g, P)\right| + \left|S(f, \mathrm{d}g, P) - \int_a^b f \,\mathrm{d}g\right| < \varepsilon
\end{aligned}$$

for all $n \ge \max(n_0, n_1)$, which completes the proof. □

The previous theorem requires that the variations of g_n are uniformly bounded. For example, this situation occurs if each g_n is nondecreasing. Then $\operatorname{var}_a^b g_n = g_n(b) - g_n(a)$ for each $n \in \mathbb{N}$, and since $\{g_n(x)\}$ was assumed to be convergent for $x = a$ and $x = b$, it follows that $\{\operatorname{var}_a^b g_n\}$ is a bounded sequence. This observation can be utilized in the next exercise.

5.7.4 Exercise. Let $\{d_k\}$ be a non-repeating sequence of points from $(a, b]$, and let $\{c_k\}$ be a sequence of nonnegative real numbers such that $\sum_{k=1}^{\infty} c_k < \infty$. Suppose that $g : [a, b] \to \mathbb{R}$ is the step function

$$g(x) = \sum_{k=1}^{\infty} c_k \chi_{[d_k, \infty)}(x) \quad \text{for } x \in [a, b].$$

Show that for every continuous function $f : [a, b] \to \mathbb{R}$, we have

$$\int_a^b f \,\mathrm{d}g = \sum_{k=1}^{\infty} c_k f(d_k).$$

Hint: Observe that $g = \lim_{n\to\infty} \sum_{k=1}^{n} g_k$, where $g_k = c_k \chi_{[d_k, \infty)}$. Show that $\int_a^b f \,\mathrm{d}g_k = c_k f(d_k)$.

Our third result is the bounded convergence theorem, which generalizes the uniform convergence theorem.

5.7.5 Theorem (BOUNDED CONVERGENCE THEOREM). *Let $g \in \mathrm{BV}([a, b])$ and let $\{f_n\}$ be a sequence of real valued functions defined on $[a, b]$ and satisfying the following conditions:*

(i) *The integrals $\int_a^b f \,\mathrm{d}g$ and $\int_a^b f_n \,\mathrm{d}g$ exist for each $n \in \mathbb{N}$.*

(ii) $\lim_{n\to\infty} f_n(t) = f(t)$ *for* $t \in [a,b]$.

(iii) *There exists a constant* $M \ge 0$ *such that* $|f_n(t)| \le M$ *for all* $n \in \mathbb{N}$ *and* $t \in [a,b]$.

Then

$$\lim_{n\to\infty} \int_a^b f_n \,\mathrm{d}g = \int_a^b f \,\mathrm{d}g.$$

Proof. By Corollary 5.3.9, the integral $\int_a^b |f_n(x) - f(x)| \,\mathrm{d}(\mathrm{var}_a^x\, g)$ exists for every $n \in \mathbb{N}$, and we have the inequality

$$\Big| \int_a^b f_n(x) \,\mathrm{d}g(x) - \int_a^b f(x) \,\mathrm{d}g(x) \Big| \le \int_a^b |f_n(x) - f(x)| \,\mathrm{d}(\mathrm{var}_a^x\, g).$$

Therefore, it is sufficient to prove that the theorem holds if the functions f_n are nonnegative, $f = 0$, and g is nondecreasing. Under these assumptions, we need to prove that

$$\lim_{n\to\infty} \int_a^b f_n \,\mathrm{d}g = 0. \tag{5.7.4}$$

Without loss of generality, assume that g is nonconstant (otherwise the statement is obvious), and denote $\gamma = g(b) - g(a)$.

Given $\varepsilon > 0$, for each $n \in \mathbb{N}$ define

$$A_n = \{t \in [a,b] \colon f_m(t) \ge \frac{\varepsilon}{3\gamma} \text{ for some } m \ge n\}.$$

It is not difficult to see that $A_{n+1} \subset A_n$ for $n \in \mathbb{N}$ and $\bigcap_n A_n = \emptyset$. Let

$$v(A_n) = \sup\{\mathrm{var}(g,E) \colon E \in \mathcal{E}(A_n)\}, \quad n \in \mathbb{N}$$

(with $v(A_n) = 0$ in case $A_n = \emptyset$). From Theorem 2.8.18 we know that $\lim_{n\to\infty} v(A_n) = 0$, thus we can choose $n_\varepsilon \in \mathbb{N}$ such that $v(A_n) < \frac{\varepsilon}{3M}$ for $n \ge n_\varepsilon$, and consequently

$$\mathrm{var}(g,E) < \frac{\varepsilon}{3M} \quad \text{for every } n \ge n_\varepsilon \text{ and } E \in \mathcal{E}(A_n). \tag{5.7.5}$$

Now, fix an arbitrary $n \ge n_\varepsilon$. By Theorem 5.1.28, $\int_a^b f_n \,\mathrm{d}g = \underline{\int_a^b} f_n \,\mathrm{d}g$. In particular, there exists a division $\boldsymbol{\alpha}$ of $[a,b]$ such that

$$\int_a^b f_n \,\mathrm{d}g - \frac{\varepsilon}{3} < \underline{S}(f_n, \mathrm{d}g, \boldsymbol{\alpha}) = \sum_{j=1}^{\nu(\boldsymbol{\alpha})} m_j\,(g(\alpha_j) - g(\alpha_{j-1})), \tag{5.7.6}$$

where $m_j = \inf_{x\in[\alpha_{j-1},\alpha_j]} f_n(x)$. Consider the sets

$$\Lambda = \{j \in \{1,\dots,\nu(\boldsymbol{\alpha})\} : m_j \geq \frac{\varepsilon}{3\gamma}\}, \qquad E = \bigcup_{j\in\Lambda} [\alpha_{j-1}, \alpha_j].$$

Clearly E is an elementary set (possibly empty) and $E \subset A_n$. Therefore, using (5.7.5) and (5.7.6) we obtain

$$\begin{aligned}\int_a^b f_n \, \mathrm{d}g &< \frac{\varepsilon}{3} + \sum_{j\in\Lambda} m_j \left(g(\alpha_j) - g(\alpha_{j-1})\right) + \sum_{j\notin\Lambda} m_j \left(g(\alpha_j) - g(\alpha_{j-1})\right) \\ &\leq \frac{\varepsilon}{3} + M \sum_{j\in\Lambda} (g(\alpha_j) - g(\alpha_{j-1})) + \frac{\varepsilon}{3\gamma} \sum_{j\notin\Lambda} (g(\alpha_j) - g(\alpha_{j-1})) \\ &\leq \frac{\varepsilon}{3} + M \operatorname{var}(g, E) + \frac{\varepsilon}{3\gamma} \left(g(b) - g(a)\right) < \varepsilon.\end{aligned}$$

In summary,

$$\int_a^b f_n \, \mathrm{d}g < \varepsilon \quad \text{for every} \quad n \geq n_\varepsilon,$$

which shows that (5.7.4) holds, concluding the proof. □

5.8 CONSEQUENCES OF RIEMANN-STIELTJES INTEGRABILITY

The goal of this section is to prove the following two assertions: 1) If the integral $\int_a^b f \, \mathrm{d}g$ exists for every finite step function g, then f is continuous. 2) If the integral $\int_a^b f \, \mathrm{d}g$ exists for every continuous function f, then g has bounded variation.

The proof of the first statement is straightforward.

5.8.1 Theorem. *If $f : [a,b] \to \mathbb{R}$ is such that the integral $\int_a^b f \, \mathrm{d}g$ exists for every finite step function $g : [a,b] \to \mathbb{R}$, then f is continuous.*

Proof. We can restrict ourselves to the (σ)-integral. Choose an arbitrary $x \in (a,b]$. Then $g = \chi_{[x,b]}$ is a finite step function which is discontinuous from the left at x. Since $(\sigma) \int_a^b f \, \mathrm{d}g$ exists, Theorem 5.1.25 implies that f is left-continuous at x. Right-continuity of f at each point $x \in [a,b)$ can be proved similarly using the function $g = \chi_{[a,x]}$. □

The second statement relies on two auxiliary lemmas.

5.8.2 Lemma. *If $a_n \geq 0$ for all $n \in \mathbb{N}$ and $\sum_{n=1}^{\infty} a_n = \infty$, then there exists a sequence $\{c_n\}$ such that*

$$c_n > 0 \ \text{ for all } n \in \mathbb{N}, \quad \lim_{n\to\infty} c_n = 0 \quad \text{and} \quad \sum_{n=1}^{\infty} c_n\, a_n = \infty.$$

Proof. The sequence $\{s_n\} = \{\sum_{k=1}^{n} a_k\}$ is nondecreasing and satisfies $\lim_{n\to\infty} s_n = \infty$. Thus, there exists an $n_0 \in \mathbb{N}$ such that $s_n > 0$ for all $n \geq n_0$. Let

$$c_n = \begin{cases} 1 & \text{if } \ n < n_0, \\ 1/s_n & \text{if } \ n \geq n_0. \end{cases}$$

Obviously, $c_n > 0$ for every $n \in \mathbb{N}$ and $\lim_{n\to\infty} c_n = 0$. On the other hand, we have

$$\sum_{k=n}^{m} c_k\, a_k = \sum_{k=n}^{m} \frac{a_k}{s_k} \geq \frac{1}{s_m} \sum_{k=n}^{m} a_k = \frac{1}{s_m}\Big(\sum_{k=1}^{m} a_k - \sum_{k=1}^{n-1} a_k\Big) = 1 - \frac{s_{n-1}}{s_m}$$

for all $m, n \in \mathbb{N}$ such that $m > n \geq n_0$. Since $\lim_{n\to\infty} s_n = \infty$, for each $n \in \mathbb{N}$ there is an $m_n > n$ such that $s_{n-1}/s_{m_n} < 1/2$. Consequently,

$$\sum_{k=n}^{m_n} c_k\, a_k > \frac{1}{2}.$$

This completes the proof of the lemma. □

5.8.3 Lemma. *Consider a function $g\colon [a,b] \to \mathbb{R}$.*

(i) *If $x_0 \in (a,b]$ and $\operatorname{var}_x^{x_0} g = \infty$ for every $x \in [a, x_0)$, then there exists an increasing sequence $\{x_k\}$ of points in $[a, x_0)$ such that*

$$\lim_{k\to\infty} x_k = x_0 \quad \text{and} \quad \sum_{k=1}^{\infty} |g(x_{k+1}) - g(x_k)| = \infty. \tag{5.8.1}$$

(ii) *If $x_0 \in [a,b)$ and $\operatorname{var}_{x_0}^{x} g = \infty$ for every $x \in (x_0, b]$, then there exists a decreasing sequence $\{x_k\}$ of points in $(x_0, b]$ satisfying* (5.8.1).

Proof. We prove the first statement and leave the second to the reader. Thus, suppose that $x_0 \in (a,b]$ and $\operatorname{var}_x^{x_0} g = \infty$ for every $x \in [a, x_0)$. We claim that

$$\sup\big\{\operatorname{var}_y^x g : x \in (y, x_0)\big\} = \infty \quad \text{for each } y \in [a, x_0). \tag{5.8.2}$$

Assume the opposite, i.e., let there be $M \in [0,\infty)$ and $y \in [a,x_0)$ such that

$$\sup\left\{\operatorname{var}_y^x g : x \in (y,x_0)\right\} \leq M. \tag{5.8.3}$$

Let $M^* = M + |g(x_0) - g(y)|$. Since $\operatorname{var}_y^{x_0} g = \infty$, we can choose a division $\{y_0, y_1, \ldots, y_m\}$ of the interval $[y, x_0]$ such that

$$\sum_{j=1}^{m} |g(y_j) - g(y_{j-1})| > 3\,M^*.$$

Observing that

$$\begin{aligned}|g(y_m) - g(y_{m-1})| &= |g(x_0) - g(y_{m-1})| \\ &\leq |g(x_0) - g(y)| + |g(y) - g(y_{m-1})| \leq M^*,\end{aligned}$$

we get

$$\sum_{j=1}^{m-1} |g(y_j) - g(y_{j-1})| > 2\,M^*,$$

and thus $\operatorname{var}_y^{y_{m-1}} g > 2\,M^*$, which contradicts (5.8.3). Hence, (5.8.2) is true.

Now, let us construct the desired sequence $\{x_k\}$. Let $u_1 = a$ and choose $u_2 \in (a, x_0)$ such that $u_2 > x_0 - 1$ and $\operatorname{var}_{u_1}^{u_2} g > 1$. If $u_1, u_2, \ldots, u_\ell \in [a, x_0)$ are already defined and satisfy

$$u_\ell \in (u_{\ell-1}, x_0) \cap (x_0 - \frac{1}{\ell - 1}, x_0) \quad \text{and} \quad \operatorname{var}_{u_{\ell-1}}^{u_\ell} g > 1,$$

find $u_{\ell+1}$ such that

$$u_{\ell+1} \in (u_\ell, x_0) \cap (x_0 - \frac{1}{\ell}, x_0) \quad \text{and} \quad \operatorname{var}_{u_\ell}^{u_{\ell+1}} g > 1.$$

The sequence $\{u_\ell\}$ is increasing and

$$\lim_{\ell\to\infty} u_\ell = x_0. \tag{5.8.4}$$

By the definition of variation, for every $\ell \in \mathbb{N}$ there is a division

$$\boldsymbol{\alpha}^\ell = \{\alpha_0^\ell, \alpha_1^\ell, \ldots, \alpha_{m_\ell}^\ell\}$$

of the interval $[u_\ell, u_{\ell+1}]$ such that

$$\sum_{j=1}^{m_\ell} |g(\alpha_j^\ell) - g(\alpha_{j-1}^\ell)| > 1$$

holds. Then

$$\sum_{\ell=1}^{\infty}\Big(\sum_{j=1}^{m_\ell}|g(\alpha_j^\ell)-g(\alpha_{j-1}^\ell)|\Big)\geq\sum_{\ell=1}^{\infty}1=\infty. \tag{5.8.5}$$

Let us reorder the elements of the sets $\boldsymbol{\alpha}^\ell$, $\ell\in\mathbb{N}$, to the sequence $\{x_k\}$ in such a way that

$$\begin{aligned} &x_1=a=\alpha_0^1,\\ &x_{k+1}=\alpha_{j+1}^\ell \quad \text{if } x_k=\alpha_j^\ell \text{ and } j<m_\ell-1,\\ &x_{k+1}=\alpha_0^{\ell+1} \quad \text{if } x_k=\alpha_{m_\ell-1}^\ell. \end{aligned}$$

By (5.8.4) and (5.8.5), the sequence $\{x_k\}$ has the required properties, and the proof of the assertion (i) is complete. □

5.8.4 Exercise. Prove the second assertion of Lemma 5.8.3.

5.8.5 Theorem. *If $g:[a,b]\to\mathbb{R}$ is such that the integral $\int_a^b f\,\mathrm{d}g$ exists for every continuous function $f:[a,b]\to\mathbb{R}$, then g has bounded variation.*

Proof. We can restrict ourselves to the (σ)-integral. By the Heine-Borel theorem and thanks to the additivity of variation with respect to intervals (cf. Theorem 2.1.14), it follows that a function $\varphi:[a,b]\to\mathbb{R}$ has bounded variation if and only if the following conditions are satisfied:

- For each $x\in(a,b]$ there is a $\delta_1\in(0,x-a)$ such that $\operatorname{var}_{x-\delta_1}^{x}\varphi<\infty$.
- For each $x\in[a,b)$ there is a $\delta_2\in(0,b-x)$ such that $\operatorname{var}_{x}^{x+\delta_2}\varphi<\infty$.

(5.8.6)

Consequently, $\operatorname{var}_a^b g=\infty$ only if there exists an $x_0\in[a,b]$ such that at least one of the conditions (5.8.6) is not satisfied for $x=x_0$. First, assume that $x_0\in(a,b]$ is such that $\operatorname{var}_x^{x_0} g=\infty$ for each $x\in[a,x_0)$. Then by the first part of Lemma 5.8.3 there is an increasing sequence $\{x_k\}$ of points in (a,x_0) such that

$$\lim_{k\to\infty}x_k=x_0 \quad\text{and}\quad \sum_{k=1}^{\infty}|g(x_{k+1})-g(x_k)|=\infty.$$

By Lemma 5.8.2, there is a sequence $\{c_k\}$ of positive numbers such that

$$\lim_{k\to\infty}c_k=0 \quad\text{and}\quad \sum_{k=1}^{\infty}c_k\,|g(x_{k+1})-g(x_k)|=\infty.$$

Now, let $\xi_k = (x_{k+1} + x_k)/2$ for each $k \in \mathbb{N}$, define

$$f(x) = \begin{cases} 0 & \text{if } \ x < x_1, \ \text{or } x \geq x_0, \ \text{or } \ x \in \{x_k\}, \\ c_k \ \operatorname{sgn}(g(x_{k+1}) - g(x_k)) & \text{if } \ x = \xi_k, \end{cases}$$

and extend the function f linearly to $[a,b]$. This implies that f is continuous on $[a,b]$. We have

$$\sum_{k=1}^{\infty} f(\xi_k)\,(g(x_{k+1}) - g(x_k)) = \infty.$$

In particular, for each $M > 0$ there is an $N_M \in \mathbb{N}$ such that

$$\sum_{k=1}^{N_M} f(\xi_k)\,(g(x_{k+1}) - g(x_k)) > M.$$

For a given $M > 0$, let

$$\boldsymbol{\alpha}_M = \{a, x_1, x_2, \ldots, x_{N_M}, x_{N_M+1}, b\}, \ \boldsymbol{\xi}_M = (a, \xi_1, \xi_2, \ldots, \xi_{N_M}, b).$$

Then $P_M = (\boldsymbol{\alpha}_M, \boldsymbol{\xi}_M)$ is a partition of $[a,b]$ and

$$S(P_M) = \sum_{k=1}^{N_M} f(\xi_k)\,(g(x_{k+1}) - g(x_k)) > M.$$

(Recall that $f(a) = f(b) = 0$.) But this means that the integral $(\sigma)\int_a^b f\,\mathrm{d}g$ cannot have a finite value.

If the latter condition from (5.8.6) is not satisfied, i.e., there is an $x_0 \in [a,b)$ such that $\operatorname{var}_{x_0}^x g = \infty$ for every $x \in (x_0, b]$, then the proof is similar, but relies on the second part of Lemma 5.8.3. □

Theorems 5.8.1 and 5.8.5 will be useful in Section 8.3, where we investigate various classes of functions and their relations to Stieltjes integrability.

5.9 MEAN VALUE THEOREMS

We now derive two mean value theorems, which apply to both kinds of the Riemann-Stieltjes integral.

5.9.1 Theorem (MEAN VALUE THEOREM). *If $f:[a,b]\to\mathbb{R}$ is continuous and $g:[a,b]\to\mathbb{R}$ is nondecreasing, then there exists an $x_0\in[a,b]$ such that*

$$\int_a^b f\,\mathrm{d}g = f(x_0)\,(g(b)-g(a)). \tag{5.9.1}$$

Proof. Corollary 5.6.2 guarantees the existence of the integral $\int_a^b f\,\mathrm{d}g$. Since g is nondecreasing on $[a,b]$, we have

$$m\,(g(b)-g(a)) \le S(P) \le M\,(g(b)-g(a))$$

for every partition P of $[a,b]$, where

$$m=\min\{f(x):x\in[a,b]\} \quad\text{and}\quad M=\max\{f(x):x\in[a,b]\}.$$

It follows that

$$m\,(g(b)-g(a)) \le \int_a^b f\,\mathrm{d}g \le M\,(g(b)-g(a)).$$

Since f is continuous, it takes on all values from the interval $[m,M]$. In particular, there is an $x_0\in[a,b]$ such that (5.9.1) holds. □

5.9.2 Theorem (SECOND MEAN VALUE THEOREM). *If $f:[a,b]\to\mathbb{R}$ is continuous and $g:[a,b]\to\mathbb{R}$ is nondecreasing, then there is an $x_0\in[a,b]$ such that*

$$\int_a^b f(x)\,g(x)\,\mathrm{d}x = g(a)\int_a^{x_0} f(x)\,\mathrm{d}x + g(b)\int_{x_0}^b f(x)\,\mathrm{d}x.$$

Proof. The function f is Riemann integrable on $[a,b]$. Let

$$h(x)=\int_a^x f(t)\,\mathrm{d}t \quad \text{for } x\in[a,b].$$

By Corollary 5.4.4 (substitution theorem), Theorem 5.5.1 (integration by parts) and Theorem 5.9.1, there is an $x_0\in[a,b]$ such that

$$\begin{aligned}\int_a^b f(x)\,g(x)\,\mathrm{d}x &= \int_a^b g\,\mathrm{d}h = h(b)\,g(b) - \int_a^b h\,\mathrm{d}g \\ &= \Big(\int_a^b f\,\mathrm{d}x\Big)\,g(b) - \Big(\int_a^{x_0} f\,\mathrm{d}x\Big)\,(g(b)-g(a)) \\ &= g(a)\int_a^{x_0} f(x)\,\mathrm{d}x + g(b)\int_{x_0}^b f(x)\,\mathrm{d}x,\end{aligned}$$

and the proof is complete. □

5.10 Other integrals of Stieltjes type

Consider functions $f, g \colon [a,b] \to \mathbb{R}$ and a division $\boldsymbol{\alpha}$ of $[a,b]$. Let

$$S_M(\boldsymbol{\alpha}) = \sum_{j=1}^{\nu(\boldsymbol{\alpha})} \frac{f(\alpha_j) + f(\alpha_{j-1})}{2} \, (g(\alpha_j) - g(\alpha_{j-1})),$$

$$S_{CL}(\boldsymbol{\alpha}) = \sum_{j=1}^{\nu(\boldsymbol{\alpha})} f(\alpha_{j-1}) \, (g(\alpha_j) - g(\alpha_{j-1})),$$

$$S_{CR}(\boldsymbol{\alpha}) = \sum_{j=1}^{\nu(\boldsymbol{\alpha})} f(\alpha_j) \, (g(\alpha_j) - g(\alpha_{j-1})),$$

and, if g is regulated and $(\boldsymbol{\alpha}, \boldsymbol{\xi})$ is a partition of $[a,b]$,

$$S_Y(\boldsymbol{\alpha}, \boldsymbol{\xi}) = \sum_{j=1}^{\nu(\boldsymbol{\alpha})} \Big(f(\alpha_{j-1}) \, \Delta^+ g(\alpha_{j-1}) + f(\xi_j) \, (g(\alpha_j -) - g(\alpha_{j-1}+)) + f(\alpha_j) \, \Delta^- g(\alpha_j) \Big).$$

If $S(\boldsymbol{\alpha}, \boldsymbol{\xi})$ in Definition 5.1.4 is replaced by $S_M(\boldsymbol{\alpha})$, $S_{CL}(\boldsymbol{\alpha})$, $S_{CR}(\boldsymbol{\alpha})$ or $S_Y(\boldsymbol{\alpha}, \boldsymbol{\xi})$, we get the definitions of the *mean integral*, *left Cauchy integral*, *right Cauchy integral*, and *Young integral*, respectively. Again, both their (δ) and (σ) variants can be considered.

5.10.1 Exercise. For each of the kind of the Stieltjes integral introduced in this chapter investigate the existence and value of the Stieltjes integral $\int_0^1 f \, \mathrm{d}g$ if $f \colon [0,1] \to \mathbb{R}$ and

$$g(x) = \begin{cases} 0 & \text{if } x \in [0, 1/2), \\ c & \text{if } x = 1/2, \\ d & \text{if } x \in (1/2, 1]. \end{cases}$$

Bibliographic remarks

Our (δ)-integral corresponds to the original Stieltjes' definition from [Stieltjes (1894)]. Sometimes it is also called the *norm integral*. The

(σ)-integral is also known as the *Moore-Pollard integral* or the *refinement Riemann-Stieltjes integral*, cf. [Pollard (1920)] and [Hildebrandt (1963)].

The Darboux-Stieltjes integral is a natural generalization of the Darboux integral, which is well known to be equivalent with the classical Riemann integral and has its origins in [Darboux (1875)].

Probably the first appearance of the pseudo-additivity condition in the context of integration theory is in [Doetsch (1935)]. Independently, the importance of pseudo-additivity for interval functions was recognized in [Getchell (1935)], as well as in [Hildebrandt (1938)]. More recently, this concept was extensively studied and utilized in [Hildebrandt (1963)].

The proof of Theorem 5.6.1 (i) is adapted from Theorem 2.17 in [Dudley, Norvaiša (2011)]. Let us mention two additional existence results for the Riemann-Stieltjes integral. Given an $\alpha \in (0,1]$, a function $f:[a,b]\to\mathbb{R}$ is called α-Hölder continuous if there exists a $K \geq 0$ such that

$$|f(x)-f(y)| \leq K\,|x-y|^{\alpha} \quad \text{for all} \quad x,y \in [a,b].$$

(The definition also makes sense for $\alpha > 1$, but the condition is satisfied only if f is constant.) A result from [L.C. Young (1936)] reads as follows: *If $f:[a,b]\to\mathbb{R}$ is α-Hölder continuous and $g:[a,b]\to\mathbb{R}$ is β-Hölder continuous with $\alpha+\beta>1$, then the integral $(\delta)\int_a^b f\,\mathrm{d}g$ exists.* The condition $\alpha+\beta>1$ is sharp: One can construct an α-Hölder continuous f and a β-Hölder continuous g such that $\alpha+\beta=1$ and none of the Riemann-Stieltjes integrals (or even the Kurzweil-Stieltjes integral treated in the next chapter) exists; see p. 63 in [Dudley, Norvaiša (1999)].

Given a $p \in (0,\infty)$, we say that a function $f:[a,b]\to\mathbb{R}$ has finite p-variation if

$$\sup_{\boldsymbol{\alpha}\in\mathscr{D}[a,b]} \Big(\sum_{j=1}^{\nu(\boldsymbol{\alpha})} |f(\alpha_j)-f(\alpha_{j-1})|^p \Big) < \infty.$$

The following result can be found in [Dudley, Norvaiša (1999)] (see Theorem 3.27 there): *Suppose that $f:[a,b]\to\mathbb{R}$ has finite p-variation and $g:[a,b]\to\mathbb{R}$ has finite q-variation, where $1/p+1/q>1$. Then the integral $(\delta)\int_a^b f\,\mathrm{d}g$ exists if f, g have no common discontinuities, and $(\sigma)\int_a^b f\,\mathrm{d}g$ exists if f, g have no common one-sided discontinuities.* Clearly, each α-Hölder continuous function has finite $1/\alpha$-variation. Thus, Young's result mentioned above follows from the theorem by Dudley and Norvaiša if we take $p=1/\alpha$ and $q=1/\beta$.

Theorem 5.7.3 is taken from [Rosentrater (2015)]. Its special case, which is known as Helly's convergence theorem, is available in Section 7 of [Helly (1912)], see also [Krabbe (1965)].

Theorem 5.7.5 (bounded convergence theorem) is usually referred to as Osgood's theorem. However, it turned out, see e.g. [Luxemburg (1971)], that Osgood in [Osgood (1897)] just rediscovered the result by C. Arzelà from [Arzelà (1885)]. Both Arzelà and Osgood formulated their results for the Riemann integral. Soon it has been accepted that Osgood's theorem is true also for the Riemann-Stieltjes integral, cf. the arguments formulated in Remark on page 976 of [Luxemburg (1971)]. To our knowledge, the first proof of bounded convergence theorem for the Riemann-Stieltjes integral was delivered in [Hildebrandt (1963)] (Theorem II.15.6). Like in [Arzelà (1885)], the proof relies on an auxiliary result known as Arzela's lemma, whose proof is rather technical unless one uses the properties of the Lebesgue measure. A more elementary proof for Riemann integral is given in [Lewin (1986)]; similar ideas were recently applied to abstract Stieltjes-type integrals in [Hanung, Monteiro, Tvrdý (2016)] and inspired the proof of Theorem 5.7.5 presented in this book. Osgood's theorem can be obtained as a simple corollary of the bounded convergence theorem for the Lebesgue-Stieltjes or Kurzweil-Stieltjes integral (see Theorem 6.8.13).

More information about the integrals introduced in Section 5.10 may be found in Section II.19 of [Hildebrandt (1963)] and in the references given there. Obviously, except the Young integral, the classes of functions integrable with respect to these integrals include all Riemann-Stieltjes integrable functions. However, not all properties of Riemann-Stieltjes integrals are maintained. For example, an analogue of Theorem 5.4.3 (substitution theorem) does not hold for the mean integral. As concerns the Young integral, the trivial example when $f \equiv 0$ and g is not regulated shows that, unlike the other integrals from Section 5.10, the Young integral in general does not cover the classical Riemann-Stieltjes integrals. On the other hand, by Theorem II.19.3.3 in [Hildebrandt (1963)] or Proposition 2.18 in [Dudley, Norvaiša (2011)], if g is regulated and the Riemann-Stieltjes integral $\int_a^b f\,\mathrm{d}g$ exists, then the corresponding Young integral also exists and has the same value. (Clearly, the assumption of regulated integrators is kept throughout the whole Section II.19.3 of [Hildebrandt (1963)].) We will return to the Young integral in Section 6.13.

Riemann-Stieltjes integrals are treated in numerous monographs devoted to real and/or functional analysis; see e.g. [Appell, Banás, Merentes

(2014)], [Dudley, Norvaiša (1999)], [Dudley, Norvaiša (2011)], [Kolmogorov, Fomin (2012)], [McLeod (1980)], [Shilov, Gurevich (1977)], or [Stroock (1994)]. We highlight the monograph [Hildebrandt (1963)], which is particularly close to the present chapter. A very nice reference on the history of Riemann-Stieltjes integrals is Chapter 7 of [Pesin (1970)]. A large collection of solved exercises involving the Riemann-Stieltjes and Darboux-Stieltjes integrals is available in [Kaczor, Nowak (2003)].

Chapter 6

Kurzweil-Stieltjes integral

6.1 Introduction

The Riemann-Stieltjes integral is simple and useful for many purposes, but it suffers from several drawbacks. For example, the assumptions of convergence theorems have to be rather strong to ensure that the limit function is integrable. Furthermore, in some applications, it is necessary to consider integrals where the integrand and the integrator have common discontinuity points, which rules out the Riemann-Stieltjes integral.

A more suitable integral is the Kurzweil-Stieltjes integral, which is the subject of the present chapter and the core topic of this book. This integral is not only more general than the well-known Lebesgue-Stieltjes integral, but its definition is quite elementary and close to the definition of the Riemann-Stieltjes integral. In comparison with the latter integral, there is no need to distinguish between the (δ) and (σ) integrals – both concepts are unified by the notions of a gauge and a δ-fine partition.

After providing the definition of the Kurzweil-Stieltjes integral, we derive a simple and useful existence result, which says that $\int_a^b f\,\mathrm{d}g$ exists if f, g are regulated and one of them has bounded variation. Our next goal will be to generalize the integration by parts and substitution formulas from the previous chapter. Then we proceed to convergence theorems for the Kurzweil-Stieltjes integral, including a generalization of the well-known monotone and dominated convergence theorems. We also explore the relation of the Kurzweil-Stieltjes integral to the Perron-Stieltjes integral, Lebesgue-Stieltjes integral, and other related concepts. Other topics treated in this chapter include properties of the indefinite integral, absolute integrability, and integration over elementary sets.

6.2 Definition and basic properties

Recall that a *division* of the interval $[a,b]$ is a finite set $\{\alpha_0, \alpha_1, \ldots, \alpha_m\}$ such that

$$a = \alpha_0 < \alpha_1 < \cdots < \alpha_m = b. \tag{6.2.1}$$

The set of all divisions of the interval $[a,b]$ is denoted by $\mathscr{D}[a,b]$. The number of subintervals induced by a division of the form (6.2.1) is denoted by $\nu(\boldsymbol{\alpha})$, i.e., we have $\alpha_{\nu(\boldsymbol{\alpha})} = b$. Finally, we denote

$$|\boldsymbol{\alpha}| = \max_{j \in \{1,\ldots,\nu(\boldsymbol{\alpha})\}} (\alpha_j - \alpha_{j-1}) \quad \text{for} \quad \boldsymbol{\alpha} \in \mathscr{D}[a,b].$$

If $\boldsymbol{\alpha}' \supset \boldsymbol{\alpha}$, we say that $\boldsymbol{\alpha}' \in \mathscr{D}[a,b]$ is a *refinement of* $\boldsymbol{\alpha} \in \mathscr{D}[a,b]$.

A *partition* of the interval $[a,b]$ is a pair $P = (\boldsymbol{\alpha}, \boldsymbol{\xi})$, where $\boldsymbol{\alpha}$ is a division of $[a,b]$, and $\boldsymbol{\xi} = (\xi_1, \ldots, \xi_{\nu(\boldsymbol{\alpha})})$ is a vector such that

$$\xi_j \in [\alpha_{j-1}, \alpha_j] \quad \text{for} \quad j \in \{1, \ldots, \nu(\boldsymbol{\alpha})\}.$$

Each point ξ_j is referred to as the *tag* of the corresponding subinterval $[\alpha_{j-1}, \alpha_j]$. For each partition $P = (\boldsymbol{\alpha}, \boldsymbol{\xi})$, we denote $\nu(P) = \nu(\boldsymbol{\alpha})$. Thus, $\nu(P)$ is the number of subintervals in P.

Let us recall also the following convention: given $a < c < b$, if $R = (\boldsymbol{\rho}, \boldsymbol{\eta})$ is a partition of $[a,c]$, and $S = (\boldsymbol{\sigma}, \boldsymbol{\zeta})$ is a partition of $[c,b]$, then $P = R \cup S$ denotes the partition of $[a,b]$ having the form $P = (\boldsymbol{\alpha}, \boldsymbol{\xi})$, with $\boldsymbol{\alpha} = \boldsymbol{\rho} \cup \boldsymbol{\sigma}$ and $\boldsymbol{\xi} = (\eta_1, \ldots, \eta_{\nu(R)}, \zeta_1, \ldots, \zeta_{\nu(S)})$.

6.2.1 Definition. A function $\delta : [a,b] \to (0, \infty)$ is called a *gauge* on $[a,b]$.

Given a gauge δ on $[a,b]$, we say that a partition $P = (\boldsymbol{\alpha}, \boldsymbol{\xi})$ of the interval $[a,b]$ is *δ-fine* if

$$[\alpha_{j-1}, \alpha_j] \subset (\xi_j - \delta(\xi_j), \xi_j + \delta(\xi_j)) \quad \text{for all } j \in \{1, \ldots, \nu(\boldsymbol{\alpha})\}. \tag{6.2.2}$$

Consider functions $f, g : [a,b] \to \mathbb{R}$ and a partition $P = (\boldsymbol{\alpha}, \boldsymbol{\xi})$ of $[a,b]$. As in the previous chapter, we use the notation

$$S(f, \mathrm{d}g, P) = \sum_{j=1}^{\nu(\boldsymbol{\alpha})} f(\xi_j)\,(g(\alpha_j) - g(\alpha_{j-1})).$$

If the functions f, g or the interval $[a,b]$ are clear from the context, we use the simpler notation $S(P)$ instead of $S(f, \mathrm{d}g, P)$.

6.2.2 Definition. Consider functions $f, g\colon [a,b] \to \mathbb{R}$. We say that the *Kurzweil-Stieltjes integral* $\int_a^b f(x)\,\mathrm{d}g(x)$ exists and equals $I \in \mathbb{R}$ if for every $\varepsilon > 0$, there exists a gauge δ_ε on $[a,b]$ such that $|S(P) - I| < \varepsilon$ for all δ_ε-fine partitions of $[a,b]$.

Often we also use the shorter notation $\int_a^b f\,\mathrm{d}g$. The function f is called the *integrand*, while g is referred to as the *integrator*.

If $a = b$, then $\int_a^a f\,\mathrm{d}g = 0$. For $b > a$, we define

$$\int_b^a f\,\mathrm{d}g = -\int_a^b f\,\mathrm{d}g.$$

In the special case when $g(x) = x$ for all $x \in [a,b]$, we speak about the Kurzweil-Henstock integral and denote it by $\int_a^b f(x)\,\mathrm{d}x$, or simply $\int_a^b f$.

The following two lemmas show that the definition of the integral is meaningful; in particular, the set of δ-fine partitions is nonempty (this result is known as the Cousin lemma), and the value of the integral is uniquely determined.

6.2.3 Lemma (COUSIN). *For every gauge $\delta\colon [a,b] \to (0,\infty)$, the set of all δ-fine partitions of $[a,b]$ is nonempty.*

Proof. Consider a gauge δ on $[a,b]$, and let M be the set of all $c \in (a,b]$ for which there exists a δ-fine partition of $[a,c]$.

The set M is nonempty, because if $c \in (a,b]$ is such that $c < a + \delta(a)$, then the partition $P = (\boldsymbol{\alpha}, \boldsymbol{\xi})$ with $\boldsymbol{\alpha} = \{a, c\}$ and $\boldsymbol{\xi} = (a)$ is a δ-fine partition of $[a,c]$.

Let $d = \sup M$ and observe that $d > a$. We claim that $d \in M$. Indeed, by the definition of supremum, there exists a $c \in (d - \delta(d), d] \cap M$. If $c = d$, then $d \in M$. Otherwise, we take a δ-fine partition $(\boldsymbol{\alpha}', \boldsymbol{\xi}')$ of $[a,c]$. Since $[c,d] \subset (d - \delta(d), d + \delta(d))$, we see that $\big(\boldsymbol{\alpha}' \cup \{d\}, (\xi_1', \ldots, \xi_{\nu(\boldsymbol{\alpha}')}', d)\big)$ is a δ-fine partition of $[a,d]$. This proves that $d \in M$.

To finish the proof, it is enough to show that $d = b$. To see this, we assume that $d < b$ and choose a partition $(\boldsymbol{\alpha}'', \boldsymbol{\xi}'')$ of $[a,d]$. Moreover, we find a $\gamma \in (d, d + \delta(d)) \cap (d, b]$. Since $[d,\gamma] \subset (d - \delta(d), d + \delta(d))$, it follows that $\big(\boldsymbol{\alpha}'' \cup \{\gamma\}, (\xi_1'', \ldots, \xi_{\nu(\boldsymbol{\alpha}'')}'', d)\big)$ is a δ-fine partition of the interval $[a,\gamma]$. Then $\gamma \in M$, which is in contradiction wit the fact that $d = \sup M$. Thus, we see that $\sup M = b$, and the proof is complete. □

6.2.4 Lemma. *If the integral* $\int_a^b f\,\mathrm{d}g$ *exists, then its value is uniquely determined.*

Proof. Assume there exist two different values I_1, $I_2 \in \mathbb{R}$ having the property mentioned in Definition 6.2.2. Let $\widetilde{\varepsilon} = \frac{1}{2}\,|I_1 - I_2|$. Then there exist gauges δ_1 and δ_2 on $[a,b]$ such that

$$|S(P) - I_1| < \widetilde{\varepsilon} \quad \text{for every } \delta_1\text{-fine partition } P \text{ of } [a,b],$$
$$|S(P) - I_2| < \widetilde{\varepsilon} \quad \text{for every } \delta_2\text{-fine partition } P \text{ of } [a,b].$$

Define a gauge δ by $\delta(x) = \min\{\delta_1(x), \delta_2(x)\}$ for all $x \in [a,b]$. Then every δ-fine partition of $[a,b]$ is both δ_1-fine and δ_2-fine. Consequently, for every δ-fine partition P of $[a,b]$ we have

$$2\,\widetilde{\varepsilon} = |I_1 - I_2| = |I_1 - S(P) + S(P) - I_2|$$
$$\le |I_1 - S(P)| + |S(P) - I_2| < 2\,\widetilde{\varepsilon}.$$

Since this is impossible, we necessarily have $I_1 = I_2$. □

6.2.5 Remark. (i) The previous proof was based on a simple fact: If two gauges δ, δ_0 on $[a,b]$ satisfy $\delta_0 \le \delta$, then every δ_0-fine partition of $[a,b]$ is also δ-fine. Therefore, if a certain statement holds for all δ_0-fine partitions of $[a,b]$, then it also holds for all δ-fine partitions of $[a,b]$. This leads to the following observation: Given an arbitrary gauge δ_0 on $[a,b]$, there is no loss of generality if we restrict Definition 6.2.2 to all gauges δ_ε such that $\delta_\varepsilon \le \delta_0$ on $[a,b]$.

(ii) Note that the concept of the Kurzweil-Stieltjes integral remains unchanged if the condition (6.2.2) in the definition of a δ-fine partition is replaced by the condition

$$[\alpha_{j-1}, \alpha_j] \subset [\xi_j - \delta(\xi_j), \xi_j + \delta(\xi_j)] \quad \text{for all } \ j \in \{1, \dots, \nu(\boldsymbol{\alpha})\},$$

which has closed intervals on the right-hand side.

If not stated otherwise, all integrals in the following text should be interpreted as Kurzweil-Stieltjes integrals.

The next result provides a Bolzano-Cauchy-type condition for the existence of the integral.

6.2.6 Theorem. *Given a pair of functions* $f, g\colon [a,b] \to \mathbb{R}$, *the integral* $\int_a^b f\,\mathrm{d}g$ *exists if and only if for each* $\varepsilon > 0$ *there is a gauge* δ_ε *on* $[a,b]$ *such* *that*

$$\big|S(P) - S(Q)\big| < \varepsilon \quad \textit{for all } \delta_\varepsilon\textit{-fine partitions } P,\ Q \textit{ of } [a,b].$$

P r o o f. If the integral $\int_a^b f\,\mathrm{d}g = I \in \mathbb{R}$ exists, then for every $\varepsilon > 0$ there is a gauge δ_ε on $[a,b]$ such that

$$|S(P) - I| < \frac{\varepsilon}{2} \quad \text{for all } \delta_\varepsilon\text{-fine partitions } P \text{ of } [a,b].$$

Hence, for every pair P, Q of δ_ε-fine partitions we have

$$|S(P) - S(Q)| \leq |S(P) - I| + |S(Q) - I| < \varepsilon.$$

To prove the converse implication, assume that for each $n \in \mathbb{N}$, there is a gauge δ_n on $[a,b]$ such that

$$\big|S(P) - S(Q)\big| < \frac{1}{n} \quad \text{for all } \delta_n\text{-fine partitions } P, Q \text{ of } [a,b].$$

Without loss of generality (cf. Remark 6.2.5), we can assume that the sequence $\{\delta_n\}$ is nonincreasing.

For each $n \in \mathbb{N}$, let P_n be an arbitrary δ_n-fine partition of $[a,b]$. Then

$$|S(P_n) - S(P_m)| < \frac{1}{k} \quad \text{for all } m, n \geq k.$$

This shows that $\{S(P_n)\}$ is a Cauchy sequence. Thus, there exists a finite limit $I = \lim_{n\to\infty} S(P_n)$.

We claim that $\int_a^b f\,\mathrm{d}g$ exists and equals I. To see this, let $\varepsilon > 0$ be given. Find $n_0 \in \mathbb{N}$ such that $1/n_0 < \varepsilon$ and $|S(P_{n_0}) - I| < \varepsilon$. Then for each δ_{n_0}-fine partition P of $[a,b]$, we have

$$|S(P) - I| \leq |S(P) - S(P_{n_0})| + |S(P_{n_0}) - I| < \frac{1}{n_0} + \varepsilon < 2\varepsilon,$$

and the proof is complete. □

The Kurzweil-Stieltjes integral is linear with respect to the integrand and the integrator.

6.2.7 Theorem. *Let $c_1, c_2 \in \mathbb{R}$ be arbitrary numbers.*

(i) *If* $f_1, f_2, g : [a,b] \to \mathbb{R}$ *are such that the integrals* $\int_a^b f_1\,\mathrm{d}g$ *and* $\int_a^b f_2\,\mathrm{d}g$ *exist, then*

$$\int_a^b (c_1 f_1 + c_2 f_2)\,\mathrm{d}g = c_1 \int_a^b f_1\,\mathrm{d}g + c_2 \int_a^b f_2\,\mathrm{d}g.$$

(ii) *If* $f, g_1, g_2 : [a,b] \to \mathbb{R}$ *are such that the integrals* $\int_a^b f\,\mathrm{d}g_1$ *and* $\int_a^b f\,\mathrm{d}g_2$ *exist, then*

$$\int_a^b f\,\mathrm{d}(c_1 g_1 + c_2 g_2) = c_1 \int_a^b f\,\mathrm{d}g_1 + c_2 \int_a^b f\,\mathrm{d}g_2.$$

Proof. We prove the first statement and leave the second as an exercise to the reader.

Let $\varepsilon > 0$ be given. By the assumption, there are gauges δ_1 and δ_2 on $[a,b]$ such that

$$\Big|S(f_1, \mathrm{d}g, P) - \int_a^b f_1\,\mathrm{d}g\Big| < \varepsilon \quad \text{for all } \delta_1\text{-fine partitions } P \text{ of } [a,b],$$
$$\Big|S(f_2, \mathrm{d}g, P) - \int_a^b f_2\,\mathrm{d}g\Big| < \varepsilon \quad \text{for all } \delta_2\text{-fine partitions } P \text{ of } [a,b].$$

Let $\delta_\varepsilon(x) = \min\{\delta_1(x), \delta_2(x)\}$ for all $x \in [a,b]$. Observe that for each partition $P = (\boldsymbol{\alpha}, \boldsymbol{\xi})$ of $[a,b]$, we have

$$S(c_1 f_1 + c_2 f_2, \mathrm{d}g, P) = c_1\, S(f_1, \mathrm{d}g, P) + c_2\, S(f_2, \mathrm{d}g, P).$$

Thus, if P is δ_ε-fine, we get

$$\begin{aligned}
&\Big|S(c_1 f_1 + c_2 f_2, \mathrm{d}g, P) - c_1 \int_a^b f_1\,\mathrm{d}g - c_2 \int_a^b f_2\,\mathrm{d}g\Big| \\
&\qquad \le |c_1|\,\Big|S(f_1, \mathrm{d}g, P) - \int_a^b f_1\,\mathrm{d}g\Big| + |c_2|\,\Big|S(f_2, \mathrm{d}g, P) - \int_a^b f_2\,\mathrm{d}g\Big| \\
&\qquad < (|c_1| + |c_2|)\,\varepsilon.
\end{aligned}$$

This shows that the first statement holds. □

6.2.8 Exercise. Prove the second statement of Theorem 6.2.7.

The next result represents another basic property of the Kurzweil-Stieltjes integral.

6.2.9 Theorem. *If the integral* $\int_a^b f\,\mathrm{d}g$ *exists and* $[c,d]\subset[a,b]$, *then the integral* $\int_c^d f\,\mathrm{d}g$ *exists as well.*

Proof. Assume that $\int_a^b f\,\mathrm{d}g$ exists and $a<c<d<b$; the cases when $a=c$ or $d=b$ are simpler and their proof is left to the reader.

By Theorem 6.2.6, there exists a gauge δ_ε on $[a,b]$ such that

$$|S(P)-S(P')|<\varepsilon \quad \text{for all } \delta_\varepsilon\text{-fine partitions } P,P' \text{ of } [a,b]. \tag{6.2.3}$$

Let us fix an arbitrary δ_ε-fine partition Q^- of $[a,c]$, and a δ_ε-fine partition Q^+ of $[d,b]$.

If Q and Q' are arbitrary δ_ε-fine partitions of $[c,d]$, consider the partitions $P=Q^-\cup Q\cup Q^+$ and $P'=Q^-\cup Q'\cup Q^+$. Then P and P' are δ_ε-fine partitions of $[a,b]$ satisfying $S(P)=S(Q^-)+S(Q)+S(Q^+)$ and $S(P')=S(Q^-)+S(Q')+S(Q^+)$. Thus, in view of (6.2.3) we have

$$|S(Q)-S(Q')|=|S(P)-S(P')|<\varepsilon.$$

The existence of the integral $\int_c^d f\,\mathrm{d}g$ now follows from Theorem 6.2.6. □

According to the previous theorem, if the integral $\int_a^b f\,\mathrm{d}g$ exists and $c\in(a,b)$, then the integrals $\int_a^c f\,\mathrm{d}g$ and $\int_c^b f\,\mathrm{d}g$ exist as well. The converse statement is also true.

6.2.10 Theorem. *Let* $f,g:[a,b]\to\mathbb{R}$ *and* $c\in(a,b)$. *If both integrals* $\int_a^c f\,\mathrm{d}g$ *and* $\int_c^b f\,\mathrm{d}g$ *exist, then the integral* $\int_a^b f\,\mathrm{d}g$ *exist as well, and*

$$\int_a^b f\,\mathrm{d}g=\int_a^c f\,\mathrm{d}g+\int_c^b f\,\mathrm{d}g.$$

Proof. Denote $I_1=\int_a^c f\,\mathrm{d}g$ and $I_2=\int_c^b f\,\mathrm{d}g$. Given an $\varepsilon>0$, there exist gauges δ'_ε on $[a,c]$ and δ''_ε on $[c,b]$ such that

$$|S(P')-I_1|<\frac{\varepsilon}{2} \quad \text{for all } \delta'_\varepsilon\text{-fine partitions } P' \text{ of } [a,c],$$
$$|S(P'')-I_2|<\frac{\varepsilon}{2} \quad \text{for all } \delta''_\varepsilon\text{-fine partitions } P'' \text{ of } [c,b].$$

Now, let the gauge $\delta_\varepsilon : [a,b] \to (0,\infty)$ be given by

$$\delta_\varepsilon(x) = \begin{cases} \min\left\{\delta'_\varepsilon(x), \frac{1}{2}(c-x)\right\} & \text{if } \; x \in [a,c), \\ \min\left\{\delta'_\varepsilon(c), \delta''_\varepsilon(c)\right\} & \text{if } \; x = c, \\ \min\left\{\delta''_\varepsilon(x), \frac{1}{2}(x-c)\right\} & \text{if } \; x \in (c,b]. \end{cases}$$

Note that

$$x + \delta_\varepsilon(x) \le x + \frac{1}{2}(c-x) < c \quad \text{if } \; x < c,$$

and

$$x - \delta_\varepsilon(x) \ge x - \frac{1}{2}(x-c) > c \quad \text{if } \; x > c.$$

Thus, $x \ne c$ always implies $c \notin (x - \delta_\varepsilon(x), x + \delta_\varepsilon(x))$. Hence, if $P = (\boldsymbol{\alpha}, \boldsymbol{\xi})$ is a δ_ε-fine partition of $[a,b]$ and $c \in [\alpha_{i-1}, \alpha_i]$ for a certain $i \in \{1, \ldots, \nu(\boldsymbol{\alpha})\}$, then necessarily $\xi_i = c$. Without loss of generality, we can assume that the tag c is shared by two adjacent subintervals in P, i.e., there exists a $k \in \{1, \ldots, \nu(\boldsymbol{\alpha})\}$ such that

$$\alpha_{k-1} < \alpha_k = \xi_k = c = \xi_{k+1} < \alpha_{k+1}.$$

Otherwise, if c is an interior point of a unique subinterval $[\alpha_{k-1}, \alpha_k]$, we can replace this interval by the pair of intervals $[\alpha_{k-1}, c]$ and $[c, \alpha_k]$ that share the same tag c. The new partition is still δ_ε-fine, and the value of the sum $S(P)$ remains unchanged.

Thus, we see that the partition P has the form $P' \cup P''$, where P' is a δ_ε-fine partition of $[a,c]$, and P'' is a δ_ε-fine partition of $[c,b]$. Since $S(P) = S(P') + S(P'')$, we get

$$\begin{aligned} |S(P) - (I_1 + I_2)| &= |S(P') + S(P'') - I_1 - I_2| \\ &\le |S(P') - I_1| + |S(P'') - I_2| < \varepsilon. \end{aligned}$$

Since P was an arbitrary δ_ε-fine partition of $[a,b]$, we see that $\int_a^b f \, \mathrm{d}g$ exists and equals $I_1 + I_2$. □

The previous proof was based on the following observation: If $c \in [a,b]$, there exists a gauge δ on $[a,b]$ which forces the point $c \in (a,b)$ to be a tag of every δ-fine partition $(\boldsymbol{\alpha}, \boldsymbol{\xi})$ of $[a,b]$. The next lemma provides a more general version of this statement.

6.2.11 Lemma. *If D is a finite set of points from $[a,b]$, there exists a gauge δ on $[a,b]$ such that for every δ-fine partition $P=(\boldsymbol{\alpha},\boldsymbol{\xi})$ of $[a,b]$ we have $D\subset\{\xi_1,\dots,\xi_{\nu(P)}\}$.*

Moreover, for each such partition P, there exists a δ-fine partition $\widetilde{P}=(\widetilde{\boldsymbol{\alpha}},\widetilde{\boldsymbol{\xi}})$ of $[a,b]$ such that $D\subset\widetilde{\boldsymbol{\alpha}}\cap\{\widetilde{\xi}_1,\dots,\widetilde{\xi}_{\nu(\widetilde{P})}\}$ and $S(\widetilde{P})=S(P)$.

Proof. Let $D=\{s_1,\dots,s_k\}$, where $a\le s_1<\dots<s_k\le b$. Consider the gauge $\delta:[a,b]\to(0,\infty)$ given by

$$\delta(x)=\begin{cases}\frac{1}{2}\min\{|x-s_j|:j\in\{1,\dots,k\}\} & \text{if } x\notin D,\\ 1, & \text{if } x\in D.\end{cases}$$

For each $j\in\{1,\dots,k\}$ we have

$$\xi+\delta(\xi)<\xi+\frac{1}{2}(s_j-\xi)<s_j \quad \text{if } \xi<s_j,$$

$$\xi-\delta(\xi)>\xi-\frac{1}{2}(s_j-\xi)>s_j \quad \text{if } \xi>s_j.$$

That is,

$$s_j\in(\xi-\delta(\xi),\xi+\delta(\xi)) \quad \text{if and only if } \xi=s_j.$$

Hence, for each δ-fine partition $P=(\boldsymbol{\alpha},\boldsymbol{\xi})$ of $[a,b]$ and for each $j\in\{1,\dots,k\}$ we have $s_j\in\{\xi_1,\dots,\xi_{\nu(P)}\}$.

We have shown that if $P=(\boldsymbol{\alpha},\boldsymbol{\xi})$ is a δ-fine partition of $[a,b]$, then $D\subset\boldsymbol{\xi}$. If $s_j\in D$ is an interior point of a certain interval $[\alpha_{i-1},\alpha_i]$, then $s_j=\xi_i$. Hence, we can split the interval $[\alpha_{i-1},\alpha_i]$ in two intervals $[\alpha_{i-1},s_j]$ and $[s_j,\alpha_i]$ with the same tag $\xi_i=s_j$. In this way, we get a new partition $\widetilde{P}=(\widetilde{\boldsymbol{\alpha}},\widetilde{\boldsymbol{\xi}})$ of $[a,b]$ such that $D\subset\widetilde{\boldsymbol{\alpha}}\cap\{\widetilde{\xi}_1,\dots,\widetilde{\xi}_{\nu(\widetilde{P})}\}$ and $S(\widetilde{P})=S(P)$. □

The next result shows that the Kurzweil-Stieltjes integral generalizes both types of the Riemann-Stieltjes integral.

6.2.12 Theorem. *If one of the Riemann-Stieltjes integrals $(\sigma)\int_a^b f\,\mathrm{d}g$ and $(\delta)\int_a^b f\,\mathrm{d}g$ exists, then the Kurzweil-Stieltjes integral $\int_a^b f\,\mathrm{d}g$ exists as well and has the same value.*

Proof. Since the existence of $(\delta)\int_a^b f\,\mathrm{d}g$ implies the existence of $(\sigma)\int_a^b f\,\mathrm{d}g$ (see Theorem 5.1.8), we can assume that the latter integral exists and equals $I\in\mathbb{R}$. Thus, given an $\varepsilon>0$, there exists a division $\boldsymbol{\alpha}_\varepsilon$ of the interval $[a,b]$ such that

$$|S(P)-I|<\varepsilon \quad \text{for every partition } P=(\boldsymbol{\alpha},\boldsymbol{\xi}) \text{ of } [a,b] \text{ satisfying } \boldsymbol{\alpha}\supset\boldsymbol{\alpha}_\varepsilon.$$

By Lemma 6.2.11, there exists a gauge δ_ε such that

$$\boldsymbol{\alpha}_\varepsilon \subset \{\xi_1, \dots, \xi_{\nu(\boldsymbol{\alpha})}\} \quad \text{for each } \delta_\varepsilon\text{-fine partition } (\boldsymbol{\alpha}, \boldsymbol{\xi}) \text{ of } [a,b].$$

Moreover, if $P = (\boldsymbol{\alpha}, \boldsymbol{\xi})$ is an arbitrary δ_ε-fine partition of $[a,b]$, there exists a partition $\widetilde{P} = (\widetilde{\boldsymbol{\alpha}}, \widetilde{\boldsymbol{\xi}})$ such that $S(P) = S(\widetilde{P})$ and $\boldsymbol{\alpha}_\varepsilon \subset \widetilde{\boldsymbol{\alpha}}$. Due to the definition of $\boldsymbol{\alpha}_\varepsilon$, we get

$$|S(P) - I| = |S(\widetilde{P}) - I| < \varepsilon.$$

Hence, the Kurzweil-Stieltjes integral $\int_a^b f \, \mathrm{d}g$ exists and equals I. □

6.2.13 Exercise. Provide a direct proof of the fact that if $(\delta) \int_a^b f \, \mathrm{d}g$ exists, then the Kurzweil-Stieltjes integral $\int_a^b f \, \mathrm{d}g$ exists as well and has the same value (avoid arguments involving the (σ) Riemann-Stieltjes integral).

In the special case when $g(x) = x$ for all $x \in [a,b]$, we know that the Kurzweil-Stieltjes integral $\int_a^b f \, \mathrm{d}g$ reduces to the Kurzweil-Henstock integral $\int_a^b f$. By Theorem 6.2.12, this integral generalizes the classical Riemann integral. As we will see, it is not difficult to construct examples of functions $f : [a,b] \to \mathbb{R}$ which are Kurzweil-Henstock integrable but not Riemann integrable.

First, let us show that each function $f : [a,b] \to \mathbb{R}$ which vanishes almost everywhere in $[a,b]$ (i.e., with the exception of a set whose Lebesgue measure is zero) is Kurzweil-Henstock integrable. Readers who are not familiar with the concept of the Lebesgue measure may skip this example or consult the beginning of Section 6.12.

6.2.14 Example. Consider a function $f : [a,b] \to \mathbb{R}$ such that the set

$$M = \{x \in [a,b] : f(x) \neq 0\}$$

has Lebesgue measure zero. We claim that the Kurzweil-Henstock integral $\int_a^b f(x) \, \mathrm{d}x$ exists and equals zero. Note that $M = \bigcup_{n \in \mathbb{N}} M_n$, where

$$M_n = \{x \in [a,b] : n - 1 < |f(x)| \le n\} \quad \text{for} \quad n \in \mathbb{N}.$$

Since $M_n \subset M$, it follows that $\mu(M_n) = 0$, where μ denotes the Lebesgue measure. Let an arbitrary $\varepsilon > 0$ be given. For each $n \in \mathbb{N}$, there is an open subset G_n of $[a,b]$ such that

$$M_n \subset G_n \quad \text{and} \quad \mu(G_n) < \frac{\varepsilon}{n \, 2^n}.$$

Now, let $\delta_\varepsilon:[a,b]\to(0,\infty)$ be a gauge satisfying

$$(x-\delta_\varepsilon(x),x+\delta_\varepsilon(x))\cap[a,b]\subset G_n \quad\text{if } x\in M_n \text{ for some } n\in\mathbb{N}.$$

If $P=(\boldsymbol{\alpha},\boldsymbol{\xi})$ is a δ_ε-fine partition of $[a,b]$, then

$$\begin{aligned}|S(P)|&\le\sum_{j=1}^{m}|f(\xi_j)|(\alpha_j-\alpha_{j-1})=\sum_{n=1}^{\infty}\sum_{j:\xi_j\in M_n}|f(\xi_j)|(\alpha_j-\alpha_{j-1})\\&\le\sum_{n=1}^{\infty}n\sum_{j:\xi_j\in M_n}(\alpha_j-\alpha_{j-1})\le\sum_{n=1}^{\infty}n\mu(G_n)<\varepsilon\sum_{n=1}^{\infty}\frac{1}{2^n}=\varepsilon.\end{aligned}$$

It follows that $\int_a^b f(x)\,\mathrm{d}x=0$.

With the help of Example 6.2.14, one can easily construct examples of Kurzweil-Henstock integrable functions which are not Riemann integrable.

6.2.15 Remark. Consider the Dirichlet function (cf. Exercise 4.1.4 (iii))

$$f_D(x)=\begin{cases}1 & \text{if } x\in\mathbb{Q},\\ 0 & \text{if } x\in\mathbb{R}\setminus\mathbb{Q}\end{cases}$$

and a nondegenerate interval $[a,b]\subset\mathbb{R}$. By Example 6.2.14, the Kurzweil-Henstock integral $\int_a^b f_D(x)\,\mathrm{d}x$ exists and equals zero. On the other hand, the Riemann integral $\int_a^b f_D(x)\,\mathrm{d}x$ does not exist. Indeed, let $\boldsymbol{\alpha}=\{\alpha_0,\alpha_1,\dots,\alpha_m\}$ be an arbitrary division of $[a,b]$. If $P=(\boldsymbol{\alpha},\boldsymbol{\xi})$ is such that all the tags in $\boldsymbol{\xi}$ are rational, we get $S(P)=b-a\neq 0$; if all the tags are irrational, we have $S(P)=0$. This shows that the Riemann integral does not exist.

The following well-known result shows that each derivative is Kurzweil-Henstock integrable (thus, the Kurzweil-Henstock integral generalizes the Newton integral when restricted to functions defined on bounded intervals).

6.2.16 Theorem. *Let $F:[a,b]\to\mathbb{R}$ be a differentiable function with $F'=f$ on $[a,b]$ (where the derivative is understood as one-sided at both endpoints). Then f is Kurzweil-Henstock integrable and $\int_a^b f(x)\,\mathrm{d}x=F(b)-F(a)$.*

Proof. Let $\varepsilon>0$ be given. For each $\xi\in[a,b]$, the relation $F'(\xi)=f(\xi)$

implies the existence of a $\delta_\varepsilon(\xi) > 0$ such that

$$|F(x) - F(\xi) - f(\xi)(x - \xi)| < \frac{\varepsilon}{b - a} |x - \xi|$$

for all $x \in [a, b] \cap (\xi - \delta_\varepsilon(\xi), \xi + \delta_\varepsilon(\xi))$.

Thus, if $P = (\boldsymbol{\alpha}, \boldsymbol{\xi})$ is an arbitrary δ_ε-fine partition of $[a, b]$, then

$$\begin{aligned}
&\big|F(\alpha_j) - F(\alpha_{j-1}) - f(\xi_j)(\alpha_j - \alpha_{j-1})\big| \\
&\quad \le \big|F(\alpha_j) - F(\xi_j) - f(\xi_j)(\alpha_j - \xi_j)\big| \\
&\quad + \big|F(\xi_j) - F(\alpha_{j-1}) - f(\xi_j)(\xi_j - \alpha_{j-1})\big| \\
&\quad < \frac{\varepsilon}{b - a} \big(|\alpha_j - \xi_j| + |\xi_j - \alpha_{j-1}|\big) = \frac{\varepsilon}{b - a} (\alpha_j - \alpha_{j-1})
\end{aligned}$$

for every $j \in \{1, \dots, \nu(\boldsymbol{\alpha})\}$. Consequently,

$$\begin{aligned}
\big|F(b) - F(a) - S(P)\big| &= \Big|\sum_{j=1}^{\nu(\boldsymbol{\alpha})} \big(F(\alpha_j) - F(\alpha_{j-1}) - f(\xi_j)(\alpha_j - \alpha_{j-1})\big)\Big| \\
&\le \sum_{j=1}^{\nu(\boldsymbol{\alpha})} \big|F(\alpha_j) - F(\alpha_{j-1}) - f(\xi_j)(\alpha_j - \alpha_{j-1})\big| \\
&< \frac{\varepsilon}{b - a} \sum_{j=1}^{m} (\alpha_j - \alpha_{j-1}) = \varepsilon.
\end{aligned}$$

This proves that $\int_a^b f(x)\,\mathrm{d}x = F(b) - F(a)$. □

Further relations between the Kurzweil-Stieltjes integral and other integrals will be explored in Sections 6.11, 6.12, and 6.13.

6.3 Existence of the integral

This section provides some sufficient conditions for the existence of the Kurzweil-Stieltjes integral $\int_a^b f\,\mathrm{d}g$. Our main goal is to show that if f, g are regulated and one of them has bounded variation, then the integral exists. Since regulated functions can be uniformly approximated by finite step functions, we begin with the simple case when one of the functions f, g is a finite step function.

6.3.1 Remark. If f is a constant function on $[a,b]$, then clearly

$$\int_a^b f\,\mathrm{d}g = f(a)\,(g(b)-g(a)) \quad \text{and} \quad \int_a^b g\,\mathrm{d}f = 0$$

for every function $g\colon [a,b]\to\mathbb{R}$.

6.3.2 Lemma. *The following relations hold for any function $f\colon [a,b]\to\mathbb{R}$:*

$$\int_a^b f\,\mathrm{d}\chi_{(\tau,b]} = f(\tau) \qquad \textit{if } \tau\in[a,b), \tag{6.3.1}$$

$$\int_a^b f\,\mathrm{d}\chi_{[\tau,b]} = f(\tau) \qquad \textit{if } \tau\in(a,b], \tag{6.3.2}$$

$$\int_a^b f\,\mathrm{d}\chi_{[a,\tau]} = -f(\tau) \qquad \textit{if } \tau\in[a,b), \tag{6.3.3}$$

$$\int_a^b f\,\mathrm{d}\chi_{[a,\tau)} = -f(\tau) \qquad \textit{if } \tau\in(a,b], \tag{6.3.4}$$

$$\int_a^b f\,\mathrm{d}\chi_{[\tau]} = 0 \qquad \textit{if } \tau\in(a,b). \tag{6.3.5}$$

Proof. It suffices to prove the relations (6.3.1) and (6.3.2); the remaining formulas then follow from Theorem 6.2.10.

a) Let $\tau\in[a,b)$ and $g=\chi_{(\tau,b]}$. Then $g\equiv 0$ on $[a,\tau]$, and $\int_a^\tau f\,\mathrm{d}g=0$. By Lemma 6.2.11, there is a gauge δ on $[\tau,b]$ such that $\tau=\alpha_0=\xi_1$ for each δ-fine partition $P=(\boldsymbol{\alpha},\boldsymbol{\xi})$ of $[\tau,b]$. Moreover, $g(\alpha_j)-g(\alpha_{j-1})=0$ for all $j\in\{2,3,\dots,\nu(P)\}$. Then

$$S(P)=f(\tau)\,(g(\alpha_1)-g(\tau))=f(\tau)$$

for each δ-fine partition $P=(\boldsymbol{\alpha},\boldsymbol{\xi})$ of $[\tau,b]$, and therefore $\int_\tau^b f\,\mathrm{d}g=f(\tau)$. Relation (6.3.1) now follows from Theorem 6.2.10.

b) Let $\tau\in(a,b]$ and $g=\chi_{[\tau,b]}$. Then $g\equiv 1$ on $[\tau,b]$, and $\int_\tau^b f\,\mathrm{d}g=0$. By Lemma 6.2.11, there is a gauge δ on $[a,\tau]$ such that $\alpha_{\nu(P)}=\xi_{\nu(P)}=\tau$ for each δ-fine partition $P=(\boldsymbol{\alpha},\boldsymbol{\xi})$ of $[a,\tau]$. Then

$$S(P)=f(\tau)\,(g(\tau)-g(\alpha_{\nu(P)-1}))=f(\tau)$$

for every δ-fine partition $P=(\boldsymbol{\alpha},\boldsymbol{\xi})$ of $[a,\tau]$, and therefore $\int_a^\tau f\,\mathrm{d}g=f(\tau)$. Relation (6.3.2) now follows from Theorem 6.2.10. □

6.3.3 Lemma. *The following relations hold for every regulated function* $g:[a,b]\to\mathbb{R}$*:*

$$\int_a^b \chi_{(\tau,b]}\,\mathrm{d}g = g(b)-g(\tau+) \quad \text{if } \tau\in[a,b), \tag{6.3.6}$$

$$\int_a^b \chi_{[\tau,b]}\,\mathrm{d}g = g(b)-g(\tau-) \quad \text{if } \tau\in(a,b], \tag{6.3.7}$$

$$\int_a^b \chi_{[a,\tau]}\,\mathrm{d}g = g(\tau+)-g(a) \quad \text{if } \tau\in[a,b), \tag{6.3.8}$$

$$\int_a^b \chi_{[a,\tau)}\,\mathrm{d}g = g(\tau-)-g(a) \quad \text{if } \tau\in(a,b], \tag{6.3.9}$$

$$\int_a^b \chi_{[\tau]}\,\mathrm{d}g = \Delta\, g(\tau) \qquad\qquad \text{if } \tau\in(a,b). \tag{6.3.10}$$

Proof. Again, we prove only the first two relations.

a) Let $\tau\in[a,b)$ and $f=\chi_{(\tau,b]}$. Then $\int_a^\tau f\,\mathrm{d}g=0$. Let $\varepsilon>0$ be given and choose an $\eta_\varepsilon>0$ be such that

$$|g(\tau+)-g(x)|<\varepsilon \quad \text{for every } x\in(\tau,\tau+\eta_\varepsilon).$$

By Lemma 6.2.11, there is a gauge δ on $[\tau,b]$ such that $\tau=\alpha_0=\xi_1$ for each δ-fine partition $P=(\boldsymbol{\alpha},\boldsymbol{\xi})$ of $[\tau,b]$. Let

$$\delta_\varepsilon(x)=\begin{cases}\min\{\eta_\varepsilon,\delta(\tau)\} & \text{if } x=\tau,\\ \delta(x), & \text{if } x\in(\tau,b].\end{cases}$$

Let an arbitrary δ_ε-fine partition $P=(\boldsymbol{\alpha},\boldsymbol{\xi})$ of $[\tau,b]$ be given. Then $\tau=\alpha_0=\xi_1$, $\alpha_1\in(\tau,\tau+\eta_\varepsilon)$, and consequently

$$\begin{aligned}&|S(P)-(g(b)-g(\tau+))|\\ &\quad=\Big|(g(b)-g(\alpha_{\nu(P)-1}))+(g(\alpha_{\nu(P)-1})-g(\alpha_{\nu(P)-2}))\\ &\qquad\quad+\dots+(g(\alpha_2)-g(\alpha_1))-(g(b)-g(\tau+))\Big|\\ &\quad=|g(\tau+)-g(\alpha_1)|<\varepsilon.\end{aligned}$$

Thus, $\int_\tau^b f\,\mathrm{d}g=g(b)-g(\tau+)$, and the relation (6.3.6) follows from Theorem 6.2.10.

b) Let $\tau \in (a,b]$ and $f = \chi_{[\tau,b]}$. Then $\int_\tau^b f\,\mathrm{d}g = g(b) - g(\tau)$. Let $\varepsilon > 0$ be given and choose an $\eta_\varepsilon > 0$ such that

$$|g(\tau-) - g(x)| < \varepsilon \quad \text{for every } x \in (\tau - \eta_\varepsilon, \tau).$$

By Lemma 6.2.11, there is a gauge δ on $[a,\tau]$ such that $\tau = \alpha_{\nu(P)} = \xi_{\nu(P)}$ for each δ-fine partition $P = (\boldsymbol{\alpha}, \boldsymbol{\xi})$ of $[a,\tau]$. Let

$$\delta_\varepsilon(x) = \begin{cases} \delta(x), & \text{if } x \in [a,\tau), \\ \min\{\eta_\varepsilon, \delta(\tau)\} & \text{if } x = \tau. \end{cases}$$

Let an arbitrary δ-fine partition $P = (\boldsymbol{\alpha}, \boldsymbol{\xi})$ of $[a,\tau]$ be given. Then $\tau = \alpha_{\nu(P)} = \xi_{\nu(P)}$, $\alpha_{\nu(P)-1} \in (\tau - \eta_\varepsilon, \tau)$, and consequently

$$\begin{aligned} \big|S(P) - (g(\tau) - g(\tau-))\big| &= \big|(g(\tau) - g(\alpha_{\nu(P)-1})) - (g(\tau) - g(\tau-))\big| \\ &= \big|g(\tau-) - g(\alpha_{\nu(P)-1})\big| < \varepsilon. \end{aligned}$$

Thus, $\int_a^\tau f\,\mathrm{d}g = g(\tau) - g(\tau-)$, and the relation (6.3.7) follows from Theorem 6.2.10. □

Since every finite step function is a linear combination of functions of the form $\chi_{(\tau,b]}$, $\chi_{[\tau,b]}$, $\chi_{[b]}$ with $\tau \in [a,b)$ (see (2.5.1)), the previous examples lead to the following statement.

6.3.4 Corollary. *Let $f, g : [a,b] \to \mathbb{R}$ satisfy one of the following conditions:*

(i) *g is a finite step function.*

(ii) *f is a finite step function and g is regulated.*

Then the integral $\int_a^b f\,\mathrm{d}g$ exists.

6.3.5 Exercise. Prove the following statement:

Let $h : [a,b] \to \mathbb{R}$ be such that $h(x) = c$ for all $x \in [a,b] \setminus \{d_1, \dots, d_n\}$. Then

$$\int_a^b f\,\mathrm{d}h = f(b)\,h(b) - f(a)\,h(a) - \big(f(b) - f(a)\big)\,c$$

holds for every function $f : [a,b] \to \mathbb{R}$.

Hint: Write $h = c + \sum_{k=1}^{n} (h(d_k) - c)\chi_{[d_k]}$ and use Lemma 6.3.2.

The following theorems provide some basic estimates for the Kurzweil-Stieltjes integral.

6.3.6 Theorem. *Let* $f, g : [a,b] \to \mathbb{R}$. *Then*

$$|S(P)| \leq \|f\| \operatorname{var}_a^b g \quad \textit{holds for each partition } P \textit{ of } [a,b].$$

If the integral $\int_a^b f \, \mathrm{d}g$ *exists, then*

$$\left| \int_a^b f \, \mathrm{d}g \right| \leq \|f\| \operatorname{var}_a^b g.$$

If the integral $\int_a^b |f(x)| \, \mathrm{d}(\operatorname{var}_a^x g)$ *exists as well, then*

$$\left| \int_a^b f \, \mathrm{d}g \right| \leq \int_a^b |f(x)| \, \mathrm{d}(\operatorname{var}_a^x g) \leq \|f\| \operatorname{var}_a^b g.$$

Proof. For every partition $P = (\boldsymbol{\alpha}, \boldsymbol{\xi})$ of $[a,b]$ we have

$$\begin{aligned} |S(P)| &\leq \sum_{j=1}^{\nu(P)} |f(\xi_j)| \, |g(\alpha_j) - g(\alpha_{j-1})| \\ &\leq \sum_{j=1}^{\nu(P)} |f(\xi_j)| \left(\operatorname{var}_{\alpha_{j-1}}^{\alpha_j} g \right) \leq \|f\| \operatorname{var}_a^b g. \end{aligned}$$

The estimates for the integral follow from the previous inequality. □

6.3.7 Theorem. *Let* $f, g : [a,b] \to \mathbb{R}$. *Then*

$$|S(P)| \leq (|f(a)| + |f(b)| + \operatorname{var}_a^b f) \, \|g\| \leq 2 \, \|f\|_{\mathrm{BV}} \, \|g\|$$

for each partition P *of* $[a,b]$. *If the integral* $\int_a^b f \, \mathrm{d}g$ *exists, then*

$$\left| \int_a^b f \, \mathrm{d}g \right| \leq \left(|f(a)| + |f(b)| + \operatorname{var}_a^b f \right) \|g\| \leq 2 \, \|f\|_{\mathrm{BV}} \, \|g\|.$$

Proof. Consider an arbitrary partition $P = (\boldsymbol{\alpha}, \boldsymbol{\xi})$ of $[a,b]$, and denote $m = \nu(P)$, $\xi_0 = a$, $\xi_{m+1} = b$. Then

$$\begin{aligned} S(P) &= f(\xi_1) \, (g(\alpha_1) - g(a)) + f(\xi_2) \, (g(\alpha_2) - g(\alpha_1)) \\ &\qquad + \cdots + f(\xi_m) \, (g(b) - g(\alpha_{m-1})) \\ &= f(b) \, g(b) - f(a) \, g(a) - (f(\xi_1) - f(a)) \, g(a) \\ &\qquad - (f(\xi_2) - f(\xi_1)) \, g(\alpha_1) - \cdots - (f(b) - f(\xi_m)) \, g(b) \\ &= f(b) \, g(b) - f(a) \, g(a) - \sum_{j=0}^{m} (f(\xi_{j+1}) - f(\xi_j)) \, g(\alpha_j), \end{aligned}$$

and the statement of the theorem follows immediately. □

Theorems 6.3.6 and 6.3.7 enable us to prove two basic convergence theorems for the Kurzweil-Stieltjes integral.

6.3.8 Theorem. *Let $f, g\colon [a,b] \to \mathbb{R}$ and $f_n\colon [a,b] \to \mathbb{R}$, $n \in \mathbb{N}$, be such that the integrals $\int_a^b f_n \,\mathrm{d}g$ exist for all $n \in \mathbb{N}$. Suppose at least one of the following conditions is satisfied:*

(i) *$g \in \mathrm{BV}([a,b])$ and $f_n \rightrightarrows f$,*

(ii) *g is bounded and $\lim_{n\to\infty} \|f_n - f\|_{\mathrm{BV}} = 0$.*

Then the integral $\int_a^b f \,\mathrm{d}g$ exists as well, and

$$\lim_{n\to\infty} \int_a^b f_n \,\mathrm{d}g = \int_a^b f \,\mathrm{d}g.$$

Proof. We claim that in both cases, the sequence $\{\int_a^b f_n \,\mathrm{d}g\}$ satisfies the Cauchy condition. In case (i), we apply Theorem 6.3.6 to get

$$\Big|\int_a^b f_n \,\mathrm{d}g - \int_a^b f_m \,\mathrm{d}g\Big| = \Big|\int_a^b (f_n - f_m) \,\mathrm{d}g\Big| \le \|f_n - f_m\| \operatorname{var}_a^b g$$

for all $m, n \in \mathbb{N}$. Since $\operatorname{var}_a^b g$ is finite and $\{f_n\}$ is uniformly convergent, the right-hand side will be arbitrarily small if m, n are sufficiently large.

In case (ii), we use Theorem 6.3.7 to get

$$\begin{aligned}\Big|\int_a^b f_n \,\mathrm{d}g - \int_a^b f_m \,\mathrm{d}g\Big| = \Big|\int_a^b (f_n - f_m) \,\mathrm{d}g\Big| &\le 2\,\|g\|\,\|f_n - f_m\|_{\mathrm{BV}} \\ &\le 2\,\|g\|\,\|f_n - f\|_{\mathrm{BV}} + 2\,\|g\|\,\|f - f_m\|_{\mathrm{BV}}\end{aligned}$$

for all $m, n \in \mathbb{N}$. Since $\|g\|$ is finite and $\|f_n - f\|_{\mathrm{BV}} \to 0$, the right-hand side will be arbitrarily small if m, n are sufficiently large.

Thus, in both cases, there exists a number $I \in \mathbb{R}$ such that

$$\lim_{n\to\infty} \int_a^b f_n \,\mathrm{d}g = I. \tag{6.3.11}$$

To show that $\int_a^b f \,\mathrm{d}g = I$, let $\varepsilon > 0$ be given. We claim there exists an $n_1 \in \mathbb{N}$ such that

$$|S(f - f_n, \mathrm{d}g, P)| < \varepsilon \quad \text{for } n \ge n_1 \text{ and every partition } P \text{ of } [a,b]. \tag{6.3.12}$$

In case (i), this follows from Theorem 6.3.6, which yields

$$|S(f - f_n, \mathrm{d}g, P)| \le \|f - f_n\| \operatorname{var}_a^b g.$$

In case (ii), we use Theorem 6.3.7 to get

$$|S(f - f_n, \mathrm{d}g, P)| \le 2\, \|f - f_n\|_{\mathrm{BV}}\, \|g\|.$$

These estimates show the validity of (6.3.12). By (6.3.11), there exists an $n_0 \ge n_1$ such that

$$|I_{n_0} - I| < \varepsilon,$$

where $I_{n_0} = \int_a^b f_{n_0}\, \mathrm{d}g$. Moreover, we choose a gauge δ_0 on $[a,b]$ such that

$$|S(f_{n_0}, \mathrm{d}g, P) - I_{n_0}| < \varepsilon \quad \text{for each } \delta_0\text{-fine partition } P \text{ of } [a,b].$$

Consequently, the estimate

$$\begin{aligned} |S(f, \mathrm{d}g, P) - I| &\le |S(f - f_{n_0}, \mathrm{d}g, P)| + |S(f_{n_0}, \mathrm{d}g, P) - I_{n_0}| + |I_{n_0} - I| \\ &< 3\varepsilon \end{aligned}$$

holds for each δ_0-fine partition P of $[a,b]$. This shows that $\int_a^b f\, \mathrm{d}g = I$, and the proof is complete. □

6.3.9 Theorem. *Let $f, g : [a,b] \to \mathbb{R}$ and $g_n : [a,b] \to \mathbb{R}$, $n \in \mathbb{N}$, be such that the integrals $\int_a^b f\, \mathrm{d}g_n$ exist for all $n \in \mathbb{N}$. Suppose at least one of the following conditions is satisfied:*

(i) *$f \in \mathrm{BV}([a,b])$ and $g_n \rightrightarrows g$,*

(ii) *f is bounded and $\lim_{n\to\infty} \operatorname{var}_a^b (g_n - g) = 0$.*

Then the integral $\int_a^b f\, \mathrm{d}g$ exists as well, and

$$\lim_{n\to\infty} \int_a^b f\, \mathrm{d}g_n = \int_a^b f\, \mathrm{d}g. \tag{6.3.13}$$

Proof. We claim that in both cases, the sequence $\{\int_a^b f\, \mathrm{d}g_n\}$ satisfies the Cauchy condition. In case (i), we apply Theorem 6.3.7 to get

$$\Big| \int_a^b f\, \mathrm{d}g_n - \int_a^b f\, \mathrm{d}g_m \Big| = \Big| \int_a^b f\, \mathrm{d}(g_n - g_m) \Big| \le 2\, \|f\|_{\mathrm{BV}}\, \|g_n - g_m\|$$

for all $m, n \in \mathbb{N}$. Since $\|f\|_{\mathrm{BV}}$ is finite and $\{g_n\}$ is uniformly convergent, the right-hand side will be arbitrarily small if m, n are sufficiently large.

In case (ii), we use Theorem 6.3.6 to get

$$\Big|\int_a^b f\,\mathrm{d}g_n - \int_a^b f\,\mathrm{d}g_m\Big| = \Big|\int_a^b f\,\mathrm{d}(g_n - g_m)\Big| \le \|f\|\ \mathrm{var}_a^b(g_n - g_m)$$
$$\le \|f\|\ \mathrm{var}_a^b\,(g_n - g) + \|f\|\ \mathrm{var}_a^b(g - g_m)$$

for all $m, n \in \mathbb{N}$. Since $\|f\|$ is finite and $\mathrm{var}_a^b(g_n - g) \to 0$, the right-hand side will be arbitrarily small if m, n are sufficiently large.

Thus, in both cases, there exists a number $I \in \mathbb{R}$ such that

$$\lim_{n\to\infty} \int_a^b f\,\mathrm{d}g_n = I. \tag{6.3.14}$$

To show that $\int_a^b f\,\mathrm{d}g = I$, let $\varepsilon > 0$ be given. We claim there exists an $n_1 \in \mathbb{N}$ such that

$$|S(f, \mathrm{d}(g - g_n), P)| < \varepsilon \text{ for } n \ge n_1 \text{ and every partition } P \text{ of } [a, b]. \tag{6.3.15}$$

In case (i), this follows from Theorem 6.3.7, which yields

$$|S(f, \mathrm{d}(g - g_n), P)| \le 2\|f\|_{\mathrm{BV}}\|g - g_n\|.$$

In case (ii), we use Theorem 6.3.6 to get

$$|S(f, \mathrm{d}(g - g_n), P)| \le \|f\|\ \mathrm{var}_a^b(g - g_n).$$

These estimates show the validity of (6.3.15). By (6.3.14), there exists an $n_0 \ge n_1$ such that

$$|I_{n_0} - I| < \varepsilon,$$

where $I_{n_0} = \int_a^b f\,\mathrm{d}g_{n_0}$. Moreover, we choose a gauge δ_0 on $[a, b]$ such that

$$|S(f, \mathrm{d}g_{n_0}, P) - I_{n_0}| < \varepsilon \quad \text{for each } \delta_0\text{-fine partition } P \text{ of } [a, b]. \tag{6.3.16}$$

Consequently, the estimate

$$|S(f, \mathrm{d}g, P) - I| \le |S(f, \mathrm{d}(g - g_{n_0}), P)| + |S(f, \mathrm{d}g_{n_0}, P) - I_{n_0}| + |I_{n_0} - I|$$
$$< 3\,\varepsilon$$

holds for each δ_0-fine partition P of $[a, b]$. This shows that $\int_a^b f\,\mathrm{d}g = I$, and the proof is complete. □

6.3.10 Remark. At first sight, it might seem that the assumptions of Theorems 6.3.8 and 6.3.9 are not completely symmetric: condition (ii) of the former theorem assumes that $\lim_{n\to\infty} \|f_n - f\|_{\mathrm{BV}} = 0$, while condition (ii) of the latter theorem requires only $\lim_{n\to\infty} \operatorname{var}_a^b(g_n - g) = 0$. However, by adding a suitable constant to each g_n, we can achieve that $g_n(a) = g(a)$ for all $n \in \mathbb{N}$. This change has no influence on the existence or value of the integrals $\int_a^b f \,\mathrm{d}g_n$, and the result is that $\lim_{n\to\infty} \|g_n - g\|_{\mathrm{BV}} = 0$.

We now prove our basic existence result for Kurzweil-Stieltjes integrals.

6.3.11 Theorem. *If $f, g : [a,b] \to \mathbb{R}$ are regulated and one of them has bounded variation, then the integral $\int_a^b f \,\mathrm{d}g$ exists.*

Proof. Assume first that $g \in \mathrm{BV}([a,b])$. By Theorem 4.1.5, there exists a sequence $\{f_n\}$ of finite step functions on $[a,b]$ such that $f_n \rightrightarrows f$. By Corollary 6.3.4, the integrals $\int_a^b f_n \,\mathrm{d}g$ exist for all $n \in \mathbb{N}$. By Theorem 6.3.8, this means that the integral $\int_a^b f \,\mathrm{d}g$ exists as well.

Next, assume that $f \in \mathrm{BV}([a,b])$. By Theorem 4.1.5, there exists a sequence $\{g_n\}$ of finite step functions on $[a,b]$ such that $g_n \rightrightarrows g$. By Corollary 6.3.4, the integrals $\int_a^b f \,\mathrm{d}g_n$ exist for all $n \in \mathbb{N}$. By Theorem 6.3.9, this means that the integral $\int_a^b f \,\mathrm{d}g$ exists as well. □

Recall from Section 2.6 that each function $h \in \mathrm{BV}([a,b])$ has a decomposition of the form $h = h^{\mathrm{C}} + h^{\mathrm{B}}$, where h^{C} is a continuous function with bounded variation (referred to as the continuous part of h), while h^{B} is a step function (the jump part of h); these functions are unique up to a constant.

If f, g are regulated and one of them has bounded variation, we know from Theorem 6.3.11 that the integral $\int_a^b f \,\mathrm{d}g$ exists. By decomposing the integrand or the integrator into the continuous and the jump part, we arrive at the following results.

6.3.12 Theorem. *If $f \in \mathrm{BV}([a,b])$, $g \in \mathrm{G}([a,b])$, D is the set of all discontinuity points of the function f in (a,b), and f^{C} is the continuous part of f with $f^{\mathrm{C}}(a) = f(a)$, then*

$$\int_a^b f \,\mathrm{d}g = \int_a^b f^{\mathrm{C}} \,\mathrm{d}g + \Delta^+ f(a)\,(g(b) - g(a+)) + \Delta^- f(b)\,\Delta^- g(b)$$
$$+ \sum_{d\in D} \Big(\Delta^- f(d)\,(g(b) - g(d-)) + \Delta^+ f(d)\,(g(b) - g(d+))\Big).$$

Proof. We focus on the case when D is infinite; the proof for a finite D is simpler and is left to the reader.

Since D is countable, we can write $D = \{d_k\}$. According to Remark 2.6.3, the functions $f^{\mathrm{B}}, f^{\mathrm{C}} : [a,b] \to \mathbb{R}$ given by

$$\begin{aligned} f^{\mathrm{B}}(x) &= \Delta^+ f(a)\, \chi_{(a,b]}(x) + \Delta^- f(b)\, \chi_{[b]}(x) \\ &\quad + \sum_{k=1}^{\infty} \big(\Delta^- f(d_k)\, \chi_{[d_k,b]}(x) + \Delta^+ f(d_k)\, \chi_{(d_k,b]}(x)\big), \\ f^{\mathrm{C}}(x) &= f(x) - f^{\mathrm{B}}(x) \end{aligned}$$

represent the jump part and the continuous part of the function f, respectively, and are chosen to satisfy $f^{\mathrm{B}}(a) = 0$ and $f^{\mathrm{C}}(a) = f(a)$.

For each $n \in \mathbb{N}$, consider the finite step function $f_n^{\mathrm{B}} : [a,b] \to \mathbb{R}$ given by

$$\begin{aligned} f_n^{\mathrm{B}}(x) &= \Delta^+ f(a)\, \chi_{(a,b]}(x) + \Delta^- f(b)\, \chi_{[b]}(x) \\ &\quad + \sum_{k=1}^{n} \big(\Delta^- f(d_k)\, \chi_{[d_k,b]}(x) + \Delta^+ f(d_k)\, \chi_{(d_k,b]}(x)\big). \end{aligned}$$

By Lemma 2.6.5, we have $\lim_{n\to\infty} \|f^{\mathrm{B}} - f_n^{\mathrm{B}}\|_{\mathrm{BV}} = 0$. Thus, Theorem 6.3.8 implies that the integral $\int_a^b f^{\mathrm{B}}\, \mathrm{d}g$ exists and

$$\int_a^b f^{\mathrm{B}}\, \mathrm{d}g = \lim_{n\to\infty} \int_a^b f_n^{\mathrm{B}}\, \mathrm{d}g.$$

By formulas (6.3.6) and (6.3.7) from Lemma 6.3.3, we have

$$\begin{aligned} \int_a^b f_n^{\mathrm{B}}\, \mathrm{d}g &= \Delta^+ f(a)\, (g(b) - g(a+)) + \Delta^- f(b)\, \Delta^- g(b) \\ &\quad + \sum_{k=1}^{n} \Big(\Delta^- f(d_k)\, (g(b) - g(d_k-)) + \Delta^+ f(d_k)\, (g(b) - g(d_k+))\Big) \end{aligned}$$

for all $n \in \mathbb{N}$. Consequently,

$$\begin{aligned} \int_a^b f^{\mathrm{B}}\, \mathrm{d}g &= \Delta^+ f(a)\, (g(b) - g(a+)) + \Delta^- f(b)\, \Delta^- g(b) \\ &\quad + \sum_{k=1}^{\infty} \Big(\Delta^- f(d_k)\, (g(b) - g(d_k-)) + \Delta^+ f(d_k)\, (g(b) - g(d_k+))\Big). \end{aligned}$$

The series on the right hand is absolutely convergent, because

$$\sum_{k=1}^{\infty} \left|\Delta^- f(d_k)\,(g(b)-g(d_k-)) + \Delta^+ f(d_k)\,(g(b)-g(d_k+))\right|$$
$$\leq 2\,\|g\| \sum_{k=1}^{\infty} \Big(|\Delta^- f(d_k)| + |\Delta^+ f(d_k)|\Big) \leq 2\,\|g\|\,(\operatorname{var}_a^b f) < \infty$$

due to Corollary 2.3.8. Thus, we can write

$$\int_a^b f^{\mathrm{B}}\,\mathrm{d}g = \Delta^+ f(a)\,(g(b)-g(a+)) + \Delta^- f(b)\,\Delta^- g(b)$$
$$+ \sum_{d\in D} \Big(\Delta^- f(d)\,(g(b)-g(d)) + \Delta^+ f(d)\,(g(b)-g(d))\Big),$$

and the proof is complete. □

6.3.13 Theorem. *If* $f \in \mathrm{G}([a,b])$, $g \in \mathrm{BV}([a,b])$, D *is the set of all discontinuity points of the function* g *in* (a,b), *and* g^{C} *is the continuous part of* g, *then*

$$\int_a^b f\,\mathrm{d}g = \int_a^b f\,\mathrm{d}g^{\mathrm{C}} + f(a)\,\Delta^+ g(a) + \sum_{d\in D} f(d)\,\Delta g(d) + f(b)\,\Delta^- g(b).$$

Proof. We focus on the case when D is infinite; the proof for a finite D is simpler and is left to the reader.

Since D is countable, we can write $D = \{d_k\}$. According to Remark 2.6.3, the functions $g^{\mathrm{B}}, g^{\mathrm{C}} : [a,b] \to \mathbb{R}$ given by

$$g^{\mathrm{B}}(x) = \Delta^+ g(a)\,\chi_{(a,b]}(x) + \Delta^- g(b)\,\chi_{[b]}(x)$$
$$+ \sum_{k=1}^{\infty} \big(\Delta^- g(d_k)\,\chi_{[d_k,b]}(x) + \Delta^+ g(d_k)\,\chi_{(d_k,b]}(x)\big),$$
$$g^{\mathrm{C}}(x) = g(x) - g^{\mathrm{B}}(x)$$

represent one possible choice of the jump part and the continuous part of the function g, respectively.

For each $n \in \mathbb{N}$, consider the function $g_n^{\mathrm{B}} : [a,b] \to \mathbb{R}$ given by

$$g_n^{\mathrm{B}}(x) = \Delta^+ g(a)\,\chi_{(a,b]}(x) + \Delta^- g(b)\,\chi_{[b]}(x)$$
$$+ \sum_{k=1}^{n} \big(\Delta^- g(d_k)\,\chi_{[d_k,b]}(x) + \Delta^+ g(d_k)\,\chi_{(d_k,b]}(x)\big).$$

By Lemma 2.6.5, we have $\lim_{n\to\infty} \|g_n^{\mathrm{B}} - g^{\mathrm{B}}\|_{\mathrm{BV}} = 0$. Thus, Theorem 6.3.9 implies that the integral $\int_a^b f\,\mathrm{d}g^{\mathrm{B}}$ exists and

$$\int_a^b f\,\mathrm{d}g^{\mathrm{B}} = \lim_{n\to\infty} \int_a^b f\,\mathrm{d}g_n^{\mathrm{B}}.$$

By formulas (6.3.1) and (6.3.2) from Lemma 6.3.2, we have

$$\int_a^b f\,\mathrm{d}g_n^{\mathrm{B}} = f(a)\,\Delta^+ g(a) + \sum_{k=1}^{n} f(d_k)\,\Delta g(d_k) + f(b)\,\Delta^- g(b)$$

for all $n \in \mathbb{N}$. Consequently,

$$\int_a^b f\,\mathrm{d}g^{\mathrm{B}} = f(a)\,\Delta^+ g(a) + \sum_{k=1}^{\infty} f(d_k)\,\Delta g(d_k) + f(b)\,\Delta^- g(b).$$

Since f is bounded, it follows from Corollary 2.3.8 that the series on the right-hand side is absolutely convergent. Thus, we can write

$$\int_a^b f\,\mathrm{d}g^{\mathrm{B}} = \Big(f(a)\,\Delta^+ g(a) + \sum_{d\in D} f(d)\,\Delta g(d) + f(b)\,\Delta^- g(b)\Big),$$

and consequently

$$\int_a^b f\,\mathrm{d}g = \int_a^b f\,\mathrm{d}g^{\mathrm{C}} + f(a)\,\Delta^+ g(a) + \sum_{d\in D} f(d)\,\Delta g(d) + f(b)\,\Delta^- g(b).$$

Although g^{B} and g^{C} are determined only up to a constant, the value of integral $\int_a^b f\,\mathrm{d}g^{\mathrm{C}}$ always remains the same. Thus, the formula does not depend on the particular choice of the decomposition of g into continuous and jump parts. □

Our final goal in this section is to determine the values of the integrals $\int_a^b h\,\mathrm{d}g$ and $\int_a^b g\,\mathrm{d}h$ in the case when h is a "nearly constant" function, i.e., it is constant with at most countably many exceptions.

6.3.14 Lemma. *Suppose that $h \in \mathrm{G}([a,b])$, $c \in \mathbb{R}$, and*

$$h(x) = c \quad \text{for } x \in [a,b] \setminus D, \tag{6.3.17}$$

where $D \subset [a,b]$ is at most countable. Then

$$h(x-) = c \ \text{ for all } x \in (a,b] \ \text{ and } \ h(x+) = c \ \text{ for all } x \in [a,b).$$

Proof. The statement evidently holds for all $x \in [a,b]$ that are not accumulation points of D. If $x \in (a,b]$ is an accumulation point of D, we can choose an increasing sequence $\{x_k\} \subset [a,x) \setminus D$ which tends to x. Clearly, $\lim_{k\to\infty} h(x_k) = c$. Since h is regulated, we necessarily have $h(x-) = c$. In a similar way, we can prove that the equality $h(x+) = c$ holds for each $x \in [a,b)$ which is an accumulation point of D. □

We remark that a function h satisfying (6.3.17) need not be regulated. To see this, consider the function

$$h(t) = \begin{cases} 1 & \text{if } t = a + \frac{1}{k},\ k \in \mathbb{N}, \\ 0 & \text{otherwise.} \end{cases}$$

More generally, it suffices to have $D = \{d_k\}$, where $\{d_k\}$ is a convergent sequence and $\lim_{k\to\infty} h(d_k) \neq c$.

If we assume that the function h in (6.3.17) has bounded variation and g is regulated, then the integrals $\int_a^b h\,\mathrm{d}g$ and $\int_a^b g\,\mathrm{d}h$ are easily calculated with the help of Theorems 6.3.12 and 6.3.13.

6.3.15 Lemma. *Suppose that $g \in \mathrm{G}([a,b])$, $h \in \mathrm{BV}([a,b])$, $c \in \mathbb{R}$, and*

$$h(x) = c \quad \text{for} \quad x \in [a,b] \setminus D,$$

where $D \subset [a,b]$ is at most countable. Then

$$\int_a^b h\,\mathrm{d}g = c\,(g(b) - g(a)) + (h(a) - c)\,\Delta^+ g(a) + \sum_{d \in D \cap (a,b)} (h(d) - c)\,\Delta g(d) + (h(b) - c)\,\Delta^- g(b),$$

$$\int_a^b g\,\mathrm{d}h = g(b)\,h(b) - g(a)\,h(a) - c\,\big(g(b) - g(a)\big).$$

Proof. The function h is a step function. Thus, the continuous part of h can be chosen to be an arbitrary constant function. If we take $h^{\mathrm{C}}(x) = h(a)$ for all $x \in [a,b]$, we can use Theorem 6.3.12 to get

$$\int_a^b h\,\mathrm{d}g = h(a)\,(g(b) - g(a)) + \Delta^+ h(a)\,(g(b) - g(a+)) + \Delta^- h(b)\,\Delta^- g(b) + \sum_{d \in D \cap (a,b)} \Big(\Delta^- h(d)\,(g(b) - g(d-)) + \Delta^+ h(d)\,(g(b) - g(d+))\Big).$$

By Lemma 6.3.14, we have

$$\Delta^+h(a) = c - h(a), \quad \Delta^-h(b) = h(b) - c,$$
$$\Delta^-h(x) = h(x) - c = -\Delta^+h(x) \quad \text{for all} \quad x \in (a,b).$$

Substituting these relations into the previous formula, we obtain

$$\begin{aligned}\int_a^b h\,\mathrm{d}g &= \int_a^b h^{\mathrm{C}}\,\mathrm{d}g + (c - h(a))\,(g(b) - g(a+)) + (h(b) - c)\,\Delta^-g(b)\\ &\quad + \sum_{d\in D\cap(a,b)} (h(d) - c)\,\big(g(b) - g(d-) - g(b) + g(d+)\big)\\ &= h(a)\,(g(b) - g(a)) + (c - h(a))\,(g(b) - g(a))\\ &\quad + (h(a) - c)\,\Delta^+g(a) + (h(b) - c)\,\Delta^-g(b)\\ &\quad + \sum_{d\in D\cap(a,b)} (h(d) - c)\,\Delta g(d),\end{aligned}$$

which proves the first statement.

Using Theorem 6.3.13, we get

$$\begin{aligned}\int_a^b g\,\mathrm{d}h &= \int_a^b g\,\mathrm{d}h^{\mathrm{C}} + g(a)\,\Delta^+h(a) + \sum_{d\in D\cap(a,b)} g(d)\,\Delta h(d) + g(b)\,\Delta^-h(b)\\ &= g(a)\,(c - h(a)) + g(b)\,(h(b) - c),\end{aligned}$$

which proves the second statement. □

We now show that the formulas from Lemma 6.3.15 are still valid in the reverse case when h is regulated and g has bounded variation.

6.3.16 Lemma. *Suppose that* $g \in \mathrm{BV}([a,b])$, $h \in \mathrm{G}([a,b])$, $c \in \mathbb{R}$, *and*

$$h(x) = c \quad \textit{for} \quad x \in [a,b] \setminus D,$$

where $D \subset [a,b]$ *is at most countable. Then*

$$\begin{aligned}\int_a^b h\,\mathrm{d}g &= c\,(g(b) - g(a)) + (h(a) - c)\,\Delta^+g(a)\\ &\quad + \sum_{d\in D\cap(a,b)} (h(d) - c)\,\Delta g(d) + (h(b) - c)\,\Delta^-g(b),\end{aligned}$$

$$\int_a^b g\,\mathrm{d}h = g(b)\,h(b) - g(a)\,h(a) - c\,(g(b) - g(a)).$$

Proof. We focus on the case when D is infinite; the proof for a finite D is simpler and is left to the reader. Since D is countable, we can write $D\cap(a,b)=\{d_k\}$.

For each $n\in\mathbb{N}$, consider the function $h_n:[a,b]\to\mathbb{R}$ given by

$$h_n(x)=c+(h(a)-c)\,\chi_{[a]}(x)+\sum_{k=1}^{n}(h(d_k)-c)\,\chi_{[d_k]}(x)+(h(b)-c)\,\chi_{[b]}(x).$$

Clearly, $h_n(a)=h(a)$ and $h_n(b)=h(b)$ for all $n\in\mathbb{N}$.

We claim that $h_n\rightrightarrows h$. To see this, notice that

$$h(x)-h_n(x)=\begin{cases} h(x)-c & \text{if } x=d_k \text{ for some } k>n,\\ 0 & \text{otherwise.}\end{cases}$$

By Lemma 6.3.14, we have

$$\Delta^- h(x)=h(x)-c \quad \text{for each } x\in(a,b).$$

Let $\varepsilon>0$ be given. By Corollary 4.1.7, the set of all $x\in(a,b)$ for which $|\Delta^- h(x)|\geq\varepsilon$ is finite. Hence, there exists an $n_\varepsilon\in\mathbb{N}$ such that

$$|\Delta^- h(d_k)|=|h(d_k)-c|<\varepsilon \quad \text{for all } k>n_\varepsilon.$$

Consequently, $\|h-h_n\|<\varepsilon$ whenever $n\geq n_\varepsilon$. This proves that $h_n\rightrightarrows h$.

Since h_n has bounded variation (because it is a finite step function), it is possible to apply Lemma 6.3.15 and calculate

$$\int_a^b h_n\,\mathrm{d}g=c\,(g(b)-g(a))+(h(a)-c)\,\Delta^+g(a)$$
$$+\sum_{k=1}^{n}(h(d_k)-c)\,\Delta g(d_k)+(h(b)-c)\,\Delta^-g(b),$$
$$\int_a^b g\,\mathrm{d}h_n=g(b)\,h(b)-g(a)h(a)-c\,(g(b)-g(a)).$$

Consequently, by Theorem 6.3.9 we have

$$\int_a^b g\,\mathrm{d}h=\lim_{n\to\infty}\int_a^b g\,\mathrm{d}h_n=g(b)\,h(b)-g(a)\,h(a)-c\,(g(b)-g(a)).$$

Similarly, by Theorem 6.3.8, we obtain

$$\begin{aligned}\int_a^b h\,\mathrm{d}g &= \lim_{n\to\infty}\int_a^b h_n\,\mathrm{d}g\\ &= c\,(g(b)-g(a)) + (h(a)-c)\,\Delta^+ g(a)\\ &\quad + \sum_{k=1}^{\infty}(h(d_k)-c)\,\Delta g(d_k) + (h(b)-c)\,\Delta^- g(b).\end{aligned}$$

The infinite series on the right-hand side is absolutely convergent, because $x \mapsto h(x) - c$ is bounded and $\sum_{k=1}^{\infty}\Delta g(d_k)$ is absolutely convergent by Corollary 2.3.8. Thus, we can write

$$\sum_{k=1}^{\infty}(h(d_k)-c)\,\Delta g(d_k) = \sum_{d\in D\cap(a,b)}(h(d)-c)\,\Delta g(d),$$

and the proof is complete. □

We conclude the present section with the following simple but useful consequence of Lemma 6.3.15.

6.3.17 Corollary. *If $f \in \mathrm{BV}([a,b])$ and $g \in \mathrm{G}([a,b])$, then*

$$\int_a^b \Delta^+ f\,\mathrm{d}g = \sum_{x\in(a,b)}\Delta^+ f(x)\,\Delta g(x) + \Delta^+ f(a)\,\Delta^+ g(a), \tag{6.3.18}$$

$$\int_a^b \Delta^- f\,\mathrm{d}g = \sum_{x\in(a,b)}\Delta^- f(x)\,\Delta g(x) + \Delta^- f(b)\,\Delta^- g(b), \tag{6.3.19}$$

with the convention that $\Delta^- f(a) = 0$, $\Delta^+ f(b) = 0$.

Proof. Since f has bounded variation, it has only finitely or countably many discontinuities. Thus, $\Delta^+ f$ and $\Delta^- f$ are zero except a finite or countable set. Moreover, by Corollary 2.3.8, the sums

$$\sum_{x\in[a,b]}|\Delta^+ f(x)| \quad\text{and}\quad \sum_{x\in[a,b]}|\Delta^- f(x)|$$

are finite. This means that $\Delta^+ f$ and $\Delta^- f$ have bounded variation. The formulas (6.3.18) and (6.3.19) are now immediate consequences of Lemma 6.3.15 with $c = 0$ and $h = \Delta^+ f$ or $h = \Delta^- f$, respectively. □

6.4 Integration by parts

The aim of this section is to prove the integration by parts theorem for the Kurzweil-Stieltjes integral. We will see that the formula is more complicated in comparison with Theorem 5.5.1 (integration by parts for the Riemann-Stieltjes integral).

We work with a fixed interval $[a,b]\subset\mathbb{R}$, and use the convention that

$$f(a-)=f(a),\quad f(b+)=f(b),$$
$$\Delta^- f(a)=\Delta^+ f(b)=0,\quad \Delta f(a)=\Delta^+ f(a),\quad \Delta f(b)=\Delta^- f(b)$$

for every regulated function $f:[a,b]\to\mathbb{R}$.

The following lemma deals with the case when one of the functions is a finite step function.

6.4.1 Lemma. *If $f:[a,b]\to\mathbb{R}$ is a finite step function and $g:[a,b]\to\mathbb{R}$ is regulated, then both the integrals $\int_a^b f\,\mathrm{d}g$ and $\int_a^b g\,\mathrm{d}f$ exist, and*

$$\int_a^b f\,\mathrm{d}g+\int_a^b g\,\mathrm{d}f=f(b)\,g(b)-f(a)\,g(a)+\sum_{a\le x\le b}\Big(\Delta^- f(x)\,\Delta^- g(x)-\Delta^+ f(x)\,\Delta^+ g(x)\Big).$$

Proof. Because f is a finite step function, there exist a division $\boldsymbol{\alpha}=\{\alpha_0,\alpha_1,\dots,\alpha_m\}$ of $[a,b]$ and real numbers $c_1,\dots,c_m\in\mathbb{R}$ such that $f(t)=c_j$ for $t\in(\alpha_{j-1},\alpha_j)$. Using Lemma 6.3.3, we obtain

$$\int_a^b f\,\mathrm{d}g=f(a)\,\Delta^+ g(a)+f(b)\,\Delta^- g(b)+\sum_{j=1}^m c_j\,(g(\alpha_j-)-g(\alpha_{j-1}+))+\sum_{j=1}^{m-1} f(\alpha_j)\,(g(\alpha_j+)-g(\alpha_j-)),$$

and, using Lemma 6.3.2,

$$\int_a^b g\,\mathrm{d}f=\sum_{j=1}^m\big(c_j\,(g(\alpha_{j-1})-g(\alpha_j))+f(\alpha_j)\,g(\alpha_j)-f(\alpha_{j-1})\,g(\alpha_{j-1})\big).$$

Therefore

$$\int_a^b f\,\mathrm{d}g + \int_a^b g\,\mathrm{d}f = f(b)\,g(b) - f(a)\,g(a)$$
$$- \sum_{j=1}^{m} c_j\,(\Delta^+ g(\alpha_{j-1}) + \Delta^- g(\alpha_j))$$
$$+ \sum_{j=1}^{m} (f(\alpha_{j-1})\,\Delta^+ g(\alpha_{j-1}) + f(\alpha_j)\,\Delta^- g(\alpha_j)).$$

Finally, we note that $\Delta^+ f(\alpha_{j-1}) = c_j - f(\alpha_{j-1})$ and $\Delta^- f(\alpha_j) = f(\alpha_j) - c_j$ for all $j \in \{1, \ldots, m\}$, and the proof is complete. □

We can now prove the main result of this section.

6.4.2 Theorem (INTEGRATION BY PARTS). *Let $f, g : [a,b] \to \mathbb{R}$ be regulated and one of them have bounded variation. Then both the integrals $\int_a^b f\,\mathrm{d}g$ and $\int_a^b g\,\mathrm{d}f$ exist, and*

$$\int_a^b f\,\mathrm{d}g + \int_a^b g\,\mathrm{d}f = f(b)\,g(b) - f(a)\,g(a)$$
$$+ \sum_{a \le x \le b} \Big(\Delta^- f(x)\,\Delta^- g(x) - \Delta^+ f(x)\,\Delta^+ g(x)\Big).$$

Proof. Since the roles of f and g are symmetric, we can assume that f is regulated and g has bounded variation. Let $\{f_n\}$ be a sequence of finite step functions which tends uniformly to f on $[a,b]$. Then by Lemma 6.4.1 we have

$$\left.\begin{aligned} &\int_a^b f_n\,\mathrm{d}g + \int_a^b g\,\mathrm{d}f_n - f_n(b)\,g(b) + f_n(a)\,g(a) \\ &\quad = \sum_{a \le x \le b} \Big(\Delta^- f_n(x)\,\Delta^- g(x) - \Delta^+ f_n(x)\,\Delta^+ g(x)\Big) \end{aligned}\right\} \tag{6.4.1}$$

for every $n \in \mathbb{N}$. By Theorems 6.3.8 and 6.3.9, the relation

$$\lim_{n\to\infty} \Big(\int_a^b f_n\,\mathrm{d}g + \int_a^b g\,\mathrm{d}f_n - f_n(b)\,g(b) + f_n(a)\,g(a)\Big)$$
$$= \int_a^b f\,\mathrm{d}g + \int_a^b g\,\mathrm{d}f - f(b)\,g(b) + f(a)\,g(a)$$

holds. Furthermore, taking into account that

$$|\Delta^+ f(t)| \le 2\,\|f\|, \quad |\Delta^- f(t)| \le 2\,\|f\|$$

$$|\Delta^+ (f_n - f)(t)| \le 2\,\|f_n - f\|, \qquad \text{and} \qquad |\Delta^- (f_n - f)(t)| \le 2\,\|f_n - f\|$$

for all $t \in [a,b]$, we obtain the estimates

$$\sum_{a\le t\le b} |\Delta^+ f(t)\,\Delta^+ g(t) - \Delta^- f(t)\,\Delta^- g(t)| \\ \le 2\,\|f\| \sum_{a\le t\le b} \Big(|\Delta^+ g(t)| + |\Delta^- g(t)|\Big) \le 2\,\|f\|\ \mathrm{var}_a^b\, g$$

and

$$\sum_{a\le t\le b} |\Delta^+ (f_n - f)(t)\,\Delta^+ g(t) - \Delta^- (f_n - f)(t)\,\Delta^- g(t)| \\ \le 2\,\|f_n - f\| \sum_{a\le t\le b} \Big(|\Delta^+ g(t)| + |\Delta^- g(t)|\Big) \le 2\,\|f_n - f\|\ \mathrm{var}_a^b\, g.$$

Consequently, the sum

$$\sum_{a\le t\le b} (\Delta^+ f(t)\,\Delta^+ g(t) - \Delta^- f(t)\,\Delta^- g(t))$$

is absolutely convergent, and

$$\lim_{n\to\infty} \sum_{a\le x\le b} \Big(\Delta^- f_n(x)\,\Delta^- g(x) - \Delta^+ f_n(x)\,\Delta^+ g(x)\Big) \\ = \sum_{a\le x\le b} \Big(\Delta^- f(x)\,\Delta^- g(x) - \Delta^+ f(x)\,\Delta^+ g(x)\Big).$$

Thus, by letting $n \to \infty$ in (6.4.1), we obtain the desired formula. □

The integration by parts formula for the Riemann-Stieltjes integral is simpler than in Theorem 6.4.2, because f and g cannot share any discontinuity points; thus, the last sum in Theorem 6.4.2 vanishes. The next exercise deals with another simple case when both f and g are left-continuous or right-continuous, respectively.

6.4.3 Exercise. Suppose that $f : [a,b] \to \mathbb{R}$ is regulated and $g : [a,b] \to \mathbb{R}$ has bounded variation. Use Theorem 6.4.2 and Corollary 6.3.17 to prove the following assertions:

(i) If f is left-continuous, then

$$\int_a^b f(x)\,\mathrm{d}g(x) + \int_a^b g(x+)\,\mathrm{d}f(x) = f(b)\,g(b) - f(a)\,g(a).$$

(ii) If f is right-continuous, then

$$\int_a^b f(x)\,\mathrm{d}g(x)+\int_a^b g(x-)\,\mathrm{d}f(x)=f(b)\,g(b)-f(a)\,g(a).$$

6.5 THE INDEFINITE INTEGRAL

The goal of this section is to obtain some basic properties of the indefinite Kurzweil-Stieltjes integral, i.e., the function $x\mapsto\int_a^x f\,\mathrm{d}g$, $x\in[a,b]$. We begin with the Saks-Henstock lemma, which is an indispensable tool to this aim.

6.5.1 Lemma (SAKS-HENSTOCK). *Let $f,g:[a,b]\to\mathbb{R}$ be such that the integral $\int_a^b f\,\mathrm{d}g$ exists. Let $\varepsilon>0$ be given and let δ be a gauge on $[a,b]$ such that*

$$\Big|S(P)-\int_a^b f\,\mathrm{d}g\Big|<\varepsilon \quad \text{for all } \delta\text{-fine partitions } P \text{ of } [a,b].$$

If $\{([s_j,t_j],\theta_j):j=1,2,\dots,n\}$ is an arbitrary system satisfying

$$\left.\begin{aligned}&a\le s_1\le\theta_1\le t_1\le s_2\le\dots\le s_n\le\theta_n\le t_n\le b,\\&[s_j,t_j]\subset(\theta_j-\delta(\theta_j),\theta_j+\delta(\theta_j)) \quad \text{for } j=1,\dots,n,\end{aligned}\right\}\tag{6.5.1}$$

then

$$\Big|\sum_{j=1}^n\Big(f(\theta_j)\,(g(t_j)-g(s_j))-\int_{s_j}^{t_j}f\,\mathrm{d}g\Big)\Big|\le\varepsilon.\tag{6.5.2}$$

Proof. Assume that the system $\{([s_j,t_j],\theta_j):j=1,2,\dots,n\}$ satisfies conditions (6.5.1). Denote $t_0=a$ and $s_{n+1}=b$.

Let $\eta>0$ and $j\in\{0,1,\dots,n\}$ be given. Assume that $t_j<s_{j+1}$. Then by Remark 6.2.5, there is a gauge δ_j on $[t_j,s_{j+1}]$ such that $\delta_j(x)\le\delta(x)$ for all $x\in[t_j,s_{j+1}]$, and a δ_j-fine partition P_j of $[t_j,s_{j+1}]$ such that

$$\Big|S(P_j)-\int_{t_j}^{s_{j+1}}f\,\mathrm{d}g\Big|<\frac{\eta}{n+1}.\tag{6.5.3}$$

If $t_j=s_{j+1}$, we set $S(P_j)=0$.

Now, consider the δ-fine partition $Q=(\boldsymbol{\beta},\boldsymbol{\eta})$ of $[a,b]$ obtained by joining the system $\{([s_j,t_j],\theta_j):j=1,2,\dots,n\}$ with the partitions $P_0,\dots,P_n$.

Then

$$S(Q) = \sum_{j=1}^{n} f(\theta_j)\,(g(t_j) - g(s_j)) + \sum_{j=0}^{n} S(P_j).$$

Hence,

$$\Big|\sum_{j=1}^{n}\Big(f(\theta_j)\,(g(t_j)-g(s_j)) - \int_{s_j}^{t_j} f\,\mathrm{d}g\Big) + \sum_{j=0}^{n}\Big(S(P_j) - \int_{t_j}^{s_{j+1}} f\,\mathrm{d}g\Big)\Big|$$
$$= \Big|S(Q) - \int_a^b f\,\mathrm{d}g\Big| < \varepsilon.$$

This together with (6.5.3) yields

$$\Big|\sum_{j=1}^{n} f(\theta_j)\,(g(t_j)-g(s_j)) - \int_{s_j}^{t_j} f\,\mathrm{d}g\Big|$$
$$\le \Big|S(Q) - \int_a^b f\,\mathrm{d}g\Big| + \Big|\sum_{j=0}^{n}\Big(S(P_j) - \int_{t_j}^{s_{j+1}} f\,\mathrm{d}g\Big)\Big| < \varepsilon + \eta.$$

Since $\eta > 0$ can be arbitrarily small, we get the inequality (6.5.2). □

Sometimes it is useful to have an estimate similar to (6.5.2), but with the absolute value inside the sum. Such estimate is easily obtained directly from the Saks-Henstock lemma.

6.5.2 Corollary. *Let $f, g : [a,b] \to \mathbb{R}$ be such that the integral $\int_a^b f\,\mathrm{d}g$ exists. Let $\varepsilon > 0$ be given and let δ be a gauge on $[a,b]$ such that*

$$\Big|S(P) - \int_a^b f\,\mathrm{d}g\Big| < \varepsilon \quad \textit{for all } \delta\textit{-fine partitions } P \textit{ of } [a,b].$$

If $\{([s_j,t_j],\theta_j) : j = 1,2,\dots,n\}$ is an arbitrary system satisfying

$$a \le s_1 \le \theta_1 \le t_1 \le s_2 \le \dots \le s_n \le \theta_n \le t_n \le b,$$
$$[s_j,t_j] \subset (\theta_j - \delta(\theta_j), \theta_j + \delta(\theta_j)) \quad \textit{for } j \in \{1,\dots,n\},$$

then

$$\sum_{j=1}^{n}\Big|f(\theta_j)\,(g(t_j)-g(s_j)) - \int_{s_j}^{t_j} f\,\mathrm{d}g\Big| \le 2\,\varepsilon. \tag{6.5.4}$$

P r o o f. Consider the sets

$$J^+ = \{j \in \{1,\dots,n\} : f(\theta_j)\,(g(t_j)-g(s_j)) - \int_{s_j}^{t_j} f\,\mathrm{d}g > 0\},$$

$$J^- = \{j \in \{1,\dots,n\} : f(\theta_j)\,(g(t_j)-g(s_j)) - \int_{s_j}^{t_j} f\,\mathrm{d}g < 0\}.$$

According to Lemma 6.5.1, we have

$$0 \le \sum_{j\in J^+} \Big(f(\theta_j)\,(g(t_j)-g(s_j)) - \int_{s_j}^{t_j} f\,\mathrm{d}g \Big) \le \varepsilon,$$

$$0 \le \sum_{j\in J^-} \Big(\int_{s_j}^{t_j} f\,\mathrm{d}g - f(\theta_j)\,(g(t_j)-g(s_j)) \Big) \le \varepsilon.$$

Adding these inequalities gives (6.5.4). □

The Saks-Henstock lemma enables us to obtain some basic properties of the indefinite Kurzweil-Stieltjes integral. The next results show that the indefinite integral need not be continuous, but it is regulated if the integrator is regulated.

6.5.3 Theorem. *Let $\int_a^b f\,\mathrm{d}g$ exist and let $c\in[a,b]$. Then*

$$\lim_{\substack{x\to c\\ x\in[a,b]}} \Big(\int_a^x f\,\mathrm{d}g + f(c)\,(g(c)-g(x)) \Big) = \int_a^c f\,\mathrm{d}g,$$

$$\lim_{\substack{x\to c\\ x\in[a,b]}} \Big(\int_x^b f\,\mathrm{d}g + f(c)\,(g(x)-g(c)) \Big) = \int_c^b f\,\mathrm{d}g.$$

Proof. We prove only the first statement, and leave the second to the reader. Let $\varepsilon>0$ be given and let δ_ε be a gauge on $[a,b]$ such that

$$\Big| S(P) - \int_a^b f\,\mathrm{d}g \Big| < \varepsilon \quad \text{for all } \delta_\varepsilon\text{-fine partitions } P \text{ of } [a,b].$$

For each $x\in(c, c+\delta_\varepsilon(c))\cap[a,b]$, the system $\{([c,x],c)\}$ satisfies the assumptions of Lemma 6.5.1, and we get

$$\Big| f(c)\,(g(x)-g(c)) - \int_c^x f\,\mathrm{d}g \Big| \le \varepsilon. \tag{6.5.5}$$

Similarly, if $x\in(c-\delta_\varepsilon(c), c)\cap[a,b]$, then, applying Lemma 6.5.1 to the system $\{([x,c],c)\}$, we get

$$\Big| f(c)\,(g(c)-g(x)) - \int_x^c f\,\mathrm{d}g \Big| \le \varepsilon.$$

Thus, inequality (6.5.5) holds for each $x \in (c-\delta_\varepsilon(c), c+\delta_\varepsilon(c)) \cap [a,b]$. Hence

$$\Big|\int_a^c f\,\mathrm{d}g - \int_a^x f\,\mathrm{d}g - f(c)\,(g(c)-g(x))\Big|$$
$$= \Big|\int_c^x f\,\mathrm{d}g - f(c)\,(g(x)-g(c))\Big| \le \varepsilon,$$

and the proof is complete. □

6.5.4 Corollary. *The following statements hold:*

(i) *If* $c \in (a,b]$, $\int_a^c f\,\mathrm{d}g$ *exists and* $\lim_{x\to c-} g(x)$ *exists as well, then*

$$\int_a^c f\,\mathrm{d}g = \lim_{x\to c-}\Big(\int_a^x f\,\mathrm{d}g\Big) + f(c)\,\Delta^- g(c).$$

(ii) *If* $c \in [a,b)$, $\int_c^b f\,\mathrm{d}g$ *exists and* $\lim_{x\to c+} g(x)$ *exists as well, then*

$$\int_c^b f\,\mathrm{d}g = \lim_{x\to c+}\Big(\int_x^b f\,\mathrm{d}g\Big) + f(c)\,\Delta^+ g(c).$$

6.5.5 Corollary. *Suppose that* $\int_a^b f\,\mathrm{d}g$ *exists,* g *is regulated on* $[a,b]$*, and let*

$$h(x) = \int_a^x f\,\mathrm{d}g \quad \textit{for} \quad x \in [a,b].$$

Then the following statements hold:

(i) h *is regulated and satisfies*

$$h(t+) = h(t) + f(t)\,\Delta^+ g(t) \quad \textit{for } t \in [a,b),$$
$$h(t-) = h(t) - f(t)\,\Delta^- g(t) \quad \textit{for } t \in (a,b].$$

(ii) *If* f *is bounded and* g *has bounded variation, then*

$$\operatorname{var}_a^b h \le \|f\|\operatorname{var}_a^b g < \infty.$$

Proof. The first assertion follows immediately from Theorem 6.5.3. To prove the second assertion, consider an arbitrary division $\boldsymbol{\alpha} = \{\alpha_0, \alpha_1, \ldots, \alpha_m\}$ of $[a,b]$. By Theorems 6.3.6 and 2.1.14 we have

$$V(h,\boldsymbol{\alpha}) = \sum_{j=1}^m |h(\alpha_j) - h(\alpha_{j-1})| = \sum_{j=1}^m \Big|\int_{\alpha_{j-1}}^{\alpha_j} f\,\mathrm{d}g\Big|$$

$$\leq \|f\| \sum_{j=1}^{m} \operatorname{var}_{\alpha_{j-1}}^{\alpha_j} g = \|f\| \operatorname{var}_a^b g < \infty,$$

and therefore the second statement holds. □

We know that if $\int_a^b f \, dg$ exists, then the integrals $\int_a^x f \, dg$ and $\int_x^b f \, dg$ exist for all $x \in [a, b]$. The next theorem provides sufficient conditions for the validity of the converse statement. (By Theorem 6.5.3, these conditions are also necessary.) Results of this type are known as Hake's theorems.

6.5.6 Theorem (HAKE). *The following statements hold:*

(i) *If $\int_a^x f \, dg$ exists for each $x \in [a, b)$ and*

$$\lim_{x \to b-} \left(\int_a^x f \, dg + f(b) \, (g(b) - g(x)) \right) = I \in \mathbb{R},$$

then $\int_a^b f \, dg = I$.

(ii) *If $\int_x^b f \, dg$ exists for each $x \in (a, b]$ and*

$$\lim_{x \to a+} \left(\int_x^b f \, dg + f(a) \, (g(x) - g(a)) \right) = I \in \mathbb{R},$$

then $\int_a^b f \, dg = I$.

Proof. We prove the first statement, and leave the second as an exercise to the reader. Let $\varepsilon > 0$ be given. Choose a $\Delta > 0$ in such a way that

$$\left| \int_a^x f \, dg + f(b) \, (g(b) - g(x)) - I \right| < \varepsilon \quad \text{for each } x \in [b - \Delta, b). \tag{6.5.6}$$

Let $\{x_k\}$ be an increasing sequence from (a, b) with $\lim_{k \to \infty} x_k = b$, and denote $x_0 = a$. For each $s \in [a, b)$, let $\kappa(s)$ be the unique number $k \in \mathbb{N}$ such that $s \in [x_{k-1}, x_k)$.

For each $k \in \mathbb{N}$, there is a gauge δ_k on $[a, x_k]$ such that

$$\left| S(P) - \int_a^{x_k} f \, dg \right| < \frac{\varepsilon}{2^k} \quad \text{for all } \delta_k\text{-fine partitions } P \text{ of } [a, x_k]. \tag{6.5.7}$$

Let δ_0 be a gauge on $[a, b)$ such that

$$\delta_0(s) \leq \delta_k(s) \quad \text{and} \quad [s - \delta_0(s), s + \delta_0(s)] \subset [a, x_k]$$

for every $k \in \mathbb{N}$ and $s \in [x_{k-1}, x_k)$.

Assume that $x \in [a,b)$ and $T=(\boldsymbol{\tau},\boldsymbol{\theta})$ is a δ_0-fine partition of $[a,x]$. We will prove that

$$\Big|S(T)-\int_a^x f\,\mathrm{d}g\Big| < \varepsilon. \tag{6.5.8}$$

To see this, let $p=\kappa(x)$, i.e., $x\in[x_{p-1},x_p)$. For every $k\in\{1,\dots,p\}$ and every $j\in\{1,\dots,\nu(T)\}$ such that $\kappa(\theta_j)=k$, we have

$$\theta_j-\delta_k(\theta_j)\le\theta_j-\delta_0(\theta_j)\le\tau_{j-1}<\tau_j\le\theta_j+\delta_0(\theta_j)\le\theta_j+\delta_k(\theta_j).$$

Taking into account (6.5.7) and applying Lemma 6.5.1 (Saks-Henstock lemma) to the system $\Big\{([\tau_{j-1},\tau_j],\theta_j): j=1,\dots,\nu(T),\ \kappa(\theta_j)=k\Big\}$, we get

$$\Big|\sum_{j:\kappa(\theta_j)=k} f(\theta_j)\,(g(\tau_j)-g(\tau_{j-1}))-\int_{\tau_{j-1}}^{\tau_j} f\,\mathrm{d}g\Big|\le\frac{\varepsilon}{2^k}\quad\text{for each } k\in\{1,\dots,p\}.$$

Consequently,

$$\begin{aligned}
&\Big|S(T)-\int_a^x f\,\mathrm{d}g\Big|\\
&\quad=\Big|\sum_{k=1}^{p}\sum_{j:\kappa(\theta_j)=k}\Big(f(\theta_j)\,(g(\tau_j)-g(\tau_{j-1}))-\int_{\tau_{j-1}}^{\tau_j} f\,\mathrm{d}g\Big)\Big|\\
&\quad\le\sum_{k=1}^{p}\Big|\sum_{j:\kappa(\theta_j)=k}\Big(f(\theta_j)\,(g(\tau_j)-g(\tau_{j-1}))-\int_{\tau_{j-1}}^{\tau_j} f\,\mathrm{d}g\Big)\Big|<\sum_{k=1}^{\infty}\frac{\varepsilon}{2^k}=\varepsilon,
\end{aligned}$$

which proves (6.5.8).

Consider the gauge δ^* on $[a,b]$ given by $\delta^*(x)=\min\{b-x,\ \delta_0(x)\}$ for $x\in[a,b)$, and $\delta^*(b)=\Delta$. Let $P=(\boldsymbol{\alpha},\boldsymbol{\xi})$ be an arbitrary δ^*-fine partition of $[a,b]$, and denote $m=\nu(P)$. Then $\xi_m=\alpha_m=b$, $\alpha_{m-1}\in(b-\Delta,b)$ and

$$\begin{aligned}
|S(P)-I|&=\Big|\sum_{j=1}^{m-1} f(\xi_j)\,(g(\alpha_j)-g(\alpha_{j-1}))+f(b)\,(g(b)-g(\alpha_{m-1}))-I\Big|\\
&\le\Big|\sum_{j=1}^{m-1} f(\xi_j)\,(g(\alpha_j)-g(\alpha_{j-1}))-\int_a^{\alpha_{m-1}} f\,\mathrm{d}g\Big|\\
&\quad+\Big|\int_a^{\alpha_{m-1}} f\,\mathrm{d}g+f(b)\,(g(b)-g(\alpha_{m-1}))-I\Big|.
\end{aligned}$$

Finally, using (6.5.8) and (6.5.6) with $x=\alpha_{m-1}$, we obtain

$$|S(P)-I|<2\,\varepsilon,$$

i.e., $\int_a^b f\,\mathrm{d}g$ exists and equals I. □

6.5.7 Exercise. Prove part (ii) of Theorem 6.5.6.

6.5.8 Example. Using Theorem 6.5.6 (Hake's theorem), we can easily obtain all formulas from Lemmas 6.3.2 and 6.3.3. For instance, the formula

$$\int_a^b f\,\mathrm{d}\chi_{[\tau,b]} = f(\tau),$$

where $\tau \in (a,b]$ and f is arbitrary, can be derived in the following way:

$$\begin{aligned}\int_a^b f\,\mathrm{d}\chi_{[\tau,b]} &= \int_a^\tau f\,\mathrm{d}\chi_{[\tau,b]}\\ &= \lim_{t\to\tau-}\Big(\int_a^t f\,\mathrm{d}\chi_{[\tau,b]} + f(\tau)(\chi_{[\tau,b]}(\tau)-\chi_{[\tau,b]}(t))\Big) = f(\tau).\end{aligned}$$

Similarly, for $\tau \in [a,b)$ and $g \in \mathrm{G}([a,b])$, we get

$$\begin{aligned}\int_a^b \chi_{[a,\tau]}\,\mathrm{d}g &= \int_a^\tau 1\,\mathrm{d}g + \int_\tau^b \chi_{[a,\tau]}\,\mathrm{d}g\\ &= g(\tau)-g(a) + \lim_{t\to\tau+}\Big(\int_t^b \chi_{[a,\tau]}\,\mathrm{d}g + g(t)-g(\tau)\Big)\\ &= g(\tau+)-g(a).\end{aligned}$$

6.5.9 Exercise. Use Theorem 6.5.6 (Hake's theorem) to prove all remaining formulas from Lemmas 6.3.2 and 6.3.3.

6.6 SUBSTITUTION

The present section is devoted to substitution theorems for the Kurzweil-Stieltjes integral. Although their statements are similar to the results obtained in Section 5.4, their applicability is much wider.

6.6.1 Theorem (SUBSTITUTION THEOREM). *Let $f,\ g,\ h\colon [a,b]\to\mathbb{R}$ be such that f is bounded and $\int_a^b g\,\mathrm{d}h$ exists. If one of the integrals*

$$\int_a^b f(x)\,\mathrm{d}\Big(\int_a^x g\,\mathrm{d}h\Big) \quad \textit{and} \quad \int_a^b fg\,\mathrm{d}h$$

exists, then the other one exists as well, and

$$\int_a^b f(x)\,\mathrm{d}\Big(\int_a^x g\,\mathrm{d}h\Big) = \int_a^b fg\,\mathrm{d}h.$$

Proof. First, notice that by Theorem 6.2.9, the function $w(x) = \int_a^x g\,\mathrm{d}h$ is defined for every $x \in [a,b]$.

Assume the integral $\int_a^b fg\,\mathrm{d}h$ exists. Let $\varepsilon > 0$ be given and let δ_ε be a gauge on $[a,b]$ such that the inequalities

$$\sum_{j=1}^{\nu(P)} \Big| f(\xi_j)\,g(\xi_j)\,(h(\alpha_j) - h(\alpha_{j-1})) - \int_{\alpha_{j-1}}^{\alpha_j} fg\,\mathrm{d}h \Big| < \varepsilon$$

and

$$\sum_{j=1}^{\nu(P)} \Big| g(\xi_j)\,(h(\alpha_j) - h(\alpha_{j-1})) - \int_{\alpha_{j-1}}^{\alpha_j} g\,\mathrm{d}h \Big| < \varepsilon$$

hold for all δ_ε-fine partitions $P = (\boldsymbol{\alpha}, \boldsymbol{\xi})$ of $[a,b]$. (Such a gauge exists by Corollary 6.5.2.)

Let a δ_ε-fine partition $P = (\boldsymbol{\alpha}, \boldsymbol{\xi})$ be given. Then

$$\begin{aligned}
&\Big| \sum_{j=1}^{\nu(P)} f(\xi_j)\,(w(\alpha_j) - w(\alpha_{j-1})) - \int_a^b fg\,\mathrm{d}h \Big| \\
&\quad \le \sum_{j=1}^{\nu(P)} \Big| f(\xi_j) \int_{\alpha_{j-1}}^{\alpha_j} g\,\mathrm{d}h - f(\xi_j)\,g(\xi_j)\,(h(\alpha_j) - g(\alpha_{j-1})) \Big| \\
&\qquad\quad + \sum_{j=1}^{\nu(P)} \Big| f(\xi_j)\,g(\xi_j)\,(h(\alpha_j) - h(\alpha_{j-1})) - \int_{\alpha_{j-1}}^{\alpha_j} fg\,\mathrm{d}h \Big| \\
&\quad \le \|f\| \sum_{j=1}^{\nu(P)} \Big| \int_{\alpha_{j-1}}^{\alpha_j} g\,\mathrm{d}h - g(\xi_j)\,(h(\alpha_j) - h(\alpha_{j-1})) \Big| \\
&\qquad\quad + \sum_{j=1}^{\nu(P)} \Big| f(\xi_j)\,g(\xi_j)\,(h(\alpha_j) - h(\alpha_{j-1})) - \int_{\alpha_{j-1}}^{\alpha_j} fg\,\mathrm{d}h \Big| \\
&\quad \le (\|f\| + 1)\,\varepsilon.
\end{aligned}$$

Hence, the integral $\int_a^b f\,\mathrm{d}w$ exists and equals $\int_a^b fg\,\mathrm{d}h$.

The remaining part of the theorem can be proved similarly, again with the help of Corollary 6.5.2. □

Let us mention two simple but useful corollaries of the substitution theorem.

6.6.2 Corollary. *Assume that* $f:[a,b]\to\mathbb{R}$ *is bounded and* $g:[a,b]\to\mathbb{R}$ *is differentiable. If one of the integrals*

$$\int_a^b f(x)\,\mathrm{d}g(x) \qquad \textit{and} \qquad \int_a^b f(x)\,g'(x)\,\mathrm{d}x$$

exists, then the other one exists as well, and

$$\int_a^b f(x)\,\mathrm{d}g(x) = \int_a^b f(x)\,g'(x)\,\mathrm{d}x.$$

6.6.3 Exercise. Prove Corollary 6.6.2.

Hint: Recall that any differentiable function $g:[a,b]\to\mathbb{R}$ is Kurzweil-Henstock integrable and $g(t)-g(a)=\int_a^t g'(x)\,\mathrm{d}x$ for all $t\in[a,b]$.

In general, the product of two Kurzweil-Henstock integrable functions need not be integrable. The next exercise provides a sufficient condition for the integrability of a product.

6.6.4 Exercise. Suppose that $f:[a,b]\to\mathbb{R}$ is Kurzweil-Henstock integrable and $g:[a,b]\to\mathbb{R}$ has bounded variation. Prove that fg is Kurzweil-Henstock integrable.

Hint: Consider the integral $\int_a^b g\,\mathrm{d}F$, where $F(t)=\int_a^t f(x)\,\mathrm{d}x$ for all $t\in[a,b]$.

The next substitution theorem is analogue of Theorem 5.4.5.

6.6.5 Theorem. *Assume that* $\phi:[c,d]\to\mathbb{R}$ *is a continuous increasing function with* $\phi(c)=a$ *and* $\phi(d)=b$. *If* $f:[a,b]\to\mathbb{R}$ *is an arbitrary function and one of the integrals*

$$\int_a^b f(x)\,\mathrm{d}g(x) \qquad \textit{and} \qquad \int_c^d f(\phi(x))\,\mathrm{d}g(\phi(x))$$

exists, then the other exists as well, and

$$\int_c^d f(\phi(x))\,\mathrm{d}g(\phi(x)) = \int_a^b f(x)\,\mathrm{d}g(x). \tag{6.6.1}$$

Proof. The assumptions imply that ϕ maps $[c,d]$ onto $[a,b]$, the inverse function ϕ^{-1} exists, it is continuous and increasing.

For a given partition $Q=(\boldsymbol{\beta},\boldsymbol{\eta})$ of $[c,d]$, let

$$\alpha_j=\phi(\beta_j) \quad \text{for } j\in\{0,\dots,\nu(Q)\},$$
$$\xi_j=\phi(\eta_j) \quad \text{for } j\in\{1,\dots,\nu(Q)\},$$

and

$$\boldsymbol{\alpha}=\{\alpha_0,\alpha_1,\dots,\alpha_{\nu(Q)}\},\ \boldsymbol{\xi}=(\xi_1,\dots,\xi_{\nu(Q)}).$$

Then $P=(\boldsymbol{\alpha},\boldsymbol{\xi})$ is partition of $[a,b]$, and we write $P=\phi(Q)$. Conversely, each partition P of $[a,b]$ has the form $P=\phi(Q)$ for a certain partition Q of $[c,d]$; we write $Q=\phi^{-1}(P)$.

Given a gauge $\widetilde{\delta}:[c,d]\to(0,\infty)$, we can find a gauge $\delta:[a,b]\to(0,\infty)$ such that

$$\phi^{-1}(\tau+\delta(\tau))<\phi^{-1}(\tau)+\widetilde{\delta}(\phi^{-1}(\tau)) \quad \text{for all } \tau\in[a,b),$$
$$\phi^{-1}(\tau-\delta(\tau))>\phi^{-1}(\tau)-\widetilde{\delta}(\phi^{-1}(\tau)) \quad \text{for all } \tau\in(a,b].$$

These conditions imply that if $P=(\boldsymbol{\alpha},\boldsymbol{\xi})$ is a δ-fine partition of $[a,b]$, then

$$\beta_j=\phi^{-1}(\alpha_j)\le\phi^{-1}(\xi_j+\delta(\xi_j))<\phi^{-1}(\xi_j)+\widetilde{\delta}(\phi^{-1}(\xi_j))=\eta_j+\widetilde{\delta}(\eta_j),$$
$$\beta_{j-1}=\phi^{-1}(\alpha_{j-1})\ge\phi^{-1}(\xi_j-\delta(\xi_j))>\phi^{-1}(\xi_j)-\widetilde{\delta}(\phi^{-1}(\xi_j))=\eta_j-\widetilde{\delta}(\eta_j)$$

for all $j\in\{1,\dots,\nu(P)\}$, i.e., $Q=\phi^{-1}(P)$ is a $\widetilde{\delta}$-fine partition of $[c,d]$. Similarly, given a gauge δ on $[a,b]$, we can find a gauge $\widetilde{\delta}$ on $[c,d]$ such that $P=\phi(Q)$ is δ-fine whenever Q is $\widetilde{\delta}$-fine.

The proof is finished by observing that the relation

$$\sum_{j=1}^{\nu(P)} f(\xi_j)\,(g(\alpha_j)-g(\alpha_{j-1}))=\sum_{j=1}^{\nu(Q)} f(\phi(\eta_j))\,(g(\phi(\beta_j))-g(\phi(\beta_{j-1})))$$

holds for every partition $P=(\boldsymbol{\alpha},\boldsymbol{\xi})$ of $[a,b]$ and $Q=\phi^{-1}(P)$. □

6.6.6 Exercise. (i) Formulate and prove an analogue of Theorem 6.6.5 for the case when ϕ is decreasing.

(ii) Formulate and prove an analogue of Theorem 5.4.6 for the Kurzweil-Stieltjes integral.

6.7 ABSOLUTE INTEGRABILITY

The Kurzweil-Stieltjes integral is a nonabsolutely convergent integral – the existence of $\int_a^b f\,\mathrm{d}g$ does not necessarily imply the existence of $\int_a^b |f|\,\mathrm{d}g$. In this section, we collect some sufficient and necessary conditions for the existence of the latter integral.

We restrict our attention to the case when g is nondecreasing. In this situation, if both $\int_a^b f\,\mathrm{d}g$ and $\int_a^b |f|\,\mathrm{d}g$ exist, we have the inequality

$$\Big|\int_a^b f\,\mathrm{d}g\Big| \le \int_a^b |f|\,\mathrm{d}g.$$

This fact follows immediately from the definition of the integral, since

$$|S(f,\mathrm{d}g,P)| \le S(|f|,\mathrm{d}g,P) \quad \text{for each partition } P \text{ of } [a,b].$$

6.7.1 Theorem. *Assume that $g\colon[a,b]\to\mathbb{R}$ is nondecreasing and $\int_a^b f\,\mathrm{d}g$ exists. Let*

$$F(x) = \int_a^x f\,\mathrm{d}g \quad \textit{for } x\in[a,b].$$

Then $\int_a^b |f|\,\mathrm{d}g$ exists if and only if F has bounded variation on $[a,b]$. In this case, we have

$$\int_a^b |f|\,\mathrm{d}g = \operatorname{var}_a^b F.$$

P r o o f. Suppose that $\int_a^b |f|\,\mathrm{d}g$ exists. If $\boldsymbol{\alpha}$ is an arbitrary division of $[a,b]$, then

$$\begin{aligned} V(F,\boldsymbol{\alpha}) &= \sum_{j=1}^{\nu(\boldsymbol{\alpha})} |F(\alpha_j)-F(\alpha_{j-1})| = \sum_{j=1}^{\nu(\boldsymbol{\alpha})} \Big|\int_{\alpha_{j-1}}^{\alpha_j} f\,\mathrm{d}g\Big| \\ &\le \sum_{j=1}^{\nu(\boldsymbol{\alpha})} \int_{\alpha_{j-1}}^{\alpha_j} |f|\,\mathrm{d}g = \int_a^b |f|\,\mathrm{d}g, \end{aligned}$$

and therefore $\operatorname{var}_a^b F$ is finite.

Conversely, suppose that F has bounded variation on $[a,b]$. Consider an arbitrary $\varepsilon>0$ and let $\boldsymbol{\alpha}$ be a division of $[a,b]$ satisfying

$$\operatorname{var}_a^b F - \varepsilon < V(F,\boldsymbol{\alpha}) \le \operatorname{var}_a^b F.$$

Observe that if $\boldsymbol{\beta}$ is a refinement of $\boldsymbol{\alpha}$, then

$$\operatorname{var}_a^b F - \varepsilon < V(F,\boldsymbol{\alpha}) \le V(F,\boldsymbol{\beta}) \le \operatorname{var}_a^b F. \tag{6.7.1}$$

Let δ be a gauge on $[a,b]$ with the following properties:

- If P is a δ-fine partition of $[a,b]$, then $\left|\int_a^b f\,\mathrm{d}g - S(f,\mathrm{d}g,P)\right| < \varepsilon$.
- If $P=(\boldsymbol{\beta},\boldsymbol{\xi})$ is a δ-fine partition of $[a,b]$, then $\boldsymbol{\alpha}\subset\{\xi_1,\dots,\xi_{\nu(P)}\}$.

The second assumption is justified by the first part of Lemma 6.2.11.

Now, if $P=(\boldsymbol{\beta},\boldsymbol{\xi})$ is a δ-fine partition of $[a,b]$, the second part of Lemma 6.2.11 guarantees the existence of a δ-fine partition $\widetilde{P}=(\widetilde{\boldsymbol{\beta}},\widetilde{\boldsymbol{\xi}})$ such that

$$S(|f|,\mathrm{d}g,\widetilde{P}) = S(|f|,\mathrm{d}g,P) \quad\text{and}\quad \boldsymbol{\alpha}\subset\widetilde{\boldsymbol{\beta}}.$$

As a result, using the fact that g is nondecreasing together with the reverse triangle inequality $||x|-|y|| \le |x-y|$ and Corollary 6.5.2, we obtain

$$\begin{aligned}\left|S(|f|,\mathrm{d}g,P) - V(F,\widetilde{\boldsymbol{\beta}})\right| &= \left|S(|f|,\mathrm{d}g,\widetilde{P}) - V(F,\widetilde{\boldsymbol{\beta}})\right| \\ &= \left|\sum_{i=1}^{\nu(\widetilde{\boldsymbol{\beta}})}\left(|f(\widetilde{\xi}_i)\,(g(\widetilde{\beta}_i)-g(\widetilde{\beta}_{i-1}))| - \Big|\int_{\widetilde{\beta}_{i-1}}^{\widetilde{\beta}_i} f\,\mathrm{d}g\Big|\right)\right| \\ &\le \sum_{i=1}^{\nu(\widetilde{\boldsymbol{\beta}})}\left|f(\widetilde{\xi}_i)\,(g(\widetilde{\beta}_i)-g(\widetilde{\beta}_{i-1})) - \int_{\widetilde{\beta}_{i-1}}^{\widetilde{\beta}_i} f\,\mathrm{d}g\right| \le 2\,\varepsilon.\end{aligned}$$

By combining this estimate with (6.7.1), we conclude that

$$\begin{aligned}&\left|S(|f|,\mathrm{d}g,P) - \operatorname{var}_a^b F\right| \\ &\qquad \le \left|S(|f|,\mathrm{d}g,\widetilde{P}) - V(F,\widetilde{\boldsymbol{\beta}})\right| + \left|V(F,\widetilde{\boldsymbol{\beta}}) - \operatorname{var}_a^b F\right| \le 3\,\varepsilon,\end{aligned}$$

which means that $\int_a^b |f|\,\mathrm{d}g$ exists and equals $\operatorname{var}_a^b F$. □

6.7.2 Example. Let $\{t_n\}$ be an increasing sequence of points from the open interval $(0,1)$ tending to $\tau\in(0,1]$. Consider the function $g\colon[0,1]\to\mathbb{R}$ given by

$$g(t) = \sum_{n=1}^{\infty} \chi_{[t_n,1]}(t)\,\frac{1}{n^2}. \tag{6.7.2}$$

It is not hard to see that g is nondecreasing and left-continuous at τ. Let

$f:[0,1]\to\mathbb{R}$ be defined as

$$f(t)=\begin{cases}(-1)^{n+1}\,n & \text{if } \ t=t_n \ \text{ for some } \ n\in\mathbb{N},\\ 0 & \text{otherwise.}\end{cases}$$

Claim 1. The integral $\int_0^1 f\,\mathrm{d}g$ exists.

Note that, for each $t\in[0,\tau)$, the integral $\int_0^t f\,\mathrm{d}g$ exists (due to Corollary 6.3.4 and using the fact that the restriction of g to the interval $[0,t]$ defines a finite step function). Moreover, by Lemma 6.3.2, we have

$$\int_0^{t_1} f\,\mathrm{d}g=f(t_1)\,g(t_1)=1,$$

and

$$\begin{aligned}\int_{t_n}^{t_{n+1}} f\,\mathrm{d}g&=\int_{t_n}^{t_{n+1}} f\,\mathrm{d}\big(g(t_n)\chi_{[t_n,t_{n+1})}+g(t_{n+1})\chi_{[t_{n+1}]}\big)\\&=f(t_{n+1})\,(g(t_{n+1})-g(t_n))=\frac{(-1)^{n+2}}{n+1},\end{aligned}$$

for each $n\in\mathbb{N}$. Since

$$\lim_{t\to\tau-}\int_0^t f\,\mathrm{d}g+f(\tau)\,\Delta^-g(\tau)=\lim_{n\to\infty}\int_0^{t_n} f\,\mathrm{d}g=\sum_{n=1}^{\infty}\frac{(-1)^{n+1}}{n}=\ln 2,$$

applying Theorem 6.5.6 (Hake's theorem), we conclude that the integral $\int_0^\tau f\,\mathrm{d}g$ exists and equals $\ln 2$. If $\tau=1$, the claim is proved. If $\tau<1$, then $\int_\tau^1 f\,\mathrm{d}g=0$ (because g is constant on $[\tau,1]$), wherefrom the proof of the claim follows.

Claim 2. $|f|$ is not integrable with respect to g on $[0,1]$.

Let $F(t)=\int_0^t f\,\mathrm{d}g$ for $t\in[0,1]$. We will show that $\operatorname{var}_0^1 F=\infty$. For every $n\in\mathbb{N}$ consider the division $D_n=\{0,t_1,t_2,\dots,t_n,1\}$ of $[0,1]$. We have

$$\begin{aligned}V(F,D_n)&=\sum_{k=1}^{n-1}\Big|\int_{t_k}^{t_{k+1}} f\,\mathrm{d}g\Big|+\Big|\int_0^{t_1} f\,\mathrm{d}g\Big|+\Big|\int_{t_n}^1 f\,\mathrm{d}g\Big|\\&=\sum_{k=1}^{n}\frac{1}{k}+\Big|\int_{t_n}^1 f\,\mathrm{d}g\Big|.\end{aligned}$$

Since $V(F,D_n)\le\operatorname{var}_0^1 F$ for every $n\in\mathbb{N}$, $\operatorname{var}_0^1 F$ cannot be finite. Thus the claim is a consequence of Theorem 6.7.1.

6.7.3 Exercise. Let $\{t_n\}$ be an increasing sequence of points from the open interval (a,b) tending to $\tau \in (a,b]$. Let $\{c_n\}$ be a sequence of nonnegative numbers such that the series $\sum c_n$ converges. Define a function $g:[a,b]\to\mathbb{R}$ by

$$g(t)=\sum_{n:t_n\le t} c_n.$$

For a given $f:[a,b]\to\mathbb{R}$ prove that:

(i) The integral $\int_a^b f\,\mathrm{d}g$ exists if and only if the series $\sum c_n\, f(t_n)$ converges and in such a case it is given by

$$\int_a^b f\,\mathrm{d}g=\sum_{n=1}^{\infty} c_n\, f(t_n).$$

(ii) The integral $\int_a^b |f|\,\mathrm{d}g$ exists if and only if the series $\sum c_n\, f(t_n)$ is absolutely convergent.

6.7.4 Theorem. *If $g:[a,b]\to\mathbb{R}$ is nondecreasing, $f,h:[a,b]\to\mathbb{R}$ are such that $\int_a^b f\,\mathrm{d}g$ and $\int_a^b h\,\mathrm{d}g$ exist and $|f(t)|\le h(t)$ for each $t\in[a,b]$, then $\int_a^b |f|\,\mathrm{d}g$ exists.*

Proof. If $[c,d]\subset[a,b]$ and P is a partition of $[c,d]$, then

$$|S(f,\mathrm{d}g,P)|\le S(h,\mathrm{d}g,P).$$

Thus, it follows from the definition of the integral that

$$\Big|\int_c^d f\,\mathrm{d}g\Big|\le\int_c^d h\,\mathrm{d}g \quad \text{for each } [c,d]\subset[a,b].$$

Let

$$F(x)=\int_a^x f\,\mathrm{d}g \quad \text{for } x\in[a,b].$$

If $\boldsymbol{\alpha}$ is an arbitrary division of $[a,b]$, we obtain the estimate

$$\begin{aligned} V(F,\alpha)&=\sum_{i=1}^{\nu(\boldsymbol{\alpha})}|F(\alpha_i)-F(\alpha_{i-1})|=\sum_{i=1}^{\nu(\boldsymbol{\alpha})}\Big|\int_{\alpha_{i-1}}^{\alpha_i} f\,\mathrm{d}g\Big| \\ &\le\sum_{i=1}^{\nu(\boldsymbol{\alpha})}\int_{\alpha_{i-1}}^{\alpha_i} h\,\mathrm{d}g=\int_a^b h\,\mathrm{d}g, \end{aligned}$$

which shows that F has bounded variation on $[a,b]$. Thus, the existence of the integral $\int_a^b |f|\,\mathrm{d}g$ follows from Theorem 6.7.1. $\square$

6.7.5 Theorem. *If $g:[a,b]\to\mathbb{R}$ is nondecreasing and $f_1, f_2:[a,b]\to\mathbb{R}$ are such that the integrals $\int_a^b f_1\,\mathrm{d}g$, $\int_a^b f_2\,\mathrm{d}g$, $\int_a^b |f_1|\,\mathrm{d}g$ and $\int_a^b |f_2|\,\mathrm{d}g$ exist, then $\int_a^b \max(f_1,f_2)\,\mathrm{d}g$ and $\int_a^b \min(f_1,f_2)\,\mathrm{d}g$ exist as well.*

Proof. By Theorem 6.7.4, the integral $\int_a^b |f_1-f_2|\,\mathrm{d}g$ exists, because

$$|f_1-f_2|\le|f_1|+|f_2|$$

and the integrals $\int_a^b (f_1-f_2)\,\mathrm{d}g$, $\int_a^b |f_1|\,\mathrm{d}g$ and $\int_a^b |f_2|\,\mathrm{d}g$ exist. Since

$$\max(f_1,f_2)=\frac{f_1+f_2+|f_1-f_2|}{2} \quad\text{and}\quad \min(f_1,f_2)=\frac{f_1+f_2-|f_1-f_2|}{2},$$

the existence of $\int_a^b \max(f_1,f_2)\,\mathrm{d}g$ and $\int_a^b \min(f_1,f_2)\,\mathrm{d}g$ follows from the linearity of the integral (i.e., from Theorem 6.2.7). □

6.8 Convergence theorems

In this section, we present additional convergence theorems for the Kurzweil-Stieltjes integral, which are applicable in situations where the results from Section 6.3 (i.e., Theorems 6.3.8 and 6.3.9) do not suffice.

We begin by introducing the concept of uniform integrability (also known as equiintegrability).

6.8.1 Definition. Consider sequences of functions $f_n:[a,b]\to\mathbb{R}$, $n\in\mathbb{N}$, and $g_n:[a,b]\to\mathbb{R}$, $n\in\mathbb{N}$. Then $\{f_n\}$ is called uniformly integrable with respect to $\{g_n\}$, if the following two conditions are satisfied:

(i) The integral $\int_a^b f_n\,\mathrm{d}g_n$ exists for each $n\in\mathbb{N}$.

(ii) For every $\varepsilon>0$, there exists a gauge δ on $[a,b]$ such that the inequality

$$\Big|\int_a^b f_n\,\mathrm{d}g_n - S(f_n,\mathrm{d}g_n,P)\Big|<\varepsilon$$

holds for each δ-fine partition P of $[a,b]$ and for every $n\in\mathbb{N}$.

In the special case when $f_n=f$ for all $n\in\mathbb{N}$, we say that f is uniformly integrable with respect to $\{g_n\}$. Symmetrically, in the case when $g_n=g$ for all $n\in\mathbb{N}$, we say that $\{f_n\}$ is uniformly integrable with respect to g.

The next result is a basic convergence theorem for the Kurzweil-Stieltjes integral. Although the assumption of uniform integrability might be difficult to verify, the result will play a key role in deriving other convergence theorems later in this section.

6.8.2 Theorem. *If $\{f_n\}$ is uniformly integrable with respect to $\{g_n\}$,*

$$\lim_{n\to\infty} f_n(t) = f(t) \quad \text{and} \quad \lim_{n\to\infty} g_n(t) = g(t) \quad \text{for all } t \in [a,b],$$

then both the integral $\int_a^b f\,\mathrm{d}g$ and the limit $\lim\limits_{n\to\infty} \int_a^b f_n\,\mathrm{d}g_n$ exist, and we have the equality

$$\int_a^b f\,\mathrm{d}g = \lim_{n\to\infty} \int_a^b f_n\,\mathrm{d}g_n.$$

Moreover, if $\{g_n\}$ is uniformly convergent to g and g is bounded, then

$$\lim_{n\to\infty} \left(\sup_{t\in[a,b]} \left| \int_a^t f_n\,\mathrm{d}g_n - \int_a^t f\,\mathrm{d}g \right| \right) = 0. \tag{6.8.1}$$

Proof. Consider an arbitrary $\varepsilon > 0$ and let δ be the corresponding gauge from Definition 6.8.1. Choose an arbitrary δ-fine partition P of $[a,b]$. Since

$$\lim_{n\to\infty} S(f_n, \mathrm{d}g_n, P) = S(f, \mathrm{d}g, P),$$

there exists an $n_0 \in \mathbb{N}$ such that

$$|S(f_m, \mathrm{d}g_m, P) - S(f_n, \mathrm{d}g_n, P)| < \varepsilon$$

holds for all $m, n \ge n_0$. Using this fact together with the definition of uniform integrability, we get

$$\begin{aligned} \left| \int_a^b f_m\,\mathrm{d}g_m - \int_a^b f_n\,\mathrm{d}g_n \right| &\le \left| \int_a^b f_m\,\mathrm{d}g_m - S(f_m, \mathrm{d}g_m, P) \right| \\ &+ |S(f_m, \mathrm{d}g_m, P) - S(f_n, \mathrm{d}g_n, P)| + \left| S(f_n, \mathrm{d}g_n, P) - \int_a^b f_n\,\mathrm{d}g_n \right| < 3\,\varepsilon \end{aligned}$$

for all $m, n \ge n_0$. In particular, $\{\int_a^b f_n\,\mathrm{d}g_n\}$ is a Cauchy sequence and thus it has a finite limit:

$$\lim_{n\to\infty} \int_a^b f_n\,\mathrm{d}g_n = L \in \mathbb{R}.$$

Now, let P be an arbitrary δ-fine partition of $[a,b]$. Choose an $n_1 \in \mathbb{N}$ such that

$$|S(f_{n_1}, \mathrm{d}g_{n_1}, P) - S(f, \mathrm{d}g, P)| < \varepsilon \quad \text{and} \quad \Big|\int_a^b f_{n_1} \, \mathrm{d}g_{n_1} - L\Big| < \varepsilon.$$

Then

$$\begin{aligned}|S(f, \mathrm{d}g, P) - L| &\le |S(f, \mathrm{d}g, P) - S(f_{n_1}, \mathrm{d}g_{n_1}, P)| \\ &+ \Big|S(f_{n_1}, \mathrm{d}g_{n_1}, P) - \int_a^b f_{n_1} \, \mathrm{d}g_{n_1}\Big| + \Big|\int_a^b f_{n_1} \, \mathrm{d}g_{n_1} - L\Big| < 3\varepsilon.\end{aligned}$$

It follows that $\int_a^b f \, \mathrm{d}g$ exists and equals L.

Assume now that $\{g_n\}$ is uniformly convergent to a bounded function g and consider an $\varepsilon > 0$, let δ be the corresponding gauge from Definition 6.8.1, and choose an arbitrary δ-fine partition $P = (\boldsymbol{\alpha}, \boldsymbol{\xi})$ of $[a,b]$.

Since $\{f_n(\xi_i)\}$ is convergent for each $i \in \{1, \dots, \nu(\boldsymbol{\alpha})\}$ and the sequence $\{g_n\}$ is uniformly bounded (because $\{g_n\}$ is convergent to g, which is bounded), there exists an $M \ge 0$ such that for all $n \in \mathbb{N}$, we have

$$|f_n(\xi_i)| \le M \quad \text{for } i \in \{1, \dots, \nu(\boldsymbol{\alpha})\} \quad \text{and} \quad |g_n(t)| \le M \quad \text{for } t \in [a,b].$$

Moreover (using also the uniform convergence of $\{g_n\}$), we can find an $n_0 \in \mathbb{N}$ such that for all $m, n \ge n_0$ and $i \in \{1, \dots, \nu(\boldsymbol{\alpha})\}$, we have

$$\begin{aligned}&|f_n(\xi_i) - f_m(\xi_i)| < \frac{\varepsilon}{M}, \\ &|g_n(t) - g_m(t)| < \frac{\varepsilon}{M} \quad \text{for } t \in [a,b], \\ &|f_n(\xi_i)\,(g_n(\alpha_i) - g_n(\alpha_{i-1})) - f_m(\xi_i)\,(g_m(\alpha_i) - g_m(\alpha_{i-1}))| < \frac{\varepsilon}{\nu(\boldsymbol{\alpha})}.\end{aligned}$$

Now, assume that $m, n \ge n_0$ and $t \in [a,b]$. Let $j \in \{1, \dots, \nu(\boldsymbol{\alpha})\}$ be such that $t \in [\alpha_{j-1}, \alpha_j]$. Then

$$\begin{aligned}&\Big|\int_a^t f_m \, \mathrm{d}g_m - \int_a^t f_n \, \mathrm{d}g_n\Big| \\ &\quad = \Big|\sum_{i=1}^{j-1} \int_{\alpha_{i-1}}^{\alpha_i} f_m \, \mathrm{d}g_m + \int_{\alpha_{j-1}}^t f_m \, \mathrm{d}g_m - \sum_{i=1}^{j-1} \int_{\alpha_{i-1}}^{\alpha_i} f_n \, \mathrm{d}g_n - \int_{\alpha_{j-1}}^t f_n \, \mathrm{d}g_n\Big| \\ &\quad \le \Big|\sum_{i=1}^{j-1} \Big(\int_{\alpha_{i-1}}^{\alpha_i} f_m \, \mathrm{d}g_m - f_m(\xi_i)\,(g_m(\alpha_i) - g_m(\alpha_{i-1}))\Big)\end{aligned}$$

$$+\int_{\alpha_{j-1}}^{t} f_m \, \mathrm{d}g_m - f_m(\xi_j)\,(g_m(t) - g_m(\alpha_{j-1}))\Big|$$
$$+\Big|\sum_{i=1}^{j-1}\Big(\int_{\alpha_{i-1}}^{\alpha_i} f_n \, \mathrm{d}g_n - f_n(\xi_i)\,(g_n(\alpha_i) - g_n(\alpha_{i-1}))\Big)$$
$$+\int_{\alpha_{j-1}}^{t} f_n \, \mathrm{d}g_n - f_n(\xi_j)\,(g_n(t) - g_n(\alpha_{j-1}))\Big|$$
$$+\sum_{i=1}^{j-1}|f_m(\xi_i)\,(g_m(\alpha_i) - g_m(\alpha_{i-1})) - f_n(\xi_i)\,(g_n(\alpha_i) - g_n(\alpha_{i-1}))|$$
$$+|f_m(\xi_j)\,(g_m(t) - g_m(\alpha_{j-1})) - f_n(\xi_j)\,(g_n(t) - g_n(\alpha_{j-1}))|.$$

We now estimate all the terms on the right-hand side. If $t \geq \xi_j$, we may apply Lemma 6.5.1 (Saks-Henstock lemma) to the system

$$\{([\alpha_{i-1}, \alpha_i], \xi_i) : i = 1, \dots, j-1\} \cup \{([\alpha_{j-1}, t], \xi_j)\}$$

and deduce that for all $\ell \in \mathbb{N}$, we have

$$\Big|\sum_{i=1}^{j-1}\Big(\int_{\alpha_{i-1}}^{\alpha_i} f_\ell \, \mathrm{d}g_\ell - f_\ell(\xi_i)\,(g_\ell(\alpha_i) - g_\ell(\alpha_{i-1}))\Big)$$
$$+\int_{\alpha_{j-1}}^{t} f_\ell \, \mathrm{d}g_\ell - f_\ell(\xi_j)\,(g_\ell(t) - g_\ell(\alpha_{j-1}))\Big| \leq \varepsilon.$$

On the other hand, if $t < \xi_j$, then applying Lemma 6.5.1 (Saks-Henstock lemma) to the systems

$$\{([\alpha_{i-1}, \alpha_i], \xi_i) : i = 1, \dots, j\} \quad \text{and} \quad \{([t, \alpha_j], \xi_j)\},$$

we get

$$\Big|\sum_{i=1}^{j-1}\Big(\int_{\alpha_{i-1}}^{\alpha_i} f_\ell \, \mathrm{d}g_\ell - f_\ell(\xi_i)\,(g_\ell(\alpha_i) - g_\ell(\alpha_{i-1}))\Big)$$
$$+\int_{\alpha_{j-1}}^{t} f_\ell \, \mathrm{d}g_\ell - f_\ell(\xi_j)\,(g_\ell(t) - g_\ell(\alpha_{j-1}))\Big|$$
$$= \Big|\sum_{i=1}^{j}\Big(\int_{\alpha_{i-1}}^{\alpha_i} f_\ell \, \mathrm{d}g_\ell - f_\ell(\xi_i)\,(g_\ell(\alpha_i) - g_\ell(\alpha_{i-1}))\Big)$$
$$-\Big(\int_{t}^{\alpha_j} f_\ell \, \mathrm{d}g_\ell - f_\ell(\xi_j)\,(g_\ell(\alpha_j) - g_\ell(t))\Big)\Big| \leq 2\,\varepsilon$$

for all $\ell \in \mathbb{N}$.

Our final two estimates are

$$\sum_{i=1}^{j-1} |f_m(\xi_i)\,(g_m(\alpha_i)-g_m(\alpha_{i-1})) - f_n(\xi_i)\,(g_n(\alpha_i)-g_n(\alpha_{i-1}))| \le \sum_{i=1}^{j-1} \frac{\varepsilon}{\nu(\boldsymbol{\alpha})} \le \varepsilon,$$

and

$$\begin{aligned} &|f_m(\xi_j)\,(g_m(t) - g_m(\alpha_{j-1})) - f_n(\xi_j)\,(g_n(t) - g_n(\alpha_{j-1}))| \\ &\quad \le |f_m(\xi_j) - f_n(\xi_j)|\,|g_m(t) - g_m(\alpha_{j-1})| \\ &\qquad + |f_n(\xi_j)|\,\big|g_m(t) - g_m(\alpha_{j-1}) - g_n(t) + g_n(\alpha_{j-1})\big| \\ &\quad < \frac{\varepsilon}{M}\,2\,M + M\,2\,\frac{\varepsilon}{M} = 4\varepsilon. \end{aligned}$$

To summarize, we have shown that

$$\Big|\int_a^t f_m\,\mathrm{d}g_m - \int_a^t f_n\,\mathrm{d}g_n\Big| \le 2\varepsilon + 2\varepsilon + \varepsilon + 4\varepsilon = 9\varepsilon$$

for all $n \ge n_0$ and $t \in [a,b]$. Thus, the sequence $\{\int_a^t f_n\,\mathrm{d}g_n\}$ is uniformly Cauchy and hence uniformly convergent on $[a,b]$. This completes the proof of (6.8.1). □

By letting $f_n \equiv 1$ for all $n \in \mathbb{N}$, it is easily seen that the uniform convergence of $\{g_n\}$ is needed to ensure the uniform convergence of the indefinite integrals in (6.8.1). The next example shows that the boundedness of the integrator g is also essential.

6.8.3 Example. Let $f_n(t) = 1/n$ for $t \in [a,b]$ and let $g : [a,b] \to \mathbb{R}$ be arbitrary with $g(a) = 0$. Then $\{f_n\}$ tends pointwise on $[a,b]$ to the zero function. Furthermore, as $S(f_n, \mathrm{d}g, P) = (1/n)\,g(b)$ for each $n \in \mathbb{N}$ and each partition P of $[a,b]$, we see that $\int_a^b f_n\,\mathrm{d}g = (1/n)\,g(b)$ for each $n \in \mathbb{N}$ and the sequence $\{f_n\}$ is uniformly integrable with respect to g. However, we claim that the indefinite integrals $F_n(t) = \int_a^t f_n\,\mathrm{d}g$ tend uniformly to the zero function if and only if g is bounded. Indeed, we have $F_n(t) = (1/n)\,g(t)$ for $t \in [a,b]$. Clearly, if g is bounded on $[a,b]$, then $\{F_n\}$ tends uniformly on $[a,b]$ to the zero function. On the other hand, if g is unbounded and an arbitrary $\varepsilon > 0$ is given, then for each $n \in \mathbb{N}$, there is a $t \in [a,b]$ such that

$$|F_n(t)| = |(1/n)\,g(t)| > \varepsilon,$$

i.e., F_n does not converge uniformly to the zero function.

The remaining results in this section provide more transparent conditions that imply uniform integrability.

6.8.4 Lemma. *If $g\colon[a,b]\to\mathbb{R}$ is a finite step function, then any sequence of real-valued functions $\{f_n\}$ is uniformly integrable with respect to g.*

6.8.5 Exercise. Prove Lemma 6.8.4.

Hint: See the proof of Lemma 6.3.2.

6.8.6 Theorem. *Assume that the sequence $\{f_n\}\subset\mathrm{BV}([a,b])$ satisfies $\sup_{n\in\mathbb{N}}\operatorname{var}_a^b f_n<\infty$ and $\lim_{n\to\infty} f_n(t)=f(t)$ for all $t\in[a,b]$. If $g\in\mathrm{G}([a,b])$, then $\{f_n\}$ is uniformly integrable with respect to g, and we have*

$$\lim_{n\to\infty}\Big(\sup_{t\in[a,b]}\Big|\int_a^t f_n\,\mathrm{d}g-\int_a^t f\,\mathrm{d}g\Big|\Big)=0.$$

Proof. In view of Theorem 6.8.2 it suffices to prove that $\{f_n\}$ is uniformly integrable with respect to g. Since g is regulated, given $\varepsilon>0$, there exists a finite step function $g^*\colon[a,b]\to\mathbb{R}$ such that $\|g^*-g\|<\varepsilon$. We know by Lemma 6.8.4 that $\{f_n\}$ is uniformly integrable with respect to g^*. Thus, there exists a gauge δ on $[a,b]$ such that

$$\Big|\int_a^b f_n\,\mathrm{d}g^*-S(f_n,\mathrm{d}g^*,P)\Big|<\varepsilon$$

for each δ-fine partition P of $[a,b]$ and for every $n\in\mathbb{N}$.

Note that the hypotheses on the sequence $\{f_n\}$ ensure that we can choose $M>0$ such that

$$|f_n(a)|+|f_n(b)|+\operatorname{var}_a^b f_n\le M\quad\text{for all } n\in\mathbb{N}.$$

Let P be a δ-fine partition of $[a,b]$. Using the inequalities above together with Theorem 6.3.7 we obtain

$$\begin{aligned}
&\Big|S(f_n,\mathrm{d}g,P)-\int_a^b f_n\,\mathrm{d}g\Big|\\
&\quad\le\Big|S(f_n,\mathrm{d}(g-g^*),P)\Big|+\Big|S(f_n,\mathrm{d}g^*,P)-\int_a^b f_n\,\mathrm{d}g^*\Big|+\Big|\int_a^b f_n\,\mathrm{d}(g^*-g)\Big|\\
&\quad\le 2\,(|f_n(a)|+|f_n(b)|+\operatorname{var}_a^b f_n)\,\|g-g^*\|+\varepsilon\\
&\quad\le\varepsilon\,(2\,M+1),
\end{aligned}$$

proving that $\{f_n\}$ is uniformly integrable with respect to g. □

6.8.7 Lemma. *Assume that the sequence* $\{g_n\} \subset \mathrm{BV}([a,b])$ *is equiregulated and satisfies* $\sup_{n\in\mathbb{N}} \operatorname{var}_a^b g_n < \infty$. *If* $f \in \mathrm{G}([a,b])$, *then* f *is uniformly integrable with respect to* $\{g_n\}$.

Proof. Denote $M = \sup_{n\in\mathbb{N}} \operatorname{var}_a^b g_n$, and let $\varepsilon > 0$ be given. By Theorem 4.1.5, we can find a division $\boldsymbol{\sigma}$ of $[a,b]$ such that

$$|f(t)-f(s)| < \varepsilon \quad \text{for } t,s \in (\sigma_{j-1},\sigma_j) \text{ and } j \in \{1,\dots,\nu(\boldsymbol{\sigma})\}. \tag{6.8.2}$$

We will show that for each $j \in \{1,\dots,\nu(\boldsymbol{\sigma})\}$ there exists a gauge δ_j on $[\sigma_{j-1},\sigma_j]$ such that the inequality

$$\left|S(f,\mathrm{d}g_n,P) - \int_{\sigma_{j-1}}^{\sigma_j} f\,\mathrm{d}g_n\right| < \frac{\varepsilon}{\nu(\boldsymbol{\sigma})} + \varepsilon \operatorname{var}_{\sigma_{j-1}}^{\sigma_j} g_n \tag{6.8.3}$$

holds for every δ_j-fine partition P of $[\sigma_{j-1},\sigma_j]$ and for every $n\in\mathbb{N}$. To this end, fix a $j \in \{1,\dots,\nu(\boldsymbol{\sigma})\}$ and consider the functions $f^*, g_n^* : [\sigma_{j-1},\sigma_j] \to \mathbb{R}$, $n \in \mathbb{N}$, given by

$$f^*(t) = \begin{cases} f(\sigma_{j-1}+) & \text{if } t = \sigma_{j-1}, \\ f(t) & \text{if } t \in (\sigma_{j-1},\sigma_j), \\ f(\sigma_j-) & \text{if } t = \sigma_j, \end{cases}$$

$$g_n^*(t) = \begin{cases} g_n(\sigma_{j-1}+), & \text{if } t = \sigma_{j-1}, \\ g_n(t), & \text{if } t \in (\sigma_{j-1},\sigma_j), \\ g_n(\sigma_j-), & \text{if } t = \sigma_j. \end{cases}$$

Clearly, f^* and g_n^* are for all $n\in\mathbb{N}$ right-continuous at σ_{j-1} and left-continuous at σ_j. Using Lemma 6.3.2 we can show that

$$\int_{\sigma_{j-1}}^{\sigma_j} f\,\mathrm{d}g_n = \int_{\sigma_{j-1}}^{\sigma_j} f\,\mathrm{d}g_n^* + f(\sigma_{j-1})\,\Delta^+ g_n(\sigma_{j-1}) + f(\sigma_j)\,\Delta^- g_n(\sigma_j).$$

Moreover, for any partition $P = (\boldsymbol{\alpha},\boldsymbol{\xi})$ of $[\sigma_{j-1},\sigma_j]$ such that $\xi_1 = \sigma_{j-1}$ and $\xi_{\nu(\boldsymbol{\alpha})} = \sigma_j$ we have

$$\begin{aligned} S(f,\mathrm{d}g_n,P) &= \sum_{\ell=2}^{\nu(\boldsymbol{\alpha})-1} f(\xi_\ell)\big(g(\alpha_\ell) - g(\alpha_{\ell_{j-1}})\big) \\ &\quad + f(\sigma_{j-1})\,(g(\alpha_1) - g(\sigma_{j-1})) + f(\sigma_j)\,(g(\sigma_j) - g(\alpha_{\nu(\boldsymbol{\alpha})-1})) \\ &= S(f,\mathrm{d}g_n^*,P) + f(\sigma_{j-1})\,\Delta^+ g_n(\sigma_{j-1}) + f(\sigma_j)\,\Delta^- g_n(\sigma_j), \end{aligned}$$

hence

$$S(f, \mathrm{d}g_n, P) - \int_{\sigma_{j-1}}^{\sigma_j} f \, \mathrm{d}g_n = S(f, \mathrm{d}g_n^*, P) - \int_{\sigma_{j-1}}^{\sigma_j} f \, \mathrm{d}g_n^*. \tag{6.8.4}$$

Due to the fact that $\{g_n\}$ is equiregulated, we can consider a gauge δ_j on $[\sigma_{j-1}, \sigma_j]$ such that $\delta_j(\sigma_{j-1})$ and $\delta_j(\sigma_j)$ satisfy

$$|\Delta^+ f(\sigma_{j-1})| \, |g_n^*(s) - g_n^*(\sigma_{j-1})| \le \frac{\varepsilon}{2\,\nu(\boldsymbol{\sigma})} \quad \text{for } s \in [\sigma_{j-1}, \sigma_{j-1} + \delta_j(\sigma_{j-1})),$$
$$|\Delta^- f(\sigma_j)| \, |g_n^*(\sigma_j) - g_n^*(s)| \le \frac{\varepsilon}{2\,\nu(\boldsymbol{\sigma})} \quad \text{for } s \in (\sigma_j - \delta_j(\sigma_j), \sigma_j],$$

for every $n \in \mathbb{N}$. By Lemma 6.2.11 we can assume that the gauge δ_j is such that every δ_j-fine partition contains the points σ_{j-1} and σ_j as tags of the first and the last subintervals, respectively. Let an arbitrary δ_j-fine partition $P = (\boldsymbol{\alpha}, \boldsymbol{\xi})$ of $[\sigma_{j-1}, \sigma_j]$ be given. Then

$$\begin{aligned}\Big|S(f, \mathrm{d}g_n^*, P) - S(f^*, \mathrm{d}g_n^*, P)\Big| &\le |f(\sigma_j) - f^*(\sigma_j)| \, |g_n^*(\sigma_j) - g_n^*(\alpha_{\nu(\boldsymbol{\alpha})-1})| \\ &+ |f(\sigma_{j-1}) - f^*(\sigma_{j-1})| \, |g_n^*(\alpha_1) - g_n^*(\sigma_{j-1})| < \frac{\varepsilon}{\nu(\boldsymbol{\sigma})}.\end{aligned}$$

It is not hard to see that $\int_{\sigma_{j-1}}^{\sigma_j} f^* \, \mathrm{d}g_n^* = \int_{\sigma_{j-1}}^{\sigma_j} f \, \mathrm{d}g_n^*$ for every $n \in \mathbb{N}$. This together with the inequality above and (6.8.4) yields

$$\begin{aligned}\Big|S(f, \mathrm{d}g_n, P) - \int_{\sigma_{j-1}}^{\sigma_j} f \, \mathrm{d}g_n\Big| &= \Big|S(f, \mathrm{d}g_n^*, P) - \int_{\sigma_{j-1}}^{\sigma_j} f^* \, \mathrm{d}g_n^*\Big| \\ &\le \Big|S(f, \mathrm{d}g_n^*, P) - S(f^*, \mathrm{d}g_n^*, P)\Big| + \Big|S(f^*, \mathrm{d}g_n^*, P) - \int_{\sigma_{j-1}}^{\sigma_j} f^* \, \mathrm{d}g_n^*\Big| \\ &< \frac{\varepsilon}{\nu(\boldsymbol{\sigma})} + \Big|S(f^*, \mathrm{d}g_n^*, P) - \int_{\sigma_{j-1}}^{\sigma_j} f^* \, \mathrm{d}g_n^*\Big|\end{aligned}$$

for every $n \in \mathbb{N}$. In order to prove (6.8.3), it is enough to show that

$$\Big|S(f^*, \mathrm{d}g_n^*, P) - \int_{\sigma_{j-1}}^{\sigma_j} f^* \, \mathrm{d}g_n^*\Big| \le \varepsilon \operatorname{var}_{\sigma_{j-1}}^{\sigma_j} g_n \quad \text{for } n \in \mathbb{N}. \tag{6.8.5}$$

Fixed $n \in \mathbb{N}$ and given $\eta > 0$, for each $\ell \in \{1, \dots, \nu(\boldsymbol{\alpha})\}$ we can choose a partition $Q_\ell = (\boldsymbol{\beta}^\ell, \boldsymbol{\eta}^\ell)$ of $[\alpha_{\ell-1}, \alpha_\ell]$ such that

$$\Big|S(f^*, \mathrm{d}g_n^*, Q_\ell) - \int_{\alpha_{\ell-1}}^{\alpha_\ell} f^* \, \mathrm{d}g_n^*\Big| < \frac{\eta}{\nu(\boldsymbol{\alpha})}. \tag{6.8.6}$$

Take $Q = Q_1 \cup \dots \cup Q_{\nu(\boldsymbol{\alpha})}$. Clearly, Q is a partition of $[\sigma_{j-1}, \sigma_j]$ and

$$S(f^*, \mathrm{d}g_n^*, P) - S(f^*, \mathrm{d}g_n^*, Q) \\ = \sum_{\ell=1}^{\nu(\boldsymbol{\alpha})} \Big(f^*(\xi_\ell)\,(g_n^*(\alpha_\ell) - g_n^*(\alpha_{\ell-1})) - S(f^*, \mathrm{d}g_n^*, Q_\ell) \Big).$$

Using (6.8.2), for each $\ell \in \{1 \dots, \nu(\boldsymbol{\alpha})\}$ we obtain

$$\big| f^*(\xi_\ell)\,(g_n^*(\alpha_\ell) - g_n^*(\alpha_{\ell-1})) - S(f^*, \mathrm{d}g_n^*, Q_\ell) \big| \\ \le \sum_{i=1}^{\nu(Q_\ell)} |f(\xi_\ell) - f(\eta_i^\ell)|\,|g_n^*(\beta_i^\ell) - g_n^*(\beta_{i-1}^\ell)| \le \varepsilon \operatorname{var}_{\alpha_{\ell-1}}^{\alpha_{\ell-1}} g_n^*,$$

and consequently

$$\big| S(f^*, \mathrm{d}g_n^*, P) - S(f^*, \mathrm{d}g_n^*, Q) \big| \le \varepsilon \operatorname{var}_{\sigma_{j-1}}^{\sigma_j} g_n^*.$$

Thus it follows from (6.8.6) that

$$\Big| S(f^*, \mathrm{d}g_n^*, P) - \int_{\sigma_{j-1}}^{\sigma_j} f^*\, \mathrm{d}g_n^* \Big| \\ \le |S(f^*, \mathrm{d}g_n^*, P) - S(f^*, \mathrm{d}g_n^*, Q)| + \Big| S(f^*, \mathrm{d}g_n^*, Q) - \int_{\sigma_{j-1}}^{\sigma_j} f^*\, \mathrm{d}g_n^* \Big| \\ < \varepsilon \operatorname{var}_{\sigma_{j-1}}^{\sigma_j} g_n^* + \eta.$$

Since $\eta > 0$ is arbitrary, we conclude that

$$\Big| S(f^*, \mathrm{d}g_n^*, P) - \int_{\sigma_{j-1}}^{\sigma_j} f^*\, \mathrm{d}g_n \Big| \le \varepsilon \operatorname{var}_{\sigma_{j-1}}^{\sigma_j} g_n^* \quad \text{for all} \quad n \in \mathbb{N},$$

and inequality (6.8.5) follows then from the fact that by Theorem 2.8.7 (iii) we have

$$\operatorname{var}_{\sigma_{j-1}}^{\sigma_j} g_n^* = \operatorname{var}(g_n^*, (\sigma_{j-1}, \sigma_j)) \\ \le \operatorname{var}(g_n, (\sigma_{j-1}, \sigma_j)) + |\Delta^+ g_n(\sigma_{j-1})| + |\Delta^- g_n(\sigma_j)| = \operatorname{var}_{\sigma_{j-1}}^{\sigma_j} g_n.$$

Altogether we have shown that for each $j \in \{1, \dots, \nu(\boldsymbol{\sigma})\}$ there exists a gauge δ_j on $[\sigma_{j-1}, \sigma_j]$ such that (6.8.3) holds. To conclude the proof, consider a gauge δ on $[a, b]$ such that $\delta(t) < \delta_j(t)$ for all $t \in [\sigma_{j-1}, \sigma_j]$, and $\boldsymbol{\sigma} \subset \{\xi_1, \dots, \xi_{\nu(P)}\}$ for each δ-fine partition $P = (\boldsymbol{\alpha}, \boldsymbol{\xi})$ of $[a, b]$. The second requirement is justified by Lemma 6.2.11, which also guarantees the existence of a δ-fine partition $\widetilde{P}=(\widetilde{\boldsymbol{\alpha}},\widetilde{\boldsymbol{\xi}})$ of $[a,b]$ such that

$$S(f,\mathrm{d}g_n,P)=S(f,\mathrm{d}g_n,\widetilde{P}) \quad \text{and} \quad \boldsymbol{\sigma}\subset\widetilde{\boldsymbol{\alpha}}.$$

Thus, we can write $\widetilde{P}=\widetilde{P}_1\cup\cdots\cup\widetilde{P}_{\nu(\boldsymbol{\sigma})}$ where, for every $j\in\{1,\dots,\nu(\boldsymbol{\sigma})\}$, $\widetilde{P}_j$ is a δ_j-fine partition of $[\sigma_{j-1},\sigma_j]$,

By (6.8.3) we obtain

$$\Big|S(f,\mathrm{d}g_n,P)-\int_a^b f\,\mathrm{d}g_n\Big|\le\sum_{j=1}^{\nu(\boldsymbol{\sigma})}\Big|S(f,\mathrm{d}g_n,\widetilde{P}_j)-\int_{\sigma_{j-1}}^{\sigma_j} f\,\mathrm{d}g_n\Big|$$

$$<\sum_{j=1}^{\nu(\boldsymbol{\sigma})}\Big(\frac{\varepsilon}{\nu(\boldsymbol{\sigma})}+\varepsilon\,\mathrm{var}_{\sigma_{j-1}}^{\sigma_j}g_n\Big)=\varepsilon\,(1+\mathrm{var}_a^b g_n)\le\varepsilon\,(1+M)$$

for every $n\in\mathbb{N}$, i.e., f is uniformly integrable with respect to $\{g_n\}$. □

6.8.8 Theorem. *Assume that the sequence $\{g_n\}\subset\mathrm{BV}([a,b])$ satisfies $\sup_{n\in\mathbb{N}}\mathrm{var}_a^b g_n<\infty$ and $g_n\rightrightarrows g$. Moreover, let $f_n:[a,b]\to\mathbb{R}$, $n\in\mathbb{N}$, be such that $\int_a^b f_n\,\mathrm{d}g_n$ exists for each $n\in\mathbb{N}$ and $f_n\rightrightarrows f$, where $f:[a,b]\to\mathbb{R}$ is regulated. Then $\{f_n\}$ is uniformly integrable with respect to $\{g_n\}$, and we have*

$$\lim_{n\to\infty}\Big(\sup_{t\in[a,b]}\Big|\int_a^t f_n\,\mathrm{d}g_n-\int_a^t f\,\mathrm{d}g\Big|\Big)=0.$$

Proof. In view of Theorem 6.8.2 it suffices to prove that $\{f_n\}$ is uniformly integrable with respect to $\{g_n\}$.

Let $\varepsilon>0$ be given. Since $\{f_n\}$ is uniformly convergent, there is $n_0\in\mathbb{N}$ such that $\|f_n-f\|<\varepsilon$ for $n\ge n_0$. Moreover, the existence of the integral $\int_a^b f_n\,\mathrm{d}g_n$ for $n\in\{1,\dots,n_0\}$ ensures that we can choose a gauge γ on $[a,b]$ such that

$$\Big|S(f_n,\mathrm{d}g_n,P)-\int_a^b f_n\,\mathrm{d}g_n\Big|<\varepsilon \tag{6.8.7}$$

holds for each γ-fine partition P of $[a,b]$ and $n=1,\dots,n_0$.

Note that the uniform convergence of the sequence $\{g_n\}$ together with Theorem 4.3.6 ensures that $\{g_n\}$ is equiregulated. The assumptions of Lemma 6.8.7 are then satisfied, implying that f is uniformly integrable with respect to $\{g_n\}$. Thus, there exists a gauge δ on $[a,b]$ such that

$$\Big|S(f,\mathrm{d}g_n,P)-\int_a^b f\,\mathrm{d}g_n\Big|<\varepsilon$$

holds for each δ-fine partition P of $[a,b]$ and for every $n\in\mathbb{N}$.

Without loss of generality we can assume that $\delta(t)\le\gamma(t)$ for $t\in[a,b]$. For this reason, if $n=1,\dots,n_0$, then (6.8.7) is satisfied for each δ-fine partition P of $[a,b]$. On the other hand, if $n>n_0$ and P is an arbitrary δ-fine partition of $[a,b]$ we have

$$\begin{aligned}\Big|S(f_n,\mathrm{d}g_n,P)-\int_a^b f_n\,\mathrm{d}g_n\Big| &\le \big|S(f_n-f,\mathrm{d}g_n,P)\big|\\ &\quad+\Big|S(f,\mathrm{d}g_n,P)-\int_a^b f\,\mathrm{d}g_n\Big|+\Big|\int_a^b (f-f_n)\,\mathrm{d}g_n\Big|\\ &< 2\,\|f_n-f\|\,\operatorname{var}_a^b g_n+\varepsilon<(2\,M+1)\,\varepsilon,\end{aligned}$$

where $M=\sup_{n\in\mathbb{N}}\operatorname{var}_a^b g_n$. In summary

$$\Big|S(f_n,\mathrm{d}g_n,P)-\int_a^b f_n\,\mathrm{d}g_n\Big|<(2\,M+1)\,\varepsilon$$

holds for each δ-fine partition P of $[a,b]$ and for every $n\in\mathbb{N}$, which proves that $\{f_n\}$ is uniformly integrable with respect to $\{g_n\}$. □

Next, we focus on convergence theorems which do not require uniform convergence of integrands and integrators. We need the following auxiliary lemma.

6.8.9 Lemma. *For each $\ell\in\mathbb{N}$, let $\mathcal{S}_\ell$ be the set of all couples $(\boldsymbol{\sigma},\boldsymbol{s})$, where $\boldsymbol{\sigma}$ is a division of $[a,b]$ and $\boldsymbol{s}=(s_1,\dots,s_{\nu(\boldsymbol{\sigma})})$ is a vector whose components are integers greater than or equal to ℓ.*

Let $g:[a,b]\to\mathbb{R}$ and let $f_n:[a,b]\to\mathbb{R}$, $n\in\mathbb{N}$, be a sequence of functions such that the integral $\int_a^b f_n\,\mathrm{d}g$ exists for each $n\in\mathbb{N}$. Moreover, assume that there are $B,C\in\mathbb{R}$ such that if $\boldsymbol{\sigma}$ is a division of $[a,b]$ and $s_1,\dots,s_{\nu(\boldsymbol{\sigma})}\in\mathbb{N}$, then

$$B\le\sum_{j=1}^{\nu(\boldsymbol{\sigma})}\int_{\sigma_{j-1}}^{\sigma_j} f_{s_j}\,\mathrm{d}g\le C. \tag{6.8.8}$$

Then for each $\varepsilon>0$ and $\ell\in\mathbb{N}$ there exist $(\boldsymbol{\pi}^\ell,\boldsymbol{p}^\ell)$, $(\boldsymbol{\rho}^\ell,\boldsymbol{r}^\ell)\in\mathcal{S}_\ell$ such that the following statements are true:

(i) *The inequalities*

$$\sum_{j=1}^{\nu(\boldsymbol{\pi}^\ell)}\int_{\pi_{j-1}^\ell}^{\pi_j^\ell} f_{p_j^\ell}\,\mathrm{d}g-\frac{\varepsilon}{2^\ell}<\sum_{j=1}^{\nu(\boldsymbol{\sigma})}\int_{\sigma_{j-1}}^{\sigma_j} f_{s_j}\,\mathrm{d}g<\sum_{j=1}^{\nu(\boldsymbol{\rho}^\ell)}\int_{\rho_{j-1}^\ell}^{\rho_j^\ell} f_{r_j^\ell}\,\mathrm{d}g+\frac{\varepsilon}{2^\ell} \tag{6.8.9}$$

hold for each $(\boldsymbol{\sigma}, \boldsymbol{s}) \in \mathcal{S}_\ell$.

(ii) *Let* $\mathcal{I} = \bigcup_{i \in I} [c_i, d_i]$ *be a finite union of nonoverlapping intervals in* $[a, b]$ *such that for each* $i \in I$ *there exists a* $j(i) \in \{1, \ldots, \nu(\boldsymbol{\pi}^\ell)\}$ *satisfying* $[c_i, d_i] \subset [\pi^\ell_{j(i)-1}, \pi^\ell_{j(i)}]$. *Then the inequality*

$$\sum_{i \in I} \int_{c_i}^{d_i} f_{p^\ell_{j(i)}} \, \mathrm{d}g - \frac{\varepsilon}{2^\ell} < \sum_{i \in I} \int_{c_i}^{d_i} f_n \, \mathrm{d}g \tag{6.8.10}$$

holds for each $n \geq \ell$.

(iii) *Let* $\mathcal{I} = \bigcup_{i \in I} [c_i, d_i]$ *be a finite union of nonoverlapping intervals in* $[a, b]$ *such that for each* $i \in I$ *there exists a* $j(i) \in \{1, \ldots, \nu(\boldsymbol{\rho}^\ell)\}$ *satisfying* $[c_i, d_i] \subset [\rho^\ell_{j(i)-1}, \rho^\ell_{j(i)}]$. *Then the inequality*

$$\sum_{i \in I} \int_{c_i}^{d_i} f_n \, \mathrm{d}g < \sum_{i \in I} \int_{c_i}^{d_i} f_{r^\ell_{j(i)}} \, \mathrm{d}g + \frac{\varepsilon}{2^\ell} \tag{6.8.11}$$

holds for each $n \geq \ell$.

P r o o f. According to (6.8.8), the values of all sums

$$\sum_{j=1}^{\nu(\boldsymbol{\sigma})} \int_{\sigma_{j-1}}^{\sigma_j} f_{s_j} \, \mathrm{d}g, \quad \text{where} \quad (\boldsymbol{\sigma}, \boldsymbol{s}) \in \mathcal{S}_\ell,$$

are contained in a bounded subset of $\mathbb{R}$. Thus, the existence of the couples $(\boldsymbol{\pi}^\ell, \boldsymbol{p}^\ell)$ and $(\boldsymbol{\rho}^\ell, \boldsymbol{r}^\ell) \in \mathcal{S}_\ell$ having the properties from statement (i) follows from the definitions of infimum and supremum.

To prove statement (ii), consider the division $\boldsymbol{\sigma} = \boldsymbol{\pi}^\ell \cup \{c_i, d_i : i \in I\}$ of $[a, b]$. For each $j \in \{1, \ldots, \nu(\boldsymbol{\sigma})\}$, there exists a $k(j) \in \{1, \ldots, \nu(\boldsymbol{\pi}^\ell)\}$ such that $[\sigma_{j-1}, \sigma_j] \subset [\pi^\ell_{k(j)-1}, \pi^\ell_{k(j)}]$. Define $\boldsymbol{s}$ as follows: If $j \in \{1, \ldots, \nu(\boldsymbol{\sigma})\}$ is such that $[\sigma_{j-1}, \sigma_j] \subset \mathcal{I}$, let $s_j = n$. Otherwise, let $s_j = p^\ell_{k(j)}$. Then $(\boldsymbol{\sigma}, \boldsymbol{s}) \in \mathcal{S}_\ell$, and relation (6.8.9) holds. Obviously,

$$\begin{aligned}\sum_{j=1}^{\nu(\boldsymbol{\sigma})} \int_{\sigma_{j-1}}^{\sigma_j} f_{s_j} \, \mathrm{d}g &= \sum_{\substack{j \in \{1, \ldots, \nu(\boldsymbol{\sigma})\} \\ [\sigma_{j-1}, \sigma_j] \subset \mathcal{I}}} \int_{\sigma_{j-1}}^{\sigma_j} f_n \, \mathrm{d}g + \sum_{\substack{j \in \{1, \ldots, \nu(\boldsymbol{\sigma})\} \\ [\sigma_{j-1}, \sigma_j] \not\subset \mathcal{I}}} \int_{\sigma_{j-1}}^{\sigma_j} f_{p^\ell_{k(j)}} \, \mathrm{d}g \\ &= \sum_{i \in I} \int_{c_i}^{d_i} f_n \, \mathrm{d}g + \sum_{\substack{j \in \{1, \ldots, \nu(\boldsymbol{\sigma})\} \\ [\sigma_{j-1}, \sigma_j] \not\subset \mathcal{I}}} \int_{\sigma_{j-1}}^{\sigma_j} f_{p^\ell_{k(j)}} \, \mathrm{d}g.\end{aligned}$$

Furthermore, for each $m \in \{1, \dots, \nu(\boldsymbol{\pi}^\ell)\}$, we have

$$[\pi^\ell_{m-1}, \pi^\ell_m] = \bigcup_{\substack{j\in\{1,\dots,\nu(\boldsymbol{\sigma})\}\\ k(j)=m}} [\sigma_{j-1}, \sigma_j],$$

and therefore

$$\begin{aligned}\sum_{j=1}^{\nu(\boldsymbol{\pi}^\ell)} \int_{\pi^\ell_{j-1}}^{\pi^\ell_j} f_{p^\ell_j}\,\mathrm{d}g &= \sum_{m=1}^{\nu(\boldsymbol{\pi}^\ell)} \sum_{\substack{j\in\{1,\dots,\nu(\boldsymbol{\sigma})\}\\ k(j)=m}} \int_{\sigma_{j-1}}^{\sigma_j} f_{p^\ell_m}\,\mathrm{d}g = \sum_{j=1}^{\nu(\sigma)} \int_{\sigma_{j-1}}^{\sigma_j} f_{p^\ell_{k(j)}}\,\mathrm{d}g \\ &= \sum_{\substack{j\in\{1,\dots,\nu(\boldsymbol{\sigma})\}\\ [\sigma_{j-1},\sigma_j]\subset\mathcal{I}}} \int_{\sigma_{j-1}}^{\sigma_j} f_{p^\ell_{k(j)}}\,\mathrm{d}g + \sum_{\substack{j\in\{1,\dots,\nu(\boldsymbol{\sigma})\}\\ [\sigma_{j-1},\sigma_j]\not\subset\mathcal{I}}} \int_{\sigma_{j-1}}^{\sigma_j} f_{p^\ell_{k(j)}}\,\mathrm{d}g \\ &= \sum_{i\in I} \int_{c_i}^{d_i} f_{p^\ell_{j(i)}}\,\mathrm{d}g + \sum_{\substack{j\in\{1,\dots,\nu(\boldsymbol{\sigma})\}\\ [\sigma_{j-1},\sigma_j]\not\subset\mathcal{I}}} \int_{\sigma_{j-1}}^{\sigma_j} f_{p^\ell_{k(j)}}\,\mathrm{d}g.\end{aligned}$$

Now, by subtracting

$$\sum_{\substack{j\in\{1,\dots,\nu(\boldsymbol{\sigma})\}\\ [\sigma_{j-1},\sigma_j]\not\subset\mathcal{I}}} \int_{\sigma_{j-1}}^{\sigma_j} f_{p^\ell_{k(j)}}\,\mathrm{d}g$$

from the first inequality in (6.8.9), we get (6.8.10).

Statement (iii) can be proved in a similar way. □

The next result is one of the main convergence theorems for the Kurzweil-Stieltjes integral. We will see that the dominated and monotone convergence theorems are straightforward consequences of this result.

6.8.10 Theorem. *Let $g \in \mathrm{BV}([a,b])$ and let $\{f_n\}$ be a sequence of real-valued functions defined on $[a,b]$ and satisfying the following conditions:*

(i) *The integral $\int_a^b f_n\,\mathrm{d}g$ exists for each $n \in \mathbb{N}$.*

(ii) $\lim_{n\to\infty} f_n(t) = f(t)$ *for $t \in [a,b]$.*

(iii) *There are $B, C \in \mathbb{R}$ such that the inequalities*

$$B \le \sum_{j=1}^{\nu(\boldsymbol{\sigma})} \int_{\sigma_{j-1}}^{\sigma_j} f_{s_j}\,\mathrm{d}g \le C$$

hold for all divisions $\boldsymbol{\sigma}$ of $[a,b]$ and all $s_1, \dots, s_{\nu(\boldsymbol{\sigma})} \in \mathbb{N}$.

Then $\{f_n\}$ *is uniformly integrable with respect to* g*, the integral* $\int_a^b f\,\mathrm{d}g$ *exists, and*

$$\int_a^b f\,\mathrm{d}g = \lim_{n\to\infty}\int_a^b f_n\,\mathrm{d}g.$$

Moreover, we have

$$\lim_{n\to\infty}\Big(\sup_{t\in[a,b]}\Big|\int_a^t f_n\,\mathrm{d}g - \int_a^t f\,\mathrm{d}g\Big|\Big) = 0.$$

Proof. According to Theorem 6.8.2, it suffices to prove that $\{f_n\}$ is uniformly integrable with respect to g. Choose an arbitrary $\varepsilon > 0$. For each $\ell\in\mathbb{N}$, let $(\boldsymbol{\pi}^\ell, \boldsymbol{p}^\ell)$ and $(\boldsymbol{\rho}^\ell, \boldsymbol{r}^\ell)$ be as in Lemma 6.8.9. Also, for each $\ell\in\mathbb{N}$, there is a gauge δ_ℓ on $[a,b]$ such that for each δ_ℓ-fine partition P of $[a,b]$, we have

$$\Big|S(f_\ell, \mathrm{d}g, P) - \int_a^b f_\ell\,\mathrm{d}g\Big| < \varepsilon, \tag{6.8.12}$$

$$\Big|S(f_{p_i^\ell}, \mathrm{d}g, P) - \int_a^b f_{p_i^\ell}\,\mathrm{d}g\Big| < \frac{\varepsilon}{2^\ell\nu(\boldsymbol{\pi}^\ell)} \quad \text{for } i\in\{1,\dots,\nu(\boldsymbol{\pi}^\ell)\}, \tag{6.8.13}$$

$$\Big|S(f_{r_i^\ell}, \mathrm{d}g, P) - \int_a^b f_{r_i^\ell}\,\mathrm{d}g\Big| < \frac{\varepsilon}{2^\ell\nu(\boldsymbol{\rho}^\ell)} \quad \text{for } i\in\{1,\dots,\nu(\boldsymbol{\rho}^\ell)\}. \tag{6.8.14}$$

Moreover, assume that $\delta_\ell(t) < \operatorname{dist}(t, (\boldsymbol{\pi}^\ell\cup\boldsymbol{\rho}^\ell)\setminus\{t\})$ holds for each $t\in[a,b]$.

Due to assumption (ii), for each $t\in[a,b]$ we can choose an $N(t)\in\mathbb{N}$ such that

$$|f(t) - f_n(t)| < \varepsilon \quad \text{for all } n \ge N(t).$$

Consider the gauge $\delta\colon[a,b]\to(0,\infty)$ given by

$$\delta(t) = \min\{\delta_1(t),\dots,\delta_{N(t)}(t)\} \quad \text{for } t\in[a,b].$$

Let $P = (\boldsymbol{\alpha}, \boldsymbol{\xi})$ be an arbitrary δ-fine partition of $[a,b]$ and $n\in\mathbb{N}$. Our goal is to obtain estimates for the terms appearing in $S(f_n, \mathrm{d}g, P)$. If $i\in\{1,\dots,\nu(\boldsymbol{\alpha})\}$ is such that $N(\xi_i)\ge n$, then $\delta(\xi_i)\le\delta_n(\xi_i)$. Thus, assumption (6.8.12) together with Lemma 6.5.1 (Saks-Henstock lemma) imply

$$\sum_{\substack{i\in\{1,\dots,\nu(\boldsymbol{\alpha})\}\\ N(\xi_i)\ge n}} f_n(\xi_i)\,(g(\alpha_i) - g(\alpha_{i-1})) \le \sum_{\substack{i\in\{1,\dots,\nu(\boldsymbol{\alpha})\}\\ N(\xi_i)\ge n}} \int_{\alpha_{i-1}}^{\alpha_i} f_n\,\mathrm{d}g + \varepsilon. \tag{6.8.15}$$

It remains to estimate those terms in $S(f_n, \mathrm{d}g, P)$ for which $N(\xi_i) < n$. To this aim, consider a fixed $\ell \in \{1, \dots, n-1\}$ together with all indices $i \in \{1, \dots, \nu(\boldsymbol{\alpha})\}$ fulfilling $N(\xi_i) = \ell$. Due to the construction of the gauge δ_ℓ, each corresponding interval $[\alpha_{i-1}, \alpha_i]$ contains at most one of the division points of $\boldsymbol{\pi}^\ell$ and, in that case, the tag ξ_i must coincide with this division point. If we split the interval $[\alpha_{i-1}, \alpha_i]$ into a pair of intervals $[\alpha_{i-1}, \xi_i]$ and $[\xi_i, \alpha_i]$ that share the same tag ξ_i, the value of $S(f_n, \mathrm{d}g, P)$ remains unchanged. Thus, without loss of generality, we can assume that for every $i \in \{1, \dots, \nu(\boldsymbol{\alpha})\}$ satisfying $N(\xi_i) = \ell$, there is a $j(i) \in \{1, \dots, \nu(\boldsymbol{\pi}^\ell)\}$ such that $[\alpha_{i-1}, \alpha_i] \subset [\pi^\ell_{j(i)-1}, \pi^\ell_{j(i)}]$. Then

$$\begin{aligned}
&\sum_{\substack{i \in \{1,\dots,\nu(\boldsymbol{\alpha})\} \\ N(\xi_i) = \ell}} f_n(\xi_i)\,(g(\alpha_i) - g(\alpha_{i-1})) \\
&\quad = \sum_{\substack{i \in \{1,\dots,\nu(\boldsymbol{\alpha})\} \\ N(\xi_i) = \ell}} (f_n(\xi_i) - f_{p^\ell_{j(i)}}(\xi_i))\,(g(\alpha_i) - g(\alpha_{i-1})) \\
&\qquad + \sum_{\substack{i \in \{1,\dots,\nu(\boldsymbol{\alpha})\} \\ N(\xi_i) = \ell}} \Big(f_{p^\ell_{j(i)}}(\xi_i)\,(g(\alpha_i) - g(\alpha_{i-1})) - \int_{\alpha_{i-1}}^{\alpha_i} f_{p^\ell_{j(i)}}\,\mathrm{d}g \Big) \\
&\qquad + \sum_{\substack{i \in \{1,\dots,\nu(\boldsymbol{\alpha})\} \\ N(\xi_i) = \ell}} \int_{\alpha_{i-1}}^{\alpha_i} f_{p^\ell_{j(i)}}\,\mathrm{d}g.
\end{aligned}$$

Since $(\boldsymbol{\pi}^\ell, \boldsymbol{p}^\ell) \in \mathcal{S}_\ell$, we also have $p^\ell_{j(i)} \geq \ell$. Hence, if $N(\xi_i) = \ell < n$, the definition of $N(\xi_i)$ gives

$$|f_n(\xi_i) - f_{p^\ell_{j(i)}}(\xi_i)| \leq |f_n(\xi_i) - f(\xi_i)| + |f(\xi_i) - f_{p^\ell_{j(i)}}(\xi_i)| < 2\,\varepsilon,$$

and consequently

$$\big|(f_n(\xi_i) - f_{p^\ell_{j(i)}}(\xi_i))\,(g(\alpha_i) - g(\alpha_{i-1}))\big| < 2\,\varepsilon\ \operatorname{var}_{\alpha_{i-1}}^{\alpha_i} g.$$

Next, we notice that

$$\begin{aligned}
&\Big| \sum_{\substack{i \in \{1,\dots,\nu(\boldsymbol{\alpha})\} \\ N(\xi_i) = \ell}} \Big(f_{p^\ell_{j(i)}}(\xi_i)\,(g(\alpha_i) - g(\alpha_{i-1})) - \int_{\alpha_{i-1}}^{\alpha_i} f_{p^\ell_{j(i)}}\,\mathrm{d}g \Big) \Big| \\
&\quad \leq \sum_{k=1}^{\nu(\boldsymbol{\pi}^\ell)} \Big| \sum_{\substack{i \in \{1,\dots,\nu(\boldsymbol{\alpha})\} \\ N(\xi_i) = \ell,\ j(i) = k}} \Big(f_{p^\ell_k}(\xi_i)\,(g(\alpha_i) - g(\alpha_{i-1})) - \int_{\alpha_{i-1}}^{\alpha_i} f_{p^\ell_k}\,\mathrm{d}g \Big) \Big|
\end{aligned}$$

$$\leq \sum_{k=1}^{\nu(\boldsymbol{\pi}^\ell)} \frac{\varepsilon}{2^\ell \nu(\boldsymbol{\pi}^\ell)} = \frac{\varepsilon}{2^\ell},$$

where the last inequality is a consequence of the assumption (6.8.13) and Lemma 6.5.1 (Saks-Henstock lemma).

To summarize, for each $\ell \in \{1, \dots, n-1\}$ the previous estimates imply

$$\sum_{\substack{i \in \{1,\dots,\nu(\boldsymbol{\alpha})\} \\ N(\xi_i)=\ell}} f_n(\xi_i)\,(g(\alpha_i) - g(\alpha_{i-1}))$$
$$< \sum_{\substack{i \in \{1,\dots,\nu(\boldsymbol{\alpha})\} \\ N(\xi_i)=\ell}} \Big(\int_{\alpha_{i-1}}^{\alpha_i} f_{p^\ell_{j(i)}}\,\mathrm{d}g + 2\,\varepsilon\, \mathrm{var}_{\alpha_{i-1}}^{\alpha_i} g \Big) + \frac{\varepsilon}{2^\ell}$$
$$< \sum_{\substack{i \in \{1,\dots,\nu(\boldsymbol{\alpha})\} \\ N(\xi_i)=\ell}} \Big(\int_{\alpha_{i-1}}^{\alpha_i} f_n\,\mathrm{d}g + 2\,\varepsilon\, \mathrm{var}_{\alpha_{i-1}}^{\alpha_i} g \Big) + 2\,\frac{\varepsilon}{2^\ell},$$

where the last inequality is a consequence of (6.8.10). Summation over $\ell \in \{1, \dots, n-1\}$ yields

$$\sum_{\substack{i \in \{1,\dots,\nu(\boldsymbol{\alpha})\} \\ N(\xi_i)<n}} f_n(\xi_i)\,(g(\alpha_i) - g(\alpha_{i-1})) < \sum_{\substack{i \in \{1,\dots,\nu(\boldsymbol{\alpha})\} \\ N(\xi_i)<n}} \int_{\alpha_{i-1}}^{\alpha_i} f_n\,\mathrm{d}g + 2\,\varepsilon\, \mathrm{var}_a^b g + 2\,\varepsilon.$$

Adding the last inequality and (6.8.15), we conclude that the estimate

$$S(f_n, \mathrm{d}g, P) = \sum_{i=1}^{\nu(\boldsymbol{\alpha})} f_n(\xi_i)\,(g(\alpha_i) - g(\alpha_{i-1})) < \int_a^b f_n\,\mathrm{d}g + 2\,\varepsilon\, \mathrm{var}_a^b g + 3\,\varepsilon$$

holds for all $n \in \mathbb{N}$. Proceeding in a similar way (using (6.8.11) and (6.8.14)), we can show that also

$$S(f_n, \mathrm{d}g, P) > \int_a^b f_n\,\mathrm{d}g - 2\,\varepsilon\, \mathrm{var}_a^b g - 3\,\varepsilon$$

holds for all $n \in \mathbb{N}$. As a consequence,

$$\Big| S(f_n, \mathrm{d}g, P) - \int_a^b f_n\,\mathrm{d}g \Big| < 2\,\varepsilon\, \mathrm{var}_a^b g + 3\,\varepsilon$$

for each δ-fine partition P of $[a,b]$ and each $n \in \mathbb{N}$, i.e., $\{f_n\}$ is uniformly integrable with respect to g. □

A straightforward consequence of Theorem 6.8.10 is the dominated convergence theorem.

6.8.11 Theorem (DOMINATED CONVERGENCE THEOREM).

Let $g\colon[a,b]\to\mathbb{R}$ be nondecreasing and let $\{f_n\}$ be a sequence of real valued functions defined on $[a,b]$ and satisfying the following conditions:

(i) *The integral $\int_a^b f_n\,\mathrm{d}g$ exists for each $n\in\mathbb{N}$.*

(ii) $\lim_{n\to\infty} f_n(t)=f(t)$ *for* $t\in[a,b]$.

(iii) *There exist functions $h_1,h_2\colon[a,b]\to\mathbb{R}$ such that the integrals $\int_a^b h_1\,\mathrm{d}g$, $\int_a^b h_2\,\mathrm{d}g$ exist and $h_1\le f_n\le h_2$ on $[a,b]$ for each $n\in\mathbb{N}$.*

Then $\{f_n\}$ is uniformly integrable with respect to g, the integral $\int_a^b f\,\mathrm{d}g$ exists and

$$\int_a^b f\,\mathrm{d}g=\lim_{n\to\infty}\int_a^b f_n\,\mathrm{d}g.$$

Moreover, we have

$$\lim_{n\to\infty}\Bigl(\sup_{t\in[a,b]}\Bigl|\int_a^t f_n\,\mathrm{d}g-\int_a^t f\,\mathrm{d}g\Bigr|\Bigr)=0.$$

Proof. The statement is a consequence of Theorem 6.8.10. To see this, let

$$B=\int_a^b h_1\,\mathrm{d}g,\qquad C=\int_a^b h_2\,\mathrm{d}g.$$

If $\boldsymbol{\sigma}$ is a division of $[a,b]$ and $s_1,\dots,s_{\nu(\boldsymbol{\sigma})}\in\mathbb{N}$, then

$$B=\sum_{j=1}^{\nu(\boldsymbol{\sigma})}\int_{\sigma_{j-1}}^{\sigma_j} h_1\,\mathrm{d}g\le\sum_{j=1}^{\nu(\boldsymbol{\sigma})}\int_{\sigma_{j-1}}^{\sigma_j} f_{s_j}\,\mathrm{d}g\le\sum_{j=1}^{\nu(\boldsymbol{\sigma})}\int_{\sigma_{j-1}}^{\sigma_j} h_2\,\mathrm{d}g=C.$$

This shows that the assumptions of Theorem 6.8.10 are satisfied, and the proof is complete. □

6.8.12 Remark. Given a nondecreasing function $g\colon[a,b]\to\mathbb{R}$ and a set $N\subset[a,b]$, we let $\mu_g(N)=\int_a^b \chi_N\,\mathrm{d}g$ if the integral exists. For example, if g is continuous at $x\in[a,b]$, then $\mu_g(\{x\})=0$. The set function μ_g is a finite measure on $[a,b]$: It is nonnegative and bounded by $g(b)-g(a)$, satisfies $\mu_g(\emptyset)=0$, and if $\{A_n\}$ is a countable collection of disjoint subsets of $[a,b]$, then $\mu_g(\bigcup_n A_n)=\sum_n \mu_g(A_n)$; this will follow from Theorem 6.8.15.

Note that μ_g coincides with the Lebesgue-Stieltjes measure corresponding to g (see Section 6.12); this is a consequence of Lemma 6.12.1 if we extend g to $\mathbb{R}$ by letting $g(t)=g(a)$ for $t<a$ and $g(t)=g(b)$ for $t>b$.

If $h:[a,b]\to\mathbb{R}$ vanishes except a set $N\subset[a,b]$ with $\mu_g(N)=0$, then $\int_a^b h\,\mathrm{d}g=0$ (this follows from Theorem 6.12.3 and the fact that the Lebesgue-Stieltjes integral $\int_{[a,b]} f\,\mathrm{d}\mu_g$ is zero). Consequently, the values of an arbitrary function $f:[a,b]\to\mathbb{R}$ on N have no influence on the existence or value of the integral $\int_a^b f\,\mathrm{d}g$.

The previous considerations imply that in Theorem 6.8.11, it suffices to assume that $\lim_{n\to\infty} f_n(t)=f(t)$ for $t\in[a,b]\setminus N$ and $h_1\le f_n\le h_2$ on $[a,b]\setminus N$, where $\mu_g(N)=0$. Other convergence theorems in this section can be modified in a similar way.

The next result is the bounded convergence theorem. In comparison with the dominated convergence theorem, the dominating hypothesis on $\{f_n\}$ is stronger. On the other hand, g is no longer assumed to be nondecreasing, but merely of bounded variation.

6.8.13 Theorem (Bounded convergence theorem).

Let $g\in\mathrm{BV}([a,b])$ and let $\{f_n\}$ be a sequence of real valued functions defined on $[a,b]$ and satisfying the following conditions:

(i) *The integral $\int_a^b f_n\,\mathrm{d}g$ exists for each $n\in\mathbb{N}$.*

(ii) $\lim\limits_{n\to\infty} f_n(t)=f(t)$ *for $t\in[a,b]$.*

(iii) *There exists a constant $M\ge 0$ such that $|f_n(t)|\le M$ for all $n\in\mathbb{N}$ and $t\in[a,b]$.*

Then $\{f_n\}$ is uniformly integrable with respect to g, the integral $\int_a^b f\,\mathrm{d}g$ exists and

$$\lim_{n\to\infty}\Big(\sup_{t\in[a,b]}\Big|\int_a^t f_n\,\mathrm{d}g-\int_a^t f\,\mathrm{d}g\Big|\Big)=0.$$

Proof. If $\boldsymbol{\sigma}$ is a division of $[a,b]$ and $s_1,\dots,s_{\nu(\boldsymbol{\sigma})}\in\mathbb{N}$, then

$$\Big|\sum_{j=1}^{\nu(\boldsymbol{\sigma})}\int_{\sigma_{j-1}}^{\sigma_j} f_{s_j}\,\mathrm{d}g\Big|\le\sum_{j=1}^{\nu(\boldsymbol{\sigma})}\Big|\int_{\sigma_{j-1}}^{\sigma_j} f_{s_j}\,\mathrm{d}g\Big|\le\sum_{j=1}^{\nu(\boldsymbol{\sigma})} M\,\mathrm{var}_{\sigma_{j-1}}^{\sigma_j} g=M\,\mathrm{var}_a^b g.$$

Hence, the assumptions of Theorem 6.8.10 are satisfied with

$$B=-M\,\mathrm{var}_a^b g,\qquad C=M\,\mathrm{var}_a^b g,$$

and the proof is complete. □

Another important consequence of Theorem 6.8.10 is the monotone convergence theorem.

6.8.14 Theorem (MONOTONE CONVERGENCE THEOREM).

Let $g\colon [a,b]\to\mathbb{R}$ be nondecreasing and $\{f_n\}$ be a sequence of functions such that the integral $\int_a^b f_n\,\mathrm{d}g$ exists for each $n\in\mathbb{N}$, and $\lim_{n\to\infty} f_n = f$ on $[a,b]$. Suppose, in addition, that one of the following conditions holds:

(i) *$\{f_n\}$ is a nondecreasing sequence and* $\displaystyle\lim_{n\to\infty}\int_a^b f_n\,\mathrm{d}g<\infty$.

(ii) *$\{f_n\}$ is a nonincreasing sequence and* $\displaystyle\lim_{n\to\infty}\int_a^b f_n\,\mathrm{d}g>-\infty$.

Then $\{f_n\}$ is uniformly integrable with respect to g, the integral $\int_a^b f\,\mathrm{d}g$ exists and

$$\int_a^b f\,\mathrm{d}g=\lim_{n\to\infty}\int_a^b f_n\,\mathrm{d}g. \tag{6.8.16}$$

Moreover, we have

$$\lim_{n\to\infty}\Big(\sup_{t\in[a,b]}\Big|\int_a^t f_n\,\mathrm{d}g-\int_a^t f\,\mathrm{d}g\Big|\Big)=0. \tag{6.8.17}$$

Proof. Suppose that the first condition holds. Let

$$B=\int_a^b f_1\,\mathrm{d}g,\qquad C=\lim_{n\to\infty}\int_a^b f_n\,\mathrm{d}g.$$

If $\boldsymbol{\sigma}$ is a division of $[a,b]$ and $s_1,\dots,s_{\nu(\boldsymbol{\sigma})}\in\mathbb{N}$, then

$$B=\sum_{j=1}^{\nu(\boldsymbol{\sigma})}\int_{\sigma_{j-1}}^{\sigma_j} f_1\,\mathrm{d}g\le\sum_{j=1}^{\nu(\boldsymbol{\sigma})}\int_{\sigma_{j-1}}^{\sigma_j} f_{s_j}\,\mathrm{d}g\le\sum_{j=1}^{\nu(\boldsymbol{\sigma})}\int_{\sigma_{j-1}}^{\sigma_j}(f_{\max(s_1,\dots,s_{\nu(\boldsymbol{\sigma})})})\,\mathrm{d}g\le C.$$

Thus, the assumptions of Theorem 6.8.10 are satisfied, and (6.8.16), (6.8.17) hold. If the second condition holds, it is enough to consider the sequence $\{-f_n\}$, which obviously satisfies the first condition. □

The monotone convergence theorem has a number of useful corollaries, which will be needed later. The first result is an analogue of Levi's theorem for term-by-term integration of infinite series.

6.8.15 Theorem. *Assume that $f_k:[a,b]\to\mathbb{R}$ is a nonnegative function for each $k\in\mathbb{N}$, $f=\sum_{k=1}^{\infty} f_k$, and $g:[a,b]\to\mathbb{R}$ is nondecreasing. If the integral $\int_a^b f_k\,\mathrm{d}g$ exists for each $k\in\mathbb{N}$ and if the sum $\sum_{k=1}^{\infty}\int_a^b f_k\,\mathrm{d}g$ is finite, then the sequence $\{s_n\}$ given by $s_n=\sum_{k=1}^{n} f_k$ on $[a,b]$ is uniformly integrable with respect to g and*

$$\int_a^b f\,\mathrm{d}g=\sum_{k=1}^{\infty}\int_a^b f_k\,\mathrm{d}g.$$

Moreover, we have

$$\lim_{n\to\infty}\Big(\sup_{t\in[a,b]}\Big|\int_a^t s_n\,\mathrm{d}g-\int_a^t f\,\mathrm{d}g\Big|\Big)=0.$$

Proof. The sequence $\{s_n\}$ is nondecreasing, the integral $\int_a^b s_n\,\mathrm{d}g$ exists for each $n\in\mathbb{N}$, and $\lim_{n\to\infty} s_n=f$ on $[a,b]$. Moreover, we have

$$\begin{aligned}\lim_{n\to\infty}\int_a^b s_n\,\mathrm{d}g&=\lim_{n\to\infty}\int_a^b\Big(\sum_{k=1}^{n} f_k\Big)\,\mathrm{d}g\\&=\lim_{n\to\infty}\sum_{k=1}^{n}\int_a^b f_k\,\mathrm{d}g=\sum_{k=1}^{\infty}\int_a^b f_k\,\mathrm{d}g<\infty.\end{aligned}$$

Thus, the statement of the present theorem follows from the monotone convergence theorem. □

6.8.16 Remark. Let $g:[a,b]\to\mathbb{R}$ be nondecreasing and $f=\sum_{k=1}^{\infty} f_k$ on $[a,b]$, where each $f_k:[a,b]\to\mathbb{R}$ is a nonnegative function such that $\int_a^b f_k\,\mathrm{d}g$ exists. If we know that $\int_a^b f\,\mathrm{d}g$ exists, then

$$\sum_{k=1}^{n}\int_a^b f_k\,\mathrm{d}g=\int_a^b\Big(\sum_{k=1}^{n} f_k\Big)\mathrm{d}g\le\int_a^b f\,\mathrm{d}g\quad\text{for each } n\in\mathbb{N}.$$

Hence, $\sum_{k=1}^{\infty}\int_a^b f_k\,\mathrm{d}g$ is finite and equals $\int_a^b f\,\mathrm{d}g$ by Theorem 6.8.15. In other words, we have established the following result:

$\int_a^b\sum_{k=1}^{\infty} f_k\,\mathrm{d}g$ exists if and only if $\sum_{k=1}^{\infty}\int_a^b f_k\,\mathrm{d}g$ is finite;

in this case, both expressions have the same value.

The monotone convergence theorem also leads to the following result.

6.8.17 Lemma. *If $g:[a,b]\to\mathbb{R}$ is nondecreasing and $\{f_n\}$ is a sequence of nonnegative functions such that the integral $\int_a^b f_n\,\mathrm{d}g$ exists for each $n\in\mathbb{N}$, then $\int_a^b(\inf_{n\in\mathbb{N}} f_n)\,\mathrm{d}g$ exists as well.*

Proof. For each $n\in\mathbb{N}$, let $h_n=\min\{f_1,\dots,f_n\}$, and note that Theorem 6.7.5 guarantees the existence of the integral $\int_a^b h_n\,\mathrm{d}g\geq 0$. The sequence $\{h_n\}$ is nonincreasing and pointwise convergent to $\inf_{k\in\mathbb{N}} f_k$. Hence, the monotone convergence theorem implies the existence of the integral $\int_a^b(\inf_{k\in\mathbb{N}} f_k)\,\mathrm{d}g$. □

Another useful corollary of the monotone convergence theorem is Fatou's lemma.

6.8.18 Lemma (FATOU'S LEMMA). *Let $g:[a,b]\to\mathbb{R}$ be nondecreasing and let $\{f_n\}$ be a sequence of functions such that the integral $\int_a^b f_n\,\mathrm{d}g$ exists for each $n\in\mathbb{N}$. Then the following statements hold:*

(i) *Assume there is a function $\varphi:[a,b]\to\mathbb{R}$ such that $\int_a^b\varphi\,\mathrm{d}g$ exists and $f_n\geq\varphi$ for each $n\in\mathbb{N}$. If $\liminf_{n\to\infty} f_n(x)<\infty$ for each $x\in[a,b]$ and $\liminf_{n\to\infty}\int_a^b f_n\,\mathrm{d}g<\infty$, then $\int_a^b(\liminf_{n\to\infty} f_n)\,\mathrm{d}g$ exists, and we have*

$$\int_a^b(\liminf_{n\to\infty} f_n)\,\mathrm{d}g\leq\liminf_{n\to\infty}\int_a^b f_n\,\mathrm{d}g.$$

(ii) *Assume there is a function $\psi:[a,b]\to\mathbb{R}$ such that $\int_a^b\psi\,\mathrm{d}g$ exists and $f_n\leq\psi$ for each $n\in\mathbb{N}$. If $\limsup_{n\to\infty} f_n(x)>-\infty$ for each $x\in[a,b]$ and $\limsup_{n\to\infty}\int_a^b f_n\,\mathrm{d}g>-\infty$, then $\int_a^b(\limsup_{n\to\infty} f_n)\,\mathrm{d}g$ exists, and we have*

$$\int_a^b(\limsup_{n\to\infty} f_n)\,\mathrm{d}g\geq\limsup_{n\to\infty}\int_a^b f_n\,\mathrm{d}g.$$

Proof. Let us prove the first statement. Without loss of generality, we can assume that $\varphi=0$ (otherwise, it is enough to consider the sequence $\{f_n-\varphi\}$). For each $n\in\mathbb{N}$, let $h_n=\inf_{k\geq n} f_k$. The sequence $\{h_n\}$ consists of nonnegative functions, it is nondecreasing, and pointwise convergent to $\liminf_{n\to\infty} f_n$. By Lemma 6.8.17, the integral $\int_a^b h_n\,\mathrm{d}g$ exists for each $n\in\mathbb{N}$. Obviously, we have

$$\int_a^b h_n\,\mathrm{d}g\leq\int_a^b f_n\,\mathrm{d}g,$$

and therefore

$$\lim_{n\to\infty} \int_a^b h_n \, dg = \liminf_{n\to\infty} \int_a^b h_n \, dg \le \liminf_{n\to\infty} \int_a^b f_n \, dg < \infty.$$

The monotone convergence theorem implies the existence of the integral $\int_a^b (\lim_{n\to\infty} h_n) \, dg$, and we get

$$\int_a^b (\liminf_{n\to\infty} f_n) \, dg = \int_a^b (\lim_{n\to\infty} h_n) \, dg = \lim_{n\to\infty} \int_a^b h_n \, dg \le \liminf_{n\to\infty} \int_a^b f_n \, dg.$$

To prove that the second statement holds, it is enough to consider the sequence $\{-f_n\}$ and apply the first statement. □

6.9 Integration over elementary sets

Up to now we have been discussing only integrals over closed intervals. In Lebesgue's integration theory, it is common to integrate over more general (measurable) sets. The generality of the Kurzweil-Stieltjes integral (in the sense of a large domain of integrable functions) seems to be the source of troubles with defining the Kurzweil-Stieltjes integral over general sets, in particular if we do not want to restrict ourselves to continuous integrators. A reasonable compromise is to limit the considerations to elementary sets, i.e., finite unions of intervals (cf. Definition 2.8.12). In this case the following definition turned out to be useful for our purposes.

6.9.1 Definition. Let $f: [a,b] \to \mathbb{R}$, $g: [a,b] \to \mathbb{R}$ and an elementary subset E of $[a,b]$ be given. The Kurzweil-Stieltjes integral of f with respect to g over E is given by

$$\int_E f \, dg = \int_a^b (f\,\chi_E) \, dg$$

provided the integral on the right-hand side exists.

According to Definition 6.9.1, the existence of the integral $\int_E f \, dg$ means that there is an $I \in \mathbb{R}$ such that for every $\varepsilon > 0$ there is a gauge δ on $[a,b]$ such that

$$\left| S(f\chi_E, dg, P) - I \right| < \varepsilon \quad \text{for all } \delta\text{-fine partitions } P \text{ of } [a,b].$$

As we will see later, the value of the integral $\int_E f \, dg$ can depend on the choice of the interval $[a,b]$ which contains E. Hence, throughout this section the interval $[a,b]$ is assumed to be fixed. Further, we use extensively the following convention: Functions g defined on the interval $[a,b]$ are supposed to be extended to some open interval containing $[a,b]$ in such a way that

$$g(a-)=g(a), \quad g(b+)=g(b), \quad \text{i.e.,} \ \ \Delta^{-}g(a)=\Delta^{+}g(b)=0.$$

The basic properties of the Kurzweil-Stieltjes integral over elementary sets given below are immediate consequences of the corresponding properties of the Kurzweil-Stieltjes integral presented in the previous sections.

6.9.2 Theorem. *Assume that E is an elementary subset of $[a,b]$, and $c_1, c_2 \in \mathbb{R}$ are arbitrary numbers.*

(i) *If $f_1, f_2, g\colon [a,b] \to \mathbb{R}$ are such that the integrals $\int_E f_1 \, \mathrm{d}g$ and $\int_E f_2 \, \mathrm{d}g$ exist, then*

$$\int_E (c_1 f_1 + c_2 f_2)\, \mathrm{d}g = c_1 \int_E f_1 \, \mathrm{d}g + c_2 \int_E f_2 \, \mathrm{d}g.$$

(ii) *If $f, g_1, g_2\colon [a,b] \to \mathbb{R}$ are such that the integrals $\int_E f \, \mathrm{d}g_1$ and $\int_E f \, \mathrm{d}g_2$ exist, then*

$$\int_E f \, \mathrm{d}(c_1 g_1 + c_2 g_2) = c_1 \int_E f \, \mathrm{d}g_1 + c_2 \int_E f \, \mathrm{d}g_2.$$

Using Theorem 6.9.2 the following additivity result of the integral with respect to elementary sets can be justified.

6.9.3 Theorem. *Let E_1 and E_2 be disjoint elementary subsets of $[a,b]$. If $f, g\colon [a,b] \to \mathbb{R}$ are such that the integrals $\int_{E_1} f \, \mathrm{d}g$ and $\int_{E_2} f \, \mathrm{d}g$ exist, then the integral $\int_{E_1 \cup E_2} f \, \mathrm{d}g$ exists as well, and*

$$\int_{E_1 \cup E_2} f \, \mathrm{d}g = \int_{E_1} f \, \mathrm{d}g + \int_{E_2} f \, \mathrm{d}g.$$

6.9.4 Remark. Note that if E is an elementary subset of $[a,b]$ and $h = 0$ on E, then obviously $\int_E h \, \mathrm{d}g = 0$ for each $g\colon [a,b] \to \mathbb{R}$. Therefore, if the integral $\int_E f_1 \, \mathrm{d}g$ exists, then the equality

$$\int_E f_1 \, \mathrm{d}g = \int_E f_2 \, \mathrm{d}g$$

holds for each function f_2 such that $f_2 = f_1$ on E.

Later in this section we will provide conditions ensuring the existence of the integral over elementary sets (see Corollary 6.9.10). For now, let us simply highlight the following consequence of Theorem 6.3.11.

6.9.5 Theorem. *Let $f \in \mathrm{G}([a,b])$ and $g \in \mathrm{BV}([a,b])$. If E is an elementary subset of $[a,b]$, then the integral $\int_E f\,\mathrm{d}g$ exists.*

Proof. It is enough to observe that $f\,\chi_E : [a,b] \to \mathbb{R}$ is a regulated function whenever $f \in \mathrm{G}([a,b])$ and E is an elementary subset of $[a,b]$. □

Now we are going to consider the properties of integrals over arbitrary bounded intervals. For degenerate intervals, i.e., singleton sets, we have the following result.

6.9.6 Theorem. *Let $\tau \in [a,b]$ and let $g : [a,b] \to \mathbb{R}$ have finite one-sided limits $g(\tau-)$ and $g(\tau+)$. Then the integral $\int_{[\tau]} f\,\mathrm{d}g$ exists and*

$$\int_{[\tau]} f\,\mathrm{d}g = f(\tau)\,\Delta g(\tau).$$

Proof. Note that $\big(f\chi_{[\tau]}\big)(t) = f(\tau)\chi_{[\tau]}(t)$ for all $t \in [a,b]$. Hence,

$$\int_{[\tau]} f\,\mathrm{d}g = \int_a^b f\chi_{[\tau]}\,\mathrm{d}g = f(\tau)\int_a^b \chi_{[\tau]}\,\mathrm{d}g$$

and the result follows from Lemma 6.3.3. □

Integration over subintervals of $[a,b]$ is described by the following assertion.

6.9.7 Theorem. *Let $f : [a,b] \to \mathbb{R}$, $g \in \mathrm{G}([a,b])$ and $a \le c < d \le b$. Then if one of the integrals*

$$\int_{[c,d]} f\,\mathrm{d}g, \quad \int_{(c,d)} f\,\mathrm{d}g, \quad \int_{[c,d)} f\,\mathrm{d}g, \quad \int_{(c,d]} f\,\mathrm{d}g, \quad \int_c^d f\,\mathrm{d}g \tag{6.9.1}$$

exists, all the others exist as well. In this case, we have the following equalities:

$$\int_{[c,d]} f\,\mathrm{d}g = \quad f(c)\,\Delta^- g(c) + \int_c^d f\,\mathrm{d}g + f(d)\,\Delta^+ g(d), \tag{6.9.2}$$

$$\int_{(c,d)} f\,\mathrm{d}g = -f(c)\,\Delta^+ g(c) + \int_c^d f\,\mathrm{d}g - f(d)\,\Delta^- g(d), \tag{6.9.3}$$

$$\int_{[c,d)} f\,\mathrm{d}g = \quad f(c)\,\Delta^- g(c) + \int_c^d f\,\mathrm{d}g - f(d)\,\Delta^- g(d), \tag{6.9.4}$$

$$\int_{(c,d]} f\,\mathrm{d}g = -f(c)\,\Delta^+ g(c) + \int_c^d f\,\mathrm{d}g + f(d)\,\Delta^+ g(d). \tag{6.9.5}$$

P r o o f. Note that

$$\int_{[c,d]} f\,\mathrm{d}g = \int_a^c (f\chi_{[c,d]})\,\mathrm{d}g + \int_c^d f\,\mathrm{d}g + \int_d^b (f\chi_{[c,d]})\,\mathrm{d}g.$$

Clearly, the first and third integral on the right-hand side exist by Lemma 6.3.3. Thus, the integral over $[c,d]$ exists if and only if the integral $\int_c^d f\,\mathrm{d}g$ exists. In this case,

$$\begin{aligned}\int_{[c,d]} f\,\mathrm{d}g &= f(c)\int_a^c \chi_{[c]}\,\mathrm{d}g + \int_c^d f\,\mathrm{d}g + f(d)\int_d^b \chi_{[d]}\,\mathrm{d}g \\ &= f(c)\,\Delta^- g(c) + \int_c^d f\,\mathrm{d}g + f(d)\,\Delta^+ g(d).\end{aligned}$$

Since both integrals $\int_{[c]} f\,\mathrm{d}g$ and $\int_{[d]} f\,\mathrm{d}g$ exist by Theorem 6.9.6, the equivalence between the existence of the integral over $[c,d]$ and the existence of the integral over the corresponding open or half-open intervals can be easily derived from the relations

$$\chi_{[c,d]} = \chi_{[c]} + \chi_{(c,d)} + \chi_{[d]} = \chi_{[c]} + \chi_{(c,d]} = \chi_{[c,d)} + \chi_{[d]}.$$

The equalities (6.9.3), (6.9.4) and (6.9.5) follow from (6.9.2) using Theorem 6.9.6. The detailed proof is left as an exercise for the reader. □

6.9.8 Remark. According to Theorem 6.9.7, if $f:[a,b]\to\mathbb{R}$, $g\in\mathrm{C}([a,b])$ and $a\le c\le d\le b$ are such that one of the integrals in (6.9.1) exists, then

$$\int_{[c,d)} f\,\mathrm{d}g = \int_{(c,d]} f\,\mathrm{d}g = \int_{(c,d)} f\,\mathrm{d}g = \int_c^d f\,\mathrm{d}g = \int_{[c,d]} f\,\mathrm{d}g.$$

We are now ready to evaluate integrals over elementary sets. To this end, we will make use of the notion of minimal decomposition of an elementary set (see Definition 2.8.12).

6.9.9 Theorem. *Let* $f:[a,b]\to\mathbb{R}$, $g\in \mathrm{G}([a,b])$ *and let* E *be an elementary subset of* $[a,b]$ *with the minimal decomposition* $\{J_k : k=1,\dots,N\}$. *If the integral* $\int_E f\,\mathrm{d}g$ *exists, then also the integrals* $\int_{J_k} f\,\mathrm{d}g$ *exist for all* $k\in\{1,\dots,N\}$ *and*

$$\int_E f\,\mathrm{d}g=\sum_{k=1}^{N}\int_{J_k} f\,\mathrm{d}g. \tag{6.9.6}$$

Proof. For $k=1,\dots,N$, let $c_k=\inf J_k$ and $d_k=\sup J_k$. By the hypothesis, the integral

$$\int_E f\,\mathrm{d}g=\int_a^b (f\chi_E)\,\mathrm{d}g$$

exists, and hence, by Theorem 6.2.9, so do all the integrals

$$\int_{c_k}^{d_k} (f\chi_E)\,\mathrm{d}g \quad \text{for} \quad k\in\{1,\dots,N\}.$$

Note that

$$\int_{c_k}^{d_k} f(\chi_E-\chi_{J_k})\,\mathrm{d}g=0, \quad \text{for} \quad k\in\{1,\dots,N\}$$

(due to the fact that $f(\chi_E-\chi_{J_k})$ vanishes on $[c_k,d_k]$) and, similarly, the integrals

$$\int_a^{c_k} (f\chi_{J_k})\,\mathrm{d}g \quad \text{and} \quad \int_{d_k}^{b} (f\chi_{J_k})\,\mathrm{d}g$$

are either zero or integrals over singleton sets (depending on whether c_k and/or d_k belong to J_k). Since all of these integrals exist, we conclude that

$$\int_{J_k} f\,\mathrm{d}g=\int_a^b (f\,\chi_{J_k})\,\mathrm{d}g$$

exist for all $k\in\{1,\dots,N\}$. Having in mind that the intervals of the minimal decomposition are pairwise disjoint, (6.9.6) follows from Theorem 6.9.3. □

Theorem 6.9.9 means that if the integral over an elementary set exists, so do the integrals over subintervals of the minimal decomposition. This implies immediately the following assertions.

6.9.10 Corollary. *If E is an elementary subset of $[a,b]$, $f:[a,b]\to\mathbb{R}$ and $g\in \mathrm{G}([a,b])$ are such that the integral $\int_E f\,\mathrm{d}g$ exists, then the integral $\int_T f\,\mathrm{d}g$ exists for every elementary subset T of E.*

In particular, if the integral $\int_a^b f\,\mathrm{d}g$ exists, then $\int_T f\,\mathrm{d}g$ exists for every elementary subset T of $[a,b]$.

Proof. Let T be an elementary subset of E and let $\{I_\ell : \ell=1,\dots,p\}$ be its minimal decomposition. Assume that $\{J_k : k=1,\dots,N\}$ is the minimal decomposition of E. Given an arbitrary $\ell\in\{1,\dots,p\}$, there exists a $k_\ell\in\{1,\dots,N\}$ such that $I_\ell\subset J_{k_\ell}$. By Theorem 6.9.9, we know that the integral $\int_a^b (f\,\chi_{J_{k_\ell}})\,\mathrm{d}g$ exists; therefore $\int_{a_\ell}^{b_\ell}(f\,\chi_{J_{k_\ell}})\,\mathrm{d}g$ also exists, where $a_\ell=\inf I_\ell$ and $b_\ell=\sup I_\ell$. Applying Theorem 6.9.7 we conclude that the integral $\int_{I_\ell}(f\,\chi_{J_{k_\ell}})\,\mathrm{d}g$ exists, while $\chi_{J_{k_\ell}}\chi_{I_\ell}=\chi_{I_\ell}$ implies that

$$\int_{I_\ell} f\,\mathrm{d}g=\int_{I_\ell}(f\,\chi_{J_{k_\ell}})\,\mathrm{d}g.$$

Since $\ell\in\{1,\dots,p\}$ is arbitrary and the intervals of the minimal decomposition are pairwise disjoint, Theorem 6.9.3 implies the existence of the integral over T. $\square$

6.9.11 Corollary. *Consider functions $f:[a,b]\to\mathbb{R}$ and $g\in\mathrm{G}([a,b])$, and let E_1 and E_2 be elementary subsets of $[a,b]$. Then the integral $\int_{E_1\cup E_2} f\,\mathrm{d}g$ exists if and only the integrals $\int_{E_1} f\,\mathrm{d}g$ and $\int_{E_2} f\,\mathrm{d}g$ exist. In this case,*

$$\int_{E_1\cup E_2} f\,\mathrm{d}g=\int_{E_1} f\,\mathrm{d}g+\int_{E_2} f\,\mathrm{d}g-\int_{E_1\cap E_2} f\,\mathrm{d}g.$$

The following estimates are, in a sense, the analogues of the estimates from Section 6.3.

6.9.12 Theorem. *Let J be a subinterval of $[a,b]$ with $c=\inf J$ and $d=\sup J$, where $c<d$. Assume that $f:[a,b]\to\mathbb{R}$ and $g\in\mathrm{BV}([a,b])$ are such that the integral $\int_J f\,\mathrm{d}g$ exists. Then the following assertions hold:*

(i) *If $J=[c,d]$, then*

$$\Big|\int_J f\,\mathrm{d}g\Big|\le\Big(\sup_{t\in J}|f(t)|\Big)\,\mathrm{var}(g,J)+|f(c)|\,|\Delta^- g(c)|+|f(d)|\,|\Delta^+ g(d)|.$$

(ii) *If $J=[c,d)$, then*

$$\Big|\int_J f\,\mathrm{d}g\Big|\le\Big(\sup_{t\in J}|f(t)|\Big)\,\mathrm{var}(g,J)+|f(c)||\Delta^- g(c)|.$$

(iii) *If* $J=(c,d]$, *then*

$$\Big|\int_J f\,\mathrm{d}g\Big| \le \Big(\sup_{t\in J}|f(t)|\Big)\ \mathrm{var}(g,J)+|f(d)||\Delta^+ g(d)|.$$

(iv) *If* $J=(c,d)$, *then*

$$\Big|\int_J f\,\mathrm{d}g\Big| \le \Big(\sup_{t\in J}|f(t)|\Big)\ \mathrm{var}(g,J).$$

Proof. (i) In the case $J=[c,d]$, the inequality is an obvious consequence of (6.9.2) and Theorem 6.3.6.

(ii) By (6.9.4) and Theorem 6.5.3 we get

$$\begin{aligned}\int_{[c,d)} f\,\mathrm{d}g &= f(c)\,\Delta^- g(c)+\int_c^d f\,\mathrm{d}g - f(d)\,\Delta^- g(d)\\ &= f(c)\,\Delta^- g(c)+\lim_{s\to d-}\int_c^s f\,\mathrm{d}g.\end{aligned}$$

This together with Theorem 2.8.5 implies

$$\begin{aligned}\Big|\int_{[c,d)} f\,\mathrm{d}g\Big| &\le |f(c)|\,|\Delta^- g(c)|+\lim_{s\to d-}\ \big(\sup_{t\in[c,s]}|f(t)|\ \mathrm{var}_c^s\, g\big)\\ &= |f(c)|\,|\Delta^- g(c)|+\Big(\sup_{t\in[c,d)}|f(t)|\Big)\ \mathrm{var}(g,[c,d)).\end{aligned}$$

(iii) To obtain the corresponding estimate in the case when $J=(c,d]$, we use (6.9.5) and follow the same arguments used in the proof of (ii).

(iv) For an arbitrary but fixed $\tau\in(c,d)$, by (6.9.3) we have

$$\begin{aligned}\int_{(c,d)} f\,\mathrm{d}g &= \int_c^d f\,\mathrm{d}g - f(c)\,\Delta^+ g(c)-f(d)\,\Delta^- g(d)\\ &= \int_c^\tau f\,\mathrm{d}g - f(c)\,\Delta^+ g(c)+\int_\tau^d f\,\mathrm{d}g - f(d)\,\Delta^- g(d).\end{aligned}$$

Thus, applying Theorem 6.5.3 we obtain

$$\int_{(c,d)} f\,\mathrm{d}g = \lim_{s\to c+}\int_s^\tau f\,\mathrm{d}g+\lim_{s\to d-}\int_\tau^s f\,\mathrm{d}g,$$

and consequently

$$\Big|\int_{(c,d)} f\,\mathrm{d}g\Big| \leq \lim_{s\to c+}\Big(\sup_{t\in[s,\tau]}|f(t)|\,\mathrm{var}_s^\tau\, g\Big) + \lim_{s\to d-}\Big(\sup_{t\in[\tau,s]}|f(t)|\,\mathrm{var}_\tau^s\, g\Big)$$
$$= \Big(\sup_{t\in(c,d)}|f(t)|\Big)\Big(\lim_{s\to c+}\mathrm{var}_s^\tau\, g + \lim_{s\to d-}\mathrm{var}_\tau^s\, g\Big).$$

The estimate in (iv) then follows from Theorem 2.8.5. □

As a consequence of Definition 2.8.13, Theorem 6.9.12 and Theorem 6.9.9 we have the following result.

6.9.13 Corollary. *If E is an elementary subset of $[a,b]$, $f:[a,b]\to\mathbb{R}$ and $g\in\mathrm{BV}([a,b])\cap\mathrm{C}([a,b])$ are such that the integral $\int_E f\,\mathrm{d}g$ exists, then*

$$\Big|\int_E f\,\mathrm{d}g\Big| \leq \Big(\sup_{t\in E}|f(t)|\Big)\,\mathrm{var}(g,E).$$

6.10 INTEGRALS OF VECTOR, MATRIX AND COMPLEX FUNCTIONS

Throughout this book, the norm of a matrix $A\in\mathscr{L}(\mathbb{R}^m,\mathbb{R}^n)$ is denoted by the symbol $|A|$ and defined by

$$|A| = \max_{j=1,\ldots,m}\sum_{i=1}^{n}|a_{i,j}|.$$

Vectors from $\mathbb{R}^n$ are identified with $n\times 1$ matrices (i.e., they are treated as column vectors). In other words, we identify the spaces $\mathbb{R}^n$ and $\mathscr{L}(\mathbb{R}^n,\mathbb{R})$. Consequently, the norm of a vector $x\in\mathbb{R}^n$ is

$$|x| = \sum_{i=1}^{n}|x_i|.$$

Clearly, we have $|A\,x|\leq|A|\,|x|$ for all $A\in\mathscr{L}(\mathbb{R}^m,\mathbb{R}^n)$, $x\in\mathbb{R}^n$. It is also known that

$$|A| = \sup\big\{|A\,x| : x\in\mathbb{R}^n,\ |x|\leq 1\big\}.$$

Let $F:[a,b]\to\mathscr{L}(\mathbb{R}^m,\mathbb{R}^p)$ and $G:[a,b]\to\mathscr{L}(\mathbb{R}^p,\mathbb{R}^n)$ be matrix functions with components

$$f_{i,k},\quad i\in\{1,\ldots,m\},\quad k\in\{1,\ldots,p\},$$

and

$$g_{k,j}, \quad k \in \{1,\dots,p\}, \quad j \in \{1,\dots,n\},$$

respectively. Then, if all the integrals

$$\int_a^b f_{i,k}\,\mathrm{d}g_{k,j} \quad i \in \{1,\dots,m\}, \quad k \in \{1,\dots,p\}, \quad j \in \{1,\dots,n\}$$

exist, the symbols

$$\int_a^b F(t)\,\mathrm{d}G(t) \quad \text{or} \quad \int_a^b F\,\mathrm{d}G$$

stand for the $m \times n$ matrix $M \in \mathscr{L}(\mathbb{R}^m, \mathbb{R}^n)$ with elements

$$m_{i,j} = \sum_{k=1}^{p} \int_a^b f_{i,k}\,\mathrm{d}g_{k,j}, \quad i = 1,\dots,m, \quad j = 1,\dots,n.$$

6.10.1 Exercise. If $\int_a^b F(t)\,\mathrm{d}G(t)$ exists and equals $M \in \mathscr{L}(\mathbb{R}^m, \mathbb{R}^n)$, verify that for each $\varepsilon > 0$, there is a gauge δ on $[a,b]$ such that the inequality

$$\Big| \sum_{\ell=1}^{\nu(P)} F(\xi_\ell)\,(G(\alpha_\ell) - G(\alpha_{\ell-1})) - M \Big| < \varepsilon$$

holds for each δ-fine partition $P = (\boldsymbol{\alpha}, \boldsymbol{\xi})$ of $[a,b]$.

Analogously, we can define the integrals $\int_a^b \mathrm{d}F\,G$, or $\int_a^b F\,\mathrm{d}G\,H$, where the values of F, G and H are matrices of appropriate dimensions.

The variation of a matrix-valued function $F\colon [a,b] \to \mathscr{L}(\mathbb{R}^m, \mathbb{R}^n)$ is defined by the same formula as in the scalar case, i.e.,

$$\operatorname{var}_a^b F = \sup_{\boldsymbol{\alpha} \in \mathscr{D}[a,b]} \sum_{j=1}^{\nu(\boldsymbol{\alpha})} |F(\alpha_j) - F(\alpha_{j-1})|.$$

One can easily verify the inequalities

$$\max_{\substack{i=1,\dots,m \\ j=1,\dots,n}} \big(\operatorname{var}_a^b f_{i,j}\big) \le \operatorname{var}_a^b F \le \sum_{i=1}^{m} \sum_{j=1}^{n} \operatorname{var}_a^b f_{i,j}.$$

This means that a matrix function $F\colon [a,b] \to \mathscr{L}(\mathbb{R}^m, \mathbb{R}^n)$ has bounded variation if and only if all its components have bounded variation. Similarly,

F is continuous or regulated if and only if all its components have the same property.

The results obtained in this and in the preceding chapter are easily generalized for matrix or vector functions. One only needs to keep in mind that matrix multiplication is in general not commutative, i.e., it is not allowed to change the order of matrix-valued functions involved in various products. For example, Theorem 6.4.2 (integration by parts) has to be formulated as follows:

If $F:[a,b]\to\mathscr{L}(\mathbb{R}^m,\mathbb{R}^p)$ is regulated and $G:[a,b]\to\mathscr{L}(\mathbb{R}^p,\mathbb{R}^n)$ has bounded variation, then both the integrals $\int_a^b F\,\mathrm{d}G$ and $\int_a^b \mathrm{d}F\,G$ exist, and we have

$$\int_a^b F\,\mathrm{d}G+\int_a^b \mathrm{d}F\,G=F(b)\,G(b)-F(a)\,G(a)$$
$$+\sum_{x\in[a,b]}\Big(\Delta^-F(x)\,\Delta^-G(x)-\Delta^+F(x)\,\Delta^+G(x)\Big).$$

Similarly, Theorem 6.6.1 (substitution theorem) can be generalized for matrix functions as follows:

Let $F:[a,b]\to\mathscr{L}(\mathbb{R}^m,\mathbb{R}^p)$ be bounded and let $G:[a,b]\to\mathscr{L}(\mathbb{R}^p,\mathbb{R}^q)$ and $H:[a,b]\to\mathscr{L}(\mathbb{R}^q,\mathbb{R}^n)$ be such that the integral $\int_a^b G\,\mathrm{d}H$ exists. If one of the integrals

$$\int_a^b F(x)\,\mathrm{d}\Big(\int_a^x G\,\mathrm{d}H\Big) \quad \textit{and} \quad \int_a^b (F\,G)\,\mathrm{d}H,$$

exists, then the other exists as well, and we have

$$\int_a^b F(x)\,\mathrm{d}\Big(\int_a^x G\,\mathrm{d}H\Big)=\int_a^b (F\,G)\,\mathrm{d}H.$$

Finally, let us consider Kurzweil-Stieltjes integrals of complex-valued functions. Given a pair of functions $f,g:[a,b]\to\mathbb{C}$ with real parts f_1,g_1 and imaginary parts f_2,g_2, we define

$$\int_a^b f\,\mathrm{d}g=\int_a^b (f_1+\mathrm{i}f_2)\,\mathrm{d}(g_1+\mathrm{i}g_2)$$
$$=\int_a^b f_1\,\mathrm{d}g_1-\int_a^b f_2\,\mathrm{d}g_2+\mathrm{i}\Big(\int_a^b f_1\,\mathrm{d}g_2+\int_a^b f_2\,\mathrm{d}g_1\Big)$$

whenever all four integrals on the right-hand side exist.

Again, most results obtained in this chapter, such as the integration by parts formula or substitution theorem, are still valid for complex-valued functions. We leave the verification of this fact up to the reader; the proofs are straightforward and based on the decomposition of complex functions into the real and imaginary parts.

6.10.2 Exercise. Given a pair of functions $f, g: [a,b] \to \mathbb{C}$, verify that if $\int_a^b f\, \mathrm{d}g$ exists and equals $I \in \mathbb{C}$, then for each $\varepsilon > 0$, there is a gauge δ on $[a,b]$ such that the inequality

$$\Big|\sum_{j=1}^{\nu(P)} f(\xi_j)\,(g(\alpha_j) - g(\alpha_{j-1})) - I\Big| < \varepsilon$$

holds for each δ-fine partition $P = (\boldsymbol{\alpha}, \boldsymbol{\xi})$ of $[a,b]$. In this context, the symbol $|z|$ denotes the absolute value of a complex number z.

6.10.3 Exercise. Let $f, g: [a,b] \to \mathbb{C}$ be such that the integral $\int_a^b f\, \mathrm{d}g$ exists. Show that

$$\overline{\int_a^b f\, \mathrm{d}g} = \int_a^b \bar{f}\, \mathrm{d}\bar{g},$$

where the symbol $\bar{z}$ denotes the complex conjugate of a complex number z.

6.11 RELATION TO THE PERRON-STIELTJES INTEGRAL

In the rest of this chapter, we explore the relations between the Kurzweil-Stieltjes integral and other concepts of integral. We begin with the *Perron-Stieltjes integral*, whose definition is based on the notion of major and minor functions.

6.11.1 Definition. Let $f, g: [a,b] \to \mathbb{R}$. We say that $M: [a,b] \to \mathbb{R}$ is a *major function* of f with respect to g if there exists a gauge $\delta: [a,b] \to (0,\infty)$ such that the inequality

$$f(\tau)\,(g(y) - g(x)) \le M(y) - M(x)$$

holds whenever $[x,y] \subset (\tau - \delta(\tau), \tau + \delta(\tau)) \cap [a,b]$ and $\tau \in [x,y]$.

Similarly, $m: [a,b] \to \mathbb{R}$ is a *minor function* of f with respect to g if there exists a gauge $\delta: [a,b] \to (0,\infty)$ such that the inequality

$$f(\tau)\,(g(y) - g(x)) \ge m(y) - m(x)$$

holds whenever $[x,y]\subset(\tau-\delta(\tau),\tau+\delta(\tau))\cap[a,b]$ and $\tau\in[x,y]$.

The set of all major functions of f with respect to g is denoted by $\mathfrak{M}(f,\mathrm{d}g)$, whereas $\mathfrak{m}(f,\mathrm{d}g)$ stands for the set of all minor functions of f with respect to g.

In the special case when $g(x)=x$ for all $x\in[a,b]$, we speak about a major function of f or a minor function of f, respectively. The next example provides some insight into the concept of major and minor functions.

6.11.2 Example. Let $F:[a,b]\to\mathbb{R}$ be a differentiable function with $F'=f$ on $[a,b]$ (where the derivative is understood as one-sided at both endpoints). We claim that for each $\varepsilon>0$, the function

$$M_\varepsilon(t)=F(t)+t\,\varepsilon,\quad t\in[a,b],$$

is a major function of f, and the function

$$m_\varepsilon(t)=F(t)-t\,\varepsilon,\quad t\in[a,b],$$

is a minor function of f.

We verify the first statement and leave the second to the reader. Choose an arbitrary $\tau\in[a,b]$ and let $\varepsilon>0$ be given. Since $F'(\tau)=f(\tau)$, there exists a $\delta(\tau)>0$ such that

$$|F(z)-F(\tau)-f(\tau)\,(z-\tau)|<\varepsilon\,|z-\tau|\quad\text{for}\ \ z\in(\tau-\delta(\tau),\tau+\delta(\tau))\cap[a,b].$$

Consequently,

$$f(\tau)\,(y-\tau)<F(y)-F(\tau)+\varepsilon\,(y-\tau)=M_\varepsilon(y)-M_\varepsilon(\tau)$$

for all $y\in[\tau,\tau+\delta(\tau))\cap[a,b]$, and

$$f(\tau)\,(\tau-x)<F(\tau)-F(x)+\varepsilon\,(\tau-x)=M_\varepsilon(\tau)-M_\varepsilon(x)$$

for all $x\in(\tau-\delta(\tau),\tau]\cap[a,b]$. By adding the previous two inequalities, we see that M_ε is a major function of f.

6.11.3 Definition. Given a pair of functions $f,g:[a,b]\to\mathbb{R}$, we define the *upper Perron-Stieltjes integral* of f with respect to g by

$$(\mathrm{PS})\overline{\int_a^b} f\,\mathrm{d}g=\inf\{M(b)-M(a):M\in\mathfrak{M}(f,\mathrm{d}g)\}$$

(with the convention that $\inf \emptyset = \infty$), and the *lower Perron-Stieltjes integral* of f with respect to g by

$$(\mathrm{PS})\underline{\int_a^b} f\,\mathrm{d}g = \sup\{m(b) - m(a) : m \in \mathfrak{m}(f, \mathrm{d}g)\}$$

(with the convention that $\sup \emptyset = -\infty$).

The next statement shows that the value of the lower integral never exceeds the value of the upper integral.

6.11.4 Lemma. *If $f, g : [a, b] \to \mathbb{R}$, then the following statements hold:*

(i) *If $M \in \mathfrak{M}(f, \mathrm{d}g)$ and $m \in \mathfrak{m}(f, \mathrm{d}g)$, there is a gauge δ on $[a, b]$ such that*

$$m(b) - m(a) \le S(P) \le M(b) - M(a)$$

for each δ-fine partition P of $[a, b]$.

(ii) $(\mathrm{PS})\underline{\int_a^b} f\,\mathrm{d}g \le (\mathrm{PS})\overline{\int_a^b} f\,\mathrm{d}g.$

Proof. To prove the first statement, let $M \in \mathfrak{M}(f, \mathrm{d}g)$ and $m \in \mathfrak{m}(f, \mathrm{d}g)$. By definition, there is a gauge δ_1 on $[a, b]$ such that

$$f(\tau)\,(g(y) - g(x)) \le M(y) - M(x)$$

whenever $[x, y] \subset (\tau - \delta_1(\tau), \tau + \delta_1(\tau)) \cap [a, b]$ and $\tau \in [x, y]$. Similarly, there is a gauge δ_2 on $[a, b]$ such that

$$f(\tau)\,(g(y) - g(x)) \ge m(y) - m(x)$$

whenever $[x, y] \subset (\tau - \delta_2(\tau), \tau + \delta_2(\tau)) \cap [a, b]$ and $\tau \in [x, y]$.

Let $\delta(x) = \min\{\delta_1(x), \delta_2(x)\}$ for all $x \in [a, b]$. Then

$$m(y) - m(x) \le f(\tau)\,(g(y) - g(x)) \le M(y) - M(x)$$

whenever $[x, y] \subset (\tau - \delta(\tau), \tau + \delta(\tau)) \cap [a, b]$. It follows that for any δ-fine partition P we have

$$m(b) - m(a) \le S(P) \le M(b) - M(a),$$

and the proof of the first statement is complete.

We have shown that the inequality

$$m(b) - m(a) \le M(b) - M(a)$$

holds for every $M \in \mathfrak{M}(f, \mathrm{d}g)$ and $m \in \mathfrak{m}(f, \mathrm{d}g)$. Hence, if $\mathfrak{M}(f, \mathrm{d}g)$ and $\mathfrak{m}(f, \mathrm{d}g)$ are nonempty, we get

$$(\mathrm{PS})\underline{\int_a^b} f\,\mathrm{d}g = \sup\{m(b) - m(a) : m \in \mathfrak{m}(f, \mathrm{d}g)\}$$

$$\leq \inf\{M(b) - M(a) : M \in \mathfrak{M}(f, \mathrm{d}g)\} = (\mathrm{PS})\overline{\int_a^b} f\,\mathrm{d}g.$$

On the other hand, if $\mathfrak{M}(f, \mathrm{d}g)$ or $\mathfrak{m}(f, \mathrm{d}g)$ is empty, it is obvious that the second statement holds. □

6.11.5 Definition. If $(\mathrm{PS})\underline{\int_a^b} f\,\mathrm{d}g = (\mathrm{PS})\overline{\int_a^b} f\,\mathrm{d}g \in \mathbb{R}$, then the common value of the upper and lower integrals is called the *Perron-Stieltjes integral* of f with respect to g, and is denoted by $(\mathrm{PS})\int_a^b f(x)\,\mathrm{d}g(x)$ or $(\mathrm{PS})\int_a^b f\,\mathrm{d}g$. In the special case when $g(x) = x$ for all $x \in [a, b]$, we speak about the Perron integral and denote it by $(\mathrm{P})\int_a^b f(x)\,\mathrm{d}x$. Similarly, the corresponding lower and upper integrals are denoted by $(\mathrm{P})\underline{\int_a^b} f(x)\,\mathrm{d}x$ and $(\mathrm{P})\overline{\int_a^b} f(x)\,\mathrm{d}x$, respectively.

We now show that the Perron-Stieltjes integral is equivalent to the Kurzweil-Stieltjes integral.

6.11.6 Theorem. *The Perron-Stieltjes integral* $(\mathrm{PS})\int_a^b f\,\mathrm{d}g$ *exists if and only if the Kurzweil-Stieltjes integral* $\int_a^b f\,\mathrm{d}g$ *exists; in that case, both integrals have the same value.*

Proof. a) Assume that $(\mathrm{PS})\int_a^b f\,\mathrm{d}g$ exists and equals $I \in \mathbb{R}$. Given an arbitrary $\varepsilon > 0$, there exist a major function $M \in \mathfrak{M}(f, \mathrm{d}g)$ and a minor function $m \in \mathfrak{m}(f, \mathrm{d}g)$ such that

$$m(b) - m(a) > I - \varepsilon, \quad M(b) - M(a) < I + \varepsilon.$$

If δ is the gauge specified in Lemma 6.11.4 (i), then the inequalities

$$I - \varepsilon < m(b) - m(a) \leq S(P) \leq M(b) - M(a) < I + \varepsilon$$

hold for every δ-fine partition P of $[a, b]$. This means that the Kurzweil-Stieltjes integral $\int_a^b f\,\mathrm{d}g$ exists and equals I.

b) Assume that $\int_a^b f\,\mathrm{d}g$ exists and equals $I \in \mathbb{R}$. Given an arbitrary $\varepsilon > 0$, there is a gauge $\delta_\varepsilon : [a, b] \to (0, \infty)$ such that

$$I - \varepsilon < S(P) < I + \varepsilon \quad \text{for all } \delta_\varepsilon\text{-fine partitions } P \text{ of } [a, b]. \qquad (6.11.1)$$

Define $M(a)=m(a)=0$, and

$$M(x)=\sup\{S(Q):Q \text{ is a } \delta_\varepsilon\text{-fine partition of } [a,x]\} \quad \text{for} \quad x\in(a,b],$$
$$m(x)=\inf\{S(Q):Q \text{ is a } \delta_\varepsilon\text{-fine partition of } [a,x]\} \quad \text{for} \quad x\in(a,b].$$

We claim that M is a major function of f with respect to g. To see this, consider an arbitrary $\tau\in[a,b]$. Using the definition of M, we get

$$M(y)\geq M(\tau)+f(\tau)\,(g(y)-g(\tau)) \quad \text{for } y\in[\tau,\tau+\delta_\varepsilon(\tau))\cap[a,b],$$

because each δ_ε-fine partition of $[a,\tau]$ can be extended to a δ_ε-fine partition P of $[a,y]$ by adjoining the interval $[\tau,y]$ and the tag τ, whose contribution to the sum $S(P)$ is $f(\tau)\,(g(y)-g(\tau))$. Analogously,

$$M(\tau)\geq M(x)+f(\tau)\,(g(\tau)-g(x)) \text{ for } x\in(\tau-\delta_\varepsilon(\tau),\tau]\cap[a,b],$$

because each δ_ε-fine partition of $[a,x]$ can be extended to a δ_ε-fine partition P of $[a,\tau]$ by adjoining the interval $[x,\tau]$ and the tag τ, whose contribution to the sum $S(P)$ is $f(\tau)\,(g(\tau)-g(x))$.

By adding the previous two inequalities, we see that

$$M(y)-M(x)\geq f(\tau)\,(g(y)-g(x))$$

whenever $[x,y]\subset(\tau-\delta_\varepsilon(\tau),\tau+\delta_\varepsilon(\tau))\cap[a,b]$ and $\tau\in[x,y]$.

Similarly, one can show that m is a minor function of f with respect to g. By (6.11.1) we have

$$I-\varepsilon\leq m(b)-m(a)\leq I+\varepsilon, \quad I-\varepsilon\leq M(b)-M(a)\leq I+\varepsilon.$$

Consequently, having in mind the definition of the upper and lower integral and Lemma 6.11.4 (ii), we see that

$$I-\varepsilon\leq(\text{PS})\underline{\int_a^b} f\,\mathrm{d}g\leq(\text{PS})\overline{\int_a^b} f\,\mathrm{d}g\leq I+\varepsilon.$$

Since $\varepsilon>0$ was arbitrary, we necessarily have

$$(\text{PS})\underline{\int_a^b} f\,\mathrm{d}g=(\text{PS})\overline{\int_a^b} f\,\mathrm{d}g=I.$$

Hence, the Perron-Stieltjes integral (PS) $\int_a^b f\,\mathrm{d}g$ exists and equals I. □

According to the previous result, we have two equivalent definitions of the same concept of integral. Depending on the situation, one definition might be easier to work with than the other one. As an illustration, let us provide an alternative proof of the fact that each derivative is Perron or Kurzweil-Henstock integrable (cf. Theorem 6.2.16).

6.11.7 Theorem. *Let $F:[a,b]\to\mathbb{R}$ be a differentiable function with $F'=f$ on $[a,b]$ (where the derivative is understood as one-sided at both endpoints). Then f is Perron integrable and* (P) $\int_a^b f(x)\,\mathrm{d}x = F(b)-F(a)$.

Proof. In Example 6.11.2, we have shown that for each $\varepsilon>0$, the function

$$M_\varepsilon(t)=F(t)+t\,\varepsilon,\quad t\in[a,b],$$

is a major function of f, and the function

$$m_\varepsilon(t)=F(t)-t\,\varepsilon,\quad t\in[a,b],$$

is a minor function of f. The definition of the lower and upper integral yields

$$\begin{aligned}(\mathrm{P})\overline{\int_a^b} f(x)\,\mathrm{d}x &\le \inf_{\varepsilon>0}\,(M_\varepsilon(b)-M_\varepsilon(a)) = \inf_{\varepsilon>0}\,(F(b)-F(a)+\varepsilon\,(b-a))\\ &= F(b)-F(a),\\ (\mathrm{P})\underline{\int_a^b} f(x)\,\mathrm{d}x &\ge \sup_{\varepsilon>0}\,(m_\varepsilon(b)-m_\varepsilon(a)) = \inf_{\varepsilon>0}\,(F(b)-F(a)-\varepsilon\,(b-a))\\ &= F(b)-F(a).\end{aligned}$$

Recalling that

$$(\mathrm{P})\underline{\int_a^b} f(x)\,\mathrm{d}x \le (\mathrm{P})\overline{\int_a^b} f(x)\,\mathrm{d}x,$$

we conclude that the value of both integrals is $F(b)-F(a)$, and the proof is complete. □

6.12 RELATION TO THE LEBESGUE-STIELTJES INTEGRAL

The goal of this section is to clarify the relationship between Kurzweil-Stieltjes integrals and Lebesgue-Stieltjes integrals. We assume some basic familiarity with measure theory and Lebesgue integration. Since the rest of the book makes no use of the results from this section, readers who are not interested in Lebesgue-Stieltjes integrals can skip this part.

First, let us recall some facts about Lebesgue-Stieltjes measures. More details can be found e.g. in Section 22 of [Yeh (2014)].

Throughout this section, we always assume that $g:\mathbb{R}\to\mathbb{R}$ is nondecreasing. Then the outer Lebesgue-Stieltjes measure of an arbitrary set $E\subset\mathbb{R}$ is given by

$$\mu_g^*(E):=\inf\Big\{\sum_{k=1}^{\infty}(g(b_k+)-g(a_k+)):E\subset\bigcup_{k=1}^{\infty}(a_k,b_k]\Big\}.$$

The outer measure is either a nonnegative real number, or ∞. The function μ_g^* is defined on the collection of all subsets of $\mathbb{R}$, but it need not be σ-additive. By restricting μ_g^* to a certain σ-algebra of sets that are called μ_g-measurable, we get the Lebesgue-Stieltjes measure μ_g, which is σ-additive. This measure has the following properties:

- Each Borel set (in particular, each open or closed set) is μ_g-measurable.
- The measures of various types of intervals are calculated as follows:

$$\begin{aligned}\mu_g([a,b]) &= g(b+)-g(a-),\\ \mu_g((a,b]) &= g(b+)-g(a+),\\ \mu_g([a,b)) &= g(b-)-g(a-),\\ \mu_g((a,b)) &= g(b-)-g(a+).\end{aligned}$$

  In particular, we have $\mu_g(\{a\})=g(a+)-g(a-)$.
- If $E\subset\mathbb{R}$ is μ_g-measurable and $\varepsilon>0$, there exists an open set G such that $E\subset G$ and $\mu_g(G\setminus E)<\varepsilon$.
- If $E\subset\mathbb{R}$ is such that $\mu_g^*(E)=0$, then E is μ_g-measurable and $\mu_g(E)=0$.

In Lebesgue's integration theory, it is common to deal with functions whose values can be not only real numbers, but also $\pm\infty$. For this reason, we set $\mathbb{R}^*=\mathbb{R}\cup\{\pm\infty\}$.

A function $f:\mathbb{R}\to\mathbb{R}^*$ is called μ_g-measurable if the set $\{x\in\mathbb{R}:f(x)>r\}$ is μ_g-measurable for each $r\in\mathbb{R}$. A function $s:\mathbb{R}\to\mathbb{R}$ is called simple if it can be expressed in the form

$$s = \sum_{j=1}^{n} c_j \chi_{A_j},$$

where $c_1, \dots, c_n$ are distinct real numbers, and $A_1, \dots, A_n$ are disjoint subsets of $\mathbb{R}$. Obviously, s is μ_g-measurable if and only if the sets $A_1, \dots, A_n$ are μ_g-measurable. If s is μ_g-measurable and nonnegative (i.e., $c_1, \dots, c_n \geq 0$), we define its Lebesgue-Stieltjes integral by the formula

$$\int_{\mathbb{R}} s \, \mathrm{d}\mu_g = \sum_{j=1}^{n} c_j \mu_g(A_j).$$

If $c_j = 0$ and $\mu_g(A_j) = \infty$ for a certain $j \in \{1, \dots, n\}$, we use the convention that $0 \cdot \infty = 0$. The value of the integral is either a nonnegative real number, or ∞.

For each nonnegative μ_g-measurable function $f : \mathbb{R} \to \mathbb{R}^*$, there exists a nondecreasing sequence of simple nonnegative functions which is convergent to f. We define

$$\int_{\mathbb{R}} f \, \mathrm{d}\mu_g = \sup \left\{ \int_{\mathbb{R}} s \, \mathrm{d}\mu_g \colon s \text{ is a simple } \mu_g\text{-measurable function, } 0 \leq s \leq f \right\}.$$

Finally, for an arbitrary μ_g-measurable function $f : \mathbb{R} \to \mathbb{R}^*$, we define

$$\int_{\mathbb{R}} f \, \mathrm{d}\mu_g = \int_{\mathbb{R}} f^+ \, \mathrm{d}\mu_g - \int_{\mathbb{R}} f^- \, \mathrm{d}\mu_g$$

whenever the difference on the right-hand side makes sense (i.e., it is not of the form $\infty - \infty$).

If $E \subset \mathbb{R}$ is a μ_g-measurable set, we let

$$\int_E f \, \mathrm{d}\mu_g = \int_{\mathbb{R}} f \chi_E \, \mathrm{d}\mu_g$$

whenever the last integral exists. Since the values of f outside E are unimportant, we can assume that f is defined only on E (and extend it to $\mathbb{R}$ in an arbitrary way).

If $g(x) = x$ for all $x \in \mathbb{R}$, then μ_g is simply the Lebesgue measure, and the Lebesgue-Stieltjes integral reduces to the ordinary Lebesgue integral.

Note that if $\int_{\mathbb{R}} f \,\mathrm{d}\mu_g$ exists and is finite, then $\int_{\mathbb{R}} |f| \,\mathrm{d}\mu_g$ also exists and is finite, because $|f|$ is μ_g-measurable, $|f| = f^+ + f^-$, and the integrals of f^+ and f^- are finite. For this reason, we say that the Lebesgue-Stieltjes integral is *absolutely convergent.*

Our goal is to prove that Lebesgue-Stieltjes integrability implies Kurzweil-Stieltjes integrability. This fact will be a fairly straightforward consequence of the next lemma.

6.12.1 Lemma. *If $E \subset [a,b]$ is a μ_g-measurable set, then the Kurzweil-Stieltjes integral $\int_a^b \chi_E \,\mathrm{d}g$ exists, and we have*

$$\int_a^b \chi_E \,\mathrm{d}g = \int_{(a,b)} \chi_E \,\mathrm{d}\mu_g + \chi_E(a)\,\Delta^+ g(a) + \chi_E(b)\,\Delta^- g(b).$$

Proof. Let $\widetilde{E} = E \cap (a,b)$ and observe that

$$\int_{(a,b)} \chi_E \,\mathrm{d}\mu_g = \int_{\mathbb{R}} \chi_{\widetilde{E}} \,\mathrm{d}\mu_g = \mu_g(\widetilde{E}).$$

Since $E \cap \{a\}$ and $E \cap \{b\}$ are either empty or singleton sets, we have

$$\int_a^b \chi_{E\cap\{a\}} \,\mathrm{d}g = \chi_E(a)\,\Delta^+ g(a), \qquad \int_a^b \chi_{E\cap\{b\}} \,\mathrm{d}g = \chi_E(b)\,\Delta^+ g(b).$$

Thus, it suffices to show that

$$\int_a^b \chi_{\widetilde{E}} \,\mathrm{d}g = \mu_g(\widetilde{E}). \tag{6.12.1}$$

Let $H = (a,b) \setminus \widetilde{E}$. Consider an arbitrary $\varepsilon > 0$. It is well known (see e.g. Lemma 22.10 in [Yeh (2014)]) that there exist open sets $G_1, G_2 \subset \mathbb{R}$ such that

$$\widetilde{E} \subset G_1, \quad \mu_g(G_1) < \mu_g(\widetilde{E}) + \varepsilon, \quad H \subset G_2, \quad \mu_g(G_2) < \mu_g(H) + \frac{\varepsilon}{2}.$$

Let δ be a gauge on $[a,b]$ with the following properties:

- $\delta(x) < \operatorname{dist}(x, \mathbb{R} \setminus G_1)$ for every $x \in \widetilde{E}$, and $\delta(x) < \operatorname{dist}(x, \mathbb{R} \setminus G_2)$ for every $x \in H$.
- $\delta(x) < x - a$ for each $x \in (a,b]$, and $\delta(x) < b - x$ for each $x \in [a,b)$.
- If $x \in (a, a+\delta(a))$, then $g(x) - g(a+) < \varepsilon/4$; if $x \in (b - \delta(b), b)$, then $g(b-) - g(x) < \varepsilon/4$.

Let $P=(\boldsymbol{\alpha},\boldsymbol{\xi})$ be a δ-fine partition of $[a,b]$. The first property from the definition of δ ensures that if $\xi_i\in\widetilde{E}$, then $[\alpha_{i-1},\alpha_i]\subset G_1$, and if $\xi_i\in H$, then $[\alpha_{i-1},\alpha_i]\subset G_2$. The second property guarantees that $\xi_1=a$ and $\xi_{\nu(P)}=b$. We can write

$$\bigcup_{\xi_i\in\widetilde{E}}[\alpha_{j-1},\alpha_j]=\bigcup_{j\in J}[\beta_{j-1},\beta_j],$$

where the right-hand side is a finite union of disjoint intervals contained in G_1 (note that the intervals on the left-hand side are nonoverlapping, but in general not disjoint). Consequently, we get

$$\begin{aligned}S(\chi_{\widetilde{E}},\mathrm{d}g,P)&=\sum_{\xi_i\in\widetilde{E}}(g(\alpha_i)-g(\alpha_{i-1}))=\sum_{j\in J}(g(\beta_j)-g(\beta_{j-1}))\\&\le\sum_{j\in J}(g(\beta_j+)-g(\beta_{j-1}-))=\sum_{j\in J}\mu_g([\beta_{j-1},\beta_j])\\&\le\mu_g(G_1)<\mu_g(\widetilde{E})+\varepsilon.\end{aligned}$$

To obtain a lower bound for $S(\chi_{\widetilde{E}},\mathrm{d}g,P)$, we write

$$\bigcup_{\xi_i\in H}[\alpha_{i-1},\alpha_i]=\bigcup_{k\in K}[\gamma_{k-1},\gamma_k],$$

where the right-hand side is a finite union of disjoint intervals contained in G_2. Recalling that $\xi_1=a\notin\widetilde{E}$ and $\xi_{\nu(P)}=b\notin\widetilde{E}$, we get

$$\begin{aligned}S(\chi_{\widetilde{E}},\mathrm{d}g,P)&=g(b)-g(a)-\sum_{\xi_i\in[a,b]\setminus\widetilde{E}}(g(\alpha_i)-g(\alpha_{i-1}))\\&=g(b)-g(a)\\&\quad-\Big(g(\alpha_1)-g(\alpha_0)+\sum_{\xi_i\in H}(g(\alpha_i)-g(\alpha_{i-1}))+g(\alpha_{\nu(P)})-g(\alpha_{\nu(P)-1})\Big)\\&=g(\alpha_{\nu(P)-1})-g(\alpha_1)-\sum_{\xi_i\in H}(g(\alpha_i)-g(\alpha_{i-1}))\\&=g(\alpha_{\nu(P)-1})-g(\alpha_1)-\sum_{k\in K}(g(\gamma_k)-g(\gamma_{k-1}))\\&>g(b-)-\frac{\varepsilon}{4}-g(a+)-\frac{\varepsilon}{4}-\sum_{k\in K}(g(\gamma_k+)-g(\gamma_{k-1}-))\\&\ge g(b-)-g(a+)-\frac{\varepsilon}{2}-\sum_{k\in K}\mu_g([\gamma_{k-1},\gamma_k])\\&\ge\mu_g((a,b))-\frac{\varepsilon}{2}-\mu_g(G_2)\ge\mu_g((a,b))-\mu_g(H)-\varepsilon=\mu_g(\widetilde{E})-\varepsilon.\end{aligned}$$

We have proved that

$$|S(\chi_{\widetilde{E}}, \mathrm{d}g, P) - \mu_g(\widetilde{E})| < \varepsilon$$

for each δ-fine partition $P = (\boldsymbol{\alpha}, \boldsymbol{\xi})$, which implies that the identity (6.12.1) holds and the proof is complete. □

Using the definition of a simple function and its Lebesgue-Stieltjes integral, we get the next result.

6.12.2 Corollary. *If $s : [a, b] \to \mathbb{R}$ is a μ_g-measurable simple function, then the Kurzweil-Stieltjes integral $\int_a^b s \,\mathrm{d}g$ exists, and we have*

$$\int_a^b s \,\mathrm{d}g = \int_{(a,b)} s \,\mathrm{d}\mu_g + s(a)\,\Delta^+ g(a) + s(b)\,\Delta^- g(b). \tag{6.12.2}$$

6.12.3 Theorem. *If $f : [a, b] \to \mathbb{R}$ and the Lebesgue-Stieltjes integral $\int_{(a,b)} f \,\mathrm{d}\mu_g$ has a finite value, then the Kurzweil-Stieltjes integral $\int_a^b f \,\mathrm{d}g$ exists, as well, and we have the relations*

$$\int_a^b f \,\mathrm{d}g = \int_{(a,b)} f \,\mathrm{d}\mu_g + f(a)\,\Delta^+ g(a) + f(b)\,\Delta^- g(b), \tag{6.12.3}$$

$$\int_{(a,b)} f \,\mathrm{d}\mu_g = \lim_{\substack{c \to a+,\\ d \to b-}} \int_c^d f \,\mathrm{d}g. \tag{6.12.4}$$

If $g(a+) = g(a)$ and $g(b+) = g(b)$, then

$$\int_a^b f \,\mathrm{d}g = \int_{(a,b]} f \,\mathrm{d}\mu_g. \tag{6.12.5}$$

If $g(a-) = g(a)$ and $g(b-) = g(b)$, then

$$\int_a^b f \,\mathrm{d}g = \int_{[a,b)} f \,\mathrm{d}\mu_g. \tag{6.12.6}$$

Proof. If the Lebesgue-Stieltjes integral exists, then f (which is considered to be zero outside $[a, b]$) is necessarily μ_g-measurable. Each of the functions f^+, f^- is nonnegative and μ_g-measurable, and therefore it is the limit of a nondecreasing sequence of nonnegative μ_g-measurable simple functions. The Kurzweil-Stieltjes and Lebesgue-Stieltjes integrals of these simple functions exist and satisfy the relation (6.12.2). Using the monotone convergence theorems for the Kurzweil-Stieltjes and Lebesgue-Stieltjes integrals, we get

$$\int_a^b f^+ \, \mathrm{d}g = \int_{(a,b)} f^+ \, \mathrm{d}\mu_g + f^+(a)\, \Delta^+ g(a) + f^+(b)\, \Delta^- g(b),$$

$$\int_a^b f^- \, \mathrm{d}g = \int_{(a,b)} f^- \, \mathrm{d}\mu_g + f^-(a)\, \Delta^+ g(a) + f^-(b)\, \Delta^- g(b),$$

which implies (6.12.3). The relation (6.12.4) then follows from Corollary 6.5.4.

If $g(a+) = g(a)$ and $g(b+) = g(b)$, observe that $\Delta^+ g(a) = 0$ and

$$f(b)\, \Delta^- g(b) = f(b)\, (g(b+) - g(b-)) = f(b)\, \mu_g(\{b\}) = \int_{\{b\}} f \, \mathrm{d}\mu_g.$$

Similarly, if $g(a-) = g(a)$ and $g(b-) = g(b)$, then $\Delta^- g(b) = 0$ and

$$f(a)\, \Delta^+ g(a) = f(a)\, (g(a+) - g(a-)) = f(a)\, \mu_g(\{a\}) = \int_{\{a\}} f \, \mathrm{d}\mu_g.$$

These facts together with (6.12.3) imply (6.12.5) and (6.12.6). □

6.12.4 Exercise. In Section 6.9 we have introduced Kurzweil-Stieltjes integrals of the form $\int_E f \, \mathrm{d}g$, where $f, g : [a,b] \to \mathbb{R}$ and E is an elementary subset of $[a,b]$. If we extend g to $\mathbb{R}$ in such a way that $g(a-) = g(a)$ and $g(b+) = g(b)$, show that if $I \subset [a,b]$ is an interval of an arbitrary type and the Lebesgue-Stieltjes integral $\int_I f \, \mathrm{d}\mu_g$ exists, then the Kurzweil-Stieltjes integral $\int_I f \, \mathrm{d}g$ exists as well and has the same value. Conclude that the same statement holds if I is replaced by an elementary set $E \subset [a,b]$.

Our next goal is to prove a partial converse to Theorem 6.12.3 and show that for nonnegative functions f, Kurzweil-Stieltjes integrability implies Lebesgue-Stieltjes integrability. We begin by establishing two auxiliary results.

6.12.5 Lemma. *Let $f : [a,b] \to \mathbb{R}$ be a nonnegative function such that the Kurzweil-Stieltjes integral $\int_a^b f \, \mathrm{d}g$ exists, and $E = \{x \in [a,b] : f(x) \geq 1\}$. Then for each $\varepsilon > 0$, there exists a μ_g-measurable set $G \subset [a,b]$ containing E and such that $\int_a^b \chi_G \, \mathrm{d}g \leq \varepsilon + \int_a^b f \, \mathrm{d}g$.*

Proof. We have $f = f\chi_{\{a\}} + f\chi_{(a,b]}$ and $E = E_1 \cup E_2$, where $E_1 = E \cap \{a\}$ and $E_2 = E \cap (a,b]$. Note that both $\int_a^b f\chi_{\{a\}}\,dg$ and $\int_a^b f\chi_{(a,b]}\,dg$ exist. We will construct measurable sets $G_1, G_2 \subset [a,b]$ such that $E_1 \subset G_1$, $E_2 \subset G_2$,

$$\int_a^b \chi_{G_1}\,\mathrm{d}g \le \frac{\varepsilon}{2} + \int_a^b f\chi_{\{a\}}\,\mathrm{d}g,$$
$$\int_a^b \chi_{G_2}\,\mathrm{d}g \le \frac{\varepsilon}{2} + \int_a^b f\chi_{(a,b]}\,\mathrm{d}g.$$

Then $G = G_1 \cup G_2$ contains E, and

$$\int_a^b \chi_G\,\mathrm{d}g \le \int_a^b (\chi_{G_1} + \chi_{G_2})\,\mathrm{d}g \le \varepsilon + \int_a^b f\,\mathrm{d}g.$$

If $f(a) < 1$, it suffices to take $G_1 = \emptyset$. If $f(a) \ge 1$, find a $c \in (a,b]$ such that $g(c) - g(a+) \le \varepsilon/2$, and let $G_1 = [a,c)$. Then $E_1 \subset \{a\} \subset G_1$ and

$$\begin{aligned}\int_a^b \chi_{G_1}\,\mathrm{d}g &= g(c-) - g(a) = g(c-) - g(a+) + g(a+) - g(a)\\ &\le g(c) - g(a+) + \frac{f(a)\,\Delta^+ g(a)}{f(a)}\\ &= g(c) - g(a+) + \frac{1}{f(a)}\int_a^b f\chi_{\{a\}}\,\mathrm{d}g\\ &\le \frac{\varepsilon}{2} + \int_a^b f\chi_{\{a\}}\,\mathrm{d}g.\end{aligned}$$

Note that g is right-continuous in $[a,b)$ with at most countably many exceptions. Hence, there exists a sequence of divisions $\{\boldsymbol{\alpha}^n\}$ of $[a,b]$ such that for each $n \in \mathbb{N}$, $\boldsymbol{\alpha}^{n+1}$ is a refinement of $\boldsymbol{\alpha}^n$, $|\boldsymbol{\alpha}^n| \le (b-a)/2^n$, and g is right-continuous at the division points $\alpha_1^n, \dots, \alpha_{\nu(\boldsymbol{\alpha})-1}^n$.

Let δ be a gauge on $[a,b]$ such that

$$\left|S(f\chi_{(a,b]}, \mathrm{d}g, P) - \int_a^b f\chi_{(a,b]}\,\mathrm{d}g\right| < \varepsilon/2$$

for each δ-fine partition P of $[a,b]$. We now construct a collection $\mathcal{I}$ of interval-point pairs of the form $((u,v],\tau)$. In the beginning, let $\mathcal{I} = \emptyset$. For each $n \in \mathbb{N}$, perform the following step: Find all intervals $[\alpha_{j-1}^n, \alpha_j^n]$ which are not contained in any interval in $\mathcal{I}$ and such that there exists a point $\tau \in [\alpha_{j-1}^n, \alpha_j^n] \cap E_2$ satisfying $[\alpha_{j-1}^n, \alpha_j^n] \subset (\tau - \delta(\tau), \tau + \delta(\tau))$; then add $((\alpha_{j-1}^n,\alpha_j^n],\tau)$ to $\mathcal{I}$. If there are several possible choices of τ for a given $[\alpha_{j-1}^n,\alpha_j^n]$, take only one of these interval-point pairs.

This procedure leads to a collection $\mathcal{I}=\{((u_k,v_k],\tau_k)\colon k\in K\}$, where $K\subset\mathbb{N}$ is either finite or countable. All intervals in $\mathcal{I}$ are pairwise disjoint, and all points τ_k satisfy $f(\tau_k)\geq 1$ (because $\tau_k\in E_2$). Moreover, we have

$$E_2\subset\bigcup_{k\in K}(u_k,v_k].$$

Indeed, if $x\in E_2$, take $n\in\mathbb{N}$ such that $(b-a)/2^n<\delta(x)$. Since x belongs to $(\alpha_{j-1}^n,\alpha_j^n]$ for a certain $j\in\{1,\ldots,\nu(\boldsymbol{\alpha}^n)\}$, and (recall that we have $|\boldsymbol{\alpha}^n|\leq(b-a)/2^n$)

$$x-\frac{b-a}{2^n}\leq\alpha_j^n-\frac{b-a}{2^n}\leq\alpha_{j-1}^n\leq\alpha_j^n\leq\alpha_{j-1}^n+\frac{b-a}{2^n}\leq x+\frac{b-a}{2^n},$$

we see that $[\alpha_{j-1}^n,\alpha_j^n]\subset(x-\delta(x),x+\delta(x))$. Thus, either $(\alpha_{j-1}^n,\alpha_j^n]$ was added to $\mathcal{I}$ in the n-th stage of its construction, or it is contained in an interval added earlier. In any case, $x\in\bigcup_{k\in K}(u_k,v_k]$.

If L is an arbitrary finite subset of K, then the collection

$$\{([u_k,v_k],\tau_k):k\in L\}$$

can be extended to a δ-fine partition P of $[a,b]$, and therefore

$$\begin{aligned}\sum_{k\in L}(g(v_k)-g(u_k))&\leq\sum_{k\in L}f(\tau_k)\,(g(v_k)-g(u_k))\\&\leq S(f\chi_{(a,b]},\mathrm{d}g,P)<\frac{\varepsilon}{2}+\int_a^b f\chi_{(a,b]}\,\mathrm{d}g.\end{aligned}$$

It follows that

$$\sum_{k\in K}(g(v_k)-g(u_k))\leq\frac{\varepsilon}{2}+\int_a^b f\chi_{(a,b]}\,\mathrm{d}g.$$

Let $G_2=\bigcup_{k\in K}(u_k,v_k]$. Since G_2 is μ_g-measurable, the integral $\int_a^b\chi_{G_2}\mathrm{d}g$ exists by Lemma 6.12.1, and we have

$$\int_a^b\chi_{G_2}\,\mathrm{d}g\leq\sum_{k\in K}\int_a^b\chi_{(u_k,v_k]}\,\mathrm{d}g.$$

If $v_k = b$, then $\int_a^b \chi_{(u_k,v_k]}\,dg = g(v_k) - g(u_k+)$. If $v_k < b$, then g is right-continuous at v_k, and

$$\int_a^b \chi_{(u_k,v_k]}\,dg = g(v_k+) - g(u_k+) = g(v_k) - g(u_k+).$$

Thus, we get

$$\int_a^b \chi_{G_2}\,dg \le \sum_{k\in K}\bigl(g(v_k) - g(u_k+)\bigr)$$
$$\le \sum_{k\in K}\bigl(g(v_k) - g(u_k)\bigr) \le \frac{\varepsilon}{2} + \int_a^b f\chi_{(a,b]}\,dg,$$

which completes the proof. □

6.12.6 Lemma. *Let $f:[a,b]\to\mathbb{R}$ be a nonnegative function such that the Kurzweil-Stieltjes integral $\int_a^b f\,dg$ exists. Then for each $\varepsilon > 0$, there exists a nonnegative μ_g-measurable function $\varphi:[a,b]\to\mathbb{R}$ satisfying $f(x) \le \varphi(x) + \varepsilon$ for all $x\in[a,b]$, and $\int_a^b \varphi\,dg \le \varepsilon + \int_a^b f\,dg$.*

Proof. Let $\varepsilon > 0$ be given. Suppose first that $f(a) = f(b) = 0$. For each $n \in \mathbb{N}\cup\{0\}$, let

$$f_n(x) = \begin{cases} 0 & \text{if } f(x) \le n\,\varepsilon,\\ f(x) - n\,\varepsilon & \text{if } f(x) \in [n\,\varepsilon, (n+1)\,\varepsilon],\\ \varepsilon & \text{if } f(x) \ge (n+1)\,\varepsilon, \end{cases}$$

and observe that

$$f(x) = \sum_{n=0}^{\infty} f_n(x) \quad \text{for } x\in[a,b]. \tag{6.12.7}$$

We claim that $\int_a^b f_n\,dg$ exists for each $n\in\mathbb{N}\cup\{0\}$. To see this, note that

$$f_n(x) = \min\{\max\{f(x) - n\,\varepsilon, 0\}, \varepsilon\} \quad \text{for } x\in[a,b].$$

Since $|f(x) - n\,\varepsilon| \le f(x) + n\,\varepsilon$ and both $\int_a^b (f - n\,\varepsilon)\,dg$ and $\int_a^b (f + n\,\varepsilon)\,dg$ exist, Theorem 6.7.4 implies that $\int_a^b |f - n\,\varepsilon|\,dg$ exists as well. According to Theorem 6.7.5, this means that $\int_a^b \max\{f - n\,\varepsilon, 0\}\,dg$ exists and consequently

$$\int_a^b \min\{\max\{f - n\,\varepsilon, 0\}, \varepsilon\}\,dg$$

exists, too; this proves the claim.

Using (6.12.7), Theorem 6.8.15 (Levi's theorem for series) and Remark 6.8.16, we get

$$\int_a^b f\,\mathrm{d}g = \sum_{n=0}^{\infty} \int_a^b f_n\,\mathrm{d}g. \tag{6.12.8}$$

For each $n \in \mathbb{N} \cup \{0\}$, Lemma 6.12.5 guarantees the existence of a μ_g-measurable set $G_n \subset [a,b]$ such that

$$\{x \in [a,b] : f_n(x) = \varepsilon\} = \{x \in [a,b] : \frac{1}{\varepsilon} f_n(x) \geq 1\} \subset G_n,$$

and

$$\int_a^b \chi_{G_n}\,\mathrm{d}g \leq \frac{1}{2^{n+1}} + \frac{1}{\varepsilon} \int_a^b f_n\,\mathrm{d}g. \tag{6.12.9}$$

Since $f(a) = f(b) = 0$, we can assume that all the sets G_n are contained in the open interval (a,b). Denote

$$\psi(x) = \varepsilon \sum_{n=0}^{\infty} \chi_{G_n}(x) \quad \text{for } x \in [a,b].$$

Since the last sum need not be convergent, the function ψ takes values in $[0,\infty]$. Using Levi's theorem for the Lebesgue-Stieltjes integral, Theorem 6.12.3 and (6.12.9), we get

$$\int_{[a,b]} \psi\,\mathrm{d}\mu_g = \varepsilon \sum_{n=0}^{\infty} \int_{[a,b]} \chi_{G_n}\,\mathrm{d}\mu_g = \varepsilon \sum_{n=0}^{\infty} \int_a^b \chi_{G_n}\,\mathrm{d}g \leq \varepsilon + \int_a^b f\,\mathrm{d}g.$$

The finiteness of the integral implies that ψ must be finite μ_g-almost everywhere in $[a,b]$, i.e., if $N = \{x \in [a,b] : \psi(x) = \infty\}$, then $\mu_g(N) = 0$. It follows that the function $\chi_N f$ is μ_g-measurable. Consequently, the function $\varphi : [a,b] \to \mathbb{R}$ given by

$$\varphi = \chi_{[a,b]\setminus N}\, \psi + \chi_N\, f = \varepsilon \sum_{n=0}^{\infty} \chi_{G_n \setminus N} + \chi_N\, f$$

is μ_g-measurable, nonnegative, and everywhere finite. Let

$$H_n = \{x \in [a,b] : n\varepsilon < f(x) < (n+1)\varepsilon\} = \{x \in [a,b] : 0 < f_n(x) < \varepsilon\}$$

for all $n \in \mathbb{N} \cup \{0\}$. Note that all the sets H_n are pairwise disjoint, and

$$f_n(x) \leq \varepsilon\,(\chi_{G_n}(x) + \chi_{H_n}(x)) \quad \text{for } x \in [a,b].$$

Thus, if $x \in [a,b] \setminus N$, we obtain

$$f(x) = \sum_{n=0}^{\infty} f_n(x) \le \varepsilon \sum_{n=0}^{\infty} \chi_{G_n}(x) + \varepsilon \sum_{n=0}^{\infty} \chi_{H_n}(x) \le \varphi(x) + \varepsilon.$$

On the other hand, if $x \in N$, then $f(x) = \varphi(x)$.

Since $\mu_g(N) = 0$, we have $\int_{(a,b)} \chi_N f \, \mathrm{d}\mu_g = 0$, and thus (by Theorem 6.12.3)

$$\int_a^b \chi_N f \, \mathrm{d}g = 0.$$

Finally, Theorem 6.8.15 (Levi's theorem for series) and (6.12.9) imply

$$\int_a^b \varphi \, \mathrm{d}g = \varepsilon \sum_{n=0}^{\infty} \int_a^b \chi_{G_n \setminus N} \, \mathrm{d}g + \int_a^b \chi_N \, f \, \mathrm{d}g \le \varepsilon \sum_{n=0}^{\infty} \int_a^b \chi_{G_n} \, \mathrm{d}g \le \varepsilon + \int_a^b f \, \mathrm{d}g.$$

Therefore, we have shown that φ has all the required properties.

In a general case when $f(a)$ and $f(b)$ are arbitrary, consider the function $f\chi_{(a,b)}$. By the previous part of the proof, there is a μ_g-measurable function $\widetilde{\varphi}$ such that

$$f\chi_{(a,b)} \le \widetilde{\varphi} + \varepsilon \quad \text{and} \quad \int_a^b \widetilde{\varphi} \, \mathrm{d}g \le \varepsilon + \int_a^b f\chi_{(a,b)} \, \mathrm{d}g.$$

Taking $\varphi = \widetilde{\varphi} + f\chi_{\{a,b\}}$, we see that

$$f = f\chi_{(a,b)} + f\chi_{\{a,b\}} \le \widetilde{\varphi} + \varepsilon + f\chi_{\{a,b\}} = \varphi + \varepsilon,$$

and $\int_a^b \varphi \, \mathrm{d}g \le \varepsilon + \int_a^b f \, \mathrm{d}g$. □

6.12.7 Theorem. *Let $g : [a,b] \to \mathbb{R}$ be given and let $f : [a,b] \to \mathbb{R}$ be a nonnegative function such that the Kurzweil-Stieltjes integral $\int_a^b f \, \mathrm{d}g$ exists. Then the Lebesgue-Stieltjes integral $\int_{(a,b)} f \, \mathrm{d}\mu_g$ also exists and is finite.*

Proof. First, assume that $\int_a^b f \, \mathrm{d}g = 0$. Define

$$E_k = \{x \in (a,b) : f(x) \ge 1/k\} \quad \text{for} \quad k \in \mathbb{N}.$$

For each $\varepsilon > 0$, Lemma 6.12.5 implies the existence of a μ_g-measurable subset G of $[a,b]$ containing the set $\{x \in [a,b] : k\, f(x) \ge 1\}$ and satisfying $\int_a^b \chi_G \, \mathrm{d}g \le \varepsilon$. Observing that $E_k \subset G \cap (a,b)$ and using Lemma 6.12.1, we get

$$\mu_g^*(E_k) \le \mu_g(G \cap (a,b)) = \int_{(a,b)} \chi_{G\cap(a,b)}\, \mathrm{d}\mu_g$$
$$= \int_a^b \chi_{G\cap(a,b)}\, \mathrm{d}g \le \int_a^b \chi_G\, \mathrm{d}g \le \varepsilon.$$

Since $\varepsilon > 0$ was arbitrary, we have $\mu_g^*(E_k) = 0$. Hence, E_k is μ_g-measurable and $\mu_g(E_k) = 0$. It follows that

$$0 = \mu_g\Big(\bigcup_{k=1}^{\infty} E_k\Big) = \mu_g(\{x \in (a,b) : f(x) > 0\}).$$

Consequently, $f\chi_{(a,b)}$ is μ_g-measurable and $\int_{(a,b)} f\, \mathrm{d}\mu_g = 0$.

Next, assume that $\int_a^b f\, \mathrm{d}g > 0$. Using Lemma 6.12.6 with $\varepsilon = 1/k$, $k \in \mathbb{N}$, we get a sequence of nonnegative μ_g-measurable functions $\{\varphi_k\}$ satisfying $f \le \varphi_k + 1/k$ and

$$\int_a^b \varphi_k\, \mathrm{d}g \le \frac{1}{k} + \int_a^b f\, \mathrm{d}g, \quad k \in \mathbb{N}.$$

Hence, $h = \liminf_{k\to\infty} \varphi_k$ is μ_g-measurable, $f \le h$, and Lemma 6.8.18 (Fatou's lemma) implies

$$\int_a^b h\, \mathrm{d}g \le \liminf_{k\to\infty} \int_a^b \varphi_k\, \mathrm{d}g \le \liminf_{k\to\infty} \Big(\frac{1}{k} + \int_a^b f\, \mathrm{d}g\Big) = \int_a^b f\, \mathrm{d}g \le \int_a^b h\, \mathrm{d}g,$$

i.e., $\int_a^b (h - f)\, \mathrm{d}g = 0$.

According to the first part of the proof, the Lebesgue-Stieltjes integral $\int_{(a,b)} (h-f)\, \mathrm{d}\mu_g$ exists and equals zero. For each $k \in \mathbb{N}$, Theorem 6.12.3 implies

$$\int_{(a,b)} \varphi_k\, \mathrm{d}\mu_g \le \int_a^b \varphi_k\, \mathrm{d}g.$$

Using Fatou's lemma for the Lebesgue-Stieltjes integral, we get

$$\int_{(a,b)} h\, \mathrm{d}\mu_g \le \liminf_{k\to\infty} \Big(\int_{(a,b)} \varphi_k\, \mathrm{d}\mu_g\Big) \le \liminf_{k\to\infty} \Big(\int_a^b \varphi_k\, \mathrm{d}g\Big) = \int_a^b h\, \mathrm{d}g,$$

which shows that $\int_{(a,b)} h\, \mathrm{d}\mu_g$ is finite. Consequently,

$$\int_{(a,b)} f\, \mathrm{d}\mu_g = \int_{(a,b)} h\, \mathrm{d}\mu_g - \int_{(a,b)} (h-f)\, \mathrm{d}\mu_g$$

exists and is finite. □

Theorems 6.12.3 and 6.12.7 show that for nonnegative functions f, the Kurzweil-Stieltjes integral $\int_a^b f\,\mathrm{d}g$ exists if and only if the Lebesgue-Stieltjes integral $\int_{(a,b)} f\,\mathrm{d}\mu_g$ exists and is finite; the relation between their values is given by the formulas (6.12.3), (6.12.4).

6.13 Relation to other Stieltjes-type integrals

Besides the Perron-Stieltjes and Lebesgue-Stieltjes integrals, there exists a wealth of other Stieltjes-type integrals. Some of them are described in the present section, including their relation to the Kurzweil-Stieltjes integral. For the sake of clarity, in what follows we write (KS) $\int_a^b f\,\mathrm{d}g$ to denote the Kurzweil-Stieltjes integral.

Young integral

Recall (see Section 5.10) that for a pair $f:[a,b]\to\mathbb{R}$ and $g\in \mathrm{G}([a,b])$, and a partition $P=(\boldsymbol{\alpha},\boldsymbol{\xi})$ of $[a,b]$, we define

$$S_Y(\boldsymbol{\alpha})=\sum_{j=1}^{\nu(P)}\Big(f(\alpha_{j-1})\,\Delta^+ g(\alpha_{j-1})+f(\xi_j)\,(g(\alpha_j-)-g(\alpha_{j-1}+)) \\ +f(\alpha_j)\,\Delta^- g(\alpha_j)\Big).$$

We say that the (σ) *Young integral* (σY) $\int_a^b f\,\mathrm{d}g$ exists and equals $I\in\mathbb{R}$ if for every $\varepsilon>0$ there is a division $\boldsymbol{\alpha}_\varepsilon$ of $[a,b]$ such that $|S_Y(P)-I|<\varepsilon$ for all partitions $P=(\boldsymbol{\alpha},\boldsymbol{\xi})$ of $[a,b]$ such that $\boldsymbol{\alpha}\supset\boldsymbol{\alpha}_\varepsilon$ and

$$\alpha_{j-1}<\xi_j<\alpha_j \quad \text{for all } \ j\in\{1,\dots,\nu(\boldsymbol{\alpha})\}.$$

Note that the sum $S_Y(P)$ can be equivalently rewritten as follows:

$$\left.\begin{aligned} S_Y(P)=f(a)\,\Delta^+g(a)+\sum_{j=1}^{\nu(P)-1} f(\alpha_j)\,\Delta\, g(\alpha_j)+f(b)\,\Delta^- g(b)\\ +\sum_{j=1}^{\nu(P)} f(\xi_j)\,(g(\alpha_j-)-g(\alpha_{j-1}+)) \end{aligned}\right\} \qquad (6.13.1)$$

It is known that the (σ) Young integral is more general than the (σ) Riemann-Stieltjes integral. For example, (σY) $\int_a^b f\,\mathrm{d}g$ exists if f is regulated and g has bounded variation (this follows from the results obtained in [Schwabik (1973a)]). Proceeding similarly as in the proof of Theorem 6.3.11, it is possible to extend this assertion as follows.

6.13.1 Theorem. *Suppose that* $f, g\colon [a,b] \to \mathbb{R}$ *are regulated and one of them has bounded variation. Then both integrals* (KS) $\int_a^b f\,\mathrm{d}g$ *and* $(\sigma\mathrm{Y})\int_a^b f\,\mathrm{d}g$ *exist and have the same value.*

The proof essentially follows the ideas of Section 6.3 and we leave it to the reader as the following (somewhat more advanced) exercise. □

6.13.2 Exercise. Prove Theorem 6.13.1 by following these steps:

(i) Verify that the formulas (6.3.1)–(6.3.10) from Lemmas 6.3.2 and 6.3.3 hold also for the (σ) Young integral.

(ii) Using Exercise 2.1.12 (ii), show that the estimate

$$|S_Y(P)| \le \|f\|\ \mathrm{var}_a^b\, g$$

holds for all partitions P of $[a,b]$ and all $f, g\colon [a,b] \to \mathbb{R}$.

(iii) For $f, g\colon [a,b] \to \mathbb{R}$ and $\alpha, \xi, \beta \in [a,b]$ such that $\alpha \le \xi \le \beta$, verify the equality

$$\begin{aligned} &f(\alpha)\,(g(\alpha+) - g(\alpha)) + f(\xi)\,(g(\beta-) - g(\alpha+)) + f(\beta)\,(g(\beta) - g(\beta-)) \\ &= (f(\alpha) - f(\xi))\,g(\alpha+) + (f(\xi) - f(\beta))\,g(\beta-) + f(\beta)\,g(\beta) - f(\alpha)\,g(\alpha). \end{aligned}$$

Having this in mind, it is easy to see that the estimate

$$|S_Y(P)| \le (|f(a)| + |f(b)| + \mathrm{var}_a^b\, f)\,\|g\|$$

holds for all partitions P of $[a,b]$.

(iv) Notice that due to the previous two steps, the estimates

$$\Big|(\sigma\mathrm{Y})\int_a^b f\,\mathrm{d}g\Big| \le \|f\|\ \mathrm{var}_a^b\, g,$$

$$\Big|(\sigma\mathrm{Y})\int_a^b f\,\mathrm{d}g\Big| \le 2\,\|f\|_{\mathrm{BV}}\,\|g\|$$

hold whenever the corresponding integrals exist.

(v) Show that the analogues of Theorems 6.3.8 and 6.3.9 hold for the (σ) Young integral.

(vi) Finally, complete the proof of Theorem 6.13.1 by proceeding as in the proof of Theorem 6.3.11.

6.13.3 Remark. The following examples show that in general, the existence of (KS) $\int_a^b f\,dg$ does not imply the existence of $(\sigma \mathrm{Y}) \int_a^b f\,dg$, and vice versa:

(i) If $g:[a,b]\to\mathbb{R}$ is regulated and

$$g(a)=g(t-)=g(s+)=g(b) \quad \text{for all} \quad t\in(a,b],\ s\in[a,b),$$

then, by (6.13.1), we have $S_Y(P)=0$ for every function $f:[a,b]\to\mathbb{R}$ and every partition P of $[a,b]$. Consequently, $(\sigma \mathrm{Y}) \int_a^b f\,dg=0$. This is no longer true for the Kurzweil-Stieltjes integral, as shown by the following example taken from [Schwabik (1973a)] (see Example 2.1 there):

Let

$$t_k=\frac{1}{k+1} \quad \text{for} \quad k\in\mathbb{N},$$

and consider the functions $f,g:[0,1]\to\mathbb{R}$ given by

$$g(t)=\begin{cases} 2^{-k} & \text{if } t=t_k \text{ for some } k\in\mathbb{N},\\ 0 & \text{for } t\in[0,1]\setminus\{t_k: k\in\mathbb{N}\},\end{cases}$$

$$f(t)=\begin{cases} 2^{k} & \text{if } t=t_k \text{ for some } k\in\mathbb{N},\\ 0 & \text{for } t\in[0,1]\setminus\{t_k: k\in\mathbb{N}\}.\end{cases}$$

Then g has bounded variation, and $g(0)=g(t-)=g(s+)=g(1)$ for $t\in(0,1]$ and $s\in[0,1)$. (Note that f is not regulated as $f(0+)$ does not exist.)

Consider an arbitrary gauge δ and let $\ell\in\mathbb{N}$ be such that $t_\ell\in(0,\delta(0))$. We can choose a δ-fine partition $P=(\boldsymbol{\alpha},\boldsymbol{\xi})$ of $[0,1]$ in such a way that

$$\alpha_0=\xi_1=0,\ \alpha_1=\xi_2=t_\ell \quad \text{and} \quad g(\alpha_j)=0 \ \text{ for } j\in\{2,\dots,\nu(\boldsymbol{\alpha})\}.$$

Then

$$S(P)=f(\xi_2)\,(g(\alpha_2)-g(\alpha_1))=-f(t_\ell)\,g(t_\ell)=-1.$$

On the other hand, we can choose a δ-fine partition $P=(\boldsymbol{\alpha},\boldsymbol{\xi})$ in such a way that $\boldsymbol{\alpha}\cap\{t_k: k\in\mathbb{N}\}=\emptyset$. As $S(P)=0$ for such a partition, it follows that the Kurzweil-Stieltjes integral (KS) $\int_a^b f\,dg$ does not exist.

(ii) If $g(x)=x$ for all $x\in[a,b]$, then the Dirichlet function f_D (see Remark 6.2.15) provides an example of a function which is not (σ) Young integrable with respect to g. Thus, the existence of the Kurzweil-Stieltjes integral does not in general imply the existence of the (σ) Young integral.

SCHWABIK'S MODIFIED KURZWEIL-STIELTJES INTEGRAL
To avoid the trouble illustrated by the first part of the above remark, Š. Schwabik introduced the notion of the *modified Kurzweil-Stieltjes integral.*

We say that the modified Kurzweil-Stieltjes integral of f with respect to g exists and equals $I \in \mathbb{R}$ if for every $\varepsilon > 0$ there is a gauge δ on $[a,b]$ such that $|S(P) - I| < \varepsilon$ for each δ-fine partition $P = (\boldsymbol{\alpha}, \boldsymbol{\xi})$ of $[a,b]$ satisfying

$$a \le \xi_1 < \alpha_1,\ \alpha_{j-1} < \xi_j < \alpha_j \text{ for } j \in \{2, \dots, \nu(\boldsymbol{\alpha}) - 1\},\ \alpha_{\nu(\boldsymbol{\alpha})-1} < \xi_{\nu(\boldsymbol{\alpha})} \le b.$$

6.13.4 Exercise. (i) If f and g are the functions from Remark 6.13.3 (i), show that the modified Kurzweil-Stieltjes integral of f with respect to g exists and equals 0.
(ii) More generally, if $g \in \mathrm{BV}([a,b])$ is such that $g(a) = g(t-) = g(s+) = g(b)$ for all $t \in (a,b]$ and $s \in [a,b)$, prove that the modified Kurzweil-Stieltjes integral of f with respect to g equals 0 for each $f : [a,b] \to \mathbb{R}$.

In Exercise 6.13.4 (ii) the assumption that g has bounded variation is essential. Indeed, [Krejčí, Kurzweil (2002)] exhibits a pair of functions f and g for which the modified Kurzweil-Stieltjes integral does not exist, while g is a regulated function defined in $[0,1]$ which vanishes except for a countable set, that is, it satisfies $g(0) = g(t-) = g(s+) = g(1)$ for $t \in (0,1]$ and $s \in [0,1)$.

So far, there exists no systematic treatment of the modified Kurzweil-Stieltjes integral. However, it is evident that this integral exists if the Kurzweil-Stieltjes integral exists; in this case, they have the same value. The applications presented in this book show that the Kurzweil-Stieltjes integral is sufficient for many purposes.

KREJČÍ'S KN-INTEGRAL
P. Krejčí introduced the so-called *KN-integral* (where "N" stands for "negligible sets"), which extends not only the (σ) Young integral but also Schwabik's modified Kurzweil-Stieltjes integral. The definition is based on a clever reduction of the set of permissible partitions.

Consider a nonempty system $\mathcal{N}$ of subsets of $[a,b]$ with the following properties:

- If $A, B \in \mathcal{N}$, then $A \cup B \in \mathcal{N}$.
- If $A \in \mathcal{N}$, then the closure of $[a,b] \setminus A$ is $[a,b]$.

Typical examples of $\mathcal{N}$ are the system of all countable subsets of $[a,b]$, or the system of all subsets of $[a,b]$ having Lebesgue measure zero.

Let $f,g\colon[a,b]\to\mathbb{R}$ be given. We say that the KN-integral of f with respect to g exists and equals $I\in\mathbb{R}$ if for every $\varepsilon>0$ there are a gauge δ on $[a,b]$ and a set $A\in\mathcal{N}$ such that $|S(P)-I|<\varepsilon$ for every δ-fine partition $P=(\boldsymbol{\alpha},\boldsymbol{\xi})$ of $[a,b]$ satisfying

$$a\le\xi_1<\alpha_1,\ \alpha_{j-1}<\xi_j<\alpha_j \text{ for } j\in\{2,\dots,\nu(\boldsymbol{\alpha})-1\},\ \alpha_{\nu(\boldsymbol{\alpha})-1}<\xi_{\nu(\boldsymbol{\alpha})}\le b$$

and

$$\alpha_1,\dots,\alpha_{\nu(\boldsymbol{\alpha})-1}\in[a,b]\setminus A.$$

Obviously, if $\mathcal{N}=\{\emptyset\}$, then the KN-integral coincides with Schwabik's modified Kurzweil-Stieltjes integral. In the general case, the existence of the modified integral implies the existence of the KN-integral. Moreover, P. Krejčí proved that if $\mathcal{N}$ is the system of all countable subsets of $[a,b]$, then the existence of the (σ)Young integral implies the existence of the KN-integral and their values coincide (to show this, one chooses A to be the set of all discontinuity point of g, which is countable because g is necessarily regulated).

DUSHNIK (INTERIOR) INTEGRAL

We say that the (σ) *Dushnik integral* $(\sigma\mathrm{D})\int_a^b f\,\mathrm{d}g$ exists and equals $I\in\mathbb{R}$ if for every $\varepsilon>0$, there is a division $\boldsymbol{\alpha}_\varepsilon$ of $[a,b]$ such that $|S(P)-I|<\varepsilon$ for all partitions $P=(\boldsymbol{\alpha},\boldsymbol{\xi})$ of $[a,b]$ satisfying $\boldsymbol{\alpha}\supset\boldsymbol{\alpha}_\varepsilon$ and $\alpha_{j-1}<\xi_j<\alpha_j$ for all $j\in\{1,\dots,\nu(\boldsymbol{\alpha})\}$.

The Dushnik integral is also known as the *interior* integral.

6.13.5 Exercise. Assume that f is the Dirichlet function and $g(x)=x$ for all $x\in[a,b]$. Show that the integral $(\sigma\mathrm{D})\int_a^b f\,\mathrm{d}g$ does not exist.

Thus, unlike the modified Kurzweil-Stieltjes integral and KN-integral, the (σ) Dushnik integral does not fully cover the Kurzweil-Stieltjes integral. Still, this concept of integral is sufficiently general for many purposes, because $(\sigma\mathrm{D})\int_a^b f\,\mathrm{d}g$ exists if f,g are regulated and one of them has bounded variation. However, the value of the integral is in general different from the value of the Kurzweil-Stieltjes integral. This is evident from the next exercise, which should be compared with Lemmas 6.3.2 and 6.3.3.

6.13.6 Exercise. (i) For any function $f\in\mathrm{G}([a,b])$, show that the following relations hold:

$$(\sigma\mathrm{D})\int_a^b f\,\mathrm{d}\chi_{(\tau,b]}=f(\tau+)\qquad \text{if } \tau\in[a,b),$$

$$(\sigma\mathrm{D})\int_a^b f\,\mathrm{d}\chi_{[\tau,b]}=f(\tau-)\qquad \text{if } \tau\in(a,b],$$

$$(\sigma\mathrm{D})\int_a^b f\,\mathrm{d}\chi_{[a,\tau]}=-f(\tau+)\quad \text{if } \tau\in[a,b),$$

$$(\sigma\mathrm{D})\int_a^b f\,\mathrm{d}\chi_{[a,\tau)}=-f(\tau-)\quad \text{if } \tau\in(a,b],$$

$$(\sigma\mathrm{D})\int_a^b f\,\mathrm{d}\chi_{[\tau]}=-\Delta f(\tau)\qquad \text{if } \tau\in(a,b).$$

(ii) For any function $g\colon[a,b]\to\mathbb{R}$, show that the following relations hold:

$$(\sigma\mathrm{D})\int_a^b \chi_{(\tau,b]}\,\mathrm{d}g=g(b)-g(\tau)\quad \text{if } \tau\in[a,b),$$

$$(\sigma\mathrm{D})\int_a^b \chi_{[\tau,b]}\,\mathrm{d}g=g(b)-g(\tau)\quad \text{if } \tau\in(a,b],$$

$$(\sigma\mathrm{D})\int_a^b \chi_{[a,\tau]}\,\mathrm{d}g=g(\tau)-g(a)\quad \text{if } \tau\in[a,b),$$

$$(\sigma\mathrm{D})\int_a^b \chi_{[a,\tau)}\,\mathrm{d}g=g(\tau)-g(a)\quad \text{if } \tau\in(a,b],$$

$$(\sigma\mathrm{D})\int_a^b \chi_{[\tau]}\,\mathrm{d}g=0\qquad\qquad \text{if } \tau\in[a,b].$$

The next result provides more information on the relation between the (σ) Dushnik and Kurzweil-Stieltjes integrals.

6.13.7 Theorem. *If $f,g\colon[a,b]\to\mathbb{R}$ are regulated and one of them has bounded variation, then both integrals $(\sigma\mathrm{D})\int_a^b f\,\mathrm{d}g$ and $(\mathrm{KS})\int_a^b g\,\mathrm{d}f$ exist, and*

$$(\sigma\mathrm{D})\int_a^b f\,\mathrm{d}g+(\mathrm{KS})\int_a^b g\,\mathrm{d}f=f(b)\,g(a)-f(a)\,g(a).\qquad (6.13.2)$$

The proof is left as an exercise to the reader.

6.13.8 Exercise. Prove Theorem 6.13.7.

Hint: Follow the steps of Exercise 6.13.2 with the (σ) Young integral replaced by the (σ) Dushnik integral. Take into account the results of Exercise 6.13.6 and observe that (6.13.2) holds if one of the functions f, g is regulated and the other is a finite step function.

By combining the relation (6.13.2) with Theorem 6.4.2 (integration by parts), we obtain the equality

$$(\sigma\mathrm{D})\int_a^b f\,\mathrm{d}g = (\mathrm{KS})\int_a^b f\,\mathrm{d}g - \sum_{a\le x\le b}\Big(\Delta^- f(x)\,\Delta^- g(x) - \Delta^+ f(x)\,\Delta^+ g(x)\Big),$$

which holds whenever $f, g\colon [a,b]\to\mathbb{R}$ are regulated and one of them has bounded variation.

INTEGRATION IN ABSTRACT SPACES

Integrals of vector and matrix functions were introduced in Section 6.10. More generally, it is possible to consider functions with values in Banach spaces. Assume that X is a Banach space equipped with a norm denoted by $\|\cdot\|_X$, and $\mathscr{L}(X)$ is the space of all continuous linear operators on X. Given functions

$$F\colon [a,b]\to\mathscr{L}(X),\quad G\colon [a,b]\to\mathscr{L}(X),\quad g\colon [a,b]\to X,$$

it is possible to define the Kurzweil-Stieltjes integrals

$$\int_a^b \mathrm{d}F\,g,\quad \int_a^b F\,\mathrm{d}g,\quad \int_a^b \mathrm{d}F\,G,\quad \int_a^b F\,\mathrm{d}G.$$

For example, $\int_a^b \mathrm{d}F\,g$ exists and equals $I\in X$ if for every $\varepsilon>0$ there is a gauge δ on $[a,b]$ such that

$$\Big\|\sum_{j=1}^{\nu(\boldsymbol{\alpha})}(F(\alpha_j)-F(\alpha_{j-1}))\,g(\xi_j) - I\Big\|_X < \varepsilon$$

for every δ-fine partition $P=(\boldsymbol{\alpha},\boldsymbol{\xi})$ of $[a,b]$. The remaining integrals can be defined in a similar way.

Most basic properties of the Kurzweil-Stieltjes integral carry over to the abstract setting, but there are some exceptions. For example, Corollary 6.5.2 holds only if the space X is finite-dimensional. This leads to certain problems connected e.g. to the substitution theorem for abstract integrals.

Given a function $f:[a,b]\to X$, we define

$$V(f,\boldsymbol{\alpha})=\sum_{j=1}^{\nu(\boldsymbol{\alpha})}\|f(\alpha_j)-f(\alpha_{j-1})\|_X$$

for each division $\boldsymbol{\alpha}$ of the interval $[a,b]$, and

$$\operatorname{var}_a^b f=\sup\{V(f,\boldsymbol{\alpha}):\boldsymbol{\alpha}\in\mathscr{D}[a,b]\}.$$

One can now show that the integrals $\int_a^b F\,\mathrm{d}G$ and $\int_a^b \mathrm{d}F\,G$ exist if $F:[a,b]\to\mathscr{L}(X)$ has bounded variation and $G:[a,b]\to\mathscr{L}(X)$ is regulated.

Instead of variation, it is possible to consider the notion of a *semivariation*, which is defined as follows: Given a function $F:[a,b]\to\mathscr{L}(X)$ and a division $\boldsymbol{\alpha}$ of $[a,b]$, let

$$V_a^b(F,\boldsymbol{\alpha})=\sup\Big\{\Big\|\sum_{j=1}^{\nu(\boldsymbol{\alpha})}(F(\alpha_j)-F(\alpha_{j-1}))\,x_j\Big\|_X\Big\},$$

where the supremum is taken over all choices of $x_j\in X$, $j\in\{1,\ldots,\nu(\boldsymbol{\alpha})\}$, satisfying $\|x_j\|_X\le 1$. Then the number

$$(\mathcal{B})\ \operatorname{var}_a^b F=\sup\{V_a^b(F,\boldsymbol{\alpha}):\ \boldsymbol{\alpha}\in\mathscr{D}[a,b]\}$$

is called the *semivariation* of the function F on $[a,b]$ (see e.g. [Hönig (1975)] and [Monteiro (2015)]). If F has finite variation, then it also has finite semivariation, but the converse statement holds only in finite-dimensional spaces. When dealing with abstract integrals, the assumption of bounded variation can be often (but not always) weakened to bounded semivariation.

BIBLIOGRAPHIC REMARKS

The Kurzweil-Stieltjes integral is a special case of the more general integral denoted by $\int_a^b \mathrm{D}U(\tau,t)$ and introduced in Kurzweil's seminal work [Kurzweil (1957)] as a tool for explaining some convergence effects in the theory of nonlinear differential equations. If $U:[a,b]\times[a,b]\to\mathbb{R}$ has the form $U(\tau,t)=f(\tau)g(t)$, then the integral $\int_a^b \mathrm{D}U(\tau,t)$ reduces to the Kurzweil-Stieltjes integral $\int_a^b f\,\mathrm{d}g$. Kurzweil explicitly used this Stieltjes integral in [Kurzweil (1958a)] to investigate generalized differential equations, including e.g. equations whose right-hand sides involve the Dirac distributions. During the 1970s, the terms Kurzweil-Stieltjes integral or Kurzweil's definition of the Perron-Stieltjes integral were already common in the works dealing with generalized differential equations (see e.g. [Schwabik (1971)] or [Schwabik, Tvrdý, Vejvoda (1979)] and the references there).

Unfortunately, monographic literature dealing with the Kurzweil-Stieltjes integral is rather scarce. Let us mention Chapter 24 of Schechter's handbook [Schechter E. (1997)] (where it is called the Henstock-Stieltjes integral), McLeod's monograph [McLeod (1980)] (where it is referred to as the gauge integral), and Chapter 2 of Dudley and Norvaiša's monograph [Dudley, Norvaiša (2011)] (where it appears as the Henstock-Kurzweil integral). The standard references on generalized ordinary differential equations, [Schwabik, Tvrdý, Vejvoda (1979)], [Schwabik (1992)], and [Kurzweil (2012)], contain substantial parts devoted to Kurzweil-Stieljtes integrals and the more general Kurzweil integrals of the form $\int_a^b \mathrm{D}U(\tau,t)$. The Kurzweil-Stieltjes integral is also a special case of Henstock's generalized Riemann integral $\int \mathrm{d}h$, where h is a point-interval function; this integral (even in the multidimensional case) is treated in [Henstock (1988)].

Lemma 6.2.3 justifies the definition of the Kurzweil-Stieljes integral. In the context of integration theory it was first mentioned in [Kurzweil (1957)]. To our knowledge, it was J. Mawhin (see Section 4.2 of [Mawhin (1992)]) who noticed that a much older paper [Cousin (1895)] dealing with complex valued functions of two variables contains the same assertion. Since then this lemma is generally called Cousin's lemma.

Some other results presented in this chapter (such as the Saks-Henstock lemma, Hake's theorem, uniform integrability, etc.) are generalizations of corresponding statements for Kurzweil-Henstock integrals given e.g. in [Bartle (2001); Gordon (1994); Lee, Výborný (2000); Schwabik, Ye (2005); Swartz (2001)]).

If we restrict ourselves to integrals $\int_a^b f\,\mathrm{d}g$ where f, g are regulated and one of them has bounded variation, then the formulas from Theorems 6.3.12 and 6.3.13, which are based on the decomposition into the continuous and the jump part, can serve as definitions of the Kurzweil-Stieltjes integral (note that the integrals $\int_a^b f^{\mathrm{C}}\,\mathrm{d}g$ in Theorem 6.3.12 and $\int_a^b f\,\mathrm{d}g^{\mathrm{C}}$ in Theorem 6.3.13 exist as Riemann-Stieltjes or Lebesgue-Stieltjes integrals). This approach can be found e.g. in [Ashordia (2016); Groh (1980); Hildebrandt (1959)].

A fairly general existence theorem for the Kurzweil-Stieltjes integral is available in [Dudley, Norvaiša (1999)] (see Theorems 4.4 and F.2 there). It is based on the notion of p-variation introduced in bibliographic remarks to Chapter 5, and reads as follows: *If* $f:[a,b]\to\mathbb{R}$ *has finite* p*-variation and* $g:[a,b]\to\mathbb{R}$ *has finite* q*-variation for some* $p,q>0$ *satisfying* $1/p+1/q>1$, *then the Kurzweil-Stieltjes integral* $\int_a^b f\,\mathrm{d}g$ *exists.* Unlike the existence result for Riemann-Stieltjes integrals mentioned in Chapter 5, there are no requirements on the continuity of f and g.

Lemma 6.5.1 (Saks-Henstock lemma) is a crucial tool for studying deeper properties of the Kurzweil-Henstock and Kurzweil-Stieltjes integrals. Sometimes it is referred to as the Henstock lemma in view of [Henstock (1946)], and its first version involving the gauge integrals was given in [Henstock (1968)]. However, R. Henstock himself pointed out that the paper [Saks (1927)] dealing with the Burkill integral was an essential inspiration for his result. For this reason, the result is nowadays called the Saks-Henstock lemma.

Theorem 6.5.6 (Hake's theorem) is a special case of Theorem 1.3.5 in [Kurzweil (1957)] (see also Theorem 17 in [Schwabik (1996)]). The origins of this result go back to [Hake (1921)], where it appeared in the context of Perron/Denjoy integration theory. Often it is referred to as one part of the Aleksandrov-Hake-Looman theorem (cf. [Aleksandrov (1924); Looman (1935)]) which states that the class of Perron (i.e., Kurzweil-Henstock) integrable functions is invariant under the Cauchy extension process. For more details see Section 6.15.4 of [Thomson (2008)] and Chapter 2 of [Pesin (1970)].

Theorem 6.6.5 can be generalized in several ways. For example, the following version of substitution theorem found its use in the application of the theory of hysteresis in economics (see [Krejčí, Lamba, Melnik, Rachinskii (2015)]): *Assume that the function* $f:[a,b]\to\mathbb{R}$ *is bounded on* $[a,b]$ *and regulated on* $[\alpha,b]$ *for every* $\alpha\in(a,b)$. *Moreover, let the function* $\phi:[a,b]\to\mathbb{R}$ *be nondecreasing on* $[a,b]$ *and let* $\phi(a)=c$, $\phi(b)=d$. *Finally, let the function* $g\in\mathrm{BV}([c,d])$ *be continuous from the right on* $[c,d)$. *Denote* $\psi(s)=\inf\{t\in[a,b]: s\le\phi(t)\}$ *for all* $s\in[c,d]$. *Then* $\int_\alpha^b f(t)\,\mathrm{d}g(\phi(t))=\int_{\phi(\alpha)}^{\phi(b)} f(\psi(s))\,\mathrm{d}g(s)$ *for all* $\alpha\in[a,b]$. Notably, the theories of Kurzweil and Kurzweil-Stieltjes integrals have contributed significantly to the investigation of hysteresis models subject to discontinuous output; see e.g. [Krejčí (2006)], [Krejčí, Lamba, Monteiro, Rachinskii (2016)] or [Liu, Krejčí, Ye (2017)].

Uniform integrability (also known as equiintegrability) is a basic concept in the study of convergence theorems for the Kurzweil-Henstock and Kurzweil-Stieltjes integrals. The definition and the first part of Theorem 6.8.2 are well known. The fact that uniform Kurzweil-Henstock integrability implies uniform convergence of indefinite integrals was observed in [Swartz (1998/99)]; Theorem 6.8.2 extends this result into the setting of Kurzweil-Stieltjes integrals. Theorem 6.8.10 is due to D. Preiss and Š. Schwabik. It is a special case of Theorem 5.5 from J. Kurzweil's book [Kurzweil (1980)], which is concerned with a more general type of integral (see also Theorem 1.28 and Remark 1.30 in [Schwabik (1992)]). In the same book, J. Kurzweil showed how to derive the dominated and monotone convergence theorems as consequences of the previously mentioned result.

Kurzweil-Stieltjes integrals over elementary sets were studied in [Hanung, Monteiro, Tvrdý (2016)] in order to establish the bounded convergence theorem for functions with values in Banach spaces.

The concept of the Perron-Stieltjes integral was introduced by A. J. Ward [Ward (1936)]; see also Section VI.8 in [Saks (1937)]. A proof of the equivalence between the Perron-Stieltjes integral and the Kurzweil-Stieltjes integral is contained in [Kurzweil (1957)]. Let us also mention the related concept of the MC-integral, which was recently introduced in [Bendová, Malý (2011)]. It has a simple definition based on the notion of an MC-derivative, and coincides with the Kurzweil-Stieltjes integral.

The proof of Theorem 6.12.7 (as well as the corresponding auxiliary lemmas), which provides the relation between Kurzweil-Stieltjes and Lebesgue-Stieltjes integrals with nonnegative integrands, follows the method described in Chapter 24 of [Schechter E. (1997)].

Integral sums of the form (6.13.1) were introduced by W. H. Young in [W.H. Young (1914)], where related (δ)-type integrals were discussed, as well. To our knowledge, a systematic study of the (σ) Young integral was initiated by T. H. Hildebrandt in [Hildebrandt (1938)]. More details are available in Section II.19 of his monograph [Hildebrandt (1963)]. Basic references on the relation between the (σ) Young integral, Kurzweil-Stieljtes integral, and the modified Kurzweil-Stieltjes integral are [Schwabik (1973a)] and [Schwabik (1973b)]. For more information about the KN integral and Dushnik integral, see [Krejčí (2003)] and [Dushnik (1931)], respectively. A noteworthy study of the Dushnik integral is contained in the monograph [Hönig (1975)] by Ch. S. Hönig, who extended its definition to functions with values in Banach spaces and subsequently used it to develop the theory of abstract Volterra-Stieltjes integral equations. A relation between the (σ) Young and (σ) Dushnik integrals analogous to Theorem 6.13.7 has been already known for a long time, cf. Theorem B in [MacNerney (1963)] or Theorem 4.7 in [Hönig (1975)]. Abstract Kurzweil-Stieltjes integrals are discussed in [Schwabik (1996)], [Schwabik (2001)] and [Monteiro, Tvrdý (2012)]. Integration of functions with values in Hilbert spaces or reflexive Banach spaces is useful in the theory of hysteresis (see e.g. [Krejčí, Laurençot (2002)] or [Krejčí, Liero (2009)]).

Chapter 7

Generalized linear differential equations

7.1 Introduction

A *generalized linear differential equation* is an integral equation of the form

$$x(t) = x(s) + \int_s^t \mathrm{d}A\, x + f(t) - f(s), \tag{7.1.1}$$

where $A\colon [a,b] \to \mathscr{L}(\mathbb{R}^n)$ is a given matrix function, and $f\colon [a,b] \to \mathbb{R}^n$ is a given vector function. A function $x\colon [a,b] \to \mathbb{R}^n$ is called a solution of equation (7.1.1) on the interval $[a,b]$ if the integral $\int_a^b \mathrm{d}A\, x$ exists, and (7.1.1) holds for all $t, s \in [a,b]$. In the special case when $f = 0$, we speak about a homogeneous equation, while the general case is referred to as the nonhomogeneous equation.

Throughout this chapter, all integrals are Kurzweil-Stieltjes integrals, and integrals of vector and matrix functions are understood in the sense of Section 6.10. As we already explained there, most properties of Kurzweil-Stieltjes integrals obtained for real functions remain valid also for vector and matrix functions; this fact should be kept in mind whenever we refer to the results of Chapter 6.

As the name suggests, generalized linear differential equations extend the concept of classical ordinary differential equations.

7.1.1 Exercise. Suppose that $P\colon [a,b] \to \mathscr{L}(\mathbb{R}^n)$ and $q\colon [a,b] \to \mathbb{R}^n$ are continuous functions. Using Theorem 6.6.1 (substitution theorem), show that $x\colon [a,b] \to \mathbb{R}^n$ is a solution of the ordinary differential equation $x'(t) = P(t)x(t) + q(t)$ on $[a,b]$ if and only if it is a solution of the generalized linear differential equation (7.1.1), where $A\colon [a,b] \to \mathscr{L}(\mathbb{R}^n)$ and

$f:[a,b]\to\mathbb{R}^n$ are given by the formulas

$$A(t)=\int_a^t P(\tau)\,\mathrm{d}\tau,\quad f(t)=\int_a^t q(\tau)\,\mathrm{d}\tau,\quad t\in[a,b].$$

The next remark provides a useful equivalent definition of solutions to generalized linear differential equations.

7.1.2 Remark. Suppose that $t_0\in[a,b]$ is given and $x:[a,b]\to\mathbb{R}^n$ satisfies the equation

$$x(t)=x(t_0)+\int_{t_0}^t \mathrm{d}A\,x+f(t)-f(t_0) \tag{7.1.2}$$

for all $t\in[a,b]$. Then for any $s\in[a,b]$, we have

$$x(s)=x(t_0)+\int_{t_0}^s \mathrm{d}A\,x+f(s)-f(t_0).$$

By subtracting this equality from (7.1.2), we find that (7.1.1) holds for all $t,s\in[a,b]$, i.e., x is a solution of equation (7.1.1). Thus, a function $x:[a,b]\to\mathbb{R}^n$ is a solution of equation (7.1.1) on $[a,b]$ if and only if it satisfies (7.1.2) for an arbitrary fixed $t_0\in[a,b]$ and all $t\in[a,b]$.

By choosing a point $t_0\in[a,b]$ and requiring that $x(t_0)=\widetilde{x}$, where $\widetilde{x}\in\mathbb{R}^n$ is a given vector, we arrive at the initial-value problem

$$x(t)=\widetilde{x}+\int_{t_0}^t \mathrm{d}A\,x+f(t)-f(t_0),\quad t\in[a,b]. \tag{7.1.3}$$

By Remark 7.1.2, it is obvious that a function $x:[a,b]\to\mathbb{R}^n$ is a solution of problem (7.1.3) if and only if it is a solution of equation (7.1.1) and satisfies $x(t_0)=\widetilde{x}$.

Our main goal in this chapter is to study the existence and uniqueness of solutions to the initial-value problem (7.1.3), as well as the continuous dependence of solutions on parameters. We will usually suppose that A is a matrix function of bounded variation, and f is a regulated vector function. To simplify notation, we introduce the following classes of functions.

7.1.3 Definition. (i) $\mathrm{G}([a,b],\mathbb{R}^n)$ is the Banach space of all regulated functions $f:[a,b]\to\mathbb{R}^n$. The norm on $\mathrm{G}([a,b],\mathbb{R}^n)$ is defined by

$$\|f\|=\sup_{t\in[a,b]}|f(t)|\quad\text{for}\quad f\in\mathrm{G}([a,b],\mathbb{R}^n),$$

where $|f(t)|$ is the norm of the vector $f(t)$ in $\mathbb{R}^n$.

(ii) $\mathrm{BV}([a,b],\mathscr{L}(\mathbb{R}^n))$ is the Banach space of all bounded variation functions $F\colon [a,b]\to\mathscr{L}(\mathbb{R}^n)$. The norm on $\mathrm{BV}([a,b],\mathscr{L}(\mathbb{R}^n))$ is defined by

$$\|F\|_{\mathrm{BV}} = |F(a)| + \operatorname{var}_a^b F \quad \text{for} \quad F \in \mathrm{BV}([a,b],\mathscr{L}(\mathbb{R}^n)),$$

where $\operatorname{var}_a^b F$ is defined as in Section 6.10 and $|F(a)|$ is the norm of the matrix $F(a)$ in $\mathscr{L}(\mathbb{R}^n)$.

The spaces $\mathrm{BV}([a,b],\mathbb{R}^n)$, $\mathrm{C}([a,b],\mathscr{L}(\mathbb{R}^n))$, $\mathrm{C}([a,b],\mathbb{R}^n)$, and the corresponding norms are defined similarly.

7.2 Differential equations with impulses

One motivation for studying generalized differential equations comes from the theory of equations with impulses. Many real-world systems involve perturbations which act for a negligible time, but significantly affect the state of the system. These phenomena are often modeled by *differential equations with impulses*, whose solutions need not be differentiable or continuous.

Impulsive systems often appear in physics (e.g., the description of clock mechanisms, oscillations of electromechanical systems, radiation of electromagnetic waves in an environment with rapidly changing parameters, stabilization of an inverted pendulum, optimal regulation using the bang-bang method), medicine (the distribution of drugs in a human body, vaccination in epidemiological models, investigation of the effect of mass measles vaccination), population dynamics (models with rapid changes in the population), or economics (trade models which admit rapid changes of prices).

The simplest instance of an impulsive process is a system of linear differential equations with linear impulses acting in finitely many preassigned points. Such a problem consists of an ordinary differential equation

$$x'(t) = P(t)\,x(t) + q(t) \tag{7.2.1}$$

together with the conditions

$$\Delta^+ x(\tau_k) = B_k\, x(\tau_k) + d_k, \quad k \in \{1,\dots,r\}, \tag{7.2.2}$$

where $P:[a,b]\to\mathscr{L}(\mathbb{R}^n)$ and $q:[a,b]\to\mathbb{R}^n$ are continuous functions, $r\in\mathbb{N}$ is the number of impulses, $a<\tau_1<\cdots<\tau_r<b$ correspond to the preassigned moments where the impulses take place, $B_1,\ldots,B_r\in\mathscr{L}(\mathbb{R}^n)$, and $d_1,\ldots,d_r\in\mathbb{R}^n$.

A regulated function $x:[a,b]\to\mathbb{R}^n$ is called a solution of the impulsive problem (7.2.1)–(7.2.2) if the restrictions of x to the intervals

$$[a,\tau_1],\ (\tau_1,\tau_2],\ldots,(\tau_{r-1},\tau_r],\ (\tau_r,b]$$

are continuously differentiable functions,

$$x'(t)=P(t)\,x(t)+q(t)\quad\text{for all } t\in[a,b]\setminus\{\tau_1,\tau_2,\ldots,\tau_r\},\tag{7.2.3}$$

and x satisfies the conditions (7.2.2). As usual, the derivatives at the endpoints are understood as one-sided derivatives.

We will now show that the problem (7.2.1)–(7.2.2) can be equivalently reformulated as a generalized linear differential equation of the form (7.1.2).

7.2.1 Theorem. *Assume that $P:[a,b]\to\mathscr{L}(\mathbb{R}^n)$ and $q:[a,b]\to\mathbb{R}^n$ are continuous, $r\in\mathbb{N}$, $a<\tau_1<\cdots<\tau_r<b$, $B_1,\ldots,B_r\in\mathscr{L}(\mathbb{R}^n)$, and $d_1,\ldots,d_r\in\mathbb{R}^n$.*

Then $x:[a,b]\to\mathbb{R}^n$ is a solution of the impulsive problem (7.2.1)–(7.2.2) if and only if it is a solution of the generalized linear differential equation

$$x(t)=x(a)+\int_a^t \mathrm{d}A\,x+f(t)-f(a),\tag{7.2.4}$$

where

$$A(t)=\int_a^t P(s)\,\mathrm{d}s+\sum_{k=1}^r \chi_{(\tau_k,b]}(t)\,B_k\quad \textit{for}\ \ t\in[a,b],$$
$$f(t)=\int_a^t q(s)\,\mathrm{d}s+\sum_{k=1}^r \chi_{(\tau_k,b]}(t)\,d_k\quad \textit{for}\ \ t\in[a,b].$$

Proof. First of all, we claim that the impulsive problem (7.2.1)–(7.2.2) is equivalent with the single equation

$$x(t)=x(a)+\int_a^t (P(s)\,x(s)+q(s))\,\mathrm{d}s+\sum_{k=1}^r \chi_{(\tau_k,b]}(t)(B_k\,x(\tau_k)+d_k).\tag{7.2.5}$$

Indeed, assume that $x:[a,b]\to\mathbb{R}^n$ is a solution of the impulsive problem (7.2.1)–(7.2.2). Consider a fixed $t\in(a,b]$. If we denote $\tau_0=a$, $\tau_{r+1}=b$, then there exists a unique $\ell\in\{1,\dots,r+1\}$ such that $t\in(\tau_{\ell-1},\tau_\ell]$, and therefore

$$x(t)=x(a)+\sum_{k=1}^{\ell-1}(x(\tau_k)-x(\tau_{k-1}+))+x(t)-x(\tau_{\ell-1}+)+\sum_{k=1}^{\ell-1}\Delta^+x(\tau_k).$$

Using (7.2.2) and (7.2.3), we obtain

$$\begin{aligned}x(t)=x(a)+\sum_{k=1}^{\ell-1}\int_{\tau_{k-1}}^{\tau_k}(P(s)\,x(s)+q(s))\,\mathrm{d}s+\int_{\tau_{\ell-1}}^{t}(P(s)\,x(s)+q(s))\,\mathrm{d}s\\ +\sum_{k=1}^{\ell-1}(B_kx(\tau_k)+d_k)=x(a)+\int_a^t(P(s)\,x(s)+q(s))\,\mathrm{d}s\\ +\sum_{k=1}^{r}\chi_{(\tau_k,b]}(t)(B_k\,x(\tau_k)+d_k),\end{aligned}$$

which shows that (7.2.5) holds for $t\in(a,b]$; the case $t=a$ is trivial. Conversely, if $x:[a,b]\to\mathbb{R}^n$ satisfies (7.2.5) for all $t\in[a,b]$, then (7.2.2) and (7.2.3) obviously hold, and x is a solution of the impulsive problem (7.2.1)–(7.2.2).

To finish the proof, it is enough to observe that

$$\int_a^t \mathrm{d}A\,x=\int_a^t P(s)\,x(s)\,\mathrm{d}s+\sum_{k=1}^{r}\chi_{(\tau_k,b]}(t)\,B_k\,x(\tau_k)\quad\text{for }\ t\in[a,b],$$

$$f(t)-f(a)=\int_a^t q(s)\,\mathrm{d}s+\sum_{k=1}^{r}\chi_{(\tau_k,b]}(t)\,d_k\quad\text{for }\ t\in[a,b].$$

The second identity is obvious, and the first one is an immediate consequence of Theorem 6.6.1 (substitution theorem) and formula (6.3.1) from Lemma 6.3.2 (see also Example 6.5.8). □

We point out that the concept of generalized linear differential equations encompasses not only linear impulsive problems, but also other types of equations, such as linear dynamic equations on time scales; see Section 8.7.

7.3 Linear operators

Before we resume our discussion of generalized linear differential equations, we collect some basic notions and results on linear operators in Banach spaces that will be needed later. More information can be found in functional analysis textbooks (see e.g. [Kolmogorov, Fomin (2012)] or [Rudin (1973)]); a basic overview is also included in the introductory part of [Schwabik, Tvrdý, Vejvoda (1979)].

Let X be a Banach space with a norm denoted by $\|\cdot\|_X$. We say that a sequence $\{x_n\}$ in X is convergent to $x \in X$ if $\lim_{n\to\infty} \|x_n - x\|_X = 0$; in this case, we write $\lim_{n\to\infty} x_n = x$, or simply $x_n \to x$.

A mapping (also called an operator) $T: X \to Y$ between two Banach spaces X, Y is called *continuous* if for each $x \in X$ and each sequence $\{x_n\}$ in X, $x_n \to x$ implies $T(x_n) \to T(x)$.

An operator $L: X \to Y$ is *linear* if

$$L(c_1 x_2 + c_2 x_2) = c_1 L(x_1) + c_2 L(x_2) \text{ holds for } x_1, x_2 \in X \quad \text{and} \quad c_1, c_2 \in \mathbb{R}.$$

If L is a linear operator, we often write $L x$ instead of $L(x)$.

We say that a linear operator $L: X \to Y$ is *bounded* if there exists a number $K \geq 0$ such that

$$\|L x\|_Y \leq K \|x\|_X \quad \text{for all } x \in X.$$

It is known that a linear operator $L: X \to Y$ is continuous if and only if it is bounded.

The set of all bounded linear operators from X into Y is denoted by $\mathscr{L}(X,Y)$. This set is a vector space when equipped with the usual operations of addition and multiplication by a real number. The norm of an operator $L \in \mathscr{L}(X,Y)$ is defined by

$$\|L\|_{\mathscr{L}(X,Y)} = \sup\{\|L x\|_Y : x \in X \quad \text{and} \quad \|x\|_X \leq 1\}.$$

The space $\mathscr{L}(X,Y)$ is a Banach space with respect to this norm.

If $X = Y$, we write $\mathscr{L}(X)$ instead of $\mathscr{L}(X,X)$. Note that for $X = \mathbb{R}^n$, each linear operator is bounded, and the space $\mathscr{L}(\mathbb{R}^n)$ coincides with the space of all $n \times n$ matrices.

The next result provides a useful sufficient condition for the invertibility of linear operators. The case when $X = \mathbb{R}^n$ is well known from linear algebra; the formulation for general Banach spaces can be found e.g. in [Taylor, Lay (1980)], Lemma IV.1.4.

7.3.1 Lemma. *If X is a Banach space, $L \in \mathscr{L}(X)$ and $\|L\|_{\mathscr{L}(X)} < 1$, then the operator $I - L$ is invertible, and*

$$\left\|(I-L)^{-1}\right\|_{\mathscr{L}(X)} \le \frac{1}{1-\|L\|_{\mathscr{L}(X)}}.$$

We say that a linear operator $L: X \to Y$ is *compact* if the image under L of every bounded subset of X is relatively compact in Y. Equivalently, L is compact if for any bounded sequence $\{x_n\}$ in X, the sequence $\{L\,x_n\}$ has a convergent subsequence. Each compact linear operator is necessarily continuous.

The following result is a generalization of one of I. Fredholm's theorems from the theory of integral equations. Its proof is included e.g. in [Dunford, Schwartz (1958,1963)] or [Schechter M. (2001)]. For our purposes, it will be a useful tool in study of solvability of generalized linear differential equations.

7.3.2 Theorem (FREDHOLM ALTERNATIVE THEOREM). *Suppose that X is a Banach space and $L \in \mathscr{L}(X)$ a compact operator. Then the equation*

$$x - L\,x = g$$

has a unique solution $x \in X$ for every $g \in X$ if and only if the corresponding homogeneous equation

$$x - L\,x = 0$$

has only the trivial solution $x = 0$.

7.4 Existence and uniqueness of solutions

The goal of this section is to study the existence and uniqueness of solutions to the initial-value problem

$$x(t) = \widetilde{x} + \int_{t_0}^{t} \mathrm{d}A\,x + f(t) - f(t_0) \quad \text{for} \quad t \in [a,b], \tag{7.4.1}$$

where $A \in \mathrm{BV}([a,b], \mathscr{L}(\mathbb{R}^n))$, $f \in \mathrm{G}([a,b], \mathbb{R}^n)$, $t_0 \in [a,b]$ and $\widetilde{x} \in \mathbb{R}^n$ are given.

The next theorem provides some information about solutions of (7.4.1).

7.4.1 Theorem. *If* $A \in \mathrm{BV}([a,b],\mathscr{L}(\mathbb{R}^n))$ *and* $f \in \mathrm{G}([a,b],\mathbb{R}^n)$, *then every solution* $x\colon[a,b]\to\mathbb{R}^n$ *of equation* (7.4.1) *is regulated on* $[a,b]$ *and satisfies the relations*

$$\Delta^- x(t) = \Delta^- A(t)x(t) + \Delta^- f(t) \quad \text{for } t \in (a,b],$$
$$\Delta^+ x(s) = \Delta^+ A(s)x(s) + \Delta^+ f(s) \quad \text{for } s \in [a,b).$$

Moreover,

$$\operatorname{var}_a^b(x-f) \le (\operatorname{var}_a^b A)\,\|x\| < \infty.$$

Proof. The fact that x is regulated, the formulas for the jumps, as well as the estimate for $\operatorname{var}_a^b(x-f)$ follow immediately from Corollary 6.5.5. □

According to Theorem 7.4.1, it is appropriate to look for solutions of equation (7.4.1) in the space $\mathrm{G}([a,b],\mathbb{R}^n)$. The basic idea is to rewrite the problem as an operator equation in $\mathrm{G}([a,b],\mathbb{R}^n)$.

Given a function $x \in \mathrm{G}([a,b],\mathbb{R}^n)$ and a point $t_0 \in [a,b]$, consider the function $\mathscr{A}_{t_0}x\colon[a,b]\to\mathbb{R}^n$ defined by

$$(\mathscr{A}_{t_0}x)(t) = \int_{t_0}^t \mathrm{d}A\,x \quad \text{for} \quad t \in [a,b]. \tag{7.4.2}$$

By Corollary 6.5.5, $\mathscr{A}_{t_0}x$ is regulated on $[a,b]$. Thus, the mapping $x \mapsto \mathscr{A}_{t_0}x$ is a linear operator on the space $\mathrm{G}([a,b],\mathbb{R}^n)$. Moreover, since (by Theorem 6.3.6) we have

$$\|\mathscr{A}_{t_0}x\| \le \big(\operatorname{var}_a^b A\big)\,\|x\| \quad \text{for all } x \in \mathrm{G}([a,b],\mathbb{R}^n),$$

it follows that $\mathscr{A}_{t_0}$ is a bounded linear operator on $\mathrm{G}([a,b],\mathbb{R}^n)$. □

Using the operator $\mathscr{A}_{t_0}$, we can rewrite the problem (7.4.1) as the equation

$$x - \mathscr{A}_{t_0}x = g, \quad \text{where} \quad g = \widetilde{x} + f - f(t_0).$$

If we show that $\mathscr{A}_{t_0}$ is a compact operator, we will be able to study the solvability of this equation using the Fredholm alternative theorem.

7.4.2 Theorem. *If* $A \in \mathrm{BV}([a,b],\mathscr{L}(\mathbb{R}^n))$ *and* $t_0 \in [a,b]$, *then the operator* $\mathscr{A}_{t_0} \in \mathscr{L}(\mathrm{G}([a,b],\mathbb{R}^n))$ *given by* (7.4.2) *is compact.*

Proof. Suppose that $\{x_n\}$ is a bounded sequence in $\mathrm{G}([a,b],\mathbb{R}^n)$, i.e., there exists a $\varkappa \in [0,\infty)$ such that

$$\|x_n\| \le \varkappa \quad \text{for every} \quad n \in \mathbb{N}.$$

We have to show that $\{\mathscr{A}_{t_0} x_n\}$ is relatively compact in $\mathrm{G}([a,b],\mathbb{R}^n)$. The function $h:[a,b] \to \mathbb{R}$ given by $h(t) = \varkappa \operatorname{var}_a^t A$ for all $t \in [a,b]$ is nondecreasing. If $n \in \mathbb{N}$ and $a \le s \le t \le b$, then

$$|(\mathscr{A}_{t_0} x_n)(t) - (\mathscr{A}_{t_0} x_n)(s)| = \left| \int_s^t \mathrm{d}A\, x_n \right| \le \|x_n\| \operatorname{var}_s^t A \le h(t) - h(s).$$

Also, for each $n \in \mathbb{N}$, we have

$$|(\mathscr{A}_{t_0} x_n)(a)| = \left| \int_{t_0}^a \mathrm{d}A\, x_n \right| \le \|x_n\| \operatorname{var}_a^{t_0} A \le \varkappa \operatorname{var}_a^b A.$$

Thus, the relative compactness of $\{\mathscr{A}_{t_0} x_n\}$ follows from Corollary 4.3.8. □

The next result is now an immediate consequence of Theorems 7.3.2 and 7.4.2.

7.4.3 Theorem. *Let $A \in \mathrm{BV}([a,b],\mathscr{L}(\mathbb{R}^n))$ and $t_0 \in [a,b]$. Then the equation*

$$x(t) - \int_{t_0}^t \mathrm{d}A\, x = g(t) \tag{7.4.3}$$

has a unique solution $x:[a,b] \to \mathbb{R}^n$ for every $g \in \mathrm{G}([a,b],\mathbb{R}^n)$ if and only if the corresponding homogeneous problem

$$x(t) - \int_{t_0}^t \mathrm{d}A\, x = 0 \tag{7.4.4}$$

has only the trivial solution $x \equiv 0$ on $[a,b]$.

Proof. Equation (7.4.3) is equivalent with the operator equation

$$x - \mathscr{A}_{t_0} x = g.$$

By Theorem 7.4.2, $\mathscr{A}_{t_0}$ is a linear compact operator on $\mathrm{G}([a,b],\mathbb{R}^n)$. Hence, the statement follows from Theorem 7.3.2. □

Now, assume that $x \in \mathrm{G}([a,b],\mathbb{R}^n)$ is a solution of equation (7.4.1), and $\tau \in (t_0, b]$. Using Theorem 7.4.1, we get

$$x(\tau) = x(\tau-) + \Delta^- x(\tau) = x(\tau-) + \Delta^- A(\tau)\, x(\tau) + \Delta^- f(\tau),$$

or equivalently

$$\big(I-\Delta^- A(\tau)\big)\, x(\tau) = x(\tau-) + \Delta^- f(\tau),$$

where I stands for the $n\times n$-identity matrix. Hence, the value of the solution x at a point $\tau\in(t_0,b]$ is uniquely determined by the limit $x(\tau-)$ if and only if the matrix $I-\Delta^- A(\tau)$ is invertible.

Similarly, if $\tau\in[a,t_0)$, Theorem 7.4.1 yields

$$x(\tau) = x(\tau+) - \Delta^+ x(\tau) = x(\tau+) - \Delta^+ A(\tau)\, x(\tau) - \Delta^+ f(\tau),$$

or equivalently

$$\big(I+\Delta^+ A(\tau)\big)\, x(\tau) = x(\tau+) - \Delta^+ f(\tau).$$

Thus, the value of the solution x at a point $\tau\in[a,t_0)$ is uniquely determined by the limit $x(\tau+)$ if and only if the matrix $I+\Delta^+ A(\tau)$ is invertible.

These considerations provide a motivation for conditions (7.4.6) and (7.4.7) in the next theorem.

7.4.4 Lemma. *Let $A\in \mathrm{BV}([a,b],\mathscr{L}(\mathbb{R}^n))$ and $t_0\in[a,b]$. Then the equation*

$$x(t) - \int_{t_0}^{t} \mathrm{d}A\, x = g(t) \tag{7.4.5}$$

has a unique solution for every function $g\in \mathrm{G}([a,b],\mathbb{R}^n)$ if and only if

$$\det\big(I-\Delta^- A(t)\big)\neq 0 \quad \textit{for all } t\in(t_0,b] \tag{7.4.6}$$

and

$$\det\big(I+\Delta^+ A(s)\big)\neq 0 \quad \textit{for all } s\in[a,t_0) \tag{7.4.7}$$

(where $(t_0,b]=\emptyset$ if $t_0=b$, and $[a,t_0)=\emptyset$ if $t_0=a$).

P r o o f. a) Assume that $t_0\in[a,b)$, A satisfies (7.4.6) and (7.4.7), and x is a solution of the homogeneous equation

$$x(t) - \int_{t_0}^{t} \mathrm{d}A\, x = 0, \quad t\in[a,b]. \tag{7.4.8}$$

We will show that x vanishes on $[t_0,b]$. Obviously, $x(t_0)=0$, and Theorem 7.4.1 implies

$$\Delta^+ x(t_0) = \Delta^+ A(t_0)\, x(t_0) = 0,$$

i.e., $x(t_0+)=0$.

Let $\alpha(t)=\operatorname{var}_{t_0}^t A$ for $t\in[t_0,b]$. Since α is nondecreasing, it has a finite limit $\alpha(t_0+)$, and there exists a $\delta\in(0,b-t_0)$ such that

$$0\le\alpha(t_0+\delta)-\alpha(t_0+)<1/2.$$

Furthermore, for all $t\in[t_0,t_0+\delta]$ we have

$$\begin{aligned}|x(t)|&\le\int_{t_0}^t|x|\,\mathrm{d}\alpha=\Delta^+\alpha(t_0)\,|x(t_0)|+\lim_{\tau\to t_0+}\int_\tau^t|x|\,\mathrm{d}\alpha\\&=\lim_{\tau\to t_0+}\int_\tau^t|x|\,\mathrm{d}\alpha\le\big(\alpha(t_0+\delta)-\alpha(t_0+)\big)\Big(\sup_{t\in[t_0,t_0+\delta]}|x(t)|\Big)\end{aligned}$$

and therefore

$$\begin{aligned}\sup_{t\in[t_0,t_0+\delta]}|x(t)|&\le\big(\alpha(t_0+\delta)-\alpha(t_0+)\big)\Big(\sup_{t\in[t_0,t_0+\delta]}|x(t)|\Big)\\&<\frac{1}{2}\Big(\sup_{t\in[t_0,t_0+\delta]}|x(t)|\Big)\end{aligned}$$

follows. However, this can be true only if $x=0$ on $[t_0,t_0+\delta]$.

Now, let $t^*=\sup\{\tau\in(t_0,b]:x=0\text{ on }[t_0,\tau]\}$. Obviously, $x=0$ on $[t_0,t^*)$, and Theorem 7.4.1 implies $0=x(t^*-)=\big(I-\Delta^-A(t^*)\big)\,x(t^*)$. However, by the assumption (7.4.6), this is possible only if $x(t^*)=0$.

Finally, assume that $t^*<b$. By the same argument that we used to prove the existence of a $\delta\in(0,b-t_0]$ such that x is zero on $[t_0,t_0+\delta]$, we can now show that there is an $\eta\in(0,b-t^*)$ such that x vanishes on $[t^*,t^*+\eta]$. This being in contradiction with the definition of t^*, we necessarily have $t^*=b$. Thus, we have proved that x vanishes on $[t_0,b]$.

Similarly, using the assumption (7.4.7), we can prove that if $t_0\in(a,b]$, then each solution x of the homogeneous equation (7.4.8) vanishes on $[a,t_0]$.

To summarize, we have proved that if (7.4.6) and (7.4.7) hold, then the homogeneous equation (7.4.8) has only the trivial solution. Consequently, by Theorem 7.4.3, equation (7.4.5) has a unique solution for each $g\in\mathrm{G}([a,b],\mathbb{R}^n)$.

b) Assume that (7.4.6) does not hold. By Lemma 7.3.1, we know that $\det\big(I-\Delta^-A(t)\big)\ne0$ if $|\Delta^-A(t)|\le1/2$. On the other hand, by Corollary 4.1.7, the reverse inequality $|\Delta^-A(t)|>1/2$ can hold only for finitely many points $t\in(t_0,b]$. Thus, the matrix $I-\Delta^-A(t)$ is not invertible for at most finite number of points $t \in (t_0, b]$, and we can choose a $t^* \in (t_0, b]$ such that

$$\det\big(I - \Delta^- A(t)\big) \neq 0 \text{ for } t \in (t_0, t^*), \quad \text{and} \quad \det\big(I - \Delta^- A(t^*)\big) = 0.$$

Now, it is well known from linear algebra that there exists a $d \in \mathbb{R}^n$ such that

$$\big(I - \Delta^- A(t^*)\big)\, c \neq d \quad \text{for every} \quad c \in \mathbb{R}^n. \tag{7.4.9}$$

Define

$$g(t) = \begin{cases} 0 & \text{if } t \neq t^*, \\ d & \text{if } t = t^*. \end{cases}$$

Then $g \in \mathrm{G}([a,b], \mathbb{R}^n)$ and $\Delta^- g(t^*) = d$. Assume that equation (7.4.5) has a solution x on $[a,b]$. Then by the first part of this proof we have $x = 0$ on $[a, t^*)$, and Theorem 7.4.1 implies

$$\big(I - \Delta^- A(t^*)\big) x(t^*) = x(t^*-) + \Delta^- g(t^*) = d.$$

This is however in contradiction with (7.4.9), and hence equation (7.4.5) cannot have a solution.

If (7.4.7) does not hold, we proceed in a similar way, find a point $t^* \in [a, t_0)$ and a function g such that

$$\big(I + \Delta^+ A(t^*)\big)\, c \neq \Delta^+ g(t^*) \quad \text{for every} \quad c \in \mathbb{R}^n,$$

and arrive at a contradiction. □

We now easily obtain the following existence and uniqueness theorem for generalized linear differential equations.

7.4.5 Theorem. *Let $A \in \mathrm{BV}([a,b], \mathscr{L}(\mathbb{R}^n))$ and $t_0 \in [a,b]$. Then the equation*

$$x(t) = \widetilde{x} + \int_{t_0}^{t} \mathrm{d}A\, x + f(t) - f(t_0) \quad \textit{for} \quad t \in [a,b],$$

has a unique solution for every function $f \in \mathrm{G}([a,b], \mathbb{R}^n)$ and every vector $\widetilde{x} \in \mathbb{R}^n$ if and only if the (7.4.6) *and* (7.4.7) *hold.*

P r o o f. The theorem is a corollary of Lemma 7.4.4, where we let

$$g(t) = \widetilde{x} + f(t) - f(t_0) \quad \text{for} \quad t \in [a,b].$$

□

7.5 A PRIORI ESTIMATES OF SOLUTIONS

Let us recall the statement of the Gronwall lemma, which plays an important role in the theory of ordinary differential equations. It is an indispensable tool in the study of uniqueness of solutions to initial-value problems, as well as continuous dependence of solutions on parameters. Its proof can be found in many textbooks dealing with ordinary differential equations, see e.g. Corollary 8.62 in [Kelley, Peterson (2010)] or Auxiliary Theorem 4.3.1 in [Kurzweil (1986)].

7.5.1 Theorem (GRONWALL LEMMA). *Assume that* u, $p\colon[a,b]\to[0,\infty)$ *are continuous,* $K\ge 0$, *and*

$$u(t)\le K+\int_a^t p(s)\,u(s)\,\mathrm{d}s \quad \text{for } t\in[a,b].$$

Then

$$u(t)\le K\,\exp\Big(\int_a^t p(s)\,\mathrm{d}s\Big) \quad \text{for } t\in[a,b].$$

For the study of generalized differential equations, we need a generalization of the Gronwall lemma involving the Kurzweil-Stieltjes integral. Its proof relies on the following auxiliary statement, which generalizes the formula

$$\int_a^b h^k(t)h'(t)\,\mathrm{d}t=\frac{h^{k+1}(b)-h^{k+1}(a)}{k+1}$$

to the case when h need not be differentiable.

7.5.2 Lemma. *Consider a function* $h\colon[a,b]\to\mathbb{R}$.

(i) *If* h *is nonnegative, nondecreasing and left-continuous, then*

$$\int_a^b h^k\,\mathrm{d}h\le\frac{h^{k+1}(b)-h^{k+1}(a)}{k+1} \quad \text{for every } k\in\mathbb{N}\cup\{0\}.$$

(ii) *If* h *is nonnegative, nonincreasing and right-continuous, then*

$$\int_a^b h^k\,\mathrm{d}h\ge\frac{h^{k+1}(b)-h^{k+1}(a)}{k+1} \quad \text{for every } k\in\mathbb{N}\cup\{0\}.$$

(iii) *If* h *is continuous and has bounded variation, then*

$$\int_a^b h^k\,\mathrm{d}h=\frac{h^{k+1}(b)-h^{k+1}(a)}{k+1} \quad \text{for every } k\in\mathbb{N}\cup\{0\}.$$

Proof. In all three cases, the existence of the integral $\int_a^b h^k \, \mathrm{d}h$ follows from Theorem 6.3.11, because both h and h^k have bounded variation.

To prove the first and second statement, we need the following observation: If h is a nonnegative function on $[a,b]$ and $g_1, g_2 : [a,b] \to \mathbb{R}$ satisfy

$$g_1(t) - g_1(s) \le g_2(t) - g_2(s) \quad \text{whenever } [s,t] \subset [a,b],$$

then $\int_a^b h \, \mathrm{d}g_1 \le \int_a^b h \, \mathrm{d}g_2$, provided that both integrals exist. This observation follows immediately from the definition of the integral.

The first statement obviously holds for $k=0$; let us assume that it holds for a certain $k \in \mathbb{N} \cup \{0\}$, and show its validity for $k+1$. Using first Theorem 6.6.1 (substitution theorem), then the induction hypothesis and our observation, and finally Theorem 6.4.2 (integration by parts), we get

$$\begin{aligned}
\int_a^b h^{k+1}(t) \, \mathrm{d}h(t) &= \int_a^b h(t) \cdot h^k(t) \, \mathrm{d}h(t) = \int_a^b h(t) \, \mathrm{d}\Big(\int_a^t h^k \, \mathrm{d}h \Big) \\
&\le \int_a^b h(t) \, \mathrm{d}\Big(\frac{h^{k+1}(t) - h^{k+1}(a)}{k+1} \Big) = \frac{1}{k+1} \int_a^b h(t) \, \mathrm{d}h^{k+1}(t) \\
&= \frac{1}{k+1} \Bigg(h^{k+2}(b) - h^{k+2}(a) - \int_a^b h^{k+1}(t) \, \mathrm{d}h(t) \\
&\qquad + \sum_{a \le t \le b} \Big(\Delta^- h(t) \, \Delta^- h^{k+1}(t) - \Delta^+ h(t) \, \Delta^+ h^{k+1}(t) \Big) \Bigg).
\end{aligned}$$

Note that the sum on the last line is nonpositive, because h and h^{k+1} are left-continuous and nondecreasing. Thus, we get

$$\int_a^b h^{k+1}(t) \, \mathrm{d}h(t) \le \frac{1}{k+1} \Big(h^{k+2}(b) - h^{k+2}(a) - \int_a^b h^{k+1}(t) \, \mathrm{d}h(t) \Big),$$

and consequently

$$\int_a^b h^{k+1} \, \mathrm{d}h \le \frac{h^{k+2}(b) - h^{k+2}(a)}{k+2},$$

which completes the proof of the first statement.

The proofs of the second and third statement are similar; it suffices to reverse the inequalities or replace them by equalities, respectively. □

We now proceed to the generalized version of the Gronwall lemma.

7.5.3 Theorem (GENERALIZED GRONWALL LEMMA). *If* $K \ge 0$, $L > 0$, $h:[a,b] \to [0,\infty)$ *is nondecreasing and left-continuous, and* $u:[a,b] \to [0,\infty)$ *satisfies*

$$u(t) \le K + L\int_a^t u(s)\,\mathrm{d}h(s) \quad \text{for } t \in [a,b], \tag{7.5.1}$$

then

$$u(t) \le K \exp\big(L\,(h(t)-h(a))\big) \quad \text{for } t \in [a,b]. \tag{7.5.2}$$

Proof. It is sufficient to show that for every $\varepsilon > 0$, the inequality

$$u(t) \le (K+\varepsilon) \exp\big(L\,(h(t)-h(a))\big) \tag{7.5.3}$$

holds for all $t \in [a,b]$. For each $\varkappa \ge 0$, denote $w_\varkappa(t) = \varkappa \exp\big(L\,(h(t)-h(a))\big)$ for $t \in [a,b]$. Choose an arbitrary fixed $\varepsilon > 0$; then (7.5.3) holds if and only if $u \le w_\varkappa$, where $\varkappa = K+\varepsilon$. To show that the second statement holds, we let $v_\varepsilon = u - w_\varkappa$ and prove that v_ε is nonpositive. We calculate

$$\begin{aligned}\int_a^t w_\varkappa\,\mathrm{d}h &= \varkappa\int_a^t \exp\big(L\,(h(s)-h(a))\big)\,\mathrm{d}h(s)\\ &= \varkappa\int_a^t \Big(\sum_{k=0}^\infty \frac{L^k}{k!}\,(h(s)-h(a))^k\Big)\,\mathrm{d}h(s) \quad \text{for } t\in[a,b].\end{aligned}$$

The infinite series is uniformly convergent; thus, by Theorem 6.3.8, we can interchange the order of the integration and summation. Using Lemma 7.5.2 (with h replaced by $h - h(a)$), we get

$$\begin{aligned}\int_a^t w_\varkappa\,\mathrm{d}h &= \varkappa\sum_{k=0}^\infty \Big(\frac{L^k}{k!}\int_a^t (h(s)-h(a))^k\Big)\,\mathrm{d}h(s)\\ &\le \varkappa\sum_{k=0}^\infty \frac{L^k\,(h(t)-h(a))^{k+1}}{(k+1)!} = \frac{\varkappa}{L}\Big(\exp(L\,(h(t)-h(a)))-1\Big)\\ &= \frac{w_\varkappa(t)-\varkappa}{L}\end{aligned}$$

for $t \in [a,b]$. This means that the function $w_\varkappa$ satisfies the inequality

$$w_\varkappa(t) \ge \varkappa + L\int_a^t w_\varkappa\,\mathrm{d}h$$

for all $t \in [a,b]$. By combining this inequality with (7.5.1), we find that

$v_\varepsilon = u - w_\varkappa$ satisfies

$$v_\varepsilon(t) \le -\varepsilon + L \int_a^t v_\varepsilon \, dh \quad \text{for } t \in [a,b]. \tag{7.5.4}$$

In particular, $v_\varepsilon(a) \le -\varepsilon < 0$. The function u is bounded, because it is nonnegative and the right-hand side of (7.5.1) is regulated, and therefore bounded from above. Since $w_\varkappa$ is also bounded, we conclude that v_ε is bounded. For $t \in (a,b]$, part (ii) of Corollary 6.5.4 implies

$$\int_a^t v_\varepsilon \, dh = v_\varepsilon(a)\, \Delta^+ h(a) + \lim_{\delta \to 0+} \int_{a+\delta}^t v_\varepsilon \, dh$$
$$\le -\varepsilon\, \Delta^+ h(a) + \|v_\varepsilon\|\,(h(t) - h(a+)) \le \|v_\varepsilon\|\,(h(t) - h(a+)).$$

Returning to (7.5.4), we obtain

$$v_\varepsilon(t) \le -\varepsilon + L \int_a^t v_\varepsilon \, dh \le -\varepsilon + L\,\|v_\varepsilon\|\,(h(t) - h(a+)) \quad \text{for } t \in (a,b].$$

Choose an $\eta > 0$ in such a way that

$$L\,\|v_\varepsilon\|\,(h(t) - h(a+)) < \varepsilon/2 \quad \text{for all } t \in [a, a+\eta].$$

Then $v_\varepsilon < 0$ on $[a, a+\eta]$.

Denote $t^* = \sup\{\tau \in [a,b] : v_\varepsilon < 0 \text{ on } [a,\tau]\}$; we see that $t^* > a$ and $v_\varepsilon < 0$ on $[a, t^*)$. Using (7.5.4), part (i) of Corollary 6.5.4 and the left-continuity of h, we get

$$v_\varepsilon(t^*) \le -\varepsilon + L \int_a^{t^*} v_\varepsilon \, dh$$
$$= -\varepsilon + L \left(v_\varepsilon(t^*)\, \Delta^- h(t^*) + \lim_{\delta \to 0+} \int_a^{t^*-\delta} v_\varepsilon \, dh \right) \le -\varepsilon < 0,$$

which proves that $v_\varepsilon < 0$ on $[a, t^*]$.

If $t^* < b$, we can repeat the previous procedure and show that there exists a $\theta \in (0, b - t^*)$ such that $v_\varepsilon < 0$ on $[a, t^* + \theta]$; this is in contradiction with the definition of t^*. Hence $t^* = b$, $v_\varepsilon < 0$ on the whole interval $[a,b]$, and therefore (7.5.3) holds. Since $\varepsilon > 0$ can be arbitrarily small, the proof is complete. □

7.5.4 Exercise. Prove the following modification of Theorem 7.5.3:

Assume that $K \geq 0$, $L > 0$, $h : [a,b] \to [0,\infty)$ *is nondecreasing and right-continuous, and* $u : [a,b] \to [0,\infty)$ *satisfies*

$$u(t) \leq K + L \int_t^b u \, \mathrm{d}h \quad \textit{for} \quad t \in [a,b].$$

Then

$$u(t) \leq K \exp\big(L\,(h(b) - h(t))\big) \quad \text{for} \quad t \in [a,b].$$

7.5.5 Exercise. Show that the classical Gronwall lemma from Theorem 7.5.1 is a consequence of the generalized Gronwall lemma from Theorem 7.5.3.

Our next goal is to obtain an a priori estimate for solutions of generalized linear differential equations, whose proof relies on the generalized Gronwall lemma and the following result.

7.5.6 Lemma. *Let* $A \in \mathrm{G}([a,b], \mathscr{L}(\mathbb{R}^n))$ *and* $t_0 \in [a,b]$. *The following statements hold:*

(i) *If A satisfies condition* (7.4.6), *then*

$$\sup_{t \in (t_0, b]} \left| (I - \Delta^- A(t))^{-1} \right| < \infty.$$

(ii) *If A satisfies condition* (7.4.7), *then*

$$\sup_{t \in [a, t_0)} \left| (I + \Delta^+ A(t))^{-1} \right| < \infty.$$

Proof. Let $t \in (t_0, b]$ be such that $|\Delta^- A(t)| \leq \frac{1}{2}$. Then, by Lemma 7.3.1, we get

$$\left| (I - \Delta^- A(t))^{-1} \right| \leq \frac{1}{1 - |\Delta^- A(t)|} < 2.$$

Corollary 4.1.7 implies that the set $\{t \in (t_0, b] : |\Delta^- A(t)| > \frac{1}{2}\}$ has at most finitely many elements, and therefore

$$\sup_{t \in (t_0, b]} \left| (I - \Delta^- A(t))^{-1} \right| < \infty.$$

The second statement can be proved similarly. □

7.5.7 Theorem. *Suppose that* $t_0 \in [a,b]$, $A \in \mathrm{BV}([a,b], \mathscr{L}(\mathbb{R}^n))$ *satisfies conditions* (7.4.6) *and* (7.4.7), $f \in \mathrm{G}([a,b], \mathbb{R}^n)$, $\widetilde{x} \in \mathbb{R}^n$, *and x is a solution* *of the equation*

$$x(t) = \widetilde{x} + \int_{t_0}^{t} \mathrm{d}A\,x + f(t) - f(t_0) \quad \textit{for } t \in [a,b]. \tag{7.5.5}$$

Then

$$|x(t)| \le c_{(A,t_0)} \left(|\widetilde{x}| + 2\,\|f\|\right) \exp\left(c_{(A,t_0)} \operatorname{var}_{t_0}^{t} A\right) \textit{ for } t \in [t_0,b],$$
$$|x(t)| \le c_{(A,t_0)} \left(|\widetilde{x}| + 2\,\|f\|\right) \exp\left(c_{(A,t_0)} \operatorname{var}_{t}^{t_0} A\right) \textit{ for } t \in [a,t_0],$$

where

$$c_{(A,t_0)} = \max\left\{1,\ \sup_{t\in(t_0,b]} \left|(I - \Delta^- A(t))^{-1}\right|,\ \sup_{t\in[a,t_0)} \left|(I + \Delta^+ A(t))^{-1}\right|\right\}. \tag{7.5.6}$$

Proof. Let us prove the first inequality. Denote

$$B(t) = \begin{cases} A(t) & \text{if } t \in [a,t_0], \\ A(t-) & \text{if } t \in (t_0,b]. \end{cases}$$

We claim that $\operatorname{var}_{t_0}^{t} B \le \operatorname{var}_{t_0}^{t} A$ for all $t \in [t_0,b]$. To see this, it is enough to take an arbitrary division $\boldsymbol{\alpha}$ of $[t_0,t]$, and observe that

$$\begin{aligned} V(B,\boldsymbol{\alpha}) &= \sum_{i=1}^{\nu(\boldsymbol{\alpha})} |B(\alpha_i) - B(\alpha_{i-1})| \\ &= |A(\alpha_1-) - A(\alpha_0)| + \sum_{i=2}^{\nu(\boldsymbol{\alpha})} |A(\alpha_i-) - A(\alpha_{i-1}-)| \\ &= \lim_{\delta\to 0+} \left(|A(\alpha_1-\delta) - A(\alpha_0)| + \sum_{i=2}^{\nu(\boldsymbol{\alpha})} |A(\alpha_i-\delta) - A(\alpha_{i-1}-\delta)|\right) \\ &\le \lim_{\delta\to 0+} \operatorname{var}_{t_0}^{t-\delta} A \le \operatorname{var}_{t_0}^{t} A. \end{aligned}$$

By Corollary 4.1.9, B is a regulated function, and we have

$$B(t+) = A(t+) \ \text{ for all } t \in [t_0,b), \ \ B(t-) = A(t-) = B(t) \ \text{ for all } t \in (t_0,b].$$

In particular, B is left-continuous on $(t_0,b]$. Note that $A(t) - B(t) = \Delta^- A(t)$ for all $t \in (t_0,b]$. Consequently, $A - B$ is nonzero in at most countably many points. Since it has bounded variation, Corollary 6.3.13 yields

$$\int_{t_0}^{t} \mathrm{d}[A-B]\,x = \Delta^+(A-B)(t_0)\,x(t_0) + \sum_{s\in(t_0,t)} \Delta(A-B)(s)\,x(s)$$
$$+\Delta^-(A-B)(t)\,x(t) = \Delta^- A(t)\,x(t) \quad \text{for} \quad t\in(t_0,b].$$

It follows from (7.5.5) that

$$\begin{aligned} x(t) &= \widetilde{x} + \int_{t_0}^{t} \mathrm{d}[A-B]\,x + \int_{t_0}^{t} \mathrm{d}B\,x + f(t) - f(t_0) \\ &= \widetilde{x} + \Delta^- A(t)\,x(t) + \int_{t_0}^{t} \mathrm{d}B\,x + f(t) - f(t_0) \quad \text{for } t\in[t_0,b], \end{aligned}$$

and therefore

$$x(t) = (I - \Delta^- A(t))^{-1}\Big(\widetilde{x} + \int_{t_0}^{t} \mathrm{d}B\,x + f(t) - f(t_0)\Big) \quad \text{for } t\in(t_0,b].$$

Having in mind that $x(t_0)=\widetilde{x}$ and $c_{(A,t_0)} \ge 1$, we easily derive the inequality

$$|x(t)| \le K + L\int_{t_0}^{t} |x|\,\mathrm{d}h \quad \text{for} \quad t\in[t_0,b],$$

where $K = c_{(A,t_0)}\Big(|\widetilde{x}| + 2\,\|f\|\Big)$, $L = c_{(A,t_0)}$ and $h(t) = \operatorname{var}_{t_0}^t B$. The function h is nondecreasing on $[t_0,b]$ and left-continuous on $(t_0,b]$, because B is left-continuous on $(t_0,b]$ (cf. Lemma 2.3.3). By applying Theorem 7.5.3 (generalized Gronwall lemma), we get

$$|x(t)| \le K\exp\big(L(h(t)-h(t_0))\big) \le c_{(A,t_0)}\,\big(|\widetilde{x}| + 2\,\|f\|\big)\exp\big(c_{(A,t_0)}\operatorname{var}_{t_0}^t A\big).$$

The proof of the second inequality is similar, and relies on the modification of the generalized Gronwall lemma from Exercise 7.5.4. □

7.5.8 Exercise. Provide a detailed proof of the second inequality in Theorem 7.5.7.

7.6 CONTINUOUS DEPENDENCE OF SOLUTIONS ON PARAMETERS

Suppose that $t_0 \in [a,b]$ and $A \in \mathrm{BV}([a,b], \mathscr{L}(\mathbb{R}^n))$ satisfy the conditions

$$\det\big(I - \Delta^- A(t)\big) \neq 0 \quad \text{for all } t \in (t_0, b], \tag{7.6.1}$$

$$\det\big(I + \Delta^+ A(s)\big) \neq 0 \quad \text{for all } s \in [a, t_0), \tag{7.6.2}$$

$\widetilde{x}, \widetilde{y} \in \mathbb{R}^n$ and $f, g \in \mathrm{G}([a,b], \mathbb{R}^n)$. Let $x : [a,b] \to \mathbb{R}^n$ be a solution of the equation

$$x(t) = \widetilde{x} + \int_{t_0}^{t} \mathrm{d}A\, x + f(t) - f(t_0), \quad t \in [a,b], \tag{7.6.3}$$

and let $y : [a,b] \to \mathbb{R}^n$ be a solution of the equation

$$y(t) = \widetilde{y} + \int_{t_0}^{t} \mathrm{d}A\, y + g(t) - g(t_0), \quad t \in [a,b]. \tag{7.6.4}$$

Then

$$x(t) - y(t) = (\widetilde{x} - \widetilde{y}) + \int_{t_0}^{t} \mathrm{d}A\,(x - y) + \big(f(t) - g(t)\big) - \big(f(t_0) - g(t_0)\big)$$

for all $t \in [a,b]$. Thus, by Theorem 7.5.7 we have

$$\|x - y\| \le c_{(A,t_0)} \big(|\widetilde{x} - \widetilde{y}| + 2\, \|f - g\|\big) \exp\big(c_{(A,t_0)} \operatorname{var}_a^b A\big), \tag{7.6.5}$$

where the number $c_{(A,t_0)} > 0$ depends only on A and t_0. We see that the distance between the solutions x, y of the equations (7.6.3) and (7.6.4) is proportional to the distance between the input data, i.e., the initial values $\widetilde{x}$, $\widetilde{y}$, as well as the right-hand sides f, g. In other words, the solution depends continuously on the choice of the initial value and the right-hand side.

Throughout this section, we study the phenomenon of continuous dependence from the following perspective: We consider sequences of equations of the form

$$x_k(t) = \widetilde{x}_k + \int_{t_0}^{t} \mathrm{d}A_k\, x_k + f_k(t) - f_k(t_0),$$

where $\widetilde{x}_k \to \widetilde{x}$, $A_k \rightrightarrows A$, and $f_k \rightrightarrows f$. We will show that under certain assumptions, the solvability of equation (7.6.3) guarantees the solvability of equations (7.6.9) for all sufficiently large $k \in \mathbb{N}$, and $x_k \rightrightarrows x$.

We need the following auxiliary result.

7.6.1 Lemma. *If A, $A_k \in \mathrm{G}([a,b],\mathbb{R}^n)$ for $k \in \mathbb{N}$ are such that $A_k \rightrightarrows A$ on $[a,b]$, then the following statements hold:*

(i) *If $I - \Delta^- A(t)$ is invertible for each $t \in (t_0,b]$, then there exists a $k_0 \in \mathbb{N}$ such that $I - \Delta^- A_k(t)$ is invertible for all $k \geq k_0$, $t \in (t_0,b]$. Moreover,*

$$\sup_{t\in(t_0,b]} \left|(I - \Delta^- A_k(t))^{-1}\right| < 2 \sup_{t\in(t_0,b]} \left|(I - \Delta^- A(t))^{-1}\right| \tag{7.6.6}$$

for each $k \geq k_0$.

(ii) *If $I + \Delta^+ A(s)$ is invertible for each $s \in [a,t_0)$, then there exists a $k_0 \in \mathbb{N}$ such that $I - \Delta^+ A_k(s)$ is invertible for all $k \geq k_0$, $t \in [a,t_0)$. Moreover,*

$$\sup_{t\in[a,t_0)} \left|(I + \Delta^+ A_k(s))^{-1}\right| < 2 \sup_{t\in[a,t_0)} \left|(I + \Delta^+ A(s))^{-1}\right| \tag{7.6.7}$$

for each $k \geq k_0$.

Proof. We prove the first statement; the proof of the second one is similar and is left to the reader.

According to Lemma 7.5.6, the quantity

$$c = \sup_{t\in(t_0,b]} \left|(I - \Delta^- A(t))^{-1}\right|$$

is finite. By Lemma 4.2.3, we have $\Delta^- A_k \rightrightarrows \Delta^- A$ on $[a,b]$. Thus, there exists a $k_0 \in \mathbb{N}$ such that

$$|\Delta^- A_k(t) - \Delta^- A(t)| < \frac{1}{4\,(c+1)} \tag{7.6.8}$$

for all $t \in [a,b]$, $k \geq k_0$. For $t \in (t_0,b]$, we can write

$$\begin{aligned} I - \Delta^- A_k(t) &= (I - \Delta^- A(t)) - (\Delta^- A_k(t) - \Delta^- A(t)) \\ &= \big(I - \Delta^- A(t)\big)\,\big(I - T_k(t)\big), \end{aligned}$$

where

$$T_k(t) = \big(I - \Delta^- A(t)\big)^{-1}\big(\Delta^- A_k(t) - \Delta^- A(t)\big).$$

To prove that $I - \Delta^- A_k(t)$ is invertible, it suffices to show that $I - T_k(t)$ is invertible. By (7.6.8) we have

$$|T_k(t)| \leq \left|\big(I - \Delta^- A(t)\big)^{-1}\right| \left|\Delta^- A_k(t) - \Delta^- A(t)\right| < \frac{1}{4}$$

for all $t \in (t_0, b]$, $k \geq k_0$. Thus, Lemma 7.3.1 guarantees that $I - T_k(t)$ and consequently also $I - \Delta^- A_k(t)$ are invertible; moreover, $\left|\left(I - T_k(t)\right)^{-1}\right| < 2$. Hence, it follows that

$$\left|\left(I - \Delta^- A_k(t)\right)^{-1}\right| \leq \left|\left(I - T_k(t)\right)^{-1}\right| \left|\left(I - \Delta^- A(t)\right)^{-1}\right| \\ < 2\left|(I - \Delta^- A(t))^{-1}\right|,$$

which proves the estimate (7.6.6). □

7.6.2 Exercise. Prove the second part of Lemma 7.6.1.

We are now able to prove the promised continuous dependence theorem for generalized linear differential equations.

7.6.3 Theorem. *Assume that* $t_0 \in [a, b]$ *and* $A \in \mathrm{BV}([a, b], \mathscr{L}(\mathbb{R}^n))$ *satisfy conditions* (7.6.1) *and* (7.6.2). *Furthermore, let* $f \in \mathrm{G}([a, b], \mathbb{R}^n)$, $\widetilde{x} \in \mathbb{R}^n$, *and let the sequences* $\{A_k\} \subset \mathrm{BV}([a, b], \mathscr{L}(\mathbb{R}^n))$, $\{f_k\} \subset \mathrm{G}([a, b], \mathbb{R}^n)$ *and* $\{x_k\} \subset \mathbb{R}^n$ *be such that*

$$A_k \rightrightarrows A, \ f_k \rightrightarrows f \ \text{ and } \ \widetilde{x}_k \to \widetilde{x} \text{ for } k \to \infty,$$
$$\sup_{k \in \mathbb{N}} \operatorname{var}_a^b A_k < \infty.$$

Then there exists a $k_0 \in \mathbb{N}$ *such that for every* $k \geq k_0$ *the equation*

$$x_k(t) = \widetilde{x}_k + \int_{t_0}^t \mathrm{d}A_k\, x_k + f_k(t) - f_k(t_0), \tag{7.6.9}$$

has a unique solution $x_k \colon [a, b] \to \mathbb{R}^n$. *Moreover,* $x_k \rightrightarrows x$, *where* $x : [a, b] \to \mathbb{R}^n$ *is the unique solution of the equation*

$$x(t) = \widetilde{x} + \int_{t_0}^t \mathrm{d}A\, x + f(t) - f(t_0) \quad \text{for } \ t \in [a, b]. \tag{7.6.10}$$

Proof. By Lemma 7.6.1, there is a $k_0 \in \mathbb{N}$ such that if $k \geq k_0$, $I - \Delta^- A_k(t)$ is invertible for all $t \in (t_0, b]$, and $I + \Delta^+ A_k(t)$ is invertible for all $t \in [a, t_0)$. Hence, Theorem 7.4.5 implies that equation (7.6.9) has a unique solution $x_k : [a, b] \to \mathbb{R}^n$ for every $k \geq k_0$, and equation (7.6.10) has a unique solution $x : [a, b] \to \mathbb{R}^n$. Put

$$w_k = (x_k - f_k) - (x - f) \quad \text{for } k \geq k_0.$$

Then $x_k - x = w_k + (f_k - f)$, and the theorem will be proved if we show that $w_k \rightrightarrows 0$. Observe that

$$w_k(t) = \widetilde{w}_k + \int_{t_0}^{t} \mathrm{d}A_k\, w_k + h_k(t) - h_k(t_0) \quad \text{for } k \ge k_0 \text{ and } t \in [a,b],$$

where $\widetilde{w}_k = (\widetilde{x}_k - f_k(t_0)) - (\widetilde{x} - f(t_0))$, and

$$h_k(t) = \int_{t_0}^{t} \mathrm{d}[A_k - A]\,(x - f) + \Big(\int_{t_0}^{t} \mathrm{d}A_k\, f_k - \int_{t_0}^{t} \mathrm{d}A\, f\Big). \tag{7.6.11}$$

If we denote $\alpha^* = \sup_{k\in\mathbb{N}} \operatorname{var}_a^b A_k$, then it follows from Theorem 7.5.7 that

$$|w_k(t)| \le c_k\, \big(|\widetilde{w}_k| + 2\,\|h_k\|\big) \exp\big(c_k \alpha^*\big) \quad \text{for } t \in [a,b] \text{ and } k \ge k_0,$$

where

$$c_k = \max\Big\{1,\ \sup_{t\in(t_0,b]} \Big|(I - \Delta^- A_k(t))^{-1}\Big|,\ \sup_{t\in[a,t_0)} \Big|(I + \Delta^+ A_k(t))^{-1}\Big|\Big\}.$$

The sequence $\{c_k\}$ is bounded, because Lemma 7.6.1 implies

$$c_k \le \max\Big\{1, 2 \sup_{t\in(t_0,b]} \Big|(I - \Delta^- A(t))^{-1}\Big|, 2 \sup_{t\in[a,t_0)} \Big|(I + \Delta^+ A(t))^{-1}\Big|\Big\}.$$

Next, notice that $\widetilde{w}_k = \widetilde{x}_k - \widetilde{x} + f(t_0) - f_k(t_0) \to 0$. Hence, to show that $w_k \rightrightarrows 0$, it is enough to prove that $h_k \rightrightarrows 0$.

By Theorem 6.8.8, we have

$$\lim_{k\to\infty} \sup_{t\in[a,b]} \Big|\int_{t_0}^{t} \mathrm{d}A_k\, f_k - \int_{t_0}^{t} \mathrm{d}A\, f\Big| = 0. \tag{7.6.12}$$

From Theorem 6.3.7, we have the estimate

$$\Big|\int_{t_0}^{t} \mathrm{d}[A_k - A]\,(x - f)\Big| \le 2\,\|A_k - A\|\,\|x - f\|_{\mathrm{BV}} \quad \text{for } t \in [a,b].$$

Since $(x - f) \in \mathrm{BV}([a,b],\mathbb{R}^n)$ by Theorem 7.4.1, it follows that

$$\lim_{k\to\infty} \sup_{t\in[a,b]} \Big|\int_{t_0}^{t} \mathrm{d}[\,A_k - A]\,(x - f)\Big| = 0. \tag{7.6.13}$$

Taking into account (7.6.12)–(7.6.13), we see from (7.6.11) that $h_k \rightrightarrows 0$, and the proof is complete. □

Our next goal is to prove a continuous dependence result also for the case when A_k need not have uniformly bounded variations. To this aim, the following auxiliary lemma is crucial.

7.6.4 Lemma. *Let $A, A_k \in \mathrm{BV}([a,b], \mathscr{L}(\mathbb{R}^n))$ for $k \in \mathbb{N}$, $t_0 \in [a,b]$. Assume that conditions* (7.6.1), (7.6.2) *are satisfied and*

$$\lim_{k\to\infty} \left(1 + \mathrm{var}_a^b A_k\right) \|A_k - A\| = 0. \tag{7.6.14}$$

Then there exist $r^* > 0$ and $k_0 \in \mathbb{N}$ such that

$$\|y\| \le r^* \left(|y(t_0)| + (1 + \mathrm{var}_a^b A_k) \sup_{t\in[a,b]} \left| y(t) - y(t_0) - \int_{t_0}^t \mathrm{d}A_k\, y \right| \right)$$

for all $y \in \mathrm{G}([a,b], \mathbb{R}^n)$ and $k \ge k_0$.

Proof. Assume that the lemma does not hold, i.e., for each $n \in \mathbb{N}$ there are $k_n \in \mathbb{N}$ and $y_n \in \mathrm{G}([a,b], \mathbb{R}^n)$ such that $\lim_{n\to\infty} k_n = \infty$ and

$$\|y_n\| > n\left(|y_n(t_0)| + (1 + \mathrm{var}_a^b A_{k_n}) \sup_{t\in[a,b]} \left| y_n(t) - y_n(t_0) - \int_{t_0}^t \mathrm{d}A_{k_n}\, y_n \right| \right).$$

We will prove that this statement leads to a contradiction. First, we rewrite the inequality as

$$\frac{1}{n} > |u_n(t_0)| + (1 + \mathrm{var}_a^b A_{k_n}) \sup_{t\in[a,b]} \left| u_n(t) - u_n(t_0) - \int_{t_0}^t \mathrm{d}A_{k_n}\, u_n \right|, \tag{7.6.15}$$

where

$$u_n(t) = \frac{y_n(t)}{\|y_n\|} \quad \text{for } t \in [a,b] \text{ and } n \in \mathbb{N}. \tag{7.6.16}$$

It follows that $|u_n(t_0)| < \frac{1}{n}$ for all $n \in \mathbb{N}$. Hence,

$$\lim_{n\to\infty} |u_n(t_0)| = 0.$$

Now, denote

$$v_n(t) = u_n(t) - u_n(t_0) - \int_{t_0}^t \mathrm{d}A_{k_n}\, u_n \quad \text{for } t \in [a,b] \text{ and } n \in \mathbb{N}. \tag{7.6.17}$$

By (7.6.15) we have

$$\left.\begin{aligned} \|v_n\| &= \frac{1}{\|y_n\|}\Big(\sup_{t\in[a,b]}\Big|y_n(t)-y_n(t_0)-\int_{t_0}^t \mathrm{d}A_{k_n}\,y_n\Big|\Big) \\ &< \frac{1}{n\,(1+\operatorname{var}_a^b A_{k_n})} \le \frac{1}{n} \qquad \text{for } n\in\mathbb{N}, \end{aligned}\right\} \tag{7.6.18}$$

and, in particular,

$$\lim_{n\to\infty}\|v_n\|=0.$$

If we denote

$$z_n(t)=u_n(t)-v_n(t) \quad \text{for } t\in[a,b] \text{ and } n\in\mathbb{N},$$

then (7.6.17) yields

$$z_n(t)=u_n(t_0)+\int_{t_0}^t \mathrm{d}A_{k_n}\,u_n \quad \text{for } t\in[a,b] \quad \text{and} \quad n\in\mathbb{N}.$$

Consequently, $z_n\in \mathrm{BV}([a,b],\mathbb{R}^n)$, $z_n(t_0)=u_n(t_0)$ and, since $\|u_n\|\le 1$ for all $n\in\mathbb{N}$, by Corollary 6.5.5 we have

$$\|z_n\|_{\mathrm{BV}}\le 2\,(1+\operatorname{var}_a^b A_{k_n}) \quad \text{for } n\in\mathbb{N}. \tag{7.6.19}$$

Now, let $n\in\mathbb{N}$ be fixed. We have

$$\begin{aligned} z_n(t) &= z_n(t_0)+\int_{t_0}^t \mathrm{d}A_{k_n}\,z_n+\int_{t_0}^t \mathrm{d}A_{k_n}\,v_n \\ &= z_n(t_0)+\int_{t_0}^t \mathrm{d}A\,z_n+\int_{t_0}^t \mathrm{d}[A_{k_n}-A]\,z_n+\int_{t_0}^t \mathrm{d}A_{k_n}\,v_n \quad \text{for } t\in[a,b], \end{aligned}$$

i.e.,

$$z_n(t)=z_n(t_0)+\int_{t_0}^t \mathrm{d}A\,z_n+h_n(t) \quad \text{for } t\in[a,b], \tag{7.6.20}$$

where

$$h_n(t)=\int_{t_0}^t \mathrm{d}[A_{k_n}-A]\,z_n+\int_{t_0}^t \mathrm{d}A_{k_n}\,v_n \quad \text{for } t\in[a,b]. \tag{7.6.21}$$

We claim that $\lim_{n\to\infty}\|h_n\|=0$. Indeed, by Theorem 6.3.7 and (7.6.19), we have

$$\begin{aligned} \sup_{t\in[a,b]}\Big|\int_{t_0}^t \mathrm{d}[A_{k_n}-A]\,z_n\Big| &\le 2\,\|A_{k_n}-A\|\,\|z_n\|_{\mathrm{BV}} \\ &\le 4\,\|A_{k_n}-A\|\,\big(1+\operatorname{var}_a^b A_{k_n}\big), \end{aligned}$$

wherefrom

$$\lim_{n\to\infty} \sup_{t\in[a,b]} \left| \int_{t_0}^{t} \mathrm{d}[A_{k_n} - A]\, z_n \right| = 0 \tag{7.6.22}$$

follows due to (7.6.14). Moreover, using Theorem 6.3.6 and (7.6.18), we get

$$\sup_{t\in[a,b]} \left| \int_{t_0}^{t} \mathrm{d}A_{k_n}\, v_n \right| \le \left(\operatorname{var}_a^b A_{k_n}\right) \|v_n\| \le \frac{1}{n}\, \frac{\operatorname{var}_a^b A_{k_n}}{(1+\operatorname{var}_a^b A_{k_n})} \le \frac{1}{n}$$

and, hence,

$$\lim_{n\to\infty} \sup_{t\in[a,b]} \left| \int_{t_0}^{t} \mathrm{d}A_{k_n}\, v_n \right| = 0. \tag{7.6.23}$$

By combining (7.6.21), (7.6.22) and (7.6.23), we obtain

$$\lim_{n\to\infty} \|h_n\| = 0.$$

Having in mind that z_n satisfies (7.6.20), applying Theorem 7.5.7 and recalling that $\lim_{n\to\infty} |z_n(t_0)| = \lim_{n\to\infty} |u_n(t_0)| = 0$, we conclude that

$$\lim_{n\to\infty} \|z_n\| \le \lim_{n\to\infty} c_{(A,t_0)} \left(|z_n(t_0)| + 2\,\|h_n\|\right) \exp\left(c_{(A,t_0)} \operatorname{var}_a^b A\right) = 0.$$

Consequently,

$$\lim_{n\to\infty} \|u_n\| = \lim_{n\to\infty} \|z_n + v_n\| \le \lim_{n\to\infty} \|z_n\| + \lim_{n\to\infty} \|v_n\| = 0.$$

However, this is impossible as $\|u_n\| = 1$ for all $n \in \mathbb{N}$ due to (7.6.16). The proof of the lemma is complete. □

7.6.5 Theorem. *Assume that* $t_0 \in [a,b]$, $A \in \mathrm{BV}([a,b], \mathscr{L}(\mathbb{R}^n))$ *satisfy conditions* (7.6.1) *and* (7.6.2). *Furthermore, let* $A_k \in \mathrm{BV}([a,b], \mathscr{L}(\mathbb{R}^n))$ *for* $k \in \mathbb{N}$ *be such that*

$$\lim_{k\to\infty} \left(1 + \operatorname{var}_a^b A_k\right) \|A_k - A\| = 0. \tag{7.6.24}$$

Also, let $f \in \mathrm{BV}([a,b], \mathbb{R}^n)$, $\widetilde{x} \in \mathbb{R}^n$, $\{f_k\} \subset \mathrm{G}([a,b], \mathbb{R}^n)$, $\{\widetilde{x}_k\} \subset \mathbb{R}^n$ *satisfy*

$$f_k \rightrightarrows f \quad \text{and} \quad \widetilde{x}_k \to \widetilde{x} \quad \text{for} \quad k \to \infty.$$

Then there exists a $k_0 \in \mathbb{N}$ *such that for every* $k \ge k_0$, *the equation*

$$x_k(t) = \widetilde{x}_k + \int_{t_0}^{t} \mathrm{d}A_k\, x_k + f_k(t) - f_k(t_0) \tag{7.6.25}$$

has a unique solution $x_k:[a,b]\to\mathbb{R}^n$. *Moreover,* $x_k\rightrightarrows x$, *where* $x\colon[a,b]\to\mathbb{R}^n$ *is the unique solution of the equation*

$$x(t)=\widetilde{x}+\int_{t_0}^{t}\mathrm{d}A\,x+f(t)-f(t_0). \tag{7.6.26}$$

Proof. First, notice that (7.6.24) implies that $A_k\rightrightarrows A$ on $[a,b]$. Therefore, by Lemma 7.6.1, there is a $k_1\in\mathbb{N}$ such that if $k\ge k_1$, then $I-\Delta^-A_k(t)$ is invertible for all $t\in(t_0,b]$, and $I+\Delta^+A_k(t)$ is invertible for all $t\in[a,t_0)$. Hence, Theorem 7.4.5 implies that equation (7.6.25) has a unique solution $x_k:[a,b]\to\mathbb{R}^n$ for every $k\ge k_1$, and equation (7.6.26) has a unique solution $x:[a,b]\to\mathbb{R}^n$. By Lemma 7.6.4 we may choose $k_0\ge k_1$ and $r^*\in(0,\infty)$ in such a way that

$$\|y\|\le r^*\left(|y(t_0)|+\left(1+\operatorname{var}_a^b A_k\right)\sup_{t\in[a,b]}\left|y(t)-y(t_0)-\int_{t_0}^{t}\mathrm{d}A_k\,y\right|\right) \tag{7.6.27}$$

holds for all $y\in\mathrm{G}([a,b],\mathbb{R}^n)$ and $k\ge k_0$.

Put $u_k=x_k-x$ for $k\ge k_0$. Then $u_k(t_0)=\widetilde{x}_k-\widetilde{x}$ and

$$\begin{aligned}&\left|u_k(t)-\int_{t_0}^{t}\mathrm{d}A_k\,u_k-(\widetilde{x}_k-\widetilde{x})\right|\\&\quad=\left|\int_{t_0}^{t}\mathrm{d}[A_k-A]\,x+(f_k(t)-f(t))-(f_k(t_0)-f(t_0))\right|\\&\quad\le 2\,\|A_k-A\|\,\|x\|_{\mathrm{BV}}+2\,\|f_k-f\|\end{aligned}$$

for $t\in[a,b]$ and $k\ge k_0$. Using (7.6.27) we deduce that the inequality

$$\|u_k\|\le r^*\left(|\widetilde{x}_k-\widetilde{x}|+\left(1+\operatorname{var}_a^b A_k\right)\left(2\,\|A_k-A\|\,\|x\|_{\mathrm{BV}}+2\,\|f_k-f\|\right)\right)$$

holds for all $k\ge k_0$. Thus, (7.6.24) together with our assumptions $\widetilde{x}_k\to\widetilde{x}$ and $f_k\rightrightarrows f$ ensures that $\lim_{k\to\infty}\|u_k\|=0$, and consequently $x_k\rightrightarrows x$. □

7.6.6 Remark. The proof of Theorem 7.6.5 could be substantially simplified and also extended to the case $f\in\mathrm{G}([a,b],\mathbb{R}^n)$ if the following assertion was true:

If $A,A_k\in\mathrm{BV}([a,b],\mathscr{L}(\mathbb{R}^n))$ *for* $n\in\mathbb{N}$ *satisfy* (7.6.24), *then the relation*

$$\lim_{k\to\infty}\left(\sup_{t\in[a,b]}\left|\int_a^t\mathrm{d}A_k\,f-\int_a^t\mathrm{d}A\,f\right|\right)=0$$

holds for each $f \in G([a,b],\mathbb{R}^n)$.

Unfortunately, this is in general false as shown by the following example.

7.6.7 Example. Let $n=1$ and $[a,b]=[0,1]$. For $k\in\mathbb{N}$, put $n_k=\lfloor k^{3/2}\rfloor+1$ (where $\lfloor x\rfloor$ is the smallest integer that does not exceed x) and $\tau_{m,k}=2^{m-n_k}$ for $m\in\{0,1,\dots,n_k\}$. Further, let

$$a_{0,k}=\frac{2^{n_k}}{k}(-1)^{n_k}, \qquad b_{0,k}=\frac{1}{k}(-1)^{n_k-1},$$

$$a_{m,k}=\frac{2^{n_k-m+1}}{k}(-1)^{n_k-m}, \quad b_{m,k}=\frac{3}{k}(-1)^{n_k-m+1}$$

for $m\in\{1,2,\dots,n_k-1\}$,

$$A_k(t)=\begin{cases}0 & \text{if } t\in[0,\tau_{0,k}],\\ a_{m,k}\,t+b_{m,k} & \text{if } t\in[\tau_{m,k},\tau_{m+1,k}] \text{ and } m\in\{0,1,\dots,n_k-1\},\end{cases}$$

$$A(t)=0 \quad \text{for} \quad t\in[0,1],$$

and

$$f(t)=\begin{cases}0 & \text{if } t=0,\\ (-1)^j\,j^{-1/4} & \text{if } t\in(2^{-j},2^{-(j-1)}] \text{ for some } j\in\mathbb{N}.\end{cases}$$

One can verify that condition (7.6.14) is satisfied, f is regulated on $[0,1]$, $\operatorname{var}_0^1 f=\infty$, and

$$\lim_{k\to\infty}\int_0^1 \mathrm{d}A_k\, f=\infty.$$

At the same time, the functions A_k and f defined above provide us with an argument explaining why Theorem 7.6.5 cannot be formulated more generally for $f\in G([a,b],\mathbb{R}^n)$. Indeed, let $t_0=0$ and $f_k(t)=f(t)$ for $t\in[0,1]$ and $k\in\mathbb{N}$. Then $x=f$ is a solution to (7.6.26) on $[0,1]$ and (7.6.25) has a solution x_k on $[0,1]$ for every $k\in\mathbb{N}$. However, it is possible to show that $\{x_k\}$ does not converge to x. The proof is quite technical and can be found in [Monteiro, Tvrdý (2014)], Example 3.4.

7.6.8 Remark. A typical case where Theorem 7.6.5 might be applicable is when $f_k=g_k+h_k$ for all $k\in\mathbb{N}$, $\{g_k\}$ is a sequence of bounded variation functions with $g_k\rightrightarrows f$, and $\{h_k\}$ is a sequence of regulated functions with $h_k\rightrightarrows 0$. Then $\{f_k\}$ is a sequence of regulated functions, whose limit has bounded variation.

7.7 FUNDAMENTAL MATRICES

As we know, the equation

$$x(t) = x(s) + \int_s^t \mathrm{d}A\,x \quad \text{for} \quad s,t \in [a,b], \tag{7.7.1}$$

represents a generalization of a homogeneous system of linear ordinary differential equations. Assume that $A \in \mathrm{BV}([a,b], \mathscr{L}(\mathbb{R}^n))$, $t_0 \in [a,b]$, and conditions (7.6.1) and (7.6.2) are satisfied. For every $\widetilde{x} \in \mathbb{R}^n$, Theorem 7.4.5 (with $f = 0$) implies that equation (7.7.1) has a unique solution $x : [a,b] \to \mathbb{R}^n$ satisfying $x(t_0) = \widetilde{x}$. According to Theorem 7.4.1, the solution x has bounded variation.

Clearly, there is a one-to-one correspondence between all solutions of (7.7.1) and their values at the point t_0. Moreover, it is easy to verify that if x, y are solutions of (7.7.1) on $[a,b]$ and $c_1, c_2 \in \mathbb{R}$, then $c_1\,x + c_2\,y$ is also a solution of (7.7.1) on $[a,b]$. These observations are summarized in the following statement.

7.7.1 Theorem. *Suppose that $t_0 \in [a,b]$ and $A \in \mathrm{BV}([a,b], \mathscr{L}(\mathbb{R}^n))$ satisfies conditions* (7.6.1) *and* (7.6.2). *Then the set of all solutions of equation* (7.7.1) *on $[a,b]$ is an n-dimensional linear subspace of* $\mathrm{BV}([a,b], \mathbb{R}^n)$.

We now introduce an analogue of the classical notion of a fundamental matrix.

7.7.2 Definition. A matrix function $X : [a,b] \to \mathscr{L}(\mathbb{R}^n)$ is called a fundamental matrix of equation (7.7.1) on the interval $[a,b]$ if

$$X(t) = X(s) + \int_s^t \mathrm{d}A\,X \qquad \text{for all } s,t \in [a,b], \tag{7.7.2}$$

and $\det X(t) \neq 0$ for at least one $t \in [a,b]$.

7.7.3 Remark. If a matrix function X satisfies the relation (7.7.2), then it is easy to verify that for any $c \in \mathbb{R}^n$, the vector function given by $x(t) = X(t)\,c$ is a solution of (7.7.1).

7.7.4 Lemma. *Assume that $t_0 \in [a,b]$ and $A \in \mathrm{BV}([a,b], \mathscr{L}(\mathbb{R}^n))$ satisfy conditions* (7.6.1) *and* (7.6.2). *Then for every matrix $\widetilde{X} \in \mathscr{L}(\mathbb{R}^n)$, there* *exists a unique matrix function* $X_{t_0} : [a,b] \to \mathscr{L}(\mathbb{R}^n)$ *such that*

$$X_{t_0}(t) = \widetilde{X} + \int_{t_0}^{t} \mathrm{d}A\, X_{t_0} \quad \textit{for all } t \in [a,b]. \tag{7.7.3}$$

Moreover, X_{t_0} *has bounded variation.*

Proof. For each $k \in \{1,\dots,n\}$, let $\widetilde{x}_k$ denote the k-th column of the matrix $\widetilde{X}$. Theorem 7.4.5 implies the existence of a unique function $x_k : [a,b] \to \mathbb{R}^n$ satisfying the equation

$$x_k(t) = \widetilde{x}_k + \int_{t_0}^{t} \mathrm{d}A\, x_k \quad \text{for} \quad t \in [a,b].$$

By Theorem 7.4.1, x_k has bounded variation on $[a,b]$. The matrix function $X_{t_0} : [a,b] \to \mathscr{L}(\mathbb{R}^n)$ whose columns are the vector functions $x_1,\dots,x_n$ is therefore a unique solution of (7.7.3) and has bounded variation. □

7.7.5 Remark. If $t_0 \in [a,b]$, $\widetilde{X} \in \mathscr{L}(\mathbb{R}^n)$ and X_{t_0} is the function defined by Lemma 7.7.4, then the function $X = X_{t_0}$ obviously satisfies (7.7.2). Therefore, X_{t_0} is a fundamental matrix of equation (7.7.1) whenever $\det \widetilde{X} \neq 0$.

For simplicity, we will now assume that

$$\left.\begin{aligned} &\det\big(I - \Delta^- A(t)\big) \neq 0 \quad \text{for every} \quad t \in (a,b], \\ &\det\big(I + \Delta^+ A(s)\big) \neq 0 \quad \text{for every} \quad s \in [a,b). \end{aligned}\right\} \tag{7.7.4}$$

7.7.6 Lemma. *If* $A \in \mathrm{BV}([a,b], \mathscr{L}(\mathbb{R}^n))$, *conditions* (7.7.4) *hold, and* X *is a fundamental matrix of* (7.7.1) *on* $[a,b]$, *then*

$$\det X(t) \neq 0 \quad \textit{for each } t \in [a,b]. \tag{7.7.5}$$

Proof. If X is a fundamental matrix of (7.7.1) and (7.7.5) does not hold, then there exist points $\tau_0, \tau_1 \in [a,b]$ such that

$$\det X(\tau_0) \neq 0 \qquad \text{and} \qquad \det X(\tau_1) = 0.$$

The latter equality implies that the columns $x_1(\tau_1), x_2(\tau_1), \dots, x_n(\tau_1)$ of the matrix $X(\tau_1)$ are linearly dependent. Hence, there are coefficients $c_1, c_2, \dots, c_n \in \mathbb{R}$, not all zero, such that

$$\sum_{k=1}^{n} c_k x_k(\tau_1) = 0.$$

Set $x(t)=\sum_{k=1}^{n} c_k\, x_k(t)$ for each $t\in[a,b]$. Then x is a solution to (7.7.1) (see Remark 7.7.3) with $x(\tau_1)=0$, i.e.,

$$x(t)=\int_{\tau_1}^{t} \mathrm{d}A\, x \quad \text{for } \ t\in[a,b].$$

However, the same equation also has the trivial solution. Since Theorem 7.4.5 guarantees uniqueness of solutions (note that (7.7.4) implies that conditions (7.6.1) and (7.6.2) are satisfied for $t_0=\tau_1$), we necessarily have $x=0$ on $[a,b]$. In particular,

$$x(\tau_0)=\sum_{k=1}^{n} c_k\, x_k(\tau_0)=0,$$

which contradicts the assumption that $\det X(\tau_0)\neq 0$. □

For functions of two variables, we use the following notation.

7.7.7 Notation. Consider a function $U:[a,b]\times[a,b]\to\mathscr{L}(\mathbb{R}^n)$. For each fixed $\tau\in[a,b]$, the symbol $U(\tau,\cdot)$ stands for the function $s\mapsto U(\tau,s)$ of a single variable $s\in[a,b]$. Similarly, the symbol $U(\cdot,\tau)$ denotes the function $t\mapsto U(t,\tau)$ of a single variable $t\in[a,b]$. Finally, we let

$$U(\tau,s+)=\lim_{\delta\to 0+} U(\tau,s+\delta), \qquad U(\tau,s-)=\lim_{\delta\to 0+} U(\tau,s-\delta),$$
$$U(t+,\tau)=\lim_{\delta\to 0+} U(t+\delta,\tau), \qquad U(t-,\tau)=\lim_{\delta\to 0+} U(t-\delta,\tau)$$

whenever the limits exist.

The following assertion follows from Lemmas 7.7.4 and 7.7.6.

7.7.8 Theorem. *If $A\in\mathrm{BV}([a,b],\mathscr{L}(\mathbb{R}^n))$ satisfies (7.7.4), then there exists a unique matrix function $U:[a,b]\times[a,b]\to\mathscr{L}(\mathbb{R}^n)$ such that*

$$U(t,s)=I+\int_{s}^{t} \mathrm{d}[A(\tau)]\, U(\tau,s) \quad \textit{for all } t,s\in[a,b]\times[a,b]. \tag{7.7.6}$$

For each $t_0\in[a,b]$, the function $U(\cdot,t_0)$ is a fundamental matrix of (7.7.1) on $[a,b]$. Furthermore, U has the following properties:

(i) $U(\cdot,s)\in\mathrm{BV}([a,b],\mathscr{L}(\mathbb{R}^n))$ *for every* $s\in[a,b]$.

(ii) $U(t,t)=I$ *for every* $t\in[a,b]$.

(iii) $\det U(t,s)\neq 0$ *for all* $t,s\in[a,b]$.

Proof. For a given $s \in [a,b]$, let $X_s \in \mathrm{BV}([a,b],\mathscr{L}(\mathbb{R}^n))$ be the unique matrix function satisfying

$$X_s(t) = I + \int_s^t \mathrm{d}A\, X_s \quad \text{for } t \in [a,b]$$

(see Lemma 7.7.4). For $t, s \in [a,b]$ set $U(t,s) = X_s(t)$, Then U satisfies equation (7.7.6) and $U(t,t) = I$ for $t \in [a,b]$. By Remark 7.7.5, this means that for every $t_0 \in [a,b]$, the function $U(\cdot, t_0)$ is a fundamental matrix of (7.7.1) on $[a,b]$. The fact that $\det U(t,s) \neq 0$ for all $t, s \in [a,b]$ follows from Lemma 7.7.6. □

7.7.9 Theorem. *Assume that* $A \in \mathrm{BV}([a,b],\mathscr{L}(\mathbb{R}^n))$ *satisfies* (7.7.4), $t_0 \in [a,b]$, $\widetilde{x} \in \mathbb{R}^n$, *and* U *is defined by Theorem* 7.7.8. *Then* $x\colon [a,b] \to \mathbb{R}^n$ *is a solution of the initial-value problem*

$$x(t) = \widetilde{x} + \int_{t_0}^t \mathrm{d}A\, x \tag{7.7.7}$$

on $[a,b]$ *if and only if*

$$x(t) = U(t,t_0)\,\widetilde{x} \quad \textit{for } t \in [a,b]. \tag{7.7.8}$$

Proof. The function x given by (7.7.8) is a solution of (7.7.7), because (7.7.6) implies

$$\int_{t_0}^t \mathrm{d}A\, x = \int_{t_0}^t \mathrm{d}[A(\tau)]\, U(\tau,s)\,\widetilde{x} = \big(U(t,t_0) - I\big)\,\widetilde{x} = x(t) - \widetilde{x} \quad \text{for } t \in [a,b].$$

By Theorem 7.4.5, this solution of (7.7.7) is necessarily unique. □

7.7.10 Definition. If $A \in \mathrm{BV}([a,b],\mathscr{L}(\mathbb{R}^n))$ satisfies (7.7.4), then the matrix function U defined by Theorem 7.7.8 is called the *Cauchy matrix* of equation (7.7.1) on $[a,b]$.

7.7.11 Corollary. *Assume that* $A \in \mathrm{BV}([a,b],\mathscr{L}(\mathbb{R}^n))$ *satisfies* (7.7.4), $t_0 \in [a,b]$, $\widetilde{X} \in \mathscr{L}(\mathbb{R}^n)$, *and* U *is the Cauchy matrix of* (7.7.1) *on* $[a,b]$. *Then a matrix function* $X\colon [a,b] \to \mathscr{L}(\mathbb{R}^n)$ *satisfies the equation*

$$X(t) = \widetilde{X} + \int_{t_0}^t \mathrm{d}A\, X \tag{7.7.9}$$

if and only if $X(t) = U(t,t_0)\widetilde{X}$ *for all* $t \in [a,b]$.

Proof. For each $k=1,\ldots,n$, let $\widetilde{x}_k$ be the k-th column of the matrix $\widetilde{X}$. A matrix function $X:[a,b]\to\mathscr{L}(\mathbb{R}^n)$ whose columns are n-vector functions $x_1,\ldots,x_n$ satisfies (7.7.9) if and only if

$$x_k(t)=\widetilde{x}_k+\int_{t_0}^{t}\mathrm{d}A\,x_k\quad\text{for all } t\in[a,b]\ \text{ and }\ k\in\{1,\ldots,n\}.$$

By Theorem 7.7.9, this happens if and only if

$$x_k(t)=U(t,t_0)\widetilde{x}_k\quad\text{for all } t\in[a,b]\ \text{ and }\ k\in\{1,\ldots,n\},$$

i.e., if and only if $X(t)=U(t,t_0)\,\widetilde{X}$ on $[a,b]$. □

7.7.12 Theorem. *If* $A\in\mathrm{BV}([a,b],\mathscr{L}(\mathbb{R}^n))$ *satisfies* (7.7.4) *and* U *is the Cauchy matrix of* (7.7.1) *on* $[a,b]$, *then the relations*

$$U(t,r)\,U(r,s)=U(t,s)\quad\textit{and}\quad U(t,r)^{-1}=U(r,t)$$

hold for all $t,s,r\in[a,b]$.

Proof. Let $r,s\in[a,b]$ be given. Using the definition of U, we have

$$\begin{aligned}U(t,s)&=I+\int_s^t\mathrm{d}[A(\tau)]\,U(\tau,s)\\&=I+\int_s^r\mathrm{d}[A(\tau)]\,U(\tau,s)+\int_r^t\mathrm{d}[A(\tau)]\,U(\tau,s)\\&=U(r,s)+\int_r^t\mathrm{d}[A(\tau)]\,U(\tau,s)\end{aligned}$$

for all $t\in[a,b]$. By Corollary 7.7.11, $U(t,s)=U(t,r)\,U(r,s)$. For $s=t$, we get $I=U(t,t)=U(t,r)\,U(r,t)$, which implies $U(t,r)^{-1}=U(r,t)$. □

7.7.13 Remark. If U is the Cauchy matrix of equation (7.7.1) on $[a,b]$, then, by Theorem 7.7.12, we have

$$U(t,s)=U(t,a)\,U(a,s)=U(t,a)U(s,a)^{-1}\quad\text{for all } t,s\in[a,b].$$

Thus, if we set $X(t)=U(t,a)$ for $t\in[a,b]$, then

$$U(t,s)=X(t)X(s)^{-1}$$

for all $t,s\in[a,b]$. Obviously, X is a fundamental matrix for (7.7.1) on $[a,b]$.

We now describe some additional properties of the Cauchy matrix.

7.7.14 Theorem. *If $A \in \mathrm{BV}([a,b], \mathscr{L}(\mathbb{R}^n))$ satisfies (7.7.4) and U is the Cauchy matrix of (7.7.1) on $[a,b]$, then there exists an $M \in (0,\infty)$ such that*

$$|U(t,s)| + \operatorname{var}_a^b U(\cdot,s) + \operatorname{var}_a^b U(t,\cdot) \le M \quad \textit{for all } t,s \in [a,b]. \qquad (7.7.10)$$

Proof. Let $e_1, \dots, e_n$ be the standard basis of $\mathbb{R}^n$, and note that $|e_k| = 1$ for each k. Denote by u_k the k-th column of the Cauchy matrix U. Then

$$u_k(t,s) = e_k + \int_s^t \mathrm{d}[A(\tau)]\, u_k(\tau,s) \quad \text{for all } t,s \in [a,b] \text{ and } k \in \{1,\dots,n\}.$$

By Theorem 7.5.7, we get

$$|u_k(t,s)| \le M_1 := c_A \exp\big(c_A \operatorname{var}_a^b A\big),$$

where

$$c_A := \max\Big\{1, \sup_{t\in(a,b]} \Big|[I - \Delta^- A(t)]^{-1}\Big|, \sup_{t\in[a,b)} \Big|[I + \Delta^+ A(t)]^{-1}\Big|\Big\}.$$

Thus,

$$|U(t,s)| = \max_{k=1,\dots,n} |u_k(t,s)| \le M_1 \quad \text{for } t,s \in [a,b]. \qquad (7.7.11)$$

Let $t_1, t_2, s \in [a,b]$ and $t_1 \le t_2$. Then

$$|u_k(t_2,s) - u_k(t_1,s)| = \Big|\int_{t_1}^{t_2} \mathrm{d}[A(\tau)]\, u_k(\tau,s)\Big| \le M_1 \operatorname{var}_{t_1}^{t_2} A$$

for each $k \in \{1,\dots,n\}$. Hence, for all $s \in [a,b]$, $k \in \{1,\dots,n\}$ and all divisions $\boldsymbol{\alpha}$ of $[a,b]$, we have

$$V(u_k(\cdot,s), \boldsymbol{\alpha}) \le M_1 \sum_{j=1}^{\nu(\boldsymbol{\alpha})} \operatorname{var}_{\alpha_{j-1}}^{\alpha_j} A = M_1 \operatorname{var}_a^b A.$$

Therefore, by denoting $M_2 = M_1 \operatorname{var}_a^b A$, we get

$$\operatorname{var}_a^b U(\cdot,s) \le \max_{k=1,\dots,n} \operatorname{var}_a^b u_k(\cdot,s) \le M_2 \quad \text{for all } s \in [a,b]. \qquad (7.7.12)$$

Let $s_1, s_2 \in [a,b]$ and $s_1 \le s_2$. Then, for all $t \in [a,b]$ and $k \in \{1,\dots,n\}$, we have

$$
\begin{aligned}
&u_k(t,s_2)-u_k(t,s_1)\\
&=\int_{s_2}^{t} \mathrm{d}[A(\tau)]\,u_k(\tau,s_2)-\int_{s_2}^{t} \mathrm{d}[A(\tau)]\,u_k(\tau,s_1)-\int_{s_1}^{s_2} \mathrm{d}[A(\tau)]\,u_k(\tau,s_1)\\
&=-\int_{s_1}^{s_2} \mathrm{d}[A(\tau)]u_k(\tau,s_1)+\int_{s_2}^{t} \mathrm{d}[A(\tau)]\big(u_k(\tau,s_2)-u_k(\tau,s_1)\big).
\end{aligned}
$$

Hence, for a fixed $k\in\{1,\dots,n\}$, the function $x(t)=u_k(t,s_2)-u_k(t,s_1)$ is a solution of the initial-value problem

$$
x(t)=-\int_{s_1}^{s_2} \mathrm{d}[A(\tau)]u_k(\tau,s_1)+\int_{s_2}^{t} \mathrm{d}A\,x
$$

on $[a,b]$. By Theorem 7.7.9 we have

$$
u_k(t,s_2)-u_k(t,s_1)=-U(t,s_2)\Big(\int_{s_1}^{s_2} \mathrm{d}[A(\tau)]u_k(\tau,s_1)\Big)
$$

for all $t\in[a,b]$, and consequently

$$
|u_k(t,s_2)-u_k(t,s_1)|\le M_1^2\,\operatorname{var}_{s_1}^{s_2} A \quad \text{for } t\in[a,b].
$$

As a result, for all $t\in[a,b]$, $k\in\{1,\dots,n\}$ and all divisions $\boldsymbol{\alpha}$ of $[a,b]$, we get

$$
V(u_k(t,\cdot),\boldsymbol{\alpha})\le M_1^2\sum_{j=1}^{\nu(\boldsymbol{\alpha})}\operatorname{var}_{\alpha_{j-1}}^{\alpha_j} A=M_1^2\,\operatorname{var}_a^b A.
$$

Therefore, by denoting $M_3=M_1^2\,\operatorname{var}_a^b A$, we conclude that

$$
\operatorname{var}_a^b U(t,\cdot)\le \max_{k=1,\dots,n}\,\operatorname{var}_a^b u_k(t,\cdot)\le M_3 \quad \text{for } t\in[a,b]. \tag{7.7.13}
$$

By (7.7.11)–(7.7.13), the theorem holds with $M=M_1+M_2+M_3$. □

7.7.15 Remark. Considering an arbitrary interval $[s_1,s_2]\subset[a,b]$ in place of $[a,b]$, one can see that the estimate

$$
\operatorname{var}_{s_1}^{s_2} U(t,\cdot)\le M_1^2\,\operatorname{var}_{s_1}^{s_2} A
$$

holds for all $t,s_1,s_2\in[a,b]$ such that $s_1\le s_2$.

7.7.16 Theorem. *If* $A\in\mathrm{BV}([a,b],\mathscr{L}(\mathbb{R}^n))$ *satisfies* (7.7.4) *and* U *is the Cauchy matrix of* (7.7.1) *on* $[a,b]$, *then*

$$\left.\begin{aligned} &U(t+,s)=\big(I+\Delta^+A(t)\big)\,U(t,s) && \textit{for } t\in[a,b),\ s\in[a,b],\\ &U(t-,s)=\big(I-\Delta^-A(t)\big)\,U(t,s) && \textit{for } t\in(a,b],\ s\in[a,b],\\ &U(t,s+)=U(t,s)\,\big(I+\Delta^+A(s)\big)^{-1} && \textit{for } t\in[a,b],\ s\in[a,b),\\ &U(t,s-)=U(t,s)\,\big(I-\Delta^-A(s)\big)^{-1} && \textit{for } t\in[a,b],\ s\in(a,b]. \end{aligned}\right\}\tag{7.7.14}$$

Proof. Notice that for all $t,s\in[a,b]$, Theorem 7.7.14 implies that the functions $U(\cdot,s)$ and $U(t,\cdot)$ have bounded variation. Hence, they are regulated and all the one-sided limits in (7.7.14) exist.

The first two relations in (7.7.14) are equivalent to the identities

$$U(t+,s)-U(t,s)=\Delta^+A(t)\,U(t,s),\quad U(t,s)-U(t-,s)=\Delta^-A(t)\,U(t,s),$$

which can be derived by applying Lemma 7.4.1 to the columns of U.

For $t\in[a,b]$ and $s\in[a,b)$, Theorem 7.7.12 yields

$$U(t,s+)\,U(s+,s)=\lim_{\delta\to 0+}U(t,s+\delta)\,U(s+\delta,s)=U(t,s),$$

and using the first statement we get

$$U(t,s+)=U(t,s)\,U(s+,s)^{-1}=U(t,s)\,\big(I+\Delta^+A(s)\big)^{-1}.$$

Similarly, for $t\in[a,b]$ and $s\in(a,b]$, Theorem 7.7.12 yields

$$U(t,s-)\,U(s-,s)=\lim_{\delta\to 0+}U(t,s-\delta)\,U(s-\delta,s)=U(t,s),$$

and using the second statement, we get

$$U(t,s-)=U(t,s)\,U(s-,s)^{-1}=U(t,s)\,\big(I-\Delta^-A(s)\big)^{-1}.$$

This completes the proof. □

7.8 Variation of constants formula

Let us return to the nonhomogeneous equation

$$x(t)=\widetilde{x}+\int_{t_0}^{t}\mathrm{d}A\,x+f(t)-f(t_0).$$

We assume that $\widetilde{x}\in\mathbb{R}^n$, $f\in G([a,b],\mathbb{R}^n)$, and $A\in BV([a,b],\mathscr{L}(\mathbb{R}^n))$ satisfies conditions (7.7.4). By Theorem 7.4.5, this initial-value problem has a unique solution on $[a,b]$, which is regulated. The aim of this section is to show that the solution can be expressed in a form resembling the *variation of constants* formula known from the theory of ordinary differential equations. To establish this result, we need several auxiliary lemmas.

7.8.1 Lemma. *Consider a function $K:[a,b]\times[a,b]\to\mathscr{L}(\mathbb{R}^n)$ such that $K(\cdot,s)$ is regulated for each $s\in[a,b]$. Moreover, suppose there exists a nondecreasing function $h:[a,b]\to\mathbb{R}$ such that*

$$|K(t,s_2)-K(t,s_1)|\le h(s_2)-h(s_1)\quad \textit{for}\quad t\in[a,b],[s_1,s_2]\subset[a,b]. \tag{7.8.1}$$

Then, for each $g\in G([a,b])$ the function $\psi:[a,b]\to\mathbb{R}$ given by

$$\psi(t)=\int_a^b d_s[K(t,s)]\,g(s)\qquad \textit{for } t\in[a,b] \tag{7.8.2}$$

is regulated, and

$$\psi(t-)=\int_a^b d_s[K(t-,s)]\,g(s)\quad \textit{for } t\in(a,b], \tag{7.8.3}$$

$$\psi(t+)=\int_a^b d_s[K(t+,s)]\,g(s)\quad \textit{for } t\in[a,b). \tag{7.8.4}$$

Proof. Let $g\in G([a,b],\mathbb{R}^n)$ and let ψ be given by (7.8.2). Obviously, ψ is then well defined on $[a,b]$, because (7.8.1) implies that $K(t,\cdot)$ has bounded variation for each $t\in[a,b]$.

First, let an arbitrary $\tau\in(a,b]$ be given and let $\{t_k\}$ be an arbitrary sequence of points from $[a,\tau)$ such that $\lim_{k\to\infty} t_k=\tau$. Having in mind (7.8.1), it is easy to verify (cf. Exercise 2.1.12) that

$$\operatorname{var}_a^b K(t_k,\cdot)\le h(b)-h(a)\quad \text{for every}\quad k\in\mathbb{N}, \tag{7.8.5}$$

$$\lim_{k\to\infty} K(t_k,s)=K(\tau-,s)\qquad \text{for every}\quad s\in[a,b] \tag{7.8.6}$$

and, by Theorem 2.7.2, also

$$\operatorname{var}_a^b K(\tau-,\cdot)\le h(b)-h(a),$$

i.e., $K(\tau-,\cdot)\in BV([a,b],\mathscr{L}(\mathbb{R}^n))$. Furthermore, thanks to (7.8.1) the convergence in (7.8.6) is by Corollary 4.3.10 uniform. Therefore,

Theorem 6.8.8, where we put

$$A_k = K(t_k, \cdot), \; A = K(\tau-, \cdot) \quad \text{and} \quad f_k = g \quad \text{for} \quad k \in \mathbb{N},$$

ensures that the relations

$$\lim_{k\to\infty} \psi(t_k) = \lim_{k\to\infty} \int_a^b \mathrm{d}_s[K(t_k, s)]\, g(s) = \int_a^b \mathrm{d}_s[K(\tau-, s)]\, g(s)$$

are true. This proves the relation (7.8.3).

The proof of the relation (7.8.4) is analogous and is left to the reader. Finally, the relations (7.8.3) and (7.8.4) imply that ψ is regulated. □

The next result is a corollary of the previous lemma.

7.8.2 Corollary. *Let $t_0 \in [a, b]$, $A \in \mathrm{BV}([a, b], \mathscr{L}(\mathbb{R}^n))$ satisfy conditions (7.7.4) and let U be the Cauchy matrix of equation (7.7.1) on $[a, b]$. Then, for each $g \in \mathrm{G}([a, b], \mathbb{R}^n)$ the function*

$$\psi(t) := \int_{t_0}^t \mathrm{d}_s[U(t, s)]\, g(s) \quad \textit{for } t \in [a, b]$$

is regulated on $[a, b]$.

Proof. The function ψ is obviously well defined on $[a, b]$. Furthermore,

$$\psi(t) = \begin{cases} \displaystyle\int_a^{t_0} \mathrm{d}_s[K_a(t, s)]\, g(s) & \text{if } t \in [a, t_0], \\[2ex] \displaystyle\int_{t_0}^b \mathrm{d}_s[K_b(t, s)]\, g(s) & \text{if } t \in [t_0, b], \end{cases}$$

where

$$K_a(t, s) = \begin{cases} -U(t, t) & \text{if } a \le s \le t \le t_0, \\ -U(t, s) & \text{if } a \le t \le s \le t_0 \end{cases}$$

and

$$K_b(t, s) = \begin{cases} U(t, s) & \text{if } t_0 \le s < t \le b, \\ U(t, t) & \text{if } t_0 \le t \le s \le b. \end{cases}$$

By Remark 7.7.15, there is a constant $M_1 \in (0, \infty)$ such that for every

$t \in [a, t_0]$ and every subinterval $[s_1, s_2]$ of $[a, t_0]$ we have

$$|K_a(t, s_2) - K_a(t, s_1)| \le \operatorname{var}_{s_1}^{s_2} K_a(t, \cdot) \le \operatorname{var}_{s_1}^{s_2} U(t, \cdot) \le M_1^2 \operatorname{var}_{s_1}^{s_2} A,$$

i.e.,

$$|K_a(t, s_2) - K_a(t, s_1)| \le h_a(s_2) - h_a(s_1) \quad \text{for } t \in [a, t_0],\ [s_1, s_2] \subset [a, t_0],$$

where $h_a(s) = M_1^2 \operatorname{var}_a^s A$ for $s \in [a, t_0]$. Similarly,

$$|K_b(t, s_2) - K_b(t, s_1)| \le h_b(s_2) - h_b(s_1) \quad \text{for } t \in [t_0, b],\ [s_1, s_2] \subset [t_0, b],$$

where $h_b(s) = M_1^2 \operatorname{var}_{t_0}^s A$ for $s \in [t_0, b]$.

Now, taking either $[a, t_0], K_a, h_a$ or $[t_0, b], K_b, h_b$ in place of $[a, b], K, h$, we can apply Lemma 7.8.1 to justify that ψ is regulated on $[a, t_0]$ or $[t_0, b]$, respectively. □

The next auxiliary assertion deals with adjusting an iterated integral, which will be useful for the proof of the main result of this section.

7.8.3 Lemma. *Let $t_0 \in [a, b]$, $A \in \mathrm{BV}([a, b], \mathscr{L}(\mathbb{R}^n))$ satisfy conditions (7.7.4) and let U be the Cauchy matrix of equation (7.7.1) on $[a, b]$. Then, for every $g \in \mathrm{G}([a, b], \mathbb{R}^n)$ and $t \in [a, b]$ we have*

$$\left.\begin{aligned} &\int_{t_0}^t \mathrm{d}[A(r)]\Big(\int_{t_0}^r \mathrm{d}_s\,[U(r, s)]\,g(s)\Big) \\ &\qquad = \int_{t_0}^t \mathrm{d}_s[U(t, s)]\,g(s) + \int_{t_0}^t \mathrm{d}[A(s)]\,g(s). \end{aligned}\right\} \tag{7.8.7}$$

Proof. The fact that U has bounded variation in both variables together with Corollary 7.8.2 imply that all integrals in (7.8.7) have a sense.

a) Given an arbitrary $t > t_0$, we will consider the interval $[t_0, t]$ fixed along the proof (the case $t \le t_0$ can be treated in similar way). Recalling that regulated functions can be uniformly approximated by finite step functions, let us first prove (7.8.7) for functions g of the form

$$g = \chi_{(\tau, t]}\,\xi, \quad g = \chi_{[\tau, t]}\,\xi, \quad g = \chi_{[t]}\,\xi, \tag{7.8.8}$$

where $\tau \in [t_0, t)$ and $\xi \in \mathbb{R}^n$ can be arbitrary. So, let $\tau \in [t_0, t)$ and $\xi \in \mathbb{R}^n$ be given and let $g = \chi_{(\tau, t]}\,\xi$. By the identity (6.3.6) from Lemma 6.3.3, we see that

$$\int_{t_0}^t \mathrm{d}[A(s)]\,g(s) = \int_\tau^t \mathrm{d}[A(s)]\,g(s) = [A(t) - A(\tau+)]\,\xi. \tag{7.8.9}$$

As in Corollary 7.8.2, let

$$\psi(r) := \int_{t_0}^{r} \mathrm{d}_s[U(r,s)]\, g(s) \quad \text{for} \quad r \in [t_0, t].$$

Using Lemma 6.3.2 and having in mind that $U(r,r) = I$ for all r, it is not difficult to verify that

$$\psi(r) = \chi_{(\tau,t]}(r)\, [I - U(r,\tau+)]\, \xi \quad \text{for} \quad r \in [t_0, t].$$

Further, applying Theorem 6.5.3, we get

$$\begin{aligned}\int_{t_0}^{t} \mathrm{d}[A(r)]\, \psi(r) &= \int_{\tau}^{t} \mathrm{d}[A(r)]\, \psi(r) = \lim_{\sigma \to \tau+} \int_{\sigma}^{t} \mathrm{d}[A(r)]\, \psi(r) \\ &= \lim_{\sigma \to \tau+} \int_{\sigma}^{t} \mathrm{d}[A(r)]\, [I - U(r,\tau+)]\, \xi,\end{aligned}$$

i.e.,

$$\int_{t_0}^{t} \mathrm{d}[A(r)] \psi(r) = (A(t) - A(\tau+))\, \xi - \lim_{\sigma \to \tau+} \int_{\sigma}^{t} \mathrm{d}[A(r)]\, U(r,\tau+)\, \xi. \quad (7.8.10)$$

Using subsequently Theorem 6.5.3, Theorem 7.7.16 and the definition of the Cauchy matrix, we deduce

$$\begin{aligned}&\lim_{\sigma \to \tau+} \int_{\sigma}^{t} \mathrm{d}[A(r)]\, U(r,\tau+)\, \xi \\ &= \Big(\int_{\tau}^{t} \mathrm{d}[A(r)]\, U(r,\tau+) - \Delta^+ A(\tau)\, U(\tau,\tau+) \Big)\, \xi \\ &= \Big(\int_{\tau}^{t} \mathrm{d}[A(r)]\, U(r,\tau)[I + \Delta^+ A(\tau)]^{-1} - \Delta^+ A(\tau)\, U(\tau,\tau)[I + \Delta^+ A(\tau)]^{-1} \Big) \xi \\ &= \Big((U(t,\tau) - I)\, [I + \Delta^+ A(\tau)]^{-1} - \Delta^+ A(\tau)\, [I + \Delta^+ A(\tau)]^{-1} \Big)\, \xi \\ &= \Big(U(t,\tau)\, [I + \Delta^+ A(\tau)]^{-1} - [I + \Delta^+ A(\tau)]\, [I + \Delta^+ A(\tau)]^{-1} \Big)\, \xi \\ &= \Big(U(t,\tau+) - I \Big)\, \xi = -\psi(t).\end{aligned}$$

Substituting this result into (7.8.10) and combining with (7.8.9), we get

$$\int_{t_0}^{t} \mathrm{d}[A(r)]\, \psi(r) = (A(t) - A(\tau+))\, \xi + \psi(t) = \psi(t) + \int_{t_0}^{t} \mathrm{d}[A(s)]\, g(s),$$

that is, (7.8.7) holds.

Analogously, we can show that (7.8.7) holds also for $t \le t_0$ and the two remaining functions g from the (7.8.8).

b) Assume that g is an arbitrary regulated function on $[a,b]$ and let $\{g_n\}$ be a sequence of finite step functions converging uniformly to g on $[t_0,t]$. By the first part of the proof we have

$$\int_{t_0}^t \mathrm{d}[A(r)]\Big(\int_{t_0}^r \mathrm{d}_s\,[U(r,s)\,]\,g_n(s)\Big) = \int_{t_0}^t \mathrm{d}_s[U(t,s)]\,g_n(s) + \int_{t_0}^t \mathrm{d}[A(s)]\,g_n(s)$$

for each $n \in \mathbb{N}$. Furthermore, by Theorem 6.3.8,

$$\lim_{n\to\infty} \int_{t_0}^t \mathrm{d}[A(s)]\,g_n(s) = \int_{t_0}^t \mathrm{d}[A(s)]\,g(s)$$

and

$$\lim_{n\to\infty} \int_{t_0}^t \mathrm{d}_s[U(t,s)]\,g_n(s) = \int_{t_0}^t \mathrm{d}_s[U(t,s)]\,g(s). \tag{7.8.11}$$

It remains to show that

$$\lim_{n\to\infty} \int_{t_0}^t \mathrm{d}[A(s)]\,\psi_n(s) = \int_{t_0}^t \mathrm{d}[A(s)]\,\psi(s), \tag{7.8.12}$$

where

$$\psi(r) = \int_{t_0}^r \mathrm{d}_s[U(r,s)]\,g(s) \quad \text{and} \quad \psi_n(r) = \int_{t_0}^r \mathrm{d}_s[U(r,s)]\,g_n(s)$$

for $n \in \mathbb{N}$ and $r \in [t_0,t]$. Note that, by Theorem 6.3.6

$$|\psi_n(r)| \le \operatorname{var}_{t_0}^r U(r,\cdot)\,\|g_n\| \quad \text{for} \quad n\in\mathbb{N} \quad \text{and} \quad r\in[t_0,t].$$

The fact that the sequence $\{g_n\}$ is uniformly bounded together with Theorem 7.7.14 implies that the sequence $\{\psi_n\}$ is also uniformly bounded. Since (7.8.11) guarantees that $\lim_{n\to\infty} \psi_n(r) = \psi(r)$ for each $r \in [t_0,t]$, Theorem 6.8.13 (bounded convergence theorem) implies that (7.8.12) holds, and the proof is complete. □

We can now state and prove the promised analogue of the classical *variation of constants formula* for solutions of nonhomogeneous generalized linear differential equations.

7.8.4 Theorem. *Assume that* $f \in \mathrm{G}([a,b],\mathbb{R}^n)$, $A \in \mathrm{BV}([a,b],\mathscr{L}(\mathbb{R}^n))$ *satisfies conditions* (7.7.4), $t_0 \in [a,b]$, $\widetilde{x} \in \mathbb{R}^n$, *and* U *is the Cauchy matrix* *of equation* (7.7.1) *on* $[a,b]$. *Then the unique solution of the equation*

$$x(t) = \widetilde{x} + \int_{t_0}^{t} \mathrm{d}A\, x + f(t) - f(t_0) \tag{7.8.13}$$

is given by the formula

$$x(t) = U(t,t_0)\,\widetilde{x} + f(t) - f(t_0) - \int_{t_0}^{t} \mathrm{d}_s[U(t,s)]\,(f(s) - f(t_0)) \quad \text{for } t \in [a,b].$$

Proof. Let an arbitrary $t \in [a,b]$ be given. By Corollary 7.8.2, the function

$$x(s) = U(s,t_0)\,\widetilde{x} + f(s) - f(t_0) - \int_{t_0}^{s} \mathrm{d}_r[U(s,r)]\,(f(r) - f(t_0))$$

is regulated on $[a,b]$. Inserting it into the integral $\int_{t_0}^{t} \mathrm{d}[A(s)]\,x(s)$, we get

$$\begin{aligned} \int_{t_0}^{t} \mathrm{d}[A(s)]\,x(s) = \int_{t_0}^{t} \mathrm{d}[A(s)]\,U(s,t_0)\,\widetilde{x} + \int_{t_0}^{t} \mathrm{d}[A(s)]\,(f(s) - f(t_0)) \\ - \int_{t_0}^{t} \mathrm{d}[A(s)] \Big(\int_{t_0}^{s} \mathrm{d}_r\,[U(s,r)]\,(f(r) - f(t_0)) \Big). \end{aligned}$$

Further, by Theorem 7.7.9 we have

$$\int_{t_0}^{t} \mathrm{d}[A(s)]\,U(s,t_0)\,\widetilde{x} = U(t,t_0)\,\widetilde{x} - \widetilde{x},$$

while Lemma 7.8.3 with $g(s) = f(s) - f(t_0)$ yields

$$\begin{aligned} &\int_{t_0}^{t} \mathrm{d}[A(r)] \Big(\int_{t_0}^{r} \mathrm{d}_s[U(r,s)]\,(f(s) - f(t_0)) \Big) \\ &\quad = \int_{t_0}^{t} \mathrm{d}_s[U(t,s)]\,(f(s) - f(t_0)) + \int_{t_0}^{t} \mathrm{d}[A(s)]\,(f(s) - f(t_0)). \end{aligned}$$

Therefore,

$$\begin{aligned} \int_{t_0}^{t} \mathrm{d}[A(s)]\,x(s) &= U(t,t_0)\,\widetilde{x} - \widetilde{x} - \int_{t_0}^{t} \mathrm{d}_s[U(t,s)]\,(f(s) - f(t_0)) \\ &= x(t) - \widetilde{x} - (f(t) - f(t_0)), \end{aligned}$$

which proves the result. □

Integration by parts leads to a different version of the variation of constants formula, which is presented in the next theorem. We make the following agreement: If $c>d$, then a sum of the form $\sum_{s\in[c,d)} h(s)$ should be interpreted as $-\sum_{s\in(d,c]} h(s)$, and the sum $\sum_{s\in(c,d]} h(s)$ should be understood as $-\sum_{s\in[d,c)} h(s)$.

7.8.5 Theorem. *Assume that* $f\in \mathrm{G}([a,b],\mathbb{R}^n)$, $A\in \mathrm{BV}([a,b],\mathscr{L}(\mathbb{R}^n))$ *satisfies conditions* (7.7.4), $t_0\in[a,b]$, $\widetilde{x}\in\mathbb{R}^n$. *Let* U *be the Cauchy matrix for* (7.7.1) *and denote* $X(t)=U(t,t_0)$ *for* $t\in[a,b]$. *Then the unique solution of equation* (7.8.13) *is given by the formula*

$$x(t)=X(t)\Bigg(\widetilde{x}+\int_{t_0}^t X(s)^{-1}\,\mathrm{d}f(s)+\sum_{s\in(t_0,t]} X(s-)^{-1}\Delta^- A(s)\,\Delta^- f(s)$$
$$-\sum_{s\in[t_0,t)} X(s+)^{-1}\Delta^+ A(s)\,\Delta^+ f(s)\Bigg) \quad \text{for } t\in[a,b].$$

Proof. Using Theorems 7.8.4 and 6.4.2 (integration by parts) together with the fact that $U(t,s)=X(t)\,X(s)^{-1}$ for all $t,s\in[a,b]$, we get

$$\begin{aligned}x(t)&=X(t)\,\widetilde{x}+f(t)-f(t_0)-X(t)\int_{t_0}^t \mathrm{d}_s X(s)^{-1}\,(f(s)-f(t_0))\\&=X(t)\,\widetilde{x}+f(t)-f(t_0)-X(t)\,X(t)^{-1}\,(f(t)-f(t_0))\\&\quad+X(t)\int_{t_0}^t X(s)^{-1}\,\mathrm{d}f(s)\\&\quad-X(t)\sum_{s\in(t_0,t]}\Delta^- X(s)^{-1}\,\Delta^- f(s)+X(t)\sum_{s\in[t_0,t)}\Delta^+ X(s)^{-1}\,\Delta^+ f(s).\end{aligned}$$

Thus, to finish the proof, it suffices to show that

$$\Delta^- X(s)^{-1}=-X(s-)^{-1}\Delta^- A(s) \quad\text{and}\quad \Delta^+ X(s)^{-1}=-X(s+)^{-1}\Delta^+ A(s).$$

These identities are simple consequences of Theorem 7.7.16:

$$\begin{aligned}\Delta^- X(s)^{-1}&=X(s)^{-1}-X(s-)^{-1}=U(t_0,s)-U(t_0,s-)\\&=U(t_0,s-)\,(U(s-,s)-I)=U(s-,t_0)^{-1}\,(I-\Delta^- A(s)-I)\\&=-X(s-)^{-1}\,\Delta^- A(s),\end{aligned}$$

$$\begin{aligned}\Delta^+ X(s)^{-1} &= X(s+)^{-1} - X(s)^{-1} = U(t_0, s+) - U(t_0, s)\\ &= U(t_0, s+)\,(I - U(s+, s)) = U(s+, t_0)^{-1}\,(I - (I + \Delta^+ A(s)))\\ &= -X(s+)^{-1}\,\Delta^+ A(s),\end{aligned}$$

and the proof is complete. □

The case when f is left-continuous and $t_0 = a$ leads to further simplification.

7.8.6 Corollary. *Assume that* $\widetilde{x} \in \mathbb{R}^n$, $f \in \mathrm{G}([a,b],\mathbb{R}^n)$ *is left-continuous and* $A \in \mathrm{BV}([a,b],\mathscr{L}(\mathbb{R}^n))$ *satisfies conditions* (7.7.4). *Let* U *be the Cauchy matrix for* (7.7.1), *denote* $X(t) = U(t,a)$ *for* $t \in [a,b]$, *and*

$$Y(s) = \begin{cases} U(a, s+), & \text{if } a \le s < b,\\ U(a,b), & \text{if } s = b.\end{cases} \tag{7.8.14}$$

Then the unique solution of the equation

$$x(t) = \widetilde{x} + \int_a^t \mathrm{d}A\, x + f(t) - f(a) \tag{7.8.15}$$

is given by

$$x(t) = X(t)\left(\widetilde{x} + \int_a^t Y(s)\,\mathrm{d}f(s)\right) \quad \text{for } t \in [a,b].$$

Proof. Using Theorem 7.8.5 and left-continuity of f, we get

$$x(t) = X(t)\Big(\widetilde{x} + \int_a^t X(s)^{-1}\,\mathrm{d}f(s) - \sum_{a \le s < t} X(s+)^{-1}\Delta^+ A(s)\,\Delta^+ f(s)\Big).$$

In the proof of Theorem 7.8.5, we have shown that

$$X(s)^{-1} = X(s+)^{-1} + X(s+)^{-1}\,\Delta^+ A(s) \quad \text{for } s \in [a,b),$$

and therefore

$$X(s)^{-1} = U(a,s+) + U(a,s+)\,\Delta^+ A(s) = Y(s) + Y(s)\,\Delta^+ A(s) \quad \text{for } s \in [a,b).$$

The left and right-hand sides are obviously equal also for $s = b$. Thus, we obtain

$$\int_a^t X(s)^{-1}\,\mathrm{d}f(s) = \int_a^t Y(s)\,\mathrm{d}f(s) + \int_a^t Y(s)\,\Delta^+ A(s)\,\mathrm{d}f(s).$$

To finish the proof, it suffices to show that

$$\int_a^t Y(s)\,\Delta^+ A(s)\,\mathrm{d}f(s) = \sum_{a\le s<t} X(s+)^{-1}\,\Delta^+ A(s)\,\Delta^+ f(s).$$

However, this is an immediate consequence of Lemma 6.3.15, left-continuity of f, and the fact that $X(s+)^{-1} = Y(s)$ for all $s \in [a,b)$. $\square$

Bibliographic remarks

Some basic references on linear as well as nonlinear generalized differential equations are [Groh (1980)], [Hildebrandt (1959)], [Hönig (1975)], [Kurzweil (1957)], [Kurzweil (1958a)], [Kurzweil (2012)], [Schwabik (1992)], [Schwabik, Tvrdý, Vejvoda (1979)], [Tvrdý (2002)].

A modification of the generalized Gronwall lemma which does not require left-continuity of the integrator is available in Chapter 22 of [Kurzweil (2012)]; a nonlinear version of the generalized Gronwall lemma can be found in Theorem 1.40 of [Schwabik (1992)].

Additional continuous dependence results were obtained in [Fraňková (1989)], [Halas, Monteiro, Tvrdý (2011)]. A condition analogous to (7.6.14) appeared in [Opial (1967)], which is concerned with the continuous dependence of solutions to linear ordinary differential equations on a parameter. Lemma 7.6.4 is an analogue of the assertion stated for ordinary differential equations in Lemma 2.5 of [Kiguradze (1988)]. Its variant in the context of functional differential equations appeared in Lemma 3.5 of [Hakl et al. (2004)]. More details concerning Example 7.6.7 may be found in [Monteiro, Tvrdý (2014)]

Differentiability of solutions with respect to initial conditions and parameters was studied in [Slavík (2013)].

Using the variation of constants formula, it is possible to investigate boundary-value problems in which one looks for a solution of a generalized linear differential equation satisfying e.g. the two-point condition

$$M\,x(a) + N\,x(b) = 0$$

where $M, N \in \mathscr{L}(\mathbb{R}^n)$. More information can be found e.g. in [Tvrdý (1991)], [Schwabik, Tvrdý (1979)], [Schwabik, Tvrdý, Vejvoda (1979)], [Schwabik (1992)].

The theory presented in this chapter can be generalized to equations where x and f takes values in a Banach space X, and A is an operator-valued function with values in $\mathscr{L}(X)$; see [Halas, Monteiro, Tvrdý (2011)], [Monteiro, Tvrdý (2013)], [Schwabik (1999b)], [Schwabik (2000)].

Chapter 8

Miscellaneous additional topics

This chapter presents several applications of the Riemann-Stieltjes and Kurzweil-Stieltjes integrals. We begin with topics related to functional analysis, such as the representation of continuous linear functionals, or the theory of distributions. Then we return to generalized linear differential equations and show how they lead to a generalization of the classical elementary functions. The final part of the chapter is devoted to the relation between Kurzweil-Stieltjes integrals, time scale integrals, and dynamic equations on time scales.

8.1 Functionals on the space of continuous functions

One of the most important tasks of functional analysis is to find explicit representations of continuous linear functionals on function spaces.

Recall that *linear functionals* on a space X are linear mappings of X into $\mathbb{R}$. The set of all linear functionals on X is a linear space when equipped with the usual operations of addition and scalar multiplication (defined pointwise). Further, if X is a Banach space equipped with a norm $\|\cdot\|_X$, then it is well known that a linear functional Φ on X is continuous if and only if it is bounded, i.e., if there is a number $K\in[0,\infty)$ such that $|\Phi(x)|\le K\,\|x\|_X$ holds for all $x\in X$. The space of continuous linear functionals on the Banach space X is denoted by X^* and is called the *dual* (or *adjoint*) *space* to X. Furthermore, X^* is a Banach space with respect to the norm given by

$$\|\Phi\|_{X^*}=\sup\{|\Phi(x)|: x\in X, \|x\|_X\le 1\} \quad \text{for} \quad \Phi\in X^*.$$

It is known that continuous linear functionals on the space $\mathrm{C}([a,b])$ are well described by means of the Riesz representation formula involving the classical Riemann-Stieltjes (δ)-integral. To present a proof of this fact we will make use of the following result from functional analysis.

8.1.1 Theorem (HAHN-BANACH). *Let X be a Banach space, and let $Y \subset X$ be its subspace. If $\Phi \in Y^*$ is a continuous linear functional on Y, then there exists a continuous linear functional $\widetilde{\Phi}$ on X such that*

$$\widetilde{\Phi}(y) = \Phi(y) \quad \textit{for} \quad y \in Y \quad \textit{and} \quad \|\widetilde{\Phi}\|_{X^*} = \|\Phi\|_{Y^*}. \tag{8.1.1}$$

Note that our formulation of the Hahn-Banach Theorem is not the most general one (see e.g. [Dunford, Schwartz (1958,1963)]). However, it is quite sufficient for our purposes.

Now we present a general form of continuous linear functionals on the space of continuous functions.

8.1.2 Theorem (RIESZ). *Φ is a continuous linear functional on $\mathrm{C}([a,b])$ if and only if there is a function $p \in \mathrm{BV}([a,b])$ such that $p(a) = 0$ and*

$$\Phi(x) = (\delta)\int_a^b x \,\mathrm{d}p \quad \textit{for any function} \quad x \in \mathrm{C}([a,b]). \tag{8.1.2}$$

In such a case, $\|\Phi\|_{(\mathrm{C}([a,b]))^*} = \operatorname{var}_a^b p$.

Proof. a) Let $x \in \mathrm{C}([a,b])$ and $p \in \mathrm{BV}([a,b])$ be given. Then by Corollary 5.6.2 the integral $(\delta)\int_a^b x \,\mathrm{d}p$ exists and, by Lemma 5.1.12, the inequality

$$\left|(\delta)\int_a^b x \,\mathrm{d}p\right| \le \left(\operatorname{var}_a^b p\right) \|x\|$$

is true. Hence, the mapping $\Phi_p : \mathrm{C}([a,b]) \to \mathbb{R}$ given by

$$\Phi_p(x) = (\delta)\int_a^b x \,\mathrm{d}p$$

is a continuous linear functional on $\mathrm{C}([a,b])$, and

$$\|\Phi_p\|_{(\mathrm{C}([a,b]))^*} \le \operatorname{var}_a^b p. \tag{8.1.3}$$

b) Let an arbitrary $\Phi \in (\mathrm{C}([a,b]))^*$ be given. Denote by X the set of all bounded functions on $[a,b]$. Obviously, X is a Banach space with respect to the supremum norm and $\mathrm{C}([a,b])$ is a closed subspace of X. For the rest of the proof, put $Y = \mathrm{C}([a,b])$.

By Theorem 8.1.1, there is a functional $\widetilde{\Phi} \in X^*$ such that $\|\widetilde{\Phi}\|_{X^*} = \|\Phi\|_{Y^*}$ and $\widetilde{\Phi}(y) = \Phi(y)$ for all $y \in Y$. Put

$$p(a) = 0 \quad \text{and} \quad p(t) = \widetilde{\Phi}(\chi_{[a,t]}) \quad \text{for } t \in (a,b]. \tag{8.1.4}$$

We will prove that $p \in \mathrm{BV}([a,b])$. To this aim, let $\boldsymbol{\alpha}$ be an arbitrary division of $[a,b]$. Denote $m = \nu(\boldsymbol{\alpha})$. Then

$$V(p,\boldsymbol{\alpha}) = \sum_{j=1}^{m} |p(\alpha_j) - p(\alpha_{j-1})| = \sum_{j=1}^{m} c_j\,(p(\alpha_j) - p(\alpha_{j-1})),$$

where $c_j = \operatorname{sgn}(p(\alpha_j) - p(\alpha_{j-1}))$ for $j \in \{1,\dots,m\}$. Hence, with respect to the definition (8.1.4), we obtain

$$\begin{aligned} V(p,\boldsymbol{\alpha}) &= c_1\,\widetilde{\Phi}(\chi_{[a,\alpha_1]}) + \sum_{j=2}^{m} c_j\,\big[\widetilde{\Phi}(\chi_{[a,\alpha_j]}) - \widetilde{\Phi}(\chi_{[a,\alpha_{j-1}]})\big] \\ &= c_1\,\widetilde{\Phi}(\chi_{[a,\alpha_1]}) + \sum_{j=2}^{m} c_j\,\widetilde{\Phi}(\chi_{(\alpha_{j-1},\alpha_j]}) \\ &= \widetilde{\Phi}\Big(c_1\,\chi_{[a,\alpha_1]} + \sum_{j=2}^{m} c_j\,\chi_{(\alpha_{j-1},\alpha_j]}\Big) = \widetilde{\Phi}(h), \end{aligned}$$

where

$$h(t) = c_1\,\chi_{[a,\alpha_1]}(t) + \sum_{j=2}^{m} c_j\,\chi_{(\alpha_{j-1},\alpha_j]}(t) \quad \text{for } t \in [a,b].$$

Obviously $\|h\|_X = \|h\| = 1$ and hence $V(p,\boldsymbol{\alpha}) \le \|\widetilde{\Phi}\|_{X^*} = \|\Phi\|_{Y^*}$ for any division $\boldsymbol{\alpha}$ of $[a,b]$. This means that

$$\operatorname{var}_a^b p \le \|\Phi\|_{Y^*}. \tag{8.1.5}$$

It remains to show that (8.1.2) is true or, in other words, $\Phi = \Phi_p$. Let $x \in Y$ and $\varepsilon > 0$ be given. Since the function x is uniformly continuous on $[a,b]$, there is a $\delta > 0$ such that

$$|x(t) - x(s)| < \varepsilon \quad \text{whenever} \quad t,s \in [a,b] \quad \text{and} \quad |t-s| < \delta.$$

Without loss of generality we can assume that $\delta > 0$ is such that

$$\Big|S(x,\mathrm{d}p,P) - (\delta)\int_a^b x\,\mathrm{d}p\Big| < \varepsilon$$

for all partitions $P=(\boldsymbol{\beta},\boldsymbol{\eta})$ of $[a,b]$ with $|\boldsymbol{\beta}|<\delta$. Let $\boldsymbol{\alpha}$ be an arbitrary division of $[a,b]$ such that $|\boldsymbol{\alpha}|<\delta$, and consider the function

$$x_{\boldsymbol{\alpha}}(t)=\begin{cases} x(\alpha_1) & \text{if } t\in[a,\alpha_1],\\ x(\alpha_j) & \text{if } t\in(\alpha_{j-1},\alpha_j] \text{ and } j\in\{2,3,\dots,\nu(\boldsymbol{\alpha})\}.\end{cases}$$

It is easy to see that $\|x-x_{\boldsymbol{\alpha}}\|_X\le\varepsilon$ and

$$x_{\boldsymbol{\alpha}}(t)=x(\alpha_1)\,\chi_{[a,\alpha_1]}(t)+\sum_{j=2}^{\nu(\boldsymbol{\alpha})}x(\alpha_j)\,\chi_{(\alpha_{j-1},\alpha_j]}(t) \quad \text{for } t\in[a,b].$$

This together with (8.1.4) implies that

$$\begin{aligned}\widetilde{\Phi}(x_{\boldsymbol{\alpha}})&=x(\alpha_1)\,\widetilde{\Phi}(\chi_{[a,\alpha_1]})+\sum_{j=2}^{\nu(\boldsymbol{\alpha})}x(\alpha_j)\,\big(\widetilde{\Phi}(\chi_{[a,\alpha_j]})-\widetilde{\Phi}(\chi_{[a,\alpha_{j-1}]})\big)\\ &=x(\alpha_1)\,\big(p(\alpha_1)-p(a)\big)+\sum_{j=2}^{\nu(\boldsymbol{\alpha})}x(\alpha_j)\,\big(p(\alpha_j)-p(\alpha_{j-1})\big)\\ &=\sum_{j=1}^{\nu(\boldsymbol{\alpha})}x(\alpha_j)\,\big(p(\alpha_j)-p(\alpha_{j-1})\big)=S(x,\mathrm{d}p,(\boldsymbol{\alpha},\boldsymbol{\xi})),\end{aligned}$$

where $\boldsymbol{\xi}=(\alpha_1,\dots,\alpha_{\nu(\boldsymbol{\alpha})})$. Therefore

$$\begin{aligned}\Big|\Phi(x)-(\delta)\int_a^b x\,\mathrm{d}p\Big|&=\Big|\widetilde{\Phi}(x)-(\delta)\int_a^b x\,\mathrm{d}p\Big|\\ &\le\Big|\widetilde{\Phi}(x)-\widetilde{\Phi}(x_{\boldsymbol{\alpha}})\Big|+\Big|\widetilde{\Phi}(x_{\boldsymbol{\alpha}})-(\delta)\int_a^b x\,\mathrm{d}p\Big|\\ &<\|\widetilde{\Phi}\|_{X^*}\,\|x-x_{\boldsymbol{\alpha}}\|_X+\Big|S(x,\mathrm{d}p,(\boldsymbol{\alpha},\boldsymbol{\xi}))-(\delta)\int_a^b x\,\mathrm{d}p\Big|\\ &<\|\widetilde{\Phi}\|_{X^*}\varepsilon+\varepsilon=\big(\|\Phi\|_{Y^*}+1\big)\,\varepsilon.\end{aligned}$$

Since $\varepsilon>0$ can be arbitrarily small, we conclude that

$$\Phi(x)=\Phi_p(x)=(\delta)\int_a^b x\,\mathrm{d}p \quad \text{for } x\in\mathrm{C}([a,b]).$$

Finally, by (8.1.3) and (8.1.5), we have $\|\Phi\|_{(\mathrm{C}([a,b]))^*}=\|\Phi\|_{Y^*}=\operatorname{var}_a^b p$. □

The correspondence

$$\Phi\in(\mathrm{C}([a,b]))^*\mapsto p\in\mathrm{BV}([a,b])$$

is not uniquely determined by the relation (8.1.2). Indeed, if p_1 and p_2 are functions such that $p_1 = p_2$ except for a countable set and (8.1.2) holds for p_1, then the relation is also satisfied for p_2. This is a consequence of the following lemma.

8.1.3 Lemma. *Let* $g \in \mathrm{BV}([a,b])$. *Then*

$$(\delta)\int_a^b f\,\mathrm{d}g = 0 \quad \textit{holds for any function } f \in \mathrm{C}([a,b]) \tag{8.1.6}$$

if and only if there is an at most countable set $D \subset (a,b)$ *such that*

$$g(t) = g(a) \quad \textit{for} \;\; t \in [a,b] \setminus D. \tag{8.1.7}$$

Proof. Without loss of generality we may assume that $g(a) = 0$.

a) Assume (8.1.7) holds. Let f be an arbitrary continuous function on $[a,b]$. We will show that

$$(\delta)\int_a^b f\,\mathrm{d}g = 0. \tag{8.1.8}$$

Clearly, if $D = \emptyset$, the integral is zero. Let $D = \{d\}$, where $d \in (a,b)$, and let $\varepsilon > 0$ be given. Choose $\delta > 0$ in such a way that

$$|f(t) - f(s)| < \varepsilon \;\text{ whenever } t, s \in [a,b] \;\text{ and }\; |t-s| < 2\,\delta.$$

Consider an arbitrary partition $P = (\boldsymbol{\alpha}, \boldsymbol{\xi})$ of $[a,b]$ such that $|\boldsymbol{\alpha}| < \delta$. Clearly, $S(f, \mathrm{d}g, P) = 0$ if $d \notin \boldsymbol{\alpha}$. Otherwise, $d = \alpha_j$ for some $j \in \{1, \dots, \nu(\boldsymbol{\alpha}) - 1\}$ and we have

$$|S(f, \mathrm{d}g, P)| = \big|f(\xi_j) - f(\xi_{j+1})\big|\,|g(d)| < \varepsilon\,\|g\|,$$

which proves that (8.1.8) holds in the case when D is a singleton set. From the linearity of the integral with respect to the integrator, it follows that (8.1.8) holds if the set D is finite.

Now, assume that D is countable, i.e., $D = \{d_k\}$. For each $n \in \mathbb{N}$, put

$$g_n(t) = \begin{cases} g(t) & \text{if } \; t \in \{d_1, d_2, \dots, d_n\}, \\ 0 & \text{if } \; t \in [a,b] \setminus \{d_1, d_2, \dots, d_n\}. \end{cases}$$

Clearly, the sequence $\{g_n\}$ is pointwise convergent to g and $\mathrm{var}_a^b\, g_n \le \mathrm{var}_a^b\, g$ for all $n \in \mathbb{N}$. Furthermore, by the previous part of the proof we have

$(\delta)\int_a^b f\,\mathrm{d}g_n = 0$ for every $n \in \mathbb{N}$. Equality (8.1.8) is then a consequence of Theorem 5.7.3.

b) Let (8.1.6) hold. Choosing $f(t) \equiv 1$, we find that $g(b) = g(a) = 0$. Put

$$f(t) = (\delta)\int_a^t g(s)\,\mathrm{d}s \quad \text{for } t \in [a,b].$$

Then $f \in \mathrm{C}([a,b])$ and, by Theorem 5.5.1 (integration by parts) and Theorem 5.4.3 (substitution theorem), we obtain

$$(\delta)\int_a^b f\,\mathrm{d}g = f(b)\,g(b) - f(a)\,g(a) - (\delta)\int_a^b g\,\mathrm{d}f = -(\delta)\int_a^b g^2(t)\,\mathrm{d}t.$$

Let $t_0 \in (a,b)$ be such that g is continuous at t_0. Then, if it were $g(t_0) \neq 0$, we could find an $\eta > 0$ such that

$$g^2(t) > 0 \quad \text{for } \ t \in (t_0 - \eta, t_0 + \eta).$$

This leads to

$$(\delta)\int_a^b f\,\mathrm{d}g = -(\delta)\int_a^b g^2(t)\,\mathrm{d}t \leq -(\delta)\int_{t_0-\eta}^{t_0-\eta} g^2(t)\,\mathrm{d}t < 0,$$

which contradicts the assumption (8.1.6). Therefore, $g(t)$ can be nonzero only if t is a discontinuity point of g. By Theorem 2.3.2 there are at most countably many such points. This shows that (8.1.7) is true. □

8.1.4 Remark. From Theorem 8.1.2 and Lemma 8.1.3 we deduce that for any continuous linear functional Φ on the space $\mathrm{C}([a,b])$, there exists a unique function $p \in \mathrm{BV}([a,b])$ such that

$$\left.\begin{aligned} &p(a) = 0, \ \ p(t+) = p(t) \quad \text{for } \ t \in (a,b) \\ &\text{and} \\ &\Phi(x) = (\delta)\int_a^b x\,\mathrm{d}p \quad \text{for any } \ x \in \mathrm{C}([a,b]). \end{aligned}\right\} \tag{8.1.9}$$

Functions $p \in \mathrm{BV}([a,b])$ which are right-continuous on (a,b) and satisfy $p(a) = 0$ are called *normalized function of bounded variation.* The set of all such functions will be denoted by the symbol $\mathrm{NBV}([a,b])$. Obviously, $\mathrm{NBV}([a,b])$ is a closed subset of $\mathrm{BV}([a,b])$. Moreover, by Theorem 8.1.2 and Lemma 8.1.3, the spaces $(\mathrm{C}([a,b]))^*$ and $\mathrm{NBV}([a,b])$ are isomorphic, i.e., the mapping

$$\Phi \in (\mathrm{C}([a,b]))^* \mapsto p \in \mathrm{NBV}([a,b]) \tag{8.1.10}$$

is one-to-one. Note that the same statement holds if $\mathrm{NBV}([a,b])$ is replaced by the space of functions of bounded variation on $[a,b]$ that are left-continuous on (a,b) and vanish at some fixed point $c\in[a,b]$.

Let $\Phi\in(\mathrm{C}([a,b]))^*$ be given and let $p\in\mathrm{NBV}([a,b])$ be determined by (8.1.10). The following theorem shows that then the equality $\|\Phi\|_{(\mathrm{C}([a,b]))^*}=\|p\|_{\mathrm{BV}}$ is true, i.e., the spaces $(\mathrm{C}([a,b]))^*$ and $\mathrm{NBV}([a,b])$ are isometrically isomorphic.

8.1.5 Theorem. *If $p\in\mathrm{NBV}([a,b])$ and $\Phi(x)=(\delta)\int_a^b x\,\mathrm{d}p$ for $x\in\mathrm{C}([a,b])$, then*

$$\|\Phi\|_{(\mathrm{C}([a,b]))^*}=\|p\|_{\mathrm{BV}}=\mathrm{var}_a^b\,p.$$

More precisely, the mapping $\Phi\in(\mathrm{C}([a,b]))^*\mapsto p\in\mathrm{NBV}([a,b])$, where p is determined by relation (8.1.9), is an isometry.

Proof. By Lemma 5.1.12 we have $\left|(\delta)\int_a^b x\,\mathrm{d}p\right|\le\left(\mathrm{var}_a^b\,p\right)\|x\|$. This means that

$$\|\Phi\|_{(\mathrm{C}([a,b]))^*}\le\|p\|_{\mathrm{BV}}. \tag{8.1.11}$$

We will prove that for every $\varepsilon>0$ there exists a function $\widetilde{x}\in\mathrm{C}([a,b])$ such that

$$\|\widetilde{x}\|=1 \quad\text{and}\quad |\Phi(\widetilde{x})|>\mathrm{var}_a^b\,p-\varepsilon. \tag{8.1.12}$$

Let an arbitrary $\varepsilon>0$ be given. Choose a division $\boldsymbol{\alpha}$ of $[a,b]$ in such a way that $\nu(\boldsymbol{\alpha})\ge 2$ and

$$V(p,\boldsymbol{\beta})>\mathrm{var}_a^b\,p-\frac{\varepsilon}{3}\quad\text{for any refinement }\ \boldsymbol{\beta}\supset\boldsymbol{\alpha}. \tag{8.1.13}$$

Set $m=\nu(\boldsymbol{\alpha})$. Recall that the right-continuity of p on (a,b) implies that the function $v(t)=\mathrm{var}_a^t\,p$ is right-continuous on (a,b) (Corollary 2.3.4). Therefore, for any $j\in\{1,\dots,m-1\}$ there is a point $t_j\in(\alpha_j,\alpha_{j+1})$ such that

$$\mathrm{var}_{\alpha_j}^{t_j}\,p=v\,(t_j)-v\,(\alpha_j)<\frac{\varepsilon}{3(m-1)}. \tag{8.1.14}$$

Put

$$\widetilde{x}(t)=\begin{cases}\mathrm{sgn}\,\big(p(\alpha_1)-p(a)\big) & \text{if }\ t\in[a,\alpha_1],\\ \mathrm{sgn}\,\big(p(\alpha_{j+1})-p(t_j)\big) & \text{if }\ t\in[t_j,\alpha_{j+1}],\ j\in\{1,\dots,m-1\},\end{cases}$$

and extend the function $\widetilde{x}$ to a continuous function on $[a,b]$ in such a way that it will be linear on the intervals $[\alpha_j,t_j]$, $j=1,\dots,m-1$. Obviously,

$\|\widetilde{x}\| = 1$. Moreover,

$$(\delta)\int_a^b \widetilde{x}\,\mathrm{d}p = |p(\alpha_1) - p(a)| + \sum_{j=1}^{m-1} |p(\alpha_{j+1}) - p(t_j)| + \sum_{j=1}^{m-1} (\delta)\int_{\alpha_j}^{t_j} \widetilde{x}\,\mathrm{d}p.$$

Since $|\widetilde{x}(t)| \le 1$ for $t \in [a,b]$, it follows by (8.1.14) that

$$\begin{aligned}\Big|(\delta)\int_a^b \widetilde{x}\,\mathrm{d}p\Big| &\ge |p(\alpha_1) - p(a)| + \sum_{j=1}^{m-1} |p(\alpha_{j+1}) - p(t_j)| - \sum_{j=1}^{m-1} \operatorname{var}_{\alpha_j}^{t_j} p \\ &\ge V(p, \widetilde{\boldsymbol{\alpha}}) - 2\sum_{j=1}^{m-1} \operatorname{var}_{\alpha_j}^{t_j} p > V(p, \widetilde{\boldsymbol{\alpha}}) - \frac{2\,\varepsilon}{3},\end{aligned}$$

where $\widetilde{\boldsymbol{\alpha}} = \{a, \alpha_1, t_1, \alpha_2 \ldots, \alpha_{m-1}, t_{m-1}, b\}$. By (8.1.13) we get

$$\Big|(\delta)\int_a^b \widetilde{x}\,\mathrm{d}p\Big| > V(p, \widetilde{\boldsymbol{\alpha}}) - \frac{2\,\varepsilon}{3} > \operatorname{var}_a^b p - \varepsilon,$$

that is, (8.1.12) holds. Hence $\sup_{\|x\|\le 1} |\Phi(x)| \ge \operatorname{var}_a^b p$, wherefrom by (8.1.11) the assertion of the theorem follows. □

In view of Theorem 8.1.5, the space $\mathrm{NBV}([a,b])$ can be identified with the space $(\mathrm{C}([a,b]))^*$.

8.1.6 Exercise. Prove the following assertion:

For a given continuous linear functional Φ on $\mathrm{C}([a,b])$, there exists a unique function $p \in \mathrm{BV}([a,b])$ such that

$$p(b) = 0,\ \ p(t-) = p(t) \quad \textit{for} \quad t \in (a,b)$$

and

$$\Phi(x) = (\delta)\int_a^b x\,\mathrm{d}p \quad \textit{for any} \quad x \in \mathrm{C}([a,b]).$$

Furthermore, in Theorem 8.1.5, the space $\mathrm{NBV}([a,b])$ can be replaced by the space of functions left-continuous on (a,b) and such that $p(b) = 0$.

Bibliographic remarks

The Riesz representation theorem, named after F. Riesz, was published in 1909 [Riesz (1909)], but more details on its proof have been given only two years later in [Riesz (1911)]. This classical result often appears in functional analysis textbooks as an illustration of the applicability of the Hahn-Banach theorem, see e.g. [Dunford, Schwartz (1958,1963)], [Schechter M.

(2001)], [Taylor, Lay (1980)]. It is also worth mentioning that an alternative proof of Riesz's theorem, based on the approximation of continuous functions by the Bernstein polynomials, can be found in [Hildebrandt (1963)].

8.2 FUNCTIONALS ON SPACES OF REGULATED FUNCTIONS

For continuous linear functionals on the space of regulated functions, an analogue of the Riesz representation (Theorem 8.1.2) requires a more general notion than the Riemann-Stieltjes integral. In this section we show that the Kurzweil-Stieltjes integral is a suitable choice for such a representation formula. To this end, we introduce the following notation:

8.2.1 Definition. We will say that $f:[a,b]\to\mathbb{R}$ is a summable function if f vanishes except for a countable set and $\sum_{a\le t\le b}|f(t)|<\infty$. For simplicity, we denote

$$s[f]=\sum_{a\le t\le b}|f(t)|.$$

It is not difficult to see that a summable function $f:[a,b]\to\mathbb{R}$ has bounded variation. The following lemmas concerning summable functions will be useful later.

8.2.2 Lemma. *If $r:[a,b]\to\mathbb{R}$ is a summable function and $x:[a,b]\to\mathbb{R}$ is bounded, then the sum*

$$\Psi_r(x)=\sum_{a\le t\le b} r(t)\,x(t) \tag{8.2.1}$$

converges and $|\Psi_r(x)|\le s[r]\,\|x\|$.

Proof. By the definition of a summable function, the sum $\sum_{a\le t\le b}|r(t)|$ is finite. Therefore

$$|\Psi_r(x)|\le\sum_{a\le t\le b}|r(t)\,x(t)|\le\|x\|\sum_{a\le t\le b}|r(t)|$$

is also finite and the result follows. □

8.2.3 Lemma. *If Φ is a continuous linear functional on $\mathrm{G}([a,b])$, then the function $r:[a,b]\to\mathbb{R}$, given by $r(t)=\Phi(\chi_{[t]})$ for $t\in[a,b]$, is summable.*

Proof. For each $n \in \mathbb{N}$, let $M_n = \{t \in [a,b] : r(t) \geq 1/n\}$. Assume that there exists $N \in \mathbb{N}$ such that M_N is infinite. Let $T \subset M_N$ be a finite set with m elements. Then $\|\chi_T\| = 1$ and

$$\Phi(\chi_T) = \sum_{t \in T} \Phi(\chi_{[t]}) = \sum_{t \in T} r(t) \geq \frac{m}{N}.$$

Note that $m \in \mathbb{N}$ can be arbitrarily large (by taking a sufficiently large set T), which contradicts the fact that Φ is bounded. Hence, M_n is finite for every $n \in \mathbb{N}$ and, consequently, the set $\{t \in [a,b] : r(t) > 0\} = \bigcup_{n=1}^{\infty} M_n$ is countable. In a similar way, we can show that $\{t \in [a,b] : r(t) < 0\}$ is also countable.

Assume $\{t_k\}$ is a non-repeating sequence of points in $[a,b]$ such that $r(t) \neq 0$ if and only if $t = t_k$ for some $k \in \mathbb{N}$. If $\{t_k\}$ is finite, then r is clearly summable. Otherwise, for each $n \in \mathbb{N}$ we get

$$\sum_{k=1}^{n} |r(t_k)| = \Big|\sum_{k=1}^{n} r(t_k)\,\lambda_k\Big| = \Big|\Phi\Big(\sum_{k=1}^{n} \lambda_k\,\chi_{[t_k]}\Big)\Big| \leq \|\Phi\|_{(\mathrm{G}([a,b]))^*},$$

where $\lambda_k = \operatorname{sgn}(r(t_k))$ for $k \in \{1, \dots, n\}$. Thus the series $\sum_{k=1}^{\infty} |r(t_k)|$ converges, and we conclude that r is summable. □

Let $p \in \mathrm{BV}([a,b])$, $q \in \mathbb{R}$, and let $r : [a,b] \to \mathbb{R}$ be a summable function. For each $x \in \mathrm{G}([a,b])$, define

$$\left.\begin{aligned} \Phi^+_{p,q,r}(x) &= q\,x(a) + \int_a^b p\,\mathrm{d}x + \sum_{a \leq t < b} r(t)\,\Delta^+ x(t), \\ \Phi^-_{p,q,r}(x) &= q\,x(a) + \int_a^b p\,\mathrm{d}x + \sum_{a < t \leq b} r(t)\,\Delta^- x(t). \end{aligned}\right\} \tag{8.2.2}$$

Lemma 8.2.2 together with the results of Chapter 6 ensures that both $\Phi^+_{p,q,r}$ and $\Phi^-_{p,q,r}$ are well-defined and linear. Moreover,

$$|\Phi^+_{p,q,r}(x)| \leq \big(|q| + |p(a)| + |p(b)| + \operatorname{var}_a^b p + 2\,s[r]\big)\,\|x\| \quad \text{for } x \in \mathrm{G}([a,b]),$$

and similarly for $\Phi^-_{p,q,r}$. In summary, for each triple (p,q,r), the identities (8.2.2) define continuous linear functionals on $\mathrm{G}([a,b])$.

The following theorem shows that all continuous linear functionals on the space of regulated functions can be described by an identity of the form (8.2.2).

8.2.4 Theorem. *If Φ is a continuous linear functional on* $\mathrm{G}([a,b])$, *then there exist* $p, \widetilde{p} \in \mathrm{BV}([a,b])$, $q \in \mathbb{R}$ *and a summable function* $r : [a,b] \to \mathbb{R}$ *such that*

$$\Phi(x) = q\,x(a) + \int_a^b p\,\mathrm{d}x - \sum_{a \le t < b} r(t)\,\Delta^+ x(t) \quad \textit{for} \quad x \in \mathrm{G}([a,b]) \tag{8.2.3}$$

and

$$\Phi(x) = q\,x(a) + \int_a^b \widetilde{p}\,\mathrm{d}x + \sum_{a < t \le b} r(t)\,\Delta^- x(t) \quad \textit{for} \quad x \in \mathrm{G}([a,b]). \tag{8.2.4}$$

P r o o f. Let $p, r : [a,b] \to \mathbb{R}$ be given by

$$p(t) = \Phi(\chi_{[t,b]}) \quad \text{and} \quad r(t) = \Phi(\chi_{[t]}) \quad \text{for} \quad t \in [a,b].$$

By Lemma 8.2.3, the function $r : [a,b] \to \mathbb{R}$ is summable. We will prove that $p \in \mathrm{BV}([a,b])$. To this end, consider an arbitrary division $\boldsymbol{\alpha}$ of $[a,b]$. Taking

$$c_j = \operatorname{sgn}(p(\alpha_{j-1}) - p(\alpha_j)) \quad \text{for} \quad j \in \{1, \dots, \nu(\boldsymbol{\alpha})\},$$

we get

$$V(p, \boldsymbol{\alpha}) = \Big| \sum_{j=1}^{\nu(\boldsymbol{\alpha})} c_j\,(p(\alpha_{j-1}) - p(\alpha_j)) \Big| = \Big| \sum_{j=1}^{\nu(\boldsymbol{\alpha})} c_j\,\Phi(\chi_{[\alpha_{j-1},\alpha_j)}) \Big| = |\Phi(h_{\boldsymbol{\alpha}})|,$$

where $h_{\boldsymbol{\alpha}} = \sum_{j=1}^{\nu(\boldsymbol{\alpha})} c_j\,\chi_{[\alpha_{j-1},\alpha_j)}$. Since $\|h_{\boldsymbol{\alpha}}\| \le 1$, we obtain the estimate $\operatorname{var}_a^b p \le \|\Phi\|_{(\mathrm{G}([a,b]))^*}$.

Further, put $q = \Phi(\chi_{[a,b]})$ and

$$\Phi_{p,q,r}(x) = q\,x(a) + \int_a^b p\,\mathrm{d}x - \sum_{a \le t < b} r(t)\,\Delta^+ x(t) \quad \text{for} \quad x \in \mathrm{G}([a,b]). \tag{8.2.5}$$

To show that $\Phi = \Phi_{p,q,r}$, it is enough to verify that

$$\Phi(x) = \Phi_{p,q,r}(x) \quad \text{for every} \quad x \in \mathrm{S}([a,b]) \tag{8.2.6}$$

(the result then follows from the continuity of both functionals Φ and $\Phi_{p,q,r}$, together with the fact that $\mathrm{S}([a,b])$ is dense in $\mathrm{G}([a,b])$). Let $x \in \mathrm{S}([a,b])$ be given by

$$x = \sum_{j=0}^{m} c_j\,\chi_{[t_j]} + \sum_{j=1}^{m} d_j\,\chi_{(t_{j-1},t_j)},$$

or equivalently,

$$x = \sum_{j=0}^{m-1} (c_j - d_{j+1})\,\chi_{[t_j]} + c_m\,\chi_{[b]} + \sum_{j=1}^{m} d_j\,\chi_{[t_{j-1},t_j)},$$

where $\{t_0, t_1, \dots, t_m\}$ is a division of $[a, b]$ and c_j, $d_j \in \mathbb{R}$ for all j. Then

$$\Phi(x) = \sum_{j=0}^{m-1} (c_j - d_{j+1})\,r(t_j) + c_m\,r(b) + \sum_{j=1}^{m} d_j\,(p(t_{j-1}) - p(t_j)).$$

On the other hand, by Lemma 6.3.2 we have

$$\int_a^b p\,\mathrm{d}x = c_m\,p(b) - c_0\,p(a) + \sum_{j=1}^{m} d_j\,(p(t_{j-1}) - p(t_j)).$$

Noting that $p(b) = r(b)$ and $\Delta^+ x(t_j) = d_{j+1} - c_j$ for $j \in \{0, \dots, m-1\}$, from the expressions above we obtain

$$\Phi(x) = c_0\,q + \int_a^b p\,\mathrm{d}x - \sum_{j=0}^{m-1} \Delta^+ x(t_j)\,r(t_j),$$

therefore (8.2.6) holds.

In order to prove (8.2.4), put $\widetilde{p}(t) = p(t) - r(t)$ for $t \in [a, b]$. Noticing that $\widetilde{p} \in \mathrm{BV}([a, b])$ and applying Lemma 6.3.15 we get

$$\int_a^b r\,\mathrm{d}x = r(a)\,\Delta^+ x(a) + \sum_{a<t<b} r(t)\,\Delta\,x(t) + r(b)\,\Delta^- x(b) \quad \text{for } x \in \mathrm{G}([a, b]).$$

This combined with (8.2.3) yields

$$\begin{aligned}\Phi(x) &= q\,x(a) + \int_a^b \widetilde{p}\,\mathrm{d}x + \int_a^b r\,\mathrm{d}x - \sum_{a \le t<b} r(t)\,\Delta^+ x(t)\\ &= q\,x(a) + \int_a^b \widetilde{p}\,\mathrm{d}x + \sum_{a<t\le b} r(t)\,\Delta^- x(t),\end{aligned}$$

that is, (8.2.4) holds. □

The identities in Theorem 8.2.4 depend on the value of the regulated function at the initial point a. With a dependence on the end point b, two other representation formulas for continuous linear functionals on $\mathrm{G}([a, b])$ can be derived.

8.2.5 Theorem. *If* Φ *is a continuous linear functional on* $\mathrm{G}([a,b])$, *then there exist* $p, \widetilde{p} \in \mathrm{BV}([a,b])$, $q \in \mathbb{R}$ *and a summable function* $r:[a,b]\to\mathbb{R}$ *such that*

$$\Phi(x) = q\,x(b) - \int_a^b p\,\mathrm{d}x - \sum_{a\le t<b} r(t)\,\Delta^+ x(t) \quad \text{for } x\in \mathrm{G}([a,b]) \tag{8.2.7}$$

and

$$\Phi(x) = q\,x(b) - \int_a^b \widetilde{p}\,\mathrm{d}x + \sum_{a<t\le b} r(t)\,\Delta^- x(t) \quad \text{for } x\in \mathrm{G}([a,b]). \tag{8.2.8}$$

8.2.6 Exercise. Prove Theorem 8.2.5.
Hint: Consider $p, \widetilde{p}:[a,b]\to\mathbb{R}$ given respectively by

$$p(t) = \begin{cases} \Phi(\chi_{[a,t)}) & \text{if } t\in(a,b],\\ 0 & \text{if } t=a, \end{cases}$$

and $\widetilde{p}(t) = p(t) + r(t)$ for $t\in[a,b]$.

In the case when the function $x\in\mathrm{G}([a,b])$ is left-continuous on $(a,b]$, the identity (8.2.4) reduces to

$$\Phi(x) = q\,x(a) + \int_a^b p\,\mathrm{d}x,$$

while (8.2.8) yields

$$\Phi(x) = q\,x(b) - \int_a^b p\,\mathrm{d}x.$$

Analogous expressions hold if we consider functions $x\in\mathrm{G}([a,b])$ right-continuous on $[a,b)$, and use identities (8.2.3) and (8.2.7). Therefore, from Theorems 8.2.4 and 8.2.5 we obtain the following representation formulas for continuous linear functionals defined on certain subspaces of $\mathrm{G}([a,b])$.

8.2.7 Theorem. (i) Φ *is a continuous linear functional on* $\mathrm{G_R}([a,b])$ *if and only if there exist* $p, \tilde{p}\in\mathrm{BV}([a,b])$ *and* $q\in\mathbb{R}$ *such that*

$$\left.\begin{aligned} \Phi(x) &= q\,x(a) + \int_a^b p\ \mathrm{d}x,\\ \Phi(x) &= q\,x(b) - \int_a^b \widetilde{p}\ \mathrm{d}x, \end{aligned}\right\} \tag{8.2.9}$$

for every $x\in\mathrm{G_R}([a,b])$.

(ii) Φ *is a continuous linear functional on* $\mathrm{G_L}([a,b])$ *if and only if there exist* $p, \widetilde{p} \in \mathrm{BV}([a,b])$ *and* $q \in \mathbb{R}$ *such that*

$$\left.\begin{aligned} \Phi(x) &= q\,x(a) + \int_a^b p\,\mathrm{d}x, \\ \Phi(x) &= q\,x(b) - \int_a^b \widetilde{p}\,\mathrm{d}x, \end{aligned}\right\} \tag{8.2.10}$$

for every $x \in \mathrm{G_L}([a,b])$.

Proof. (i) Given $p, \widetilde{p} \in \mathrm{BV}([a,b])$ and $q \in \mathbb{R}$, it is not difficult to see that each identity in (8.2.9) defines a continuous linear functional Φ on $\mathrm{G_R}([a,b])$.

Now, consider an arbitrary linear bounded functional Φ on $\mathrm{G_R}([a,b])$ and, for $x \in \mathrm{G}([a,b])$, define

$$\widetilde{x}(t) = \begin{cases} x(t+) & \text{if } t \in [a,b), \\ x(b) & \text{if } t = b, \end{cases}$$

and $\widetilde{\Phi}(x) = \Phi(\widetilde{x})$. By Corollary 4.1.9, $\widetilde{x} \in \mathrm{G_R}([a,b])$, and hence the mapping $\widetilde{\Phi} : \mathrm{G}([a,b]) \to \mathbb{R}$ is well defined and linear. Furthermore,

$$|\widetilde{\Phi}(x)| \le \|\Phi\|_{(\mathrm{G_R}([a,b]))^*} \|\widetilde{x}\| \le \|\Phi\|_{(\mathrm{G_R}([a,b]))^*} \|x\| \quad \text{for } x \in \mathrm{G}([a,b]),$$

that is, $\widetilde{\Phi}$ is a continuous linear functional on $\mathrm{G}([a,b])$. Applying Theorems 8.2.4 and 8.2.5, the equalities in (8.2.9) follow from the fact that $\widetilde{\Phi} = \Phi$ on $\mathrm{G_R}([a,b])$.

The proof of (ii) is analogous. □

Next we show that Theorem 8.2.7 (i) gives rise to an isomorphism between $\mathrm{BV}([a,b])$ and the dual space of $\mathrm{G_R}([a,b])$.

8.2.8 Theorem. *For* $p \in \mathrm{BV}([a,b])$, *let*

$$\Phi_p(x) = p(a)\,x(a) + \int_a^b p\,\mathrm{d}x \quad \textit{for} \ \ x \in \mathrm{G_R}([a,b]).$$

Then, the mapping

$$p \in \mathrm{BV}([a,b]) \mapsto \Phi_p \in (\mathrm{G_R}([a,b]))^* \tag{8.2.11}$$

is an isomorphism.

Proof. Clearly, (8.2.11) defines a linear mapping and, by Theorem 8.2.7 (i), it is also surjective. Now, consider $p \in \mathrm{BV}([a,b])$ such that $\Phi_p \equiv 0$. Since by Lemma 6.3.2 we have

$$\Phi_p(\chi_{[a,b]}) = p(a),$$
$$\Phi_p(\chi_{[\tau,b]}) = p(\tau) \quad \text{if} \quad \tau \in (a,b],$$

we conclude that $p \equiv 0$, showing that the mapping is one-to-one.

Finally, from Theorem 6.3.7 it follows that

$$\|\Phi_p\|_{(\mathrm{G_R}([a,b]))^*} \leq 2\,|p(a)| + |p(b)| + \mathrm{var}_a^b\, p \leq 3\,\|p\|_{\mathrm{BV}},$$

which implies that (8.2.11) is continuous. □

Similarly, one can show that $\mathrm{BV}([a,b])$ is isomorphic to the dual space of $\mathrm{G_L}([a,b])$.

8.2.9 Theorem. *For* $p \in \mathrm{BV}([a,b])$, *let*

$$\Phi_p(x) = p(b)\,x(b) - \int_a^b p \;\mathrm{d}x, \quad x \in \mathrm{G_L}([a,b]).$$

Then, the mapping

$$p \in \mathrm{BV}([a,b]) \mapsto \Phi_p \in (\mathrm{G_L}([a,b]))^*$$

is an isomorphism.

8.2.10 Exercise. Prove Theorem 8.2.9.

Regarding continuous linear functionals on $\mathrm{G_{reg}}([a,b])$ we have the following statement.

8.2.11 Theorem. Φ *is a continuous linear functional on* $\mathrm{G_{reg}}([a,b])$ *if and only if there exists* $p \in \mathrm{BV}([a,b])$ *such that* $\Phi = \Phi_p$, *where*

$$\Phi_p(x) = p(a)\,x(a) + \int_a^b p \;\mathrm{d}x \quad \text{for} \quad x \in \mathrm{G_{reg}}([a,b]).$$

8.2.12 Exercise. Prove Theorem 8.2.11.
Hint: Consider the function $p\colon [a,b] \to \mathbb{R}$ given by

$$p(t) = \begin{cases} \Phi(\chi_{[a,b]}) & \text{if } t = a, \\ \Phi(\frac{1}{2}\chi_{[t]} + \chi_{(t,b]}) & \text{if } t \in (a,b). \end{cases}$$

Bibliographic remarks

The material in the present section revises the results contained in [Schwabik (1988)], [Tvrdý (1989)] and [Tvrdý (2002)]. Representation of continuous linear functionals on the space of regulated functions by means of other Stieltjes-type integrals has been also investigated in [Kaltenborn (1934)], [Hönig (1975)], where the Dushnik integral was used, and in [Hildebrandt (1966)], [Brokate, Krejčí (2003)], where the Young integral was used. It is worth highlighting that the representations given by (8.2.10) differ from the one presented in [Hönig (1975)] not only in the integral used but also in its form. According to [Hönig (1975)], a functional $\Phi \in (\mathrm{G_L}([a,b]))^*$ can be described by the equality

$$\Phi(x) = (\sigma\mathrm{D})\int_a^b x\,\mathrm{d}p \quad \text{for } x \in \mathrm{G_L}([a,b]),$$

where $p \in \mathrm{BV}([a,b])$, $p(a) = 0$, and the integral is understood as the Dushnik (σ)-integral, cf. Section 6.13. On the other hand, the representation by means of the Kurzweil-Stieltjes integral, besides adding an extra term, has regulated functions x in the role of integrators.

8.3 Adjoint classes of Kurzweil-Stieltjes integrable functions

In mathematical analysis, to understand classes of functions which are integrable (in some sense) is fairly crucial for applications to differential equations. When the integration process is of the Stieltjes type though, there are two possible ways of addressing the question of integrability. First, for a fixed integrator g, we can ask for which functions f the integral $\int f\,\mathrm{d}g$ exists. Second, for a fixed integrand f, we can ask for which class of functions g the integral $\int f\,\mathrm{d}g$ exists. To put it another way, in Stieltjes-type integration a class of functions $\mathcal{B}$ determines a class $\mathcal{A}$ so that the integral $\int f\,\mathrm{d}g$ exists provided $f \in \mathcal{A}$ and $g \in \mathcal{B}$. Related notion is that of adjoint classes defined as follows:

8.3.1 Definition. Let $\mathcal{A}$ and $\mathcal{B}$ be two classes of functions defined on $[a,b]$, and consider an integration process T. We say that $\mathcal{A}$ is *adjoint with* $\mathcal{B}$ *regarding the integral* T if the following three conditions are satisfied:

(i) The integral ${}^{(T)}\!\!\int_a^b f\,\mathrm{d}g$ exists for every $f \in \mathcal{A}$ and every $g \in \mathcal{B}$.

(ii) If the integral ${}^{(T)}\!\!\int_a^b f\,\mathrm{d}g$ exists for every $g \in \mathcal{B}$, then $f \in \mathcal{A}$.

(iii) If the integral ${}^{(T)}\!\!\int_a^b f\,\mathrm{d}g$ exists for every $f \in \mathcal{A}$, then $g \in \mathcal{B}$.

Notice that in Definition 8.3.1 the order of sets $\mathcal{A}$ and $\mathcal{B}$ is important as it refers to the classes of integrands and integrators, respectively. Further, notice also that if $\mathcal{A}$ and $\mathcal{B}$ are adjoint with respect to a given integral T, then neither $\mathcal{A}$ nor $\mathcal{B}$ can be enlarged so that the integrability condition (i) is preserved. Moreover, conditions (ii) and (iii) mean that the integrability with respect to a given integral can be regarded as a tool to characterize some particular properties of functions.

In light of the definition above, we can see that Corollary 5.6.2, Theorems 5.8.5 and 5.8.1 imply that $\mathrm{C}([a,b])$ and $\mathrm{BV}([a,b])$ are adjoint classes of Riemann-Stieltjes integrable functions. From this, we know that the existence of the Riemann-Stieltjes integral with respect to every function of bounded variation ensures continuity. Obviously, we cannot expect such a property to hold in the theory of Kurzweil-Stieltjes integration. Indeed, the following result, which partially generalizes Lemma 6.3.16, indicates a whole class of discontinuous functions for which the Kurzweil-Stieltjes integral with respect to functions of bounded variation always exists.

8.3.2 Theorem. *Let $D = \{d_k\} \subset [a,b]$, $c \in \mathbb{R}$ and let $f : [a,b] \to \mathbb{R}$ be bounded and such that*

$$f(t) = c \quad \text{for } t \in [a,b] \setminus D.$$

Then the integral $\int_a^b f\,\mathrm{d}g$ exists for all $g \in \mathrm{BV}([a,b])$.

Proof. If D is finite, the assertion of the corollary is obvious. So, let D be infinite. For each $n \in \mathbb{N}$, put $D_n = \{d_1, d_2, \dots, d_n\}$ and define

$$f_n(t) = \begin{cases} c & \text{if } t \in [a,b] \setminus D_n, \\ f(t) & \text{if } t \in D_n. \end{cases}$$

Since f_n is a finite step function, the integral $\int_a^b f_n\,\mathrm{d}g$ exists for every $g \in \mathrm{BV}([a,b])$ (see Corollary 6.3.4). It is easy to see that the sequence $\{f_n\}$ is uniformly bounded by $K = \max\{\|f\|, |c|\}$ and $f_n(t)$ tends to $f(t)$

for every $t \in [a,b]$. Hence the result follows from Theorem 6.8.13 (bounded convergence theorem). □

In this section we discuss the adjoint classes of Kurzweil-Stieltjes integrable functions in two directions. First, we show that the class of regulated functions, $\mathrm{G}([a,b])$, is not adjoint with the class of functions of bounded variation, $\mathrm{BV}([a,b])$, regarding the Kurzweil-Stieltjes integral. In the sequel, we analyze the reverse order, that is, $\mathrm{BV}([a,b])$ and $\mathrm{G}([a,b])$, showing that it yields a pair of adjoint classes of Kurzweil-Stieltjes integrable functions.

The next example shows that condition (ii) of Definition 8.3.1 fails to be true when $\mathcal{A} = \mathrm{G}([a,b])$ and $\mathcal{B} = \mathrm{BV}([a,b])$.

8.3.3 Example. Let $c \in \mathbb{R}$ and consider a function $f : [0,1] \to \mathbb{R}$ given by

$$f(t) = c \quad \text{for } t \in [a,b] \setminus D,$$

where $D = \{d_k\} \subset [a,b]$ is infinite and such that $\lim_{k\to\infty} d_k = d \notin D$ while $f(d_k)$ does not converge to c. Thus, f is not regulated, but by Theorem 8.3.2 the integral $\int_a^b f \, \mathrm{d}g$ exists for all $g \in \mathrm{BV}([a,b])$.

The example above leads us to conclude that $\mathrm{G}([a,b])$ is not adjoint with $\mathrm{BV}([a,b])$ regarding the Kurzweil-Stieltjes integral. Yet, we might wonder whether condition (iii) of Definition 8.3.1 holds for these classes of functions. As we will see in the following example, the answer is again negative.

8.3.4 Example. Let $g : [0,1] \to \mathbb{R}$ be given by

$$g(t) = \begin{cases} \frac{1}{k} & \text{if } t = \frac{1}{k} \text{ for some } k \in \mathbb{N} \text{ such that } k \geq 2, \\ 0 & \text{otherwise.} \end{cases} \tag{8.3.1}$$

Claim 1. $\operatorname{var}_a^b g = \infty$.

Note that $\Delta^- g(\frac{1}{k}) = -\Delta^+ g(\frac{1}{k}) = \dfrac{1}{k}$ for $k \in \mathbb{N}$ and hence

$$\sum_{0 \leq t < 1} |\Delta^+ g(t)| + \sum_{0 < t \leq 1} |\Delta^- g(t)| = 2 \sum_{k=1}^{\infty} \frac{1}{k} = \infty.$$

On the other hand, if we had $g \in \mathrm{BV}([0,1])$, then by Corollary 2.3.8 we would have

$$\sum_{0\le t<1} |\Delta^+ g(t)| + \sum_{0<t\le 1} |\Delta^- g(t)| \le \operatorname{var}_0^1 g.$$

Therefore $\operatorname{var}_0^1 g = \infty$.

Claim 2. $\int_0^1 f\,\mathrm{d}g = 0$ *for every* $f:[0,1]\to\mathbb{R}$.

Let $f:[0,1]\to\mathbb{R}$ and $\varepsilon>0$ be given. Put $D=\{0\}\cup\{\frac{1}{k}: k\in\mathbb{N},\ k\ge 2\}$ and define

$$\delta(t)=\begin{cases}\operatorname{dist}(t,D) & \text{if } t\notin D,\\ \operatorname{dist}(t,D\setminus\{t\}) & \text{if } t=\frac{1}{k} \text{ for some } k\in\mathbb{N} \text{ such that } k\ge 2,\\ \eta, & \text{if } t=0,\end{cases}$$

where $\eta>0$ is such that

$$|f(0)|\,|g(s)|<\varepsilon \quad \text{for} \quad s\in(0,\eta) \tag{8.3.2}$$

(such a number exists as $g(0+)=g(0)=0$). Let $P=(\boldsymbol{\alpha},\boldsymbol{\xi})$ be a δ-fine partition of $[0,1]$. Due to the definition of δ we have $\xi_1=0$ and for each $j\in\{2,\dots,\nu(\boldsymbol{\alpha})\}$ the subinterval $[\alpha_{j-1},\alpha_j]$ contains at most one point of D. Moreover, it is not difficult to see that if $\alpha_j=\xi_j=\frac{1}{k}$ for some $j\in\{2,\dots,\nu(\boldsymbol{\alpha})-1\}$ and $k\ge 2$, then $\xi_{j+1}=\xi_j=\frac{1}{k}$. Hence, in such case,

$$\begin{aligned}&f(\xi_j)\,(g(\alpha_j)-g(\alpha_{j-1}))+f(\xi_{j+1})\,(g(\alpha_{j+1})-g(\alpha_j))\\ &\quad = f(\tfrac{1}{k})\,(g(\alpha_{j+1})-g(\alpha_{j-1}))=0,\end{aligned}$$

since none of the points α_{j-1} and α_{j+1} belong to $D\setminus\{0\}$. Let $\widetilde{P}$ be the partition obtained from P by combining two adjacent subintervals whenever they have a common tag. All subintervals determined by $\widetilde{P}$, except the first one, contribute nothing to $S(f,\mathrm{d}g,\widetilde{P})$. Therefore

$$\big|S(f,\mathrm{d}g,P)\big| = \big|S(f,\mathrm{d}g,\widetilde{P})\big| = |f(0)\,g(\alpha_1)|<\varepsilon,$$

where the last inequality is due to (8.3.2). This proves Claim 2.

Concerning the function g in (8.3.1), it is worth highlighting that, unlike what we observed for the integration in the Kurzweil sense, the Riemann-Stieltjes integral fails to exist even for all continuous functions. This is a direct consequence of the fact that $\mathrm{C}([a,b])$ and $\mathrm{BV}([a,b])$ are adjoint with respect to the Riemann-Stieltjes integral. (See Theorem 5.8.5 for details.)

The example above not only invalidates condition (iii) for the classes $\mathrm{G}([a,b])$ and $\mathrm{BV}([a,b])$, but also shows that there is no class of functions $\mathcal{A}$ adjoint with $\mathrm{BV}([a,b])$ regarding the Kurzweil-Stieltjes integral. Nevertheless, as we will see in the following theorem, integrability with respect to functions of bounded variation ensures boundedness of the integrator.

8.3.5 Theorem. *If $f\colon[a,b]\to\mathbb{R}$ is a function such that the integral $\int_a^b f\,\mathrm{d}g$ exists for every $g\in\mathrm{BV}([a,b])$, then f is bounded.*

Proof. For contradiction, assume that there exists an unbounded function f satisfying the hypothesis. Without loss of generality, assume that f is unbounded from above (otherwise, consider the function $-f$). Thus, there exist $c\in[a,b]$ and a sequence $\{t_n\}$ in $[a,b]$ such that

$$\lim_{n\to\infty} t_n=c \quad\text{and}\quad \lim_{n\to\infty} f(t_n)=+\infty.$$

Note that the intersection of $\{t_n: n\in\mathbb{N}\}$ with at least one of the intervals $[a,c)$, $(c,b]$ is an infinite set. Thus, we can assume that the sequence $\{t_n\}$ is monotone and $\{f(t_n)\}$ is increasing, with $f(t_n)>0$ for all $n\in\mathbb{N}$. Denote

$$y_0=0, \qquad y_n=\frac{1}{f(t_n)}-\frac{1}{f(t_{n+1})}, \qquad s_n=\sum_{k=0}^{n} y_k, \quad \text{for } n\in\mathbb{N}.$$

If $\{t_n\}$ increases to c, define $g\colon[a,b]\to\mathbb{R}$ by

$$g(t)=\begin{cases} 0 & \text{if } t\in[a,t_1],\\ s_{n-1} & \text{if } t\in[t_n,t_{n+1}),\ n\in\mathbb{N},\\ \lim\limits_{n\to\infty} s_n & \text{if } t\in[c,b]. \end{cases}$$

Note that g is continuous at c and has bounded variation, because it is monotone. By hypothesis, the integral $\int_a^b f\,\mathrm{d}g$ exists; thus, using Corollary 6.5.4 we can write

$$\int_a^c f\,\mathrm{d}g=\lim_{t\to c-}\int_a^t f\,\mathrm{d}g+f(c)\,\Delta^{-}g(c)=\lim_{n\to\infty}\int_a^{t_n} f\,\mathrm{d}g. \tag{8.3.3}$$

For each $k\in\mathbb{N}$, calculating the integral we obtain

$$\begin{aligned}\int_{t_k}^{t_{k+1}} f\,\mathrm{d}g&=\int_{t_k}^{t_{k+1}} f\,\mathrm{d}\big(s_{k-1}\chi_{[t_k,t_{k+1})}+s_k\chi_{[t_{k+1}]}\big)\\ &=f(t_{k+1})(s_k-s_{k-1})=f(t_{k+1})\,y_k\end{aligned}$$

(see Lemma 6.3.2), and consequently

$$\int_a^{t_n} f\,dg = \int_a^{t_1} f\,dg + \sum_{k=1}^{n-1} \int_{t_k}^{t_{k+1}} f\,dg = \sum_{k=1}^{n-1} f(t_{k+1})\,y_k. \tag{8.3.4}$$

We claim that $\sum_{n=1}^{\infty} f(t_{n+1})\,y_n$ diverges. Indeed, since the sequence $\{f(t_n)\}$ is increasing and unbounded, for each $n \in \mathbb{N}$ we can choose $m_n \in \mathbb{N}$, such that

$$m_n > n \quad \text{and} \quad \frac{f(t_n)}{f(t_{m_n+1})} < \frac{1}{2}.$$

Therefore,

$$\begin{aligned}\sum_{k=n}^{m_n} f(t_{k+1})\,y_k &= \sum_{k=n}^{m_n} \frac{f(t_{k+1}) - f(t_k)}{f(t_k)} \geq \frac{1}{f(t_{m_n+1})} \sum_{k=n}^{m_n} (f(t_{k+1}) - f(t_k)) \\ &= 1 - \frac{f(t_n)}{f(t_{m_n+1})} > \frac{1}{2}.\end{aligned}$$

This means that $\sum_{n=1}^{\infty} f(t_{n+1})\,y_n$ does not satisfy the Cauchy condition, and consequently the series diverges. Having this in mind, the equality (8.3.4) together with (8.3.3) contradicts the existence of the integral $\int_a^c f\,dg$.

If, on the other hand, $\{t_n\}$ decreases to c, we redefine g accordingly and using similar argument we get that $\int_c^b f\,dg$ equals a divergent series; again a contradiction. In summary, we conclude that a function satisfying the hypothesis must be bounded. □

Now we turn our attention to the following question: Is $\mathrm{BV}([a,b])$ adjoint with $\mathrm{G}([a,b])$ regarding the Kurzweil-Stieltjes integral? We know from Theorem 6.3.11 that condition (i) of Definition 8.3.1 is satisfied for these classes of functions. The verification of the remaining ones is contained in Theorems 8.3.6 and 8.3.8.

8.3.6 Theorem. *If $f:[a,b] \to \mathbb{R}$ is a function such that the integral $\int_a^b f\,dg$ exists for every $g \in \mathrm{G}([a,b])$, then $f \in \mathrm{BV}([a,b])$.*

Proof. For contradiction, assume that there exists a function f such that $\mathrm{var}_a^b f = \infty$, while the integral $\int_a^b f\,dg$ exists for each $g \in \mathrm{G}([a,b])$. Recall that a function belongs to $\mathrm{BV}([a,b])$ if and only if each point of $[a,b]$ has a neighborhood on which the variation is finite. Hence, as $f \notin \mathrm{BV}([a,b])$, there must exist a point c which satisfies either $c \in (a,b]$ and $\mathrm{var}_t^c f = \infty$ for every $t \in [a,c)$, or $c \in [a,b)$ and $\mathrm{var}_c^t f = \infty$ for every $t \in (c,b]$.

By Lemma 5.8.3 (i), in the former case there is an increasing sequence $\{t_k\}$ in (a,c) such that

$$\lim_{k\to\infty} t_k = c \quad \text{and} \quad \sum_{k=1}^{\infty} |f(t_{k+1}) - f(t_k)| = \infty$$

and further, due to Lemma 5.8.2, we can choose a sequence $\{y_k\}$ of positive numbers in such a way that

$$\lim_{k\to\infty} y_k = 0 \quad \text{and} \quad \sum_{k=1}^{\infty} y_k |f(t_{k+1}) - f(t_k)| = \infty. \tag{8.3.5}$$

Let $\lambda_k = \operatorname{sgn}\big(f(t_k) - f(t_{k+1})\big)$ for $k \in \mathbb{N}$. Define

$$g(t) = \begin{cases} y_k\,\lambda_k & \text{if } t \in (t_k, t_{k+1}) \text{ and } k \in \mathbb{N}, \\ 0 & \text{otherwise.} \end{cases}$$

Clearly, $g \in \mathrm{G}([a,b])$ and $g(c-) = g(c) = 0$. Since $\int_a^b f\,\mathrm{d}g$ exists, the integral also exists in every subinterval of $[a,b]$. In particular, $\int_a^{t_1} f\,\mathrm{d}g = 0$ and for each $k \in \mathbb{N}$, by Lemma 6.3.2, we obtain

$$\int_{t_k}^{t_{k+1}} f\,\mathrm{d}g = y_k\,\lambda_k \int_{t_k}^{t_{k+1}} f\,\mathrm{d}\big[\chi_{(t_k,t_{k+1})}\big] = y_k\,\lambda_k\,\big(f(t_k) - f(t_{k+1})\big),$$

that is,

$$\int_{t_k}^{t_{k+1}} f\,\mathrm{d}g = y_k\,|f(t_{k+1}) - f(t_k)|.$$

Due to the convergence of the sequence $\{t_n\}$ and Corollary 6.5.4, we can write

$$\begin{aligned} \int_a^c f\,\mathrm{d}g &= \lim_{t\to c-} \int_a^t f\,\mathrm{d}g + f(c)\,\Delta^- g(c) = \lim_{n\to\infty} \int_a^{t_n} f\,\mathrm{d}g \\ &= \int_a^{t_1} f\,\mathrm{d}g + \lim_{n\to\infty} \sum_{k=1}^{n-1} \int_{t_k}^{t_{k+1}} f\,\mathrm{d}g = \lim_{n\to\infty} \sum_{k=1}^{n-1} y_k |f(t_{k+1}) - f(t_k)|. \end{aligned}$$

Hence, in view of (8.3.5), $\int_a^c f\,\mathrm{d}g$ diverges, which is a contradiction. Therefore we conclude that a function satisfying the hypothesis must have a bounded variation.

In the case when there is $c \in [a,b)$ such that $\operatorname{var}_c^t f = \infty$ for every $t \in (c,b]$, the proof is similar, but relies on part (ii) of Lemma 5.8.3. □

We remark that, in the theorem above, G([a, b]) cannot be replaced by the class C([a, b]). In other words, the existence of the Kurzweil–Stieltjes integral with respect to continuous integrators does not ensure bounded variation. To illustrate this fact we can again make use of the function given in (8.3.1).

8.3.7 Example. Let $f\colon [0,1] \to \mathbb{R}$ be given by

$$f(t) = \begin{cases} \frac{1}{k} & \text{if } t = \frac{1}{k} \text{ for some } k \in \mathbb{N} \text{ such that } k \geq 2, \\ 0 & \text{otherwise.} \end{cases}$$

We know that $f \notin \mathrm{BV}([0,1])$ (see Example 8.3.1). Let us prove that $\int_0^1 f\,\mathrm{d}g = 0$ for every $g \in C[0,1]$. Consider an arbitrary $\varepsilon > 0$. Since g is uniformly continuous, for each $k \in \mathbb{N}$ there exists $\rho_k > 0$ such that

$$\omega_{Q_k}(g) < \frac{\varepsilon}{k}, \quad Q_k = (\tfrac{1}{k} - \rho_k, \tfrac{1}{k} + \rho_k) \cap [0,1],$$

where $\omega_{Q_k}(g)$ denotes the modulus of oscillation of g on Q_k. Put

$$D = \{0\} \cup \{\frac{1}{k} : k \in \mathbb{N},\, k \geq 2\}$$

and define a gauge δ on $[0,1]$ by

$$\delta(t) = \begin{cases} \operatorname{dist}(t, D) & \text{if } t \notin D, \\ \min\{\rho_k\,,\, \operatorname{dist}(t, D \setminus \{t\})\} & \text{if } t = \frac{1}{k},\ k \in \mathbb{N},\ k \geq 2, \\ 1 & \text{if } t = 0. \end{cases}$$

Let $P = (\boldsymbol{\alpha}, \boldsymbol{\xi})$ be a δ-fine partition of $[0,1]$. Then, for each $j \in \{2, \dots, \nu(\boldsymbol{\alpha})\}$, the subinterval $[\alpha_{j-1}, \alpha_j]$ contains at most one point of D. Let Λ be the set of all $j \in \{1, \dots, \nu(\boldsymbol{\alpha})\}$ such that $\xi_j = \frac{1}{k_j}$ for some $k_j \in \mathbb{N}$. Clearly, $[\alpha_{j-1}, \alpha_j] \subset Q_{k_j}$ for every $j \in \Lambda$. Therefore,

$$|S(f, \mathrm{d}g, P)| = \Big| \sum_{j \in \Lambda} f(\xi_j)\,(g(\alpha_j) - g(\alpha_{j-1})) \Big| \leq \sum_{j \in \Lambda} \frac{1}{k_j}\, \omega_{Q_{k_j}}(g) < \varepsilon \sum_{n=1}^{\infty} \frac{1}{n^2}$$

wherefrom it follows that $\int_0^1 f\,\mathrm{d}g = 0.$

To conclude that BV([a, b]) is adjoint with G([a, b]) it remains to verify also condition (iii). As we will see, we need not test the integrability of all functions of the space BV([a, b]) but only a particular type of finite step functions.

8.3.8 Theorem. *Let* $g\colon[a,b]\to\mathbb{R}$. *If, for every interval* J, *the integral* $\int_a^b \chi_J \, \mathrm{d}g$ *exists, then* $g\in \mathrm{G}([a,b])$.

Proof. Given $c\in[a,b)$ we will prove that $g(c+)$ exists. To this aim, let $f=\chi_{(c,b]}$ and let $\varepsilon>0$ be given. Thus, we can choose a gauge δ on $[a,b]$ such that

$$\Big|S(f,\mathrm{d}g,P)-\int_a^b f\,\mathrm{d}g\Big|<\frac{\varepsilon}{2} \quad \text{for every } \delta\text{-fine partition } P \text{ of } [a,b]. \qquad (8.3.6)$$

In view of Lemma 6.2.11 we can assume that the gauge δ is such that every δ-fine partition contains c as the tag of some subinterval.

Fix an arbitrary δ-fine partition $P=(\boldsymbol{\alpha},\boldsymbol{\xi})$ of $[a,b]$ and let an index $\ell\in\{1,\dots,\nu(\boldsymbol{\alpha})\}$ be such that $\xi_\ell=c$. For each $t\in(c,\alpha_\ell)$, define the partition

$$\widetilde{P}_t=P_t^-\cup P_t\cup Q_t\cup P_t^+,$$

where

$$\begin{aligned} P_t &=(\{\alpha_{\ell-1},t\},(c)),\\ P_t^- &=(\{\alpha_0,\dots,\alpha_{\ell-1}\},(\xi_1,\dots,\xi_{\ell-1})),\\ P_t^+ &=(\{\alpha_\ell,\dots,\alpha_{\nu(\boldsymbol{\alpha})}\},(\xi_{\ell+1},\dots,\xi_{\nu(\boldsymbol{\alpha})})), \end{aligned}$$

and Q_k is an arbitrary δ-fine partition of $[t,\alpha_\ell]$. Thus, $\widetilde{P}_t$ is a δ-fine partition of $[a,b]$ and a simple calculation shows that

$$S(f,\mathrm{d}g,\widetilde{P}_t)=S(f,\mathrm{d}g,Q_t)+S(f,\mathrm{d}g,P_t^+)=g(b)-g(t).$$

This together with (8.3.6) implies that for each pair $u,v\in(c,\alpha_\ell)$ we get

$$|g(u)-g(v)|=|S(f,\mathrm{d}g,\widetilde{P}_u)-S(f,\mathrm{d}g,\widetilde{P}_v)|<\varepsilon,$$

showing that the Cauchy condition for the existence of $g(c+)$ is satisfied. Analogously, we can show the existence of $g(c-)$ for every $c\in(a,b]$. □

Theorems 8.3.6 and 8.3.8 lead to the following corollary.

8.3.9 Corollary. $\mathrm{BV}([a,b])$ *is adjoint with* $\mathrm{G}([a,b])$ *regarding the Kurzweil-Stieltjes integral.*

BIBLIOGRAPHIC REMARKS

The results on adjoint classes of Kurzweil-Stieltjes integrable functions are taken from the paper [Monteiro (2017)], while Definition 8.3.1 is inspired by the concept of adjoint classes introduced in [Chen (1998/99a)].

Adjoint classes of integrable functions have been studied in connection to a variety of integration processes in a handful of papers. Concerning the Riemann-Stieltjes integral, we highlight the basic result contained in Hildebrandt's monograph [Hildebrandt (1963)] (see Chapter II, section 12 there) and also the work of H. Chen in [Chen (1998/99a)]. Generalizations of the Riemann-Stieltjes integral have been considered in [Chen (1998/99b)], [Chen (2000/01)] and [Sworowski (2007)].

Theorem 8.3.8 corresponds to Proposition 2.60 in [Dudley, Norvaiša (2011)]; however, herein we presented a slightly different proof.

8.4 DISTRIBUTIONS

In this section we outline some applications of the Kurzweil-Stieltjes integral in the theory of distributions, which are understood in the sense of L. Schwartz. Let us recall some of the basic notions and definitions.

8.4.1 Definition. The symbol $\mathfrak{D}[a,b]$ stands for the set of functions $\varphi\colon \mathbb{R}\to\mathbb{R}$, which are infinitely differentiable and such that $\varphi^{(k)}(t)=0$ for all $t\in\mathbb{R}\setminus(a,b)$ and $k\in\mathbb{N}\cup\{0\}$. Functions from $\mathfrak{D}[a,b]$ are called *test functions* on $[a,b]$.

The set $\mathfrak{D}[a,b]$ is a linear space when equipped with the usual operations of addition and scalar multiplication. We say that a sequence φ_n converges to φ in $\mathfrak{D}[a,b]$ if and only if

$$\lim_{n\to\infty}\|\varphi_n^{(k)}-\varphi^{(k)}\|=0 \quad \text{for any} \quad k\in\mathbb{N}\cup\{0\}.$$

With the topology induced by the notion of convergence above, $\mathfrak{D}[a,b]$ is a topological vector space.

Typical examples of test functions on $[a,b]$ are given by:

$$\varphi_{c,d}(t)=\begin{cases}\exp\left(\frac{1}{c-t}+\frac{1}{t-d}\right) & \text{for } t\in(c,d),\\ 0 & \text{for } t\in\mathbb{R}\setminus(c,d),\end{cases} \tag{8.4.1}$$

where $[c,d]$ is an arbitrary closed subinterval of $[a,b]$.

8.4.2 Definition. Continuous linear functionals on the topological vector space $\mathfrak{D}[a,b]$ are called *distributions* on $[a,b]$, and the set of all distributions on $[a,b]$ is denoted by the symbol $\mathfrak{D}^*[a,b]$.

In other words, the set $\mathfrak{D}^*[a,b]$ is the dual space to $\mathfrak{D}[a,b]$.

For a given distribution $f \in \mathfrak{D}^*[a,b]$ and a test function $\varphi \in \mathfrak{D}[a,b]$, the value $f(\varphi)$ is traditionally denoted by $\langle f, \varphi \rangle$.

8.4.3 Remark. Let $\mathrm{L}([a,b])$ denote the space of all functions $f\colon [a,b] \to \mathbb{R}$ whose Lebesgue integral $\int_a^b f(t)\,\mathrm{d}t$ exists and is finite. Given $f \in \mathrm{L}([a,b])$ put

$$\langle f, \varphi \rangle = \int_a^b f(t)\,\varphi(t)\,\mathrm{d}t \quad \text{for } \varphi \in \mathfrak{D}[a,b].$$

The mapping $\varphi \mapsto \langle f, \varphi \rangle$ defines a distribution on $[a,b]$, which will be also denoted by the symbol f. We say that the distribution f is determined by the function f.

The null element of the space $\mathfrak{D}^*[a,b]$ is the distribution that maps each test function to zero. Notice that this distribution is determined by an arbitrary measurable function f which vanishes almost everywhere in $[a,b]$. In particular, if $f \in \mathrm{G}([a,b])$, then $f = 0 \in \mathfrak{D}^*[a,b]$ if and only if $f(t-) = f(s+) = 0$ for all $t \in (a,b]$, $s \in [a,b)$. Likewise, if $f \in \mathrm{G}_{\mathrm{reg}}([a,b])$, then $f = 0 \in \mathfrak{D}^*[a,b]$ if and only if $f(t) = 0$ for all $t \in [a,b]$. Consequently, if $g \in \mathrm{L}([a,b])$, then there is at most one function $f \in \mathrm{G}_{\mathrm{reg}}([a,b])$ such that $f = g$ a.e. on $[a,b]$. Furthermore, for $f, g \in \mathrm{L}([a,b])$, the equality $f = g$ holds in the sense of $\mathfrak{D}^*[a,b]$ if and only if $f = g$ a.e. on $[a,b]$.

8.4.4 Definition. For a given distribution $f \in \mathfrak{D}^*[a,b]$, its *distributional derivative* f' is defined by $\langle f', \varphi \rangle = -\langle f, \varphi' \rangle$ for $\varphi \in \mathfrak{D}[a,b]$.

Similarly, for each $k \in \mathbb{N}$, we define

$$\langle f^{(k)}, \varphi \rangle = (-1)^k \langle f, \varphi^{(k)} \rangle \quad \text{for } \varphi \in \mathfrak{D}[a,b].$$

Note that distributional derivatives of absolutely continuous functions are determined by their classical derivatives.

8.4.5 Example. Given an arbitrary $\tau \in (a,b)$, by δ_τ we denote the *Dirac δ-distribution* (concentrated at τ) defined by

$$\langle \delta_\tau, \varphi \rangle = \varphi(\tau) \quad \text{for every } \varphi \in \mathfrak{D}[a,b].$$

One can show that δ_τ corresponds to the distributional derivative of the *Heaviside function* given by $h_\tau(t) = H(t-\tau)$, $t \in [a,b]$, where

$$H(t) = \begin{cases} 0 & \text{for } t < 0, \\ \frac{1}{2} & \text{for } t = 0, \\ 1 & \text{for } t > 0. \end{cases}$$

Indeed, by Theorem 6.4.2 (integration by parts) and Corollary 6.6.2, as well as relations (6.3.1) and (6.3.5), we get

$$\langle h'_\tau, \varphi\rangle = -\langle h_\tau, \varphi'\rangle = -\int_a^b h_\tau(t)\,\varphi'(t)\,\mathrm{d}t = -\int_a^b h_\tau\,\mathrm{d}\varphi = \int_a^b \varphi\,\mathrm{d}h_\tau = \varphi(\tau),$$

which shows that $\delta_\tau = h'_\tau$ in the sense of distributions.

8.4.6 Theorem. *Let $f \in \mathrm{L}([a,b])$. Then its distributional derivative f' is the zero distribution if and only if there is a constant $c \in \mathbb{R}$ such that $f(t) = c$ for almost all $t \in [a,b]$.*

Proof. Let $f(t) = c$ for almost all $t \in [a,b]$ and $\varphi \in \mathfrak{D}[a,b]$, then

$$\langle f', \varphi\rangle = -\langle c, \varphi'\rangle = -c\int_a^b \varphi'(s)\,\mathrm{d}s = -c\,(\varphi(b) - \varphi(a)) = 0.$$

Conversely, assume that distributional derivative f' is the zero distribution. Given an arbitrary test function $\varphi \in \mathfrak{D}[a,b]$, let

$$\rho(t) = \begin{cases} \displaystyle\int_a^t \big(\varphi(s) - a_0\,\Theta(s)\big)\,\mathrm{d}s & \text{for } t \in [a,b], \\ 0 & \text{for } t \in \mathbb{R}\setminus[a,b], \end{cases}$$

where

$$a_0 = \int_a^b \varphi(s)\,\mathrm{d}s, \quad \Theta(t) = \frac{\varphi_{a,b}(t)}{\displaystyle\int_a^b \varphi_{a,b}(s)\,\mathrm{d}s},$$

and $\varphi_{a,b}$ is the function given by (8.4.1). Then

$$\int_a^b \Theta(s)\,\mathrm{d}s = 1$$

wherefrom it follows easily that $\rho(a) = \rho(b) = 0$, i.e., $\rho \in \mathfrak{D}[a,b]$. Furthermore,

$$\rho'(t) = \varphi(t) - a_0\,\Theta(t) \quad \text{for } t \in [a,b].$$

Hence $0=-\langle f',\rho\rangle=\langle f,\rho'\rangle=\langle f,\varphi\rangle-\Big(\int_a^b\varphi(s)\,\mathrm{d}s\Big)\,\langle f,\Theta\rangle$. Therefore, by letting $c=\langle f,\Theta\rangle$, we get

$$\langle f,\varphi\rangle=\Big(\int_a^b\varphi(s)\,\mathrm{d}s\Big)\,\langle f,\Theta\rangle=\int_a^b c\,\varphi(s)\,\mathrm{d}s$$

for any $\varphi\in\mathfrak{D}[a,b]$. Thus, $f=c$ in the sense of distributions. □

8.4.7 Exercise. For $f\in\mathrm{L}([a,b])$ and $k\in\mathbb{N}$ show that $f^{(k)}=0\in\mathfrak{D}^*[a,b]$ if and only if there are $c_0,c_1,\dots,c_{k-1}\in\mathbb{R}$ such that

$$f(t)=c_0+c_1\,t+\cdots+c_{k-1}\,t^{k-1}\quad\text{for almost all}\quad t\in[a,b].$$

An important problem of the theory of distributions is a proper definition of the product of two distributions. The following classical definitions apply only to very special kinds of distributions.

8.4.8 Definition. (i) If $f,g\in\mathrm{L}([a,b])$ are such that $f\,g\in\mathrm{L}([a,b])$, then

$$\langle f\,g,\varphi\rangle=\int_a^b f(t)\,g(t)\,\varphi(t)\,\mathrm{d}t\quad\text{for}\quad\varphi\in\mathfrak{D}[a,b].$$

(ii) If $f\in\mathfrak{D}^*[a,b]$ and $g:[a,b]\to\mathbb{R}$ is infinitely differentiable on $[a,b]$, then $\langle fg,\varphi\rangle=\langle f,g\varphi\rangle$.

To deal with differential equations with distributional coefficients, it is useful to have a reasonable definition of the distributions $f\,g'$ and $f'\,g$, where $f\in\mathrm{G}([a,b])$ and $g\in\mathrm{BV}([a,b])$. Obviously, Definition 8.4.8 does not cover such cases. To formulate proper definitions for such couples, the Kurzweil-Stieltjes integral turns to be very helpful.

8.4.9 Definition. If $f\in\mathrm{G}([a,b])$ and $g\in\mathrm{BV}([a,b])$, then we define

$$\langle f'\,g,\varphi\rangle=\int_a^b\varphi\,g\;\mathrm{d}f\qquad\text{and}\qquad\langle f\,g',\varphi\rangle=\int_a^b\varphi f\;\mathrm{d}g\quad\text{for}\quad\varphi\in\mathfrak{D}[a,b].$$

8.4.10 Theorem. *Let $f\in\mathrm{G}([a,b])$ and $g\in\mathrm{BV}([a,b])$ be such that*

$$\Delta^+f(t)\,\Delta^+g(t)=\Delta^-f(t)\,\Delta^-g(t)\quad\textit{for every}\quad t\in(a,b).\tag{8.4.2}$$

Then $(f\,g)'=f\,g'+f'\,g$.

P r o o f. Using Definition 8.4.4 together with Theorem 6.4.2 (integration by parts), Theorem 6.6.1 (substitution theorem) and Corollary 6.6.2, for $\varphi \in \mathfrak{D}[a,b]$ we obtain

$$\begin{aligned}\langle (f\,g)', \varphi\rangle &= -\langle f\,g, \varphi'\rangle \\ &= -\int_a^b f(t)\,g(t)\,\varphi'(t)\,\mathrm{d}t = -\int_a^b f(t)\,g(t)\,\mathrm{d}\varphi(t) \\ &= \int_a^b \varphi(t)\,\mathrm{d}\big(f(t)\,g(t) - f(a)\,g(a)\big) \\ &= \int_a^b \varphi(t)\,\mathrm{d}\Big(\int_a^t g\,\mathrm{d}f + \int_a^t f\,\mathrm{d}g\Big) \\ &= \int_a^b \varphi\,g\,\mathrm{d}f + \int_a^b \varphi\,f\,\mathrm{d}g = \langle f'\,g, \varphi\rangle + \langle f\,g', \varphi\rangle,\end{aligned}$$

which completes the proof. □

8.4.11 Remark. Let $f \in \mathrm{G}([a,b])$ and $g \in \mathrm{BV}([a,b])$. Then the condition (8.4.2) is obviously satisfied e.g. in the following cases:

(i) For each $t \in (a,b)$, at least one of f, g is left-continuous at t, and at least one of f, g is right-continuous at t.

(ii) For each $t \in (a,b)$, we have $f(t) = (f(t-) + f(t+))/2$ and $g(t) = (g(t-) + g(t+))/2$.

8.4.12 Exercise.

(i) Let $\tau \in (a,b)$, $\varkappa \in \mathbb{R}$ and

$$g(t) = \begin{cases} 0 & \text{if } t < \tau, \\ \varkappa & \text{if } t = \tau, \\ 1 & \text{if } t > \tau. \end{cases}$$

Prove that $\int_a^b \varphi\,g\,\mathrm{d}g = \varphi(\tau)\,\varkappa$ for any $\varphi \in \mathrm{BV}([a,b])$.
Hint: Using Exercise 6.3.5, determine the values of the integrals $\int_a^\tau \varphi\,g\,\mathrm{d}g$ and $\int_\tau^b \varphi\,g\,\mathrm{d}g$.

(ii) If h_τ and δ_τ are the Heaviside and Dirac distributions concentrated at a point $\tau \in (a,b)$, show that $h_\tau\,\delta_\tau = \delta_\tau/2$.

Bibliographic remarks

More information about distributions can be found e.g. in [Rudin (1973)], [Friedlander, Joshi (1999)], [Halperin (1952)] and, of course, in the monograph by L. Schwartz [Schwartz (1998)]. The extended definition of the product of distributions by means of the Kurzweil-Stieltjes integral has shown its applicability in [Pelant, Tvrdý (1993)] and [Tvrdý (1994)].

In [Talvila (2008)], E. Talvila introduced the distributional Denjoy integral which, roughly speaking, inverts distributional derivatives of continuous functions. Later, in [Talvila (2009)], this notion was extended so as to integrate distributional derivatives of regulated functions. These distributional integrals are both very general (including, for example, the Henstock-Kurzweil integral), and they have been the background of several papers related to distributional differential equations; e.g. [Heikkilä (2012)], [Malý (2014)], [Monteiro, Satco (2016)], [Ye, Liu (2016)].

8.5 Generalized elementary functions

The aim of this section is to show that the theory of generalized linear differential equations from Chapter 7 can be used to extend the definitions of the exponential, hyperbolic and trigonometric functions.

One possible way of introducing the classical exponential function is to define it as the unique solution of the initial-value problem

$$z'(t) = z(t), \quad z(0) = 1.$$

More generally, for every continuous function p defined on the real line, the initial-value problem $z'(t) = p(t)\,z(t)$, $z(t_0) = 1$, which can be written in the equivalent integral form

$$z(t) = 1 + \int_{t_0}^{t} p(s)\,z(s)\,\mathrm{d}s, \tag{8.5.1}$$

has the unique solution $z(t) = e^{\int_{t_0}^{t} p(s)\,\mathrm{d}s}$. Using Theorem 6.6.1 (substitution theorem), we can rewrite equation (8.5.1) as the generalized linear differential equation

$$z(t) = 1 + \int_{t_0}^{t} z(s)\,\mathrm{d}P(s) \tag{8.5.2}$$

with $P(s)=\int_{t_0}^{s} p$. In this section, we study equation (8.5.2) for an arbitrary function P with bounded variation (not necessarily differentiable or continuous). The solution of this equation will be called the generalized exponential function and denoted by $e_{\mathrm{d}P}$. If P is a real function, then $e_{\mathrm{d}P}$ is simply a special case of the Cauchy matrix U introduced in Definition 7.7.10 with $n=1$. In this scalar case, we will be able to obtain a much more detailed information about solutions of equation (8.5.2) than in the n-dimensional case studied in Section 7.7. Moreover, we will focus on the more general case when P is a complex-valued function. To this end, we need an existence and uniqueness theorem for generalized linear differential equations with complex coefficients.

8.5.1 Theorem. *Consider a function $P\colon[a,b]\to\mathbb{C}$, which has bounded variation on $[a,b]$. Let $t_0\in[a,b]$ and assume that $1+\Delta^+P(t)\neq 0$ for every $t\in[a,t_0)$, and $1-\Delta^-P(t)\neq 0$ for every $t\in(t_0,b]$. Then, for every $\widetilde{z}\in\mathbb{C}$, there exists a unique function $z\colon[a,b]\to\mathbb{C}$ such that*

$$z(t)=\widetilde{z}+\int_{t_0}^{t} z(s)\,\mathrm{d}P(s) \quad \text{for } t\in[a,b]. \tag{8.5.3}$$

The function z has bounded variation on $[a,b]$. If P and $\widetilde{z}$ are real, then z is real as well.

Proof. We decompose all complex quantities into real and imaginary parts as follows: $P=P_1+\mathrm{i}\,P_2$, $z=z_1+\mathrm{i}\,z_2$, and $\widetilde{z}=\widetilde{z}_1+\mathrm{i}\,\widetilde{z}_2$. Now, we see that equation (8.5.3) is equivalent to the following system of two equations with real coefficients:

$$z_1(t)=\widetilde{z}_1+\int_{t_0}^{t} z_1(s)\,\mathrm{d}P_1(s)-\int_{t_0}^{t} z_2(s)\,\mathrm{d}P_2(s),$$

$$z_2(t)=\widetilde{z}_2+\int_{t_0}^{t} z_1(s)\,\mathrm{d}P_2(s)+\int_{t_0}^{t} z_2(s)\,\mathrm{d}P_1(s).$$

The system can be also written in the vector form

$$u(t)=\widetilde{u}+\int_{t_0}^{t} \mathrm{d}[A(s)]\,u(s) \quad \text{for } t\in[a,b]$$

with

$$\widetilde{u}=\begin{pmatrix}\widetilde{z}_1\\ \widetilde{z}_2\end{pmatrix},\quad u(t)=\begin{pmatrix}z_1(t)\\ z_2(t)\end{pmatrix},\quad A(t)=\begin{pmatrix}P_1(t) & -P_2(t)\\ P_2(t) & P_1(t)\end{pmatrix} \quad \text{for } t\in[a,b].$$

Since P has bounded variation on $[a,b]$, it is clear that A has the same property. The condition $1+\Delta^+P(t)\neq 0$ implies

$$1+\Delta^+P_1(t)\neq 0 \quad \text{or} \quad \Delta^+P_2(t)\neq 0 \quad \text{for} \quad t\in[a,t_0),$$

and similarly $1-\Delta^-P(t)\neq 0$ implies

$$1-\Delta^-P_1(t)\neq 0 \quad \text{or} \quad \Delta^-P_2(t)\neq 0 \quad \text{for} \quad t\in(t_0,b].$$

In view of this, we have

$$\det(I+\Delta^+A(t))=(1+\Delta^+P_1(t))^2+(\Delta^+P_2(t))^2\neq 0 \quad \text{for} \quad t\in[a,t_0),$$
$$\det(I-\Delta^-A(t))=(1-\Delta^-P_1(t))^2+(\Delta^-P_2(t))^2\neq 0 \quad \text{for} \quad t\in(t_0,b].$$

Hence, existence and uniqueness of solution to equation (8.5.3) follows from Theorem 7.4.5. By Theorem 7.4.1, the solution has bounded variation on $[a,b]$.

If P and $\widetilde{z}$ are real, the equation for z_2 simplifies to $z_2(t)=\int_{t_0}^{t} z_2\,\mathrm{d}P_1$, whose solution is identically zero and therefore z is real. □

The previous existence and uniqueness theorem guarantees that the definition of the generalized exponential function is meaningful.

8.5.2 Definition. Consider a function $P\colon[a,b]\to\mathbb{C}$, which has bounded variation on $[a,b]$. Let $t_0\in[a,b]$ and assume that $1+\Delta^+P(t)\neq 0$ for every $t\in[a,t_0)$, and $1-\Delta^-P(t)\neq 0$ for every $t\in(t_0,b]$. Then we define the generalized exponential function $t\mapsto e_{\mathrm{d}P}(t,t_0)$, $t\in[a,b]$, as the unique solution $z\colon[a,b]\to\mathbb{C}$ of the generalized linear differential equation

$$z(t)=1+\int_{t_0}^{t} z(s)\,\mathrm{d}P(s).$$

The next theorem summarizes the basic properties of the generalized exponential functions. Some of them were already established in Section 7.7 in the context of Cauchy matrices, but we repeat them here for reader's convenience. In each of the eight statements, we assume that the function P is such that all exponentials appearing in the given identity are defined. For example, in statement (v), it is necessary to assume that

$$1+\Delta^+P(\tau)\neq 0 \quad \text{for every} \quad \tau\in[a,\max\{s,r\}),$$
$$1-\Delta^-P(\tau)\neq 0 \quad \text{for every} \quad \tau\in(\min\{s,t\},b].$$

8.5.3 Theorem. *Let $P\colon[a,b]\to\mathbb{C}$ be a function with bounded variation. The generalized exponential function has the following properties:*

(i) *If P is constant, then $e_{\mathrm{d}P}(t,t_0)=1$ for every $t\in[a,b]$.*

(ii) $e_{\mathrm{d}P}(t,t)=1$ *for every $t\in[a,b]$.*

(iii) *The function $t\mapsto e_{\mathrm{d}P}(t,t_0)$ is regulated on $[a,b]$ and satisfies*

$$\begin{aligned}
\Delta^+ e_{\mathrm{d}P}(t,t_0)&=\Delta^+P(t)\,e_{\mathrm{d}P}(t,t_0) && \text{for } t\in[a,b),\\
\Delta^- e_{\mathrm{d}P}(t,t_0)&=\Delta^-P(t)\,e_{\mathrm{d}P}(t,t_0) && \text{for } t\in(a,b],\\
e_{\mathrm{d}P}(t+,t_0)&=(1+\Delta^+P(t))\,e_{\mathrm{d}P}(t,t_0) && \text{for } t\in[a,b),\\
e_{\mathrm{d}P}(t-,t_0)&=(1-\Delta^-P(t))\,e_{\mathrm{d}P}(t,t_0) && \text{for } t\in(a,b].
\end{aligned}$$

(iv) *The function $t\mapsto e_{\mathrm{d}P}(t,t_0)$ has bounded variation on $[a,b]$.*

(v) $e_{\mathrm{d}P}(t,s)\,e_{\mathrm{d}P}(s,r)=e_{\mathrm{d}P}(t,r)$ *for every $t,s,r\in[a,b]$.*

(vi) $e_{\mathrm{d}P}(t,s)=e_{\mathrm{d}P}(s,t)^{-1}$ *for every t, $s\in[a,b]$.*

(vii) $\overline{e_{\mathrm{d}P}(t,t_0)}=e_{\mathrm{d}\overline{P}}(t,t_0)$ *for every $t\in[a,b]$, where $\overline{z}$ denotes the complex conjugate of $z\in\mathbb{C}$.*

(viii) *If P is continuous, then $e_{\mathrm{d}P}(t,t_0)=e^{P(t)-P(t_0)}$ for every $t\in[a,b]$.*

Proof. The first two statements are obvious. Statement (iii) is a consequence of Corollary 6.5.5 (extended to complex-valued case), and statement (iv) follows from Theorem 8.5.1.

To prove statement (v), note that, given arbitrary $r,s\in[a,b]$, we have

$$\begin{aligned}
e_{\mathrm{d}P}(t,r)&=1+\int_r^t e_{\mathrm{d}P}(\tau,r)\,\mathrm{d}P(\tau)\\
&=1+\int_r^s e_{\mathrm{d}P}(\tau,r)\,\mathrm{d}P(\tau)+\int_s^t e_{\mathrm{d}P}(\tau,r)\,\mathrm{d}P(\tau)\\
&=e_{\mathrm{d}P}(s,r)+\int_s^t e_{\mathrm{d}P}(\tau,r)\,\mathrm{d}P(\tau)
\end{aligned}$$

for every $t\in[a,b]$. Hence, the function $y(t)=e_{\mathrm{d}P}(t,r)$ is a solution on $[a,b]$ of the generalized linear differential equation

$$x(t)=\widetilde{x}+\int_s^t x(s)\,\mathrm{d}P(s),\qquad \text{where } \ \widetilde{x}=e_{\mathrm{d}P}(s,r).$$

On the other hand, it is not hard to see that $z(t)=e_{\mathrm{d}P}(t,s)\,\widetilde{x}$ for $t\in[a,b]$

is also a solution of the same equation. By the uniqueness of solutions, we have $y(t) = z(t)$ for all $t \in [a, b]$, which proves statement (v).

Statement (vi) is a direct consequence of statement (v). Indeed, for $t, s \in [a, b]$, we obtain

$$e_{\mathrm{d}P}(t, s)\, e_{\mathrm{d}P}(s, t) = e_{\mathrm{d}P}(t, t) = 1.$$

By the definition of the exponential function, we have

$$e_{\mathrm{d}P}(t, t_0) = 1 + \int_{t_0}^{t} e_{\mathrm{d}P}(s, t_0)\, \mathrm{d}P(s).$$

Taking the complex conjugate of both sides, we get

$$\overline{e_{\mathrm{d}P}(t, t_0)} = 1 + \int_{t_0}^{t} \overline{e_{\mathrm{d}P}(s, t_0)}\, \mathrm{d}\overline{P}(s),$$

which proves statement (vii).

To prove statement (viii), assume that P is a continuous function with bounded variation. Let $\widetilde{P}(t) = P(t) - P(t_0)$ and $z(t) = e^{\widetilde{P}(t)}$ for all $t \in [a, b]$. Using the uniform convergence theorem (Theorem 5.7.1) and the third part of Lemma 7.5.2 (which holds also for complex functions), we get

$$\begin{aligned} 1 + \int_{t_0}^{t} z(s)\, \mathrm{d}P(s) &= 1 + \int_{t_0}^{t} z(s)\, \mathrm{d}\widetilde{P}(s) \\ &= 1 + \int_{t_0}^{t} \Big(\sum_{k=0}^{\infty} \frac{\widetilde{P}(s)^k}{k!} \Big)\, \mathrm{d}\widetilde{P}(s) = 1 + \sum_{k=0}^{\infty} \Big(\frac{1}{k!} \int_{t_0}^{t} \widetilde{P}(s)^k\, \mathrm{d}\widetilde{P}(s) \Big) \\ &= 1 + \sum_{k=0}^{\infty} \frac{1}{k!} \frac{\widetilde{P}(t)^{k+1} - \widetilde{P}(t_0)^{k+1}}{k+1} = 1 + \sum_{k=0}^{\infty} \frac{\widetilde{P}(t)^{k+1}}{(k+1)!} = z(t) \end{aligned}$$

for all $t \in [a, b]$. It follows that $e_{\mathrm{d}P}(t, t_0) = z(t) = e^{P(t) - P(t_0)}$. □

As an immediate consequence of parts (iii) and (vi) of Theorem 8.5.3, we see that the generalized exponential function is regulated with respect to both arguments; this fact will be used in the proof of the next result.

We now derive an explicit formula for the value of $e_{\mathrm{d}P}(t, t_0)$. It is sufficient to focus on the case when $t > t_0$; the formula for the case $t < t_0$ then follows easily from the identity $e_{\mathrm{d}P}(t, t_0) = e_{\mathrm{d}P}(t_0, t)^{-1}$.

The next theorem involves infinite products of (possibly complex) numbers. Recall that if $\sum_{k=1}^{\infty} a_k$ is an absolutely convergent series of complex numbers, then the infinite product $\prod_{k=1}^{\infty} (1 + a_k)$ is also absolutely convergent; in particular, the product converges to the same value after an arbitrary rearrangement of the sequence $\{a_k\}$. Now, suppose that $P\colon [a,b]\to\mathbb{R}$ is a function of bounded variation with infinitely many discontinuity points in (a,b), which are arranged in a sequence $\{s_k\}$. By Theorem 2.3.6, the sum $\sum_{k=1}^{\infty}(|\Delta^+P(s_k)|+|\Delta^-P(s_k)|)$ is finite. Hence, the products $\prod_{k=1}^{\infty}(1+\Delta^+P(s_k))$ and $\prod_{k=1}^{\infty}(1-\Delta^-P(s_k))$ are absolutely convergent. Since the order of the factors is unimportant, we introduce a similar convention as in Remark 2.3.7, and write

$$\prod_{x\in(a,b)}(1+\Delta^+P(x))=\prod_{k=1}^{\infty}(1+\Delta^+P(s_k)),$$

$$\prod_{x\in(a,b)}(1-\Delta^-P(x))=\prod_{k=1}^{\infty}(1-\Delta^-P(s_k)).$$

The symbols

$$\prod_{x\in[a,b)},\quad \prod_{x\in(a,b]},\quad \text{and}\quad \prod_{x\in[a,b]}$$

should be understood in an analogous way.

8.5.4 Theorem. *If $t>t_0$, then*

$$e_{\mathrm{d}P}(t,t_0)=\frac{e^{P(t-)-P(t_0+)}}{e^{\sum_{\tau\in(t_0,t)}\Delta P(\tau)}}\,\frac{\prod_{\tau\in[t_0,t)}(1+\Delta^+P(\tau))}{\prod_{\tau\in(t_0,t]}(1-\Delta^-P(\tau))}. \tag{8.5.4}$$

Proof. We begin by verifying the formula in the case when P has only finitely many discontinuities in $[t_0,t]$. Let $\boldsymbol{\alpha}$ be a division of $[t_0,t]$ containing all discontinuity points of P. Then P is continuous on each interval (α_{k-1},α_k), where $k\in\{1,\dots,\nu(\boldsymbol{\alpha})\}$. Parts (ii), (iii), (v), (vi), (viii) of Theorem 8.5.3 imply

$$\begin{aligned}
e_{\mathrm{d}P}(t,t_0)&=\prod_{k=1}^{\nu(\boldsymbol{\alpha})}e_{\mathrm{d}P}(\alpha_k,\alpha_{k-1})\\
&=\prod_{k=1}^{\nu(\boldsymbol{\alpha})}e_{\mathrm{d}P}(\alpha_k,\alpha_k-)\,e_{\mathrm{d}P}(\alpha_k-,\alpha_{k-1}+)\,e_{\mathrm{d}P}(\alpha_{k-1}+,\alpha_{k-1})\\
&=\prod_{k=1}^{\nu(\boldsymbol{\alpha})}(1-\Delta^-P(\alpha_k))^{-1}\,e^{P(\alpha_k-)-P(\alpha_{k-1}+)}\,(1+\Delta^+P(\alpha_{k-1}))\\
&=\frac{e^{P(t-)-P(t_0+)}}{e^{\sum_{k=1}^{\nu(\boldsymbol{\alpha})-1}\Delta P(\alpha_k)}}\,\frac{\prod_{k=1}^{\nu(\boldsymbol{\alpha})}(1+\Delta^+P(\alpha_{k-1}))}{\prod_{k=1}^{\nu(\boldsymbol{\alpha})}(1-\Delta^-P(\alpha_k))},
\end{aligned}$$

which agrees with (8.5.4).

Now, assume that P has infinitely many discontinuities in $[t_0,t]$, and let $D=\{s_k\}$ be a sequence of all discontinuity points contained in (t_0,t). According to Theorem 2.6.1 and its proof (which remains valid for complex functions), we have the decomposition $P=P^{\mathrm{C}}+P^{\mathrm{B}}$, where

$$\begin{aligned}P^{\mathrm{B}}(\tau)&=P(t_0)+\Delta^+P(t_0)\,\chi_{(t_0,t]}(\tau)+\Delta^-P(t)\,\chi_{[t]}(\tau)\\&\quad+\sum_{k=1}^{\infty}\Big(\Delta^+P(s_k)\,\chi_{(s_k,t]}(\tau)+\Delta^-P(s_k)\,\chi_{[s_k,t]}(\tau)\Big)\quad\text{for}\quad\tau\in[t_0,t],\end{aligned}$$

is the jump part of P, and $P^{\mathrm{C}}=P-P^{\mathrm{B}}$ is the continuous part of P. For each $n\in\mathbb{N}$, let $P_n=P^{\mathrm{C}}+P_n^{\mathrm{B}}$, where

$$\begin{aligned}P_n^{\mathrm{B}}(\tau)&=P(t_0)+\Delta^+P(t_0)\,\chi_{(t_0,t]}(\tau)+\Delta^-P(t)\,\chi_{[t]}(\tau)\\&\quad+\sum_{k=1}^{n}\Big(\Delta^+P(s_k)\,\chi_{(s_k,t]}(\tau)+\Delta^-P(s_k)\,\chi_{[s_k,t]}(\tau)\Big)\quad\text{for}\quad\tau\in[t_0,t].\end{aligned}$$

For each $n\in\mathbb{N}$, we have $\Delta^+P_n(t_0)=\Delta^+P(t_0)$ and $\Delta^-P_n(t)=\Delta^-P(t)$; moreover, $\Delta^+P_n(s_k)=\Delta^+P(s_k)$ and $\Delta^-P_n(s_k)=\Delta^-P(s_k)$ for all $k\le n$. Since the discontinuities of P_n are contained in the finite set $\{t_0,s_1,\dots,s_n,t\}$, we obtain

$$\begin{aligned}e_{\mathrm{d}P_n}(t,t_0)&=\frac{e^{P_n(t-)-P_n(t_0+)}}{e^{\sum_{\tau\in(t_0,t)}\Delta P_n(\tau)}}\,\frac{1+\Delta^+P_n(t_0)}{1-\Delta^-P_n(t)}\,\frac{\prod_{\tau\in(t_0,t)}(1+\Delta^+P_n(\tau))}{\prod_{\tau\in(t_0,t)}(1-\Delta^-P_n(\tau))}\\&=\frac{e^{P_n(t-)-P_n(t_0+)}}{e^{\sum_{k=1}^{n}\Delta P(s_k)}}\,\frac{1+\Delta^+P(t_0)}{1-\Delta^-P(t)}\,\frac{\prod_{k=1}^{n}(1+\Delta^+P(s_k))}{\prod_{k=1}^{n}(1-\Delta^-P(s_k))}.\end{aligned}$$

According to the proof of Lemma 2.6.5, the sequence $\{P_n\}$ is convergent to P in the BV norm. It follows that the variations of P_n, $n\in\mathbb{N}$, are uniformly bounded, and $\{P_n\}$ is uniformly convergent to P. Hence, Theorem 7.6.3 (continuous dependence of solutions to generalized linear differential equations with respect to the right-hand side) implies

$$\begin{aligned}e_{\mathrm{d}P}(t,t_0)&=\lim_{n\to\infty}e_{\mathrm{d}P_n}(t,t_0)\\&=\frac{e^{P(t-)-P(t_0+)}}{e^{\sum_{k=1}^{\infty}\Delta P(s_k)}}\,\frac{1+\Delta^+P(t_0)}{1-\Delta^-P(t)}\,\frac{\prod_{k=1}^{\infty}(1+\Delta^+P(s_k))}{\prod_{k=1}^{\infty}(1-\Delta^-P(s_k))}.\end{aligned}$$

The series $\sum_{k=1}^{\infty} \Delta P(s_k)$ and the two infinite products on the right-hand side are absolutely convergent, which proves that formula (8.5.4) holds. □

8.5.5 Corollary. *If $P:[t_0,t]\to\mathbb{R}$ has bounded variation, is left-continuous, and $1+\Delta^+P(\tau)>0$ for all $\tau\in[t_0,t)$, then*

$$0<e_{dP}(t,t_0)\le e^{P(t)-P(t_0)}.$$

Proof. Using the formula from Theorem 8.5.4 and left-continuity of P, we get

$$\begin{aligned} e_{dP}(t,t_0) &= \frac{e^{P(t)-P(t_0+)}\prod_{\tau\in[t_0,t)}(1+\Delta^+P(\tau))}{\exp\left(\sum_{\tau\in(t_0,t)}\Delta^+P(\tau)\right)} \\ &= \frac{e^{P(t)-P(t_0)}\prod_{\tau\in[t_0,t)}(1+\Delta^+P(\tau))}{\exp\left(\sum_{\tau\in[t_0,t)}\Delta^+P(\tau)\right)} \\ &= \frac{e^{P(t)-P(t_0)}\prod_{\tau\in[t_0,t)}(1+\Delta^+P(\tau))}{\prod_{\tau\in[t_0,t)}\exp\left(\Delta^+P(\tau)\right)}. \end{aligned}$$

Consequently, if $1+\Delta^+P(\tau)>0$ for all $\tau\in[t_0,t)$, then

$$0<\frac{\prod_{\tau\in[t_0,t)_{\mathbb{T}}}(1+\Delta^+P(t))}{\prod_{\tau\in[t_0,t)_{\mathbb{T}}}\exp\left(\Delta^+P(t)\right)}\le 1,$$

and the proof is complete. □

In the next theorem, we prove that the product of two exponentials e_{dP} and e_{dQ} equals the exponential of a certain function denoted by $P\oplus Q$. We have $P\oplus Q=P+Q$ if P, Q are continuous, but the general definition of $P\oplus Q$ is more complicated and takes into account the jumps of P and Q. We make the following agreement: If $c>d$, then a sum of the form $\sum_{s\in[c,d)} h(s)$ should be interpreted as $-\sum_{s\in(d,c]} h(s)$, and the sum $\sum_{s\in(c,d]} h(s)$ should be understood as $-\sum_{s\in[d,c)} h(s)$.

8.5.6 Theorem. *Assume that $P,Q:[a,b]\to\mathbb{C}$ have bounded variation and*

$$(1+\Delta^+P(t))\,(1+\Delta^+Q(t))\neq 0 \quad \textit{for every} \;\; t\in[a,t_0),$$
$$(1-\Delta^-P(t))\,(1-\Delta^-Q(t))\neq 0 \quad \textit{for every} \;\; t\in(t_0,b].$$

Then

$$e_{\mathrm{d}P}(t,t_0)\, e_{\mathrm{d}Q}(t,t_0) = e_{\mathrm{d}(P\oplus Q)}(t,t_0), \quad t\in[a,b],$$

where

$$(P\oplus Q)(t) = P(t) + Q(t) + \int_{t_0}^{t} \Delta^+ Q(s)\,\mathrm{d}P(s) - \int_{t_0}^{t} \Delta^- P(s)\,\mathrm{d}Q(s), \quad (8.5.5)$$

with the convention that $\Delta^+Q(t)=0$, $\Delta^-P(t_0)=0$. *Equivalently, we have*

$$(P\oplus Q)(t) = P(t) + Q(t) + \sum_{s\in[t_0,t)} \Delta^+Q(s)\,\Delta^+P(s) - \sum_{s\in(t_0,t]} \Delta^-Q(s)\,\Delta^-P(s).$$

Proof. By Corollary 6.3.17, we have

$$\begin{aligned}
&\int_{t_0}^{t} \Delta^+ Q(s)\,\mathrm{d}P(s) - \int_{t_0}^{t} \Delta^- P(s)\,\mathrm{d}Q(s)\\
&\quad = \sum_{s\in(t_0,t)} \big(\Delta^+Q(s)\,\Delta P(s) - \Delta^-P(s)\,\Delta Q(s)\big)\\
&\qquad\quad + \Delta^+Q(t_0)\,\Delta^+P(t_0) - \Delta^-P(t)\,\Delta^-Q(t)\\
&\quad = \sum_{s\in(t_0,t)} \big(\Delta^+Q(s)\,(\Delta^+P(s)+\Delta^-P(s))\\
&\qquad\quad - \Delta^-P(s)\,(\Delta^+Q(s)+\Delta^-Q(s))\big)\\
&\qquad\quad + \Delta^+Q(t_0)\,\Delta^+P(t_0) - \Delta^-P(t)\,\Delta^-Q(t)\\
&\quad = \sum_{s\in[t_0,t)} \Delta^+Q(s)\,\Delta^+P(s) - \sum_{s\in(t_0,t]} \Delta^-Q(s)\,\Delta^-P(s).
\end{aligned}$$

Hence, the two formulas for $P\oplus Q$ are equivalent. For $t\in[a,b]$, let

$$R(t) = \sum_{s\in(t_0,t]} \Delta^-Q(s)\,\Delta^-P(s) \quad\text{and}\quad T(t) = \sum_{s\in[t_0,t)} \Delta^+Q(s)\,\Delta^+P(s).$$

Using the definition of variation together with Corollary 2.3.8, it is not difficult to check that R,T have bounded variation. Furthermore, the definitions of R,T imply

$$\begin{aligned}
&\Delta^-R(t) = \Delta^-Q(t)\,\Delta^-P(t), \quad \Delta^+R(t)=0,\\
&\Delta^-T(t)=0, \quad \Delta^+T(t) = \Delta^+Q(t)\,\Delta^+P(t).
\end{aligned}$$

In view of this, it is clear that $P \oplus Q$ has bounded variation on $[a,b]$. Moreover,

$$\begin{aligned} 1-\Delta^-(P\oplus Q)(t) &= 1-\Delta^- P(t)-\Delta^- Q(t)-\Delta^- T(t)+\Delta^- R(t) \\ &= 1-\Delta^- P(t)-\Delta^- Q(t)+\Delta^- Q(t)\,\Delta^- P(t) \\ &= \big(1-\Delta^- P(t)\big)\big(1-\Delta^- Q(t)\big)\neq 0 \quad \text{for } t\in(t_0,b]. \end{aligned}$$

Proceeding in a similar way, one can show that

$$1+\Delta^+(P\oplus Q)(t)=\big(1+\Delta^+ P(t)\big)\big(1+\Delta^+ Q(t)\big)\neq 0 \quad \text{for} \quad t\in[a,t_0).$$

Therefore, the exponential function $t\mapsto e_{\mathrm{d}(P\oplus Q)}(t,t_0)$ is defined.

For $t\in[a,b]$, Theorem 6.4.2 (integration by parts) gives

$$\begin{aligned} &e_{\mathrm{d}P}(t,t_0)\,e_{\mathrm{d}Q}(t,t_0)=e_{\mathrm{d}P}(t_0,t_0)\,e_{\mathrm{d}Q}(t_0,t_0) \\ &\quad+\int_{t_0}^t e_{\mathrm{d}P}(s,t_0)\,\mathrm{d}[e_{\mathrm{d}Q}(s,t_0)]+\int_{t_0}^t e_{\mathrm{d}Q}(s,t_0)\,\mathrm{d}[e_{\mathrm{d}P}(s,t_0)] \\ &\quad+\sum_{s\in[t_0,t)}\Delta^+ e_{\mathrm{d}P}(s,t_0)\,\Delta^+ e_{\mathrm{d}Q}(s,t_0)-\sum_{s\in(t_0,t]}\Delta^- e_{\mathrm{d}P}(s,t_0)\,\Delta^- e_{\mathrm{d}Q}(s,t_0). \end{aligned}$$

Let us examine the terms on the right-hand side. Obviously,

$$e_{\mathrm{d}P}(t_0,t_0)e_{\mathrm{d}Q}(t_0,t_0)=1.$$

Using Theorem 6.6.1 (substitution theorem), we have

$$\begin{aligned} \int_{t_0}^t e_{\mathrm{d}P}(s,t_0)\,\mathrm{d}[e_{\mathrm{d}Q}(s,t_0)] &= \int_{t_0}^t e_{\mathrm{d}P}(s,t_0)\,\mathrm{d}\Big[1+\int_{t_0}^s e_{\mathrm{d}Q}(u,t_0)\,\mathrm{d}Q(u)\Big] \\ &= \int_{t_0}^t e_{\mathrm{d}P}(s,t_0)e_{\mathrm{d}Q}(s,t_0)\,\mathrm{d}Q(s), \end{aligned}$$

and

$$\begin{aligned} \int_{t_0}^t e_{\mathrm{d}Q}(s,t_0)\,\mathrm{d}[e_{\mathrm{d}P}(s,t_0)] &= \int_{t_0}^t e_{\mathrm{d}Q}(s,t_0)\,\mathrm{d}\Big[1+\int_{t_0}^s e_{\mathrm{d}P}(u,t_0)\,\mathrm{d}P(u)\Big] \\ &= \int_{t_0}^t e_{\mathrm{d}Q}(s,t_0)\,e_{\mathrm{d}P}(s,t_0)\,\mathrm{d}P(s). \end{aligned}$$

Finally, by performing an algebraic manipulation and using subsequently Corollary 6.3.17, part (iii) of Theorem 8.5.3 and finally Theorem 6.6.1 (substitution theorem), we get

$$\sum_{s\in[t_0,t)}\Delta^+ e_{\mathrm{d}P}(s,t_0)\,\Delta^+ e_{\mathrm{d}Q}(s,t_0)-\sum_{s\in(t_0,t]}\Delta^- e_{\mathrm{d}P}(s,t_0)\,\Delta^- e_{\mathrm{d}Q}(s,t_0)$$

$$
\begin{aligned}
&= \Delta^+ e_{\mathrm{d}P}(t_0,t_0)\,\Delta^+ e_{\mathrm{d}Q}(t_0,t_0) \\
&\qquad + \sum_{s\in(t_0,t)} [\Delta^+ e_{\mathrm{d}P}(s,t_0) + \Delta^- e_{\mathrm{d}P}(s,t_0)]\,\Delta^+ e_{\mathrm{d}Q}(s,t_0) \\
&\qquad - \sum_{s\in(t_0,t)} \Delta^- e_{\mathrm{d}P}(s,t_0)\,[\Delta^- e_{\mathrm{d}Q}(s,t_0) + \Delta^+ e_{\mathrm{d}Q}(s,t_0)] \\
&\qquad - \Delta^- e_{\mathrm{d}P}(t,t_0)\,\Delta^- e_{\mathrm{d}Q}(t,t_0)) \\
&= \Delta^+ e_{\mathrm{d}P}(t_0,t_0)\,\Delta^+ e_{\mathrm{d}Q}(t_0,t_0) \\
&\qquad + \sum_{s\in(t_0,t)} \Delta e_{\mathrm{d}P}(s,t_0)\,\Delta^+ e_{\mathrm{d}Q}(s,t_0) \\
&\qquad - \sum_{s\in(t_0,t)} \Delta^- e_{\mathrm{d}P}(s,t_0)\,\Delta e_{\mathrm{d}Q}(s,t_0) - \Delta^- e_{\mathrm{d}P}(t,t_0)\,\Delta^- e_{\mathrm{d}Q}(t,t_0)) \\
&= \int_{t_0}^{t} \Delta^+ e_{\mathrm{d}Q}(s,t_0)\,\mathrm{d}[e_{\mathrm{d}P}(s,t_0)] - \int_{t_0}^{t} \Delta^- e_{\mathrm{d}P}(s,t_0)\,\mathrm{d}[e_{\mathrm{d}Q}(s,t_0)] \\
&= \int_{t_0}^{t} \Delta^+ Q(s) e_{\mathrm{d}Q}(s,t_0)\,e_{\mathrm{d}P}(s,t_0)\,\mathrm{d}P(s) \\
&\qquad - \int_{t_0}^{t} \Delta^- P(s) e_{\mathrm{d}P}(s,t_0)\,e_{\mathrm{d}Q}(s,t_0)\,\mathrm{d}Q(s) \\
&= \int_{t_0}^{t} e_{\mathrm{d}Q}(s,t_0)\,e_{\mathrm{d}P}(s,t_0)\,\mathrm{d}\Big[\int_{t_0}^{s} \Delta^+ Q(u)\,\mathrm{d}P(u) - \int_{t_0}^{s} \Delta^- P(u)\,\mathrm{d}Q(u)\Big].
\end{aligned}
$$

By combining the previous results, we obtain

$$
e_{\mathrm{d}P}(t,t_0)\,e_{\mathrm{d}Q}(t,t_0) = 1 + \int_{t_0}^{t} e_{\mathrm{d}P}(s,t_0)\,e_{\mathrm{d}Q}(s,t_0)\,\mathrm{d}\,[(P\oplus Q)(s)]\,,
$$

with $P\oplus Q$ given by (8.5.5). □

8.5.7 Exercise. Obviously, the binary operation $\oplus$ introduced in the previous theorem is commutative. Verify that

$$
\begin{aligned}
((P\oplus Q)\oplus R)(t) &= (P\oplus(Q\oplus R))(t) = P(t)+Q(t)+R(t) \\
&\quad + \sum_{s\in[t_0,t)} \Big(\Delta^+ Q(s)\,\Delta^+ P(s) + \Delta^+ R(s)\,\Delta^+ P(s) + \Delta^+ Q(s)\,\Delta^+ R(s)\Big) \\
&\quad + \sum_{s\in[t_0,t)} \Delta^+ P(s)\,\Delta^+ Q(s)\,\Delta^+ R(s) \\
&\quad - \sum_{s\in(t_0,t]} \Big(\Delta^- Q(s)\,\Delta^- P(s) + \Delta^- R(s)\,\Delta^- P(s) + \Delta^- Q(s)\,\Delta^- R(s)\Big)
\end{aligned}
$$

$$+\sum_{s\in(t_0,t]} \Delta^- P(s)\,\Delta^- Q(s)\,\Delta^- R(s),$$

i.e., the operation $\oplus$ is also associative.

The next result shows that the reciprocal value of an exponential function is again an exponential function.

8.5.8 Theorem. *Assume that $P\colon[a,b]\to\mathbb{C}$ has bounded variation and*

$$1+\Delta^+ P(t)\neq 0 \quad \textit{for every} \quad t\in[a,b),$$
$$1-\Delta^- P(t)\neq 0 \quad \textit{for every} \quad t\in(a,b].$$

Then

$$(e_{\mathrm{d}P}(t,t_0))^{-1}=e_{\mathrm{d}(\ominus P)}(t,t_0) \quad \textit{for} \quad t\in[a,b], \tag{8.5.6}$$

where

$$(\ominus P)(t)=-P(t)+\sum_{s\in[t_0,t)}\frac{(\Delta^+ P(s))^2}{1+\Delta^+ P(s)}-\sum_{s\in(t_0,t]}\frac{(\Delta^- P(s))^2}{1-\Delta^- P(s)}.$$

Proof. For $t\in[a,b]$, we have $(\ominus P)(t)=-P(t)+R_1(t)-R_2(t)$, where

$$R_1(t)=\sum_{s\in[t_0,t)}\frac{(\Delta^+ P(s))^2}{1+\Delta^+ P(s)}, \quad R_2(t)=\sum_{s\in(t_0,t]}\frac{(\Delta^- P(s))^2}{1-\Delta^- P(s)}.$$

These functions have bounded variation on $[a,b]$ and satisfy

$$\begin{aligned} &\Delta^- R_1(t)=0, && \Delta^+ R_1(t)=\frac{(\Delta^+ P(t))^2}{1+\Delta^+ P(t)},\\ &\Delta^+ R_2(t)=0, && \Delta^- R_2(t)=\frac{(\Delta^- P(t))^2}{1-\Delta^- P(t)}. \end{aligned} \tag{8.5.7}$$

Thus, $\ominus P$ has bounded variation on $[a,b]$ and

$$1+\Delta^+(\ominus P)(t)=1-\Delta^+ P(t)+\frac{(\Delta^+ P(t))^2}{1+\Delta^+ P(t)}=\frac{1}{1+\Delta^+ P(t)}\neq 0 \quad \text{for } t\in[a,t_0),$$
$$1-\Delta^-(\ominus P)(t)=1+\Delta^- P(t)+\frac{(\Delta^- P(t))^2}{1-\Delta^- P(t)}=\frac{1}{1-\Delta^- P(t)}\neq 0 \quad \text{for } t\in(t_0,b],$$

which implies that the exponential function $e_{\mathrm{d}(\ominus P)}$ is defined.

Using the relations (8.5.7) together with the definition of $\oplus$ given in Theorem 8.5.6, we obtain

$$
\begin{aligned}
(P\oplus(\ominus P))(t) &= P(t)-P(t)+R_1(t)-R_2(t)\\
&\quad+\sum_{s\in[t_0,t)}\Delta^+(-P+R_1-R_2)(s)\,\Delta^+P(s)\\
-\sum_{s\in(t_0,t]}\Delta^-(-P+R_1-R_2)(s)\,\Delta^-P(s) &= R_1(t)-R_2(t)\\
&\quad+\sum_{s\in[t_0,t)}\Big(-(\Delta^+P(s))^2+\frac{(\Delta^+P(s))^3}{1+\Delta^+P(s)}\Big)\\
&\quad-\sum_{s\in(t_0,t]}\Big(-(\Delta^-P(s))^2-\frac{(\Delta^-P(s))^3}{1-\Delta^-P(s)}\Big)\\
=R_1(t)-R_2(t)-\sum_{s\in[t_0,t)}\frac{(\Delta^+P(s))^2}{1+\Delta^+P(s)}&+\sum_{s\in(t_0,t]}\frac{(\Delta^-P(s))^2}{1-\Delta^-P(s)}=0.
\end{aligned}
$$

Now, it follows from Theorem 8.5.6 and part (i) of Theorem 8.5.3 that

$$e_{\mathrm{d}P}(t,t_0)\,e_{\mathrm{d}(\ominus P)}(t,t_0)=e_{\mathrm{d}(P\oplus(\ominus P))}(t,t_0)=1,$$

which proves the relation (8.5.6). □

Our next goal is to investigate the sign of the generalized exponential function corresponding to a real function P.

8.5.9 Theorem. *Consider a function $P\colon[a,b]\to\mathbb{R}$, which has bounded variation and*

$$1+\Delta^+P(t)\neq 0\quad\textit{for every}\quad t\in[a,b),$$

$$1-\Delta^-P(t)\neq 0\quad\textit{for every}\quad t\in(a,b].$$

Then, for every $t_0\in[a,b]$, the following statements hold:

(i) $e_{\mathrm{d}P}(t,t_0)\neq 0$ *for all* $t\in[a,b]$.

(ii) *If* $t\in[a,b)$ *and* $1+\Delta^+P(t)<0$, *then* $e_{\mathrm{d}P}(t,t_0)\,e_{\mathrm{d}P}(t+,t_0)<0$.

(iii) *If* $t\in(a,b]$ *and* $1-\Delta^-P(t)<0$, *then* $e_{\mathrm{d}P}(t,t_0)\,e_{\mathrm{d}P}(t-,t_0)<0$.

(iv) *If* $t\in(a,b)$, $1+\Delta^+P(t)>0$ *and* $1-\Delta^-P(t)>0$, *then* $t\mapsto e_{\mathrm{d}P}(t,t_0)$ *does not change sign in the neighborhood of* t.

Proof. If $e_{dP}(t,t_0)=0$ for a certain $t\in[a,b]$, we can use Theorem 8.5.3 to obtain

$$1=e_{dP}(t_0,t_0)=e_{dP}(t_0,t)\,e_{dP}(t,t_0)=0,$$

which is a contradiction. This proves statement (i). Statements (ii) and (iii) follow immediately from part (iii) of Theorem 8.5.3. Finally, if $1+\Delta^+P(t)>0$ and $1-\Delta^-P(t)>0$, then $e_{dP}(t+,t_0)$ and $e_{dP}(t-,t_0)$ have the same sign as $e_{dP}(t,t_0)$, which proves (iv). □

According to the previous theorem, the exponential function changes sign at all points t such that $1+\Delta^+P(t)<0$ or $1-\Delta^-P(t)<0$. Since P has bounded variation, we conclude that the interval $[a,b]$ can contain only finitely many points where the exponential function changes its sign.

The next theorem describes the class of all real functions P for which the generalized exponential function remains positive.

8.5.10 Theorem. *Let $\mathcal{P}_+$ be the class consisting of all functions $P:[a,b]\to\mathbb{R}$ that have bounded variation and satisfy $1+\Delta^+P(t)>0$ for every $t\in[a,b)$, and $1-\Delta^-P(t)>0$ for every $t\in(a,b]$. Then the following statements hold:*

(i) *$P\in\mathcal{P}_+$ if and only if $1+\Delta^+P(t)\neq 0$ for every $t\in[a,b)$, $1-\Delta^-P(t)\neq 0$ for every $t\in(a,b]$, and the inequality $e_{dP}(t,t_0)>0$ holds for all $t,t_0\in[a,b]$.*

(ii) *If $P,Q\in\mathcal{P}_+$, then $P\oplus Q\in\mathcal{P}_+$.*

(iii) *If $P\in\mathcal{P}_+$, then $\ominus P\in\mathcal{P}_+$.*

Proof. The first statement follows from Theorem 8.5.9 and the fact that $e_{dP}(t_0,t_0)$ is positive. Statement (ii) is a consequence of the formulas

$$1-\Delta^-(P\oplus Q)(t)=\left(1-\Delta^-P(t)\right)\left(1-\Delta^-Q(t)\right),$$

$$1+\Delta^+(P\oplus Q)(t)=\left(1+\Delta^+P(t)\right)\left(1+\Delta^+Q(t)\right),$$

which were obtained in the proof of Theorem 8.5.6. Similarly, the last statement is a consequence of the formulas

$$1+\Delta^+(\ominus P)(t)=\frac{1}{1+\Delta^+P(t)},\quad 1-\Delta^-(\ominus P)(t)=\frac{1}{1-\Delta^-P(t)},$$

which were obtained in the proof of Theorem 8.5.8. □

Using the exponential function, we can now introduce the generalized hyperbolic functions.

8.5.11 Definition. Consider a function $P\colon [a,b]\to\mathbb{C}$, which has bounded variation on $[a,b]$. Let $t_0\in[a,b]$ and assume that

$$1-(\Delta^+P(t))^2\neq 0 \quad \text{for} \quad t\in[a,t_0),$$
$$1-(\Delta^-P(t))^2\neq 0 \quad \text{for} \quad t\in(t_0,b].$$

Then we define the generalized hyperbolic functions $\cosh_{\mathrm{d}P}$ and $\sinh_{\mathrm{d}P}$ by the formulas

$$\cosh_{\mathrm{d}P}(t,t_0)=\frac{e_{\mathrm{d}P}(t,t_0)+e_{\mathrm{d}(-P)}(t,t_0)}{2} \quad \text{for} \quad t\in[a,b],$$
$$\sinh_{\mathrm{d}P}(t,t_0)=\frac{e_{\mathrm{d}P}(t,t_0)-e_{\mathrm{d}(-P)}(t,t_0)}{2} \quad \text{for} \quad t\in[a,b].$$

Note that the condition $1-(\Delta^+P(t))^2\neq 0$ is equivalent to

$$(1+\Delta^+P(t))(1+\Delta^+(-P)(t))\neq 0,$$

and $1-(\Delta^-P(t))^2\neq 0$ is equivalent to

$$(1-\Delta^-P(t))(1-\Delta^-(-P)(t))\neq 0.$$

Therefore, $e_{\mathrm{d}P}$ and $e_{\mathrm{d}(-P)}$ are well defined.

Obviously, the two hyperbolic functions are real if P is real, and for $P(s)=s$, we obtain the classical hyperbolic functions:

$$\cosh_{\mathrm{d}P}(t,t_0)=\cosh(t-t_0), \quad \sinh_{\mathrm{d}P}(t,t_0)=\sinh(t-t_0).$$

More generally, if P is continuous, then

$$\cosh_{\mathrm{d}P}(t,t_0)=\cosh(P(t)-P(t_0)) \quad \text{and} \quad \sinh_{\mathrm{d}P}(t,t_0)=\sinh(P(t)-P(t_0)).$$

In the next theorem, we obtain the analogues of the well-known formulas

$$(\cosh z)'=\sinh z, \quad (\sinh z)'=\cosh z \quad \text{and} \quad \cosh^2 z-\sinh^2=1.$$

8.5.12 Theorem. *Consider a function $P\colon [a,b]\to\mathbb{C}$, which has bounded variation and*

$$1-(\Delta^+P(t))^2\neq 0 \quad \textit{for every } t\in[a,t_0),$$
$$1-(\Delta^-P(t))^2\neq 0 \quad \textit{for every } t\in(t_0,b].$$

The generalized hyperbolic functions have the following properties:

(i) $\cosh_{\mathrm{d}P}(t_0,t_0)=1$, $\sinh_{\mathrm{d}P}(t_0,t_0)=0$.

(ii) $\cosh_{\mathrm{d}P}(t,t_0)=1+\int_{t_0}^{t}\sinh_{\mathrm{d}P}(s,t_0)\,\mathrm{d}P(s)$ *for* $t\in[a,b]$.

(iii) $\sinh_{\mathrm{d}P}(t,t_0)=\int_{t_0}^{t}\cosh_{\mathrm{d}P}(s,t_0)\,\mathrm{d}P(s)$ *for* $t\in[a,b]$.

(iv) $\cosh^2_{\mathrm{d}P}(t,t_0)-\sinh^2_{\mathrm{d}P}(t,t_0)=e_{\mathrm{d}Q}(t,t_0)$ *for* $t\in[a,b]$, *where*

$$Q(t)=(P\oplus(-P))(t)=\int_{t_0}^{t}(\Delta^-P(s)-\Delta^+P(s))\,\mathrm{d}P(s)$$
$$=\sum_{s\in(t_0,t]}(\Delta^-P(s))^2-\sum_{s\in[t_0,t)}(\Delta^+P(s))^2$$

with the convention that $\Delta^+P(t)=0$, $\Delta^-P(t_0)=0$.

Proof. The first statement is obvious. Using the definition of the generalized exponential function, we obtain

$$\cosh_{\mathrm{d}P}(t,t_0)=\frac{1}{2}\Big(1+\int_{t_0}^{t}e_{\mathrm{d}P}(s,t_0)\,\mathrm{d}P(s)+1+\int_{t_0}^{t}e_{\mathrm{d}(-P)}(s,t_0)\,\mathrm{d}(-P)(s)\Big)$$
$$=1+\frac{1}{2}\int_{t_0}^{t}(e_{\mathrm{d}P}(s,t_0)-e_{\mathrm{d}(-P)}(s,t_0))\,\mathrm{d}P(s)=1+\int_{t_0}^{t}\sinh_{\mathrm{d}P}(s,t_0)\,\mathrm{d}P(s),$$

which proves statement (ii). Similarly we ca prove statement (iii). To verify the last one, observe that

$$\cosh^2_{\mathrm{d}P}(t,t_0)-\sinh^2_{\mathrm{d}P}(t,t_0)$$
$$=\Big(\frac{e_{\mathrm{d}P}(t,t_0)+e_{\mathrm{d}(-P)}(t,t_0)}{2}\Big)^2-\Big(\frac{e_{\mathrm{d}P}(t,t_0)-e_{\mathrm{d}(-P)}(t,t_0)}{2}\Big)^2$$
$$=e_{\mathrm{d}P}(t,t_0)e_{\mathrm{d}(-P)}(t,t_0)=e_{\mathrm{d}(P\oplus(-P))}(t,t_0).$$

From Theorem 8.5.6, we have

$$(P\oplus(-P))(t)=-\int_{t_0}^{t}\Delta^+P(s)\,\mathrm{d}P(s)+\int_{t_0}^{t}\Delta^-P(s)\,\mathrm{d}P(s)$$
$$=-\sum_{s\in[t_0,t)}(\Delta^+P(s))^2+\sum_{s\in(t_0,t]}(\Delta^-P(s))^2,$$

which completes the proof. □

Finally, we introduce the generalized trigonometric functions.

8.5.13 Definition. Consider a function $P\colon [a,b] \to \mathbb{C}$, which has bounded variation on $[a,b]$. Let $t_0 \in [a,b]$ and assume that

$$1-(\Delta^+ P(t))^2 \neq 0 \quad \text{for every } t \in [a,t_0),$$

$$1-(\Delta^- P(t))^2 \neq 0 \quad \text{for every } t \in (t_0,b].$$

Then we define the generalized trigonometric functions $\cos_{\mathrm{d}P}$ and $\sin_{\mathrm{d}P}$ by the formulas

$$\cos_{\mathrm{d}P}(t,t_0) = \frac{e_{\mathrm{d}(\mathrm{i}\,P)}(t,t_0) + e_{\mathrm{d}(-\mathrm{i}\,P)}(t,t_0)}{2} = \cosh_{\mathrm{d}(iP)}(t,t_0) \qquad \text{for } t \in [a,b],$$

$$\sin_{\mathrm{d}P}(t,t_0) = \frac{e_{\mathrm{d}(\mathrm{i}\,P)}(t,t_0) - e_{\mathrm{d}(-\mathrm{i}\,P)}(t,t_0)}{2\mathrm{i}} = -i\,\sinh_{\mathrm{d}(iP)}(t,t_0) \quad \text{for } t \in [a,b].$$

Note that the condition $1+(\Delta^+ P(t))^2 \neq 0$ is equivalent to

$$(1+\Delta^+(\mathrm{i}\,P)(t))(1+\Delta^+(-\mathrm{i}\,P)(t)) \neq 0,$$

and $1+(\Delta^- P(t))^2 \neq 0$ is equivalent to

$$(1-\Delta^-(\mathrm{i}\,P)(t))(1-\Delta^-(-\mathrm{i}\,P)(t)) \neq 0.$$

Therefore, $e_{\mathrm{d}(\mathrm{i}\,P)}$ and $e_{\mathrm{d}(-\mathrm{i}\,P)}$ are well defined. If P is a real function, both conditions are always satisfied.

Again, it is easy to see that for $P(s)=s$, our definitions coincide with the classical trigonometric functions:

$$\cos_{\mathrm{d}P}(t,t_0) = \cos(t-t_0), \quad \sin_{\mathrm{d}P}(t,t_0) = \sin(t-t_0).$$

Also, if P is continuous, then

$$\cos_{\mathrm{d}P}(t,t_0) = \cos(P(t)-P(t_0)) \quad \text{and} \quad \sin_{\mathrm{d}P}(t,t_0) = \sin(P(t)-P(t_0)).$$

If P is real, the trigonometric functions are real as well: By part (vii) of Theorem 8.5.3,

$$e_{\mathrm{d}(\mathrm{i}\,P)} + e_{\mathrm{d}(-\mathrm{i}\,P)} = e_{\mathrm{d}(\mathrm{i}\,P)} + \overline{e_{\mathrm{d}(\mathrm{i}\,P)}},$$

which is purely real. Similarly,

$$e_{\mathrm{d}(\mathrm{i}\,P)} - e_{\mathrm{d}(-\mathrm{i}\,P)} = e_{\mathrm{d}(\mathrm{i}\,P)} - \overline{e_{\mathrm{d}(\mathrm{i}\,P)}},$$

which is purely imaginary.

We now derive the analogues of the well-known formulas

$$(\cos z)' = -\sin z, \quad (\sin z)' = \cos z \quad \text{and} \quad \cos^2 z + \sin^2 = 1.$$

8.5.14 Theorem. *Consider a function $P\colon [a,b] \to \mathbb{C}$, which has bounded variation and*

$$1 + (\Delta^+ P(t))^2 \neq 0 \quad \textit{for every } t \in [a, t_0),$$
$$1 + (\Delta^- P(t))^2 \neq 0 \quad \textit{for every } t \in (t_0, b].$$

The generalized trigonometric functions have the following properties:

(i) $\cos_{\mathrm{d}P}(t_0, t_0) = 1$, $\sin_{\mathrm{d}P}(t_0, t_0) = 0$.

(ii) $\cos_{\mathrm{d}P}(t, t_0) = 1 - \int_{t_0}^t \sin_{\mathrm{d}P}(s, t_0)\,\mathrm{d}P(s)$ *for* $t \in [a, b]$.

(iii) $\sin_{\mathrm{d}P}(t, t_0) = \int_{t_0}^t \cos_{\mathrm{d}P}(s, t_0)\,\mathrm{d}P(s)$ *for* $t \in [a, b]$.

(iv) $\cos^2_{\mathrm{d}P}(t, t_0) + \sin^2_{\mathrm{d}P}(t, t_0) = e_{\mathrm{d}Q}(t, t_0)$ *for* $t \in [a, b]$, *where*

$$Q(t) = (\mathrm{i}\,P \oplus (-\mathrm{i}\,P))(t) = \int_{t_0}^t (\Delta^+ P(s) - \Delta^- P(s))\,\mathrm{d}P(s)$$
$$= \sum_{s \in [t_0, t)} (\Delta^+ P(s))^2 - \sum_{s \in (t_0, t]} (\Delta^- P(s))^2$$

with the convention that $\Delta^+ P(t) = 0$, $\Delta^- P(t_0) = 0$.

Proof. The first statement is obvious. Using the definition of the generalized exponential function, we get

$$\cos_{\mathrm{d}P}(t, t_0) = \frac{1}{2}\left(1 + \int_{t_0}^t e_{\mathrm{d}(\mathrm{i}\,P)}(s, t_0)\,\mathrm{d}[\mathrm{i}\,P(s)] + 1 + \int_{t_0}^t e_{\mathrm{d}(-\mathrm{i}\,P)}(s, t_0)\,\mathrm{d}[-\mathrm{i}\,P(s)]\right)$$
$$= 1 + \frac{\mathrm{i}}{2}\int_{t_0}^t (e_{\mathrm{d}(\mathrm{i}\,P)}(s, t_0) - e_{\mathrm{d}(-\mathrm{i}\,P)}(s, t_0))\,\mathrm{d}P(s)$$
$$= 1 + \frac{\mathrm{i}}{2}\int_{t_0}^t (2\mathrm{i}\ \sin_{\mathrm{d}P}(s, t_0))\,\mathrm{d}P(s) = 1 - \int_{t_0}^t \sin_{\mathrm{d}P}(s, t_0)\,\mathrm{d}P(s),$$

which proves statement (ii). The proof of statement (iii) is similar. To prove the last one, we observe that

$$\begin{aligned}
&\cos^2_{\mathrm{d}P}(t,t_0)+\sin^2_{\mathrm{d}P}(t,t_0)\\
&=\Big(\frac{e_{\mathrm{d}(\mathrm{i}\,P)}(t,t_0)+e_{\mathrm{d}(-\mathrm{i}\,P)}(t,t_0)}{2}\Big)^2+\Big(\frac{e_{\mathrm{d}(\mathrm{i}\,P)}(t,t_0)-e_{\mathrm{d}(-\mathrm{i}\,P)}(t,t_0)}{2\mathrm{i}}\Big)^2\\
&=e_{\mathrm{d}(\mathrm{i}\,P)}(t,t_0)\,e_{\mathrm{d}(-\mathrm{i}\,P)}(t,t_0)=e_{\mathrm{d}(\mathrm{i}\,P\oplus(-\mathrm{i}\,P))}(t,t_0).
\end{aligned}$$

From Theorem 8.5.6, we have

$$\begin{aligned}
(\mathrm{i}\,P\oplus(-\mathrm{i}\,P))(t)&=\int_{t_0}^{t}\Delta^+P(s)\,\mathrm{d}P(s)-\int_{t_0}^{t}\Delta^-P(s)\,\mathrm{d}P(s)\\
&=\sum_{s\in[t_0,t)}(\Delta^+P(s))^2-\sum_{s\in(t_0,t]}(\Delta^-P(s))^2,
\end{aligned}$$

which completes the proof. □

Bibliographic remarks

Most of the material in the present section is taken from the paper [Monteiro, Slavík (2014)]. The explicit formula (8.5.4) for the solution of a scalar generalized linear differential equation can be found in [Groh (1980)]; a slightly different but equivalent formula is available in [Kurzweil (2012)], where it is discussed in connection with the Gronwall inequality. The result presented there can be formulated as follows:

Assume that $K>0$, $h:[a,b]\to[0,\infty)$ is nondecreasing, $\Delta^-h(t)<1$ for all $t\in(a,b]$, and $\Delta^+h(t)<1$ for all $t\in[a,b)$. Moreover, suppose that $u:[a,b]\to[0,\infty)$ satisfies

$$u(t)\le K+\int_a^t u\,\mathrm{d}h \quad \textit{for} \quad t\in[a,b].$$

Then $u(t)\le K\,e_{\mathrm{d}h}(t,a)$ for all $t\in[a,b]$. Clearly, if h is left-continuous, then the last inequality combined with the estimate from Corollary 8.5.5 lead to the Gronwall inequality presented in Theorem 7.5.3. A similar result can be also found in [Groh (1980)], where the condition $\Delta^+h(t)<1$ is omitted, but u is assumed to have bounded variation.

8.6 Integration on time scales

The time scale calculus, which originated in the work of S. Hilger [Hilger (1990)], is a popular tool that provides a unification of the continuous and discrete calculus. It is concerned with functions $f:\mathbb{T}\to\mathbb{R}$, where $\mathbb{T}$ is a *time scale* – an arbitrary nonempty closed set $\mathbb{T}\subset\mathbb{R}$. As we will see, the choice $\mathbb{T}=\mathbb{R}$ leads to the classical continuous calculus, while $\mathbb{T}=\mathbb{Z}$ corresponds to the discrete calculus. Another frequently studied time scale is $\mathbb{T}=q^{\mathbb{Z}}=\{q^n:n\in\mathbb{Z}\}$, where $q>1$; this leads to the quantum calculus. The basic operations of the time scale calculus are the Δ-derivative, ∇-derivative, Δ-integral, and ∇-integral. The main goal of this section is to show that both types of integrals on time scales are special cases of the Kurzweil-Stieltjes integral.

Let us start by introducing some basic notation. If $t\in\mathbb{T}$ and $t<\sup\mathbb{T}$, we denote

$$\sigma(t)=\inf\{s\in\mathbb{T}:s>t\},\qquad \mu(t)=\sigma(t)-t.$$

Moreover, if $t=\sup\mathbb{T}<\infty$, we define $\sigma(t)=t$, $\mu(t)=0$. The functions $\sigma:\mathbb{T}\to\mathbb{T}$ and $\mu:\mathbb{T}\to[0,\infty)$ are referred to as the *forward jump operator* and *forward graininess*, respectively. If $t\in\mathbb{T}$ satisfies $\sigma(t)>t$, we say that t is *right-scattered*; otherwise, if $\sigma(t)=t$ and $t<\sup\mathbb{T}$, then t is called *right-dense.*

Similarly, if $t\in\mathbb{T}$ and $t>\inf\mathbb{T}$, let

$$\rho(t)=\sup\{s\in\mathbb{T}:s<t\},\qquad \nu(t)=t-\rho(t).$$

Moreover, if $t=\inf\mathbb{T}>-\infty$, we define $\rho(t)=t$, $\nu(t)=0$. The functions $\rho:\mathbb{T}\to\mathbb{T}$ and $\nu:\mathbb{T}\to[0,\infty)$ are called the *backward jump operator* and *backward graininess*, respectively. If $t\in\mathbb{T}$ satisfies $\rho(t)<t$, we say that t is *left-scattered*; otherwise, if $\rho(t)=t$ and $t>\inf\mathbb{T}$, then t is called *left-dense.*

For an arbitrary pair of real numbers $a<b$, we use the following notation:

$$[a,b]_{\mathbb{T}}=[a,b]\cap\mathbb{T},\ [a,b)_{\mathbb{T}}=[a,b)\cap\mathbb{T},\ (a,b]_{\mathbb{T}}=(a,b]\cap\mathbb{T},\ (a,b)_{\mathbb{T}}=(a,b)\cap\mathbb{T}.$$

These sets are referred to as the time scale intervals, and the subscript $\mathbb{T}$ helps to distinguish them from ordinary intervals.

A function $f:\mathbb{T}\to\mathbb{R}$ is called *rd-continuous*, if it is continuous at all right-dense points and regulated on $\mathbb{T}$. Similarly, f is called *ld-continuous*, if it is continuous at all left-dense points and regulated on $\mathbb{T}$.

Given a function $f:\mathbb{T}\to\mathbb{R}$, we can introduce the Δ-derivative and the ∇-derivative of f at a point $t\in\mathbb{T}$. Although our main interest lies in integration theory, we include the definitions of both derivatives and some of their properties (for more details, see [Bohner, Peterson (2001); Bohner, Guseinov (2003)]).

8.6.1 Definition. Consider a function $f:\mathbb{T}\to\mathbb{R}$ and a point $t\in\mathbb{T}$.

(i) Suppose that $t<\sup\mathbb{T}$, or $t=\sup\mathbb{T}$ and $\rho(t)=t$. We say that the Δ – derivative $f^{\Delta}(t)$ exists and equals $D\in\mathbb{R}$, if for every $\varepsilon>0$, there is a $\delta>0$ such that

$$|f(\sigma(t))-f(s)-D(\sigma(t)-s)|<\varepsilon|\sigma(t)-s|$$

for all $s\in(t-\delta,t+\delta)_{\mathbb{T}}$.

(ii) Suppose that $t>\inf\mathbb{T}$, or $t=\inf\mathbb{T}$ and $\sigma(t)=t$. We say that the ∇ – derivative $f^{\nabla}(t)$ exists and equals $D\in\mathbb{R}$, if for every $\varepsilon>0$, there is a $\delta>0$ such that

$$|f(\rho(t))-f(s)-D(\rho(t)-s)|<\varepsilon|\rho(t)-s|$$

for all $s\in(t-\delta,t+\delta)_{\mathbb{T}}$.

The following remarks should help to clarify the meaning of both derivatives, as well as the difference between them:

- If $\mathbb{T}=\mathbb{R}$, then $f^{\Delta}(t)=f^{\nabla}(t)=f'(t)$, i.e., both derivatives coincide with the classical derivative.
- If $\mathbb{T}=\mathbb{Z}$, then $f^{\Delta}(t)=f(t+1)-f(t)$ and $f^{\nabla}(t)=f(t)-f(t-1)$, i.e., the Δ- and ∇-derivative reduce to the forward difference and backward difference, respectively.
- More generally, if $t\in\mathbb{T}$ satisfies $\sigma(t)>t$, then

$$f^{\Delta}(t)=\frac{f(\sigma(t))-f(t)}{\sigma(t)-t}=\frac{f(\sigma(t))-f(t)}{\mu(t)}.$$

Similarly, if $t\in\mathbb{T}$ satisfies $\rho(t)<t$, then

$$f^{\nabla}(t)=\frac{f(t)-f(\rho(t))}{t-\rho(t)}=\frac{f(t)-f(\rho(t))}{\nu(t)}.$$

We now proceed to Δ- and ∇-integrals of a function $f:[a,b]_{\mathbb{T}}\to\mathbb{R}$, which are in a certain sense inverse operations to the Δ- and ∇-derivatives. As in the classical calculus, there exist definitions in the spirit of Newton, Riemann, Lebesgue, and Kurzweil. Moreover, these integrals also have their Stieltjes-type counterparts. We are primarily interested in Kurzweil integrals, but it is instructive to begin with Riemann integrals.

In the rest of this section, we always assume that $a, b \in \mathbb{T}$. A partition of $[a,b]_{\mathbb{T}}$ is a partition $P=(\boldsymbol{\alpha},\boldsymbol{\xi})$ of $[a,b]$ such that $\boldsymbol{\alpha}\subset\mathbb{T}$ and the components of the vector $\boldsymbol{\xi}$ are elements of $\mathbb{T}$. In certain situations we might also make use of the alternative interpretation of a partition as a collection of interval-point pairs (see Remark 5.1.2). We keep the notation used earlier in this book and write

$$S(P)=\sum_{j=1}^{\nu(\boldsymbol{\alpha})} f(\xi_j)(\alpha_j-\alpha_{j-1}).$$

(Throughout this section, the symbol ν has two different meanings: $\nu(\boldsymbol{\alpha})$ denotes the number of subintervals in a division $\boldsymbol{\alpha}$, while $\nu(t)$ is the backward graininess at a point $t\in\mathbb{T}$. The meaning of ν will be always clear from the context.)

The definition of the classical Riemann integral relies on partitions of $[a,b]$ such that the distance of each two successive division points does not exceed a certain $\delta>0$. However, if we replace $[a,b]$ by the time scale interval $[a,b]_{\mathbb{T}}$, such a partition need not exist. For this reason, the Riemann Δ- and ∇-integrals involve a modified type of partitions that are described in the next definition.

8.6.2 Definition. Given a $\delta>0$, the symbol $\mathcal{P}_\delta([a,b]_{\mathbb{T}})$ will denote all partitions $P=(\boldsymbol{\alpha},\boldsymbol{\xi})$ of $[a,b]_{\mathbb{T}}$ such that for each $i\in\{1,\dots,\nu(\boldsymbol{\alpha})\}$ we have either $\alpha_i-\alpha_{i-1}\le\delta$, or $\sigma(\alpha_{i-1})=\alpha_i$.

We are now able to define the Riemann Δ- and ∇-integrals.

8.6.3 Definition. Consider a function $f:[a,b]_{\mathbb{T}}\to\mathbb{R}$.

(i) We say that the Riemann Δ-integral $\int_a^b f(t)\,\Delta t$ exists and equals $I\in\mathbb{R}$, if for every $\varepsilon>0$, there is a $\delta>0$ such that $|S(P)-I|<\varepsilon$ for all partitions $P\in\mathcal{P}_\delta([a,b]_{\mathbb{T}})$ satisfying $\xi_i\in[\alpha_{i-1},\rho(\alpha_i)]_{\mathbb{T}}$ for each $i\in\{1,\dots,\nu(\boldsymbol{\alpha})\}$.

(ii) We say that the Riemann ∇-integral $\int_a^b f(t)\,\nabla t$ exists and equals $I\in\mathbb{R}$, if for every $\varepsilon>0$, there is a $\delta>0$ such that $|S(P)-I|<\varepsilon$ for all partitions $P\in\mathcal{P}_\delta([a,b]_{\mathbb{T}})$ satisfying $\xi_i\in[\sigma(\alpha_{i-1}),\alpha_i]_{\mathbb{T}}$ for each $i\in\{1,\dots,\nu(\boldsymbol{\alpha})\}$.

A complete theory of Riemann Δ- and ∇-integrals can be found in [Bohner, Guseinov (2003)]; here we limit ourselves to several remarks:

- If $\mathbb{T}=\mathbb{R}$, then
$$\int_a^b f(t)\,\Delta t=\int_a^b f(t)\,\nabla t=\int_a^b f(t)\,\mathrm{d}t,$$
i.e., both integrals coincide with the classical Riemann integral.

- The requirement $\xi_i\in[\alpha_{i-1},\rho(\alpha_i)]_{\mathbb{T}}$ in the definition of the Δ-integral means that if α_i is left-scattered, then it cannot serve as a tag for $[\alpha_{i-1},\alpha_i]_{\mathbb{T}}$. Thus, if $t\in\mathbb{T}$ satisfies $\sigma(t)>t$, then the only possible partition $P=(\boldsymbol{\alpha},\boldsymbol{\xi})$ that is taken into account in the definition of $\int_t^{\sigma(t)}f(t)\,\Delta t$ is $t=\alpha_0=\xi_1<\alpha_1=\sigma(t)$. Therefore,
$$\int_t^{\sigma(t)} f(t)\,\Delta t=f(t)\,(\sigma(t)-t)=f(t)\,\mu(t).$$
Similarly, the requirement $\xi_i\in[\sigma(\alpha_{i-1}),\alpha_i]_{\mathbb{T}}$ in the definition of the ∇-integral ensures that if α_{i-1} is right-scattered, then it cannot serve as a tag for $[\alpha_{i-1},\alpha_i]_{\mathbb{T}}$. Hence, if $t\in\mathbb{T}$ satisfies $\rho(t)<t$, the only possible partition $P=(\boldsymbol{\alpha},\boldsymbol{\xi})$ appearing in the definition of $\int_{\rho(t)}^t f(t)\,\nabla t$ is $\rho(t)=\alpha_0<\xi_1=\alpha_1=t$. Therefore,
$$\int_{\rho(t)}^t f(t)\,\nabla t=f(t)\,(t-\rho(t))=f(t)\,\nu(t).$$

- If $\mathbb{T}=\mathbb{Z}$, then $\int_a^b f(t)\,\Delta t=\sum_{t=a}^{b-1} f(t)$ and $\int_a^b f(t)\,\nabla t=\sum_{t=a+1}^{b} f(t)$ (see the previous remark; for $\delta\le 1$, there is only one possible choice for the partition of $[a,b]_{\mathbb{T}}$ appearing in Definition 8.6.3).

- If $f:[a,b]_{\mathbb{T}}\to\mathbb{R}$ is regulated, then it is Riemann Δ- and ∇-integrable, and the indefinite integrals
$$F_1(t)=\int_a^t f(s)\,\Delta s \quad\text{and}\quad F_2(t)=\int_a^t f(s)\,\nabla s,\qquad t\in[a,b]_{\mathbb{T}},$$
are continuous. Moreover, if f is rd-continuous, then $F_1^{\Delta}(t)=f(t)$ for all $t\in[a,b)_{\mathbb{T}}$; if f is ld-continuous, then $F_2^{\nabla}(t)=f(t)$ for all $t\in(a,b]_{\mathbb{T}}$.

- If F is continuous, $F^{\Delta}=f$ on $[a,b)_{\mathbb{T}}$ and f is Δ-integrable, then $\int_a^b f(t)\,\Delta t=F(b)-F(a)$. Similarly, if F is continuous, $F^{\nabla}=f$ on $(a,b]_{\mathbb{T}}$ and f is ∇-integrable, then $\int_a^b f(t)\,\nabla t=F(b)-F(a)$.

We now turn our attention to Kurzweil Δ- and ∇-integrals. As in the classical case, they have several advantages over the Riemann (or Lebesgue) integrals: the class of integrable functions is larger, and the assumptions of the fundamental theorem of calculus are less restrictive (each function which is a Δ- or ∇-derivative is Kurzweil Δ- or ∇-integrable). In time scale calculus, the concepts of a gauge on $[a,b]_{\mathbb{T}}$ and a δ-fine partition of $[a,b]_{\mathbb{T}}$ take the following form:

8.6.4 Definition. Consider a pair of functions $\delta_L, \delta_R : [a,b]_{\mathbb{T}} \to (0,\infty)$. Then $\delta = (\delta_L, \delta_R)$ is called a Δ-gauge on $[a,b]_{\mathbb{T}}$ if $\delta_R(t) \geq \mu(t)$ for all $t \in [a,b)_{\mathbb{T}}$, and a ∇-gauge on $[a,b]_{\mathbb{T}}$ if $\delta_L(t) \geq \nu(t)$ for all $t \in (a,b]_{\mathbb{T}}$. If $\delta = (\delta_L, \delta_R)$ is either a Δ-gauge or a ∇-gauge on $[a,b]_{\mathbb{T}}$, a partition $P = (\boldsymbol{\alpha}, \boldsymbol{\xi})$ of $[a,b]_{\mathbb{T}}$ is called δ-fine if

$$[\alpha_{j-1}, \alpha_j] \subset [\xi_j - \delta_L(\xi_j), \xi_j + \delta_R(\xi_j)] \quad \text{for all } j = 1, \ldots, \nu(\boldsymbol{\alpha}). \tag{8.6.1}$$

8.6.5 Remark. We emphasize that the interval on the right-hand side of (8.6.1) is closed and cannot be replaced by an open interval. This is in contrast to Definition 6.2.1, where it does not matter whether we choose an open or closed interval (see Remark 6.2.5) – both choices lead to the same definition of the Kurzweil-Stieltjes integral.

Instead of introducing the Kurzweil Δ- and ∇-integrals, we choose a more general approach in the spirit of this book, and define the Kurzweil-Stieltjes Δ- and ∇-integrals of a function $f : [a,b]_{\mathbb{T}} \to \mathbb{R}$ with respect to a function $g : [a,b]_{\mathbb{T}} \to \mathbb{R}$. As in the rest of the book, if $P = (\boldsymbol{\alpha}, \boldsymbol{\xi})$ is a partition of $[a,b]_{\mathbb{T}}$, we write

$$S(f, \mathrm{d}g, P) = \sum_{j=1}^{\nu(\boldsymbol{\alpha})} f(\xi_j)\,(g(\alpha_j) - g(\alpha_{j-1})).$$

8.6.6 Definition. Consider a pair of functions $f, g : [a,b]_{\mathbb{T}} \to \mathbb{R}$.

(i) We say that the Kurzweil-Stieltjes Δ-integral $\int_a^b f(t)\,\Delta g(t)$ exists and equals $I \in \mathbb{R}$, if for every $\varepsilon > 0$, there is a Δ-gauge δ on $[a,b]_{\mathbb{T}}$ such that

$$|S(f, \mathrm{d}g, P) - I| < \varepsilon$$

for all δ-fine partitions P of $[a,b]_{\mathbb{T}}$.

(ii) We say that the Kurzweil-Stieltjes ∇-integral $\int_a^b f(t)\,\nabla g(t)$ exists and equals $I \in \mathbb{R}$, if for every $\varepsilon > 0$, there is a ∇-gauge δ on $[a,b]_{\mathbb{T}}$

such that

$$|S(f, \mathrm{d}g, P) - I| < \varepsilon$$

for all δ-fine partitions P of $[a,b]_{\mathbb{T}}$.

If $g(t) = t$ for all $t \in [a,b]_{\mathbb{T}}$, the two integrals are referred to as the Kurzweil Δ-integral and ∇-integral, and they are denoted by $\int_a^b f(t)\,\Delta t$ and $\int_a^b f(t)\,\nabla t$, respectively. Let us show that Riemann integrability implies Kurzweil integrability.

8.6.7 Theorem. *Consider a function $f : [a,b]_{\mathbb{T}} \to \mathbb{R}$.*

(i) *If the Riemann Δ-integral $\int_a^b f(t)\,\Delta t$ exists, then the Kurzweil Δ-integral $\int_a^b f(t)\,\Delta t$ also exists and has the same value.*

(ii) *If the Riemann ∇-integral $\int_a^b f(t)\,\nabla t$ exists, then the Kurzweil ∇-integral $\int_a^b f(t)\,\nabla t$ also exists and has the same value.*

Proof. We prove the first statement, and leave the second up to the reader. Suppose that the Riemann Δ-integral $I = \int_a^b f(t)\,\Delta t$ exists, and choose an arbitrary $\varepsilon > 0$. By definition, there is a $\delta > 0$ such that

$$|S(P) - I| < \varepsilon$$

for all partitions $P = (\boldsymbol{\alpha}, \boldsymbol{\xi}) \in \mathcal{P}_\delta([a,b]_{\mathbb{T}})$ such that $\xi_i \in [\alpha_{i-1}, \rho(\alpha_i)]_{\mathbb{T}}$ for all $i \in \{1, \ldots, \nu(\boldsymbol{\alpha})\}$. We define a Δ-gauge $\widetilde{\delta} = (\widetilde{\delta}_L, \widetilde{\delta}_R)$ on $[a,b]_{\mathbb{T}}$ as follows:

$$\widetilde{\delta}_L(t) = \begin{cases} \nu(t)/2 & \text{if } t \in (a,b]_{\mathbb{T}} \text{ and } \rho(t) < t, \\ \delta & \text{otherwise,} \end{cases}$$

$$\widetilde{\delta}_R(t) = \begin{cases} \mu(t) & \text{if } t \in [a,b)_{\mathbb{T}} \text{ and } \sigma(t) > t, \\ \delta & \text{otherwise.} \end{cases}$$

Now, consider an arbitrary $\widetilde{\delta}$-fine partition $P = (\boldsymbol{\alpha}, \boldsymbol{\xi})$ of $[a,b]_{\mathbb{T}}$. By splitting each interval-point pair $([\alpha_{i-1}, \alpha_i], \xi_i)$ into $([\alpha_{i-1}, \xi_i], \xi_i)$ and $([\xi_i, \alpha_i], \xi_i)$, we get a partition $P' = (\boldsymbol{\alpha}', \boldsymbol{\xi}')$ which is still $\widetilde{\delta}$-fine and satisfies $S(P) = S(P')$. By the definition a $\widetilde{\delta}$-fine partition, we have

$$[\alpha'_{j-1}, \alpha'_j] \subset [\xi'_j - \widetilde{\delta}_L(\xi'_j), \xi'_j + \widetilde{\delta}_R(\xi'_j)] \quad \text{for all } j = 1, \ldots, \nu(\boldsymbol{\alpha}').$$

It follows from the construction of P' that for each $j = 1, \ldots, \nu(\boldsymbol{\alpha}')$, either $\xi'_j = \alpha'_{j-1}$, or $\xi'_j = \alpha'_j$. In the latter case, we have $\alpha'_{j-1} \in [\xi'_j - \widetilde{\delta}_L(\xi'_j), \xi'_j)_{\mathbb{T}}$.

Observe that if $\rho(\xi'_j) < \xi'_j$, then the interval $[\xi'_j - \widetilde{\delta}_L(\xi'_j), \xi'_j)_{\mathbb{T}}$ is empty (by the definition of $\widetilde{\delta}_L$). This shows that the tag ξ'_j cannot coincide with the right endpoint α'_j if it is left-scattered, i.e., we always have $\xi'_j \in [\alpha_{j-1}, \rho(\alpha_j)]_{\mathbb{T}}$.

Finally, observe that the inequality $\alpha'_j - \alpha'_{j-1} > \delta$ can be true only if $\xi'_j = \alpha'_{j-1}$ and $\alpha'_j = \sigma(\alpha'_{j-1})$; this shows that $P' \in \mathcal{P}_\delta([a,b]_{\mathbb{T}})$. Consequently,

$$|S(P) - I| = |S(P') - I| < \varepsilon,$$

i.e., the Kurzweil Δ-integral exists and equals the Riemann Δ-integral. □

Instead of developing a theory of Kurzweil-Stieltjes Δ- and ∇-integrals, we show that they are in fact special cases of the Kurzweil-Stieltjes integral from Definition 6.2.2. This means that all basic properties of the Kurzweil-Stieltjes Δ- and ∇-integrals can be obtained as corollaries of the results from Chapter 6.

First of all, we describe two possible ways of extending a function defined on the time scale interval $[a,b]_{\mathbb{T}}$ to the full interval $[a,b]$. For each $t \in [a,b]$, let

$$t^* = \inf\{s \in [a,b]_{\mathbb{T}} : s \geq t\}, \quad t_* = \sup\{s \in [a,b]_{\mathbb{T}} : s \leq t\}.$$

Note that if $t \in [a,b]_{\mathbb{T}}$, then $t^* = t_* = t$; otherwise, t_* and t^* are elements of $[a,b]_{\mathbb{T}}$ satisfying $t_* < t < t^*$. Now, for an arbitrary function $g : [a,b]_{\mathbb{T}} \to \mathbb{R}$, we define the extension $g^* : [a,b] \to \mathbb{R}$ by

$$g^*(t) = g(t^*) \quad \text{for all } t \in [a,b], \tag{8.6.2}$$

and the extension $g_* : [a,b] \to \mathbb{R}$ by

$$g_*(t) = g(t_*) \quad \text{for all } t \in [a,b]. \tag{8.6.3}$$

These two extensions appear in the statement of the next result.

8.6.8 Theorem. *Consider a pair of functions $f, g : [a,b]_{\mathbb{T}} \to \mathbb{R}$. Let $\widetilde{f} : [a,b] \to \mathbb{R}$ be an arbitrary function such that $\widetilde{f}(t) = f(t)$ for all $t \in [a,b]_{\mathbb{T}}$. Then the following statements hold:*

(i) *The integral $\int_a^b f(t)\,\Delta g(t)$ exists if and only if the integral $\int_a^b \widetilde{f}(t)\,\mathrm{d}g^*(t)$ exists; in this case, both integrals have the same value.*

(ii) *The integral $\int_a^b f(t)\,\nabla g(t)$ exists if and only if the integral $\int_a^b \widetilde{f}(t)\,\mathrm{d}g_*(t)$ exists; in this case, both integrals have the same value.*

Proof. We prove only the first statement; the proof of the second one is similar, and is left to the reader. We will repeatedly use the fact that if $P=(\boldsymbol{\alpha},\boldsymbol{\xi})$ is a partition of $[a,b]_{\mathbb{T}}$, then all the division points and tags are elements of $\mathbb{T}$, and therefore

$$\begin{aligned}S(\widetilde{f},\mathrm{d}g^*,P)&=\sum_{j=1}^{\nu(\boldsymbol{\alpha})}\widetilde{f}(\xi_j)\,(g^*(\alpha_j)-g^*(\alpha_{j-1}))\\&=\sum_{j=1}^{\nu(\boldsymbol{\alpha})}f(\xi_j)\,(g(\alpha_j)-g(\alpha_{j-1}))=S(f,\mathrm{d}g,P).\end{aligned}$$

Suppose first that $\int_a^b\widetilde{f}(t)\mathrm{d}g^*(t)$ exists. Given an arbitrary $\varepsilon>0$, there is a gauge $\widetilde{\delta}\colon[a,b]\to(0,\infty)$ such that $|S(\widetilde{f},\mathrm{d}g^*,P)-\int_a^b\widetilde{f}(t)\mathrm{d}g^*(t)|<\varepsilon$ for each $\widetilde{\delta}$-fine partition P of $[a,b]$. Now, let $\delta=(\delta_L,\delta_R)$ be a Δ-gauge on $[a,b]_{\mathbb{T}}$ given by $\delta_L(t)=\widetilde{\delta}(t)/2$ and $\delta_R(t)=\max(\widetilde{\delta}(t)/2,\mu(t))$ for all $t\in[a,b]_{\mathbb{T}}$. Let $P=(\boldsymbol{\alpha},\boldsymbol{\xi})$ be an arbitrary δ-fine partition of $[a,b]_{\mathbb{T}}$. Note that P need not be $\widetilde{\delta}$-fine: there might exist $i\in\{1,\dots,\nu(\boldsymbol{\alpha})\}$ such that $\alpha_i\ge\xi_i+\widetilde{\delta}(\xi_i)$. In this case, we necessarily have $\delta_R(\xi_i)=\mu(\xi_i)>0$, $\alpha_i=\sigma(\xi_i)$, and therefore $g^*(t)=\alpha_i$ for all $t\in(\xi_i,\alpha_i]$. After replacing each such interval-point pair $([\alpha_{i-1},\alpha_i],\xi_i)$ by the interval-point pair $([\alpha_{i-1},\xi_i+\widetilde{\delta}(\xi_i)/2],\xi_i)$ and an arbitrary $\widetilde{\delta}$-fine partition of $[\xi_i+\widetilde{\delta}(\xi_i)/2,\alpha_i]$, we get a partition P' of $[a,b]$, which is $\widetilde{\delta}$-fine and satisfies $S(\widetilde{f},\mathrm{d}g^*,P')=S(\widetilde{f},\mathrm{d}g^*,P)$. Hence, we have

$$\Big|S(f,\mathrm{d}g,P)-\int_a^b\widetilde{f}(t)\mathrm{d}g^*(t)\Big|=\Big|S(\widetilde{f},\mathrm{d}g^*,P')-\int_a^b\widetilde{f}(t)\mathrm{d}g^*(t)\Big|<\varepsilon,$$

which proves that $\int_a^b f(t)\,\Delta g(t)$ exists and equals $\int_a^b\widetilde{f}(t)\mathrm{d}g^*(t)$.

Conversely, assume that $\int_a^b f(t)\,\Delta g(t)$ exists. Given an arbitrary $\varepsilon>0$, there is a Δ-gauge $\delta=(\delta_L,\delta_R)$ on $[a,b]_{\mathbb{T}}$ such that

$$|S(f,\mathrm{d}g,P)-\int_a^b f(t)\,\Delta g(t)|<\varepsilon\quad\text{for each }\ \delta\text{-fine partition }P\text{ of }[a,b]_{\mathbb{T}}.$$

Now, let $\widetilde{\delta}\colon[a,b]\to(0,\infty)$ be a gauge given by

$$\widetilde{\delta}(t)=\begin{cases}\min(\delta_L(t),\sup[t,t+\delta_R(t)]_{\mathbb{T}}-t) & \text{if }\ t\in(a,b)_{\mathbb{T}},\\ \sup[a,a+\delta_R(a)]_{\mathbb{T}}-a & \text{if }\ t=a,\\ \delta_L(b) & \text{if }\ t=b,\\ \operatorname{dist}(t,[a,b]_{\mathbb{T}}) & \text{if }\ t\in[a,b]\setminus\mathbb{T}.\end{cases}$$

Let $P=(\boldsymbol{\alpha},\boldsymbol{\xi})$ be an arbitrary $\widetilde{\delta}$-fine partition of $[a,b]$. Note that P need not be a partition of $[a,b]_{\mathbb{T}}$ (i.e., the division points and tags need not be elements of $\mathbb{T}$). However, if we show that there exists a δ-fine partition P' of $[a,b]_{\mathbb{T}}$ such that $S(f,\mathrm{d}g,P')=S(\widetilde{f},\mathrm{d}g^*,P)$, then

$$\left|S(\widetilde{f},\mathrm{d}g^*,P)-\int_a^b f(t)\,\Delta g(t)\right|=\left|S(f,\mathrm{d}g,P)-\int_a^b f(t)\,\Delta g(t)\right|<\varepsilon,$$

and the proof will be complete.

Thus, suppose that P is not a partition of $[a,b]_{\mathbb{T}}$ (otherwise take $P'=P$). Since $\alpha_0=a\in\mathbb{T}$, there exists the smallest $i\in\{1,\dots,\nu(\boldsymbol{\alpha})\}$ such that $\alpha_{i-1}\in\mathbb{T}$, and at least one of α_i, ξ_i is not an element of $\mathbb{T}$. However, since the definition of $\widetilde{\delta}$ guarantees that we have either $\xi_i\in\mathbb{T}$, or $\xi_i\notin\mathbb{T}$ and $[\alpha_{i-1},\alpha_i]\cap\mathbb{T}=\emptyset$, we necessarily have $\xi_i\in[a,b)_{\mathbb{T}}$ and $\alpha_i\notin\mathbb{T}$. We now modify the partition P as follows:

- Replace the interval-point pair $([\alpha_{i-1},\alpha_i],\xi_i)$ by $([\alpha_{i-1},\alpha_i^*],\xi_i)$.
- Remove all interval-point pairs $([\alpha_{j-1},\alpha_j],\xi_j)$ for which the inclusion $[\alpha_{j-1},\alpha_j]\subset[\alpha_i,\alpha_i^*]$ is true.
- If there is an interval-point pair $([\alpha_{j-1},\alpha_j],\xi_j)$ such that

$$\alpha_i\le\alpha_{j-1}<\alpha_i^*<\alpha_j,$$

  replace it by $([\alpha_i^*,\alpha_j],\xi_j)$.

Denote the partition obtained in this way by $\widetilde{P}$. Because g^* is constant on $[\alpha_i,\alpha_i^*]$, it is clear that $S(\widetilde{f},\mathrm{d}g^*,P')=S(\widetilde{f},\mathrm{d}g^*,P)$. Since

$$\alpha_i<\xi_i+\widetilde{\delta}(\xi_i)\le\sup[\xi_i,\xi_i+\delta_R(\xi_i)]_{\mathbb{T}},$$

it follows from the definition of α_i^* that $\alpha_i^*\le\sup[\xi_i,\xi_i+\delta_R(\xi_i)]_{\mathbb{T}}$, and consequently

$$[\alpha_{i-1},\alpha_i^*]\subset[\xi_i-\delta_L(\xi_i),\xi_i+\delta_R(\xi_i)].$$

If necessary, we can repeat the procedure we have just described with P replaced by $\widetilde{P}$. The procedure does not increase the total number of division points, and decreases the number of division points that are not elements of $\mathbb{T}$. Thus, after finitely many steps, we obtain the desired δ-fine partition P' of $[a,b]_{\mathbb{T}}$ such that $S(f,\mathrm{d}g,P')=S(\widetilde{f},\mathrm{d}g^*,P')=S(\widetilde{f},\mathrm{d}g^*,P)$, and the proof is complete. $\square$

8.6.9 Corollary. *Consider a function* $f:[a,b]_{\mathbb{T}} \to \mathbb{R}$. *Let* $\widetilde{f}:[a,b] \to \mathbb{R}$ *be an arbitrary function such that* $\widetilde{f}(t)=f(t)$ *for all* $t \in [a,b]_{\mathbb{T}}$. *Also, let* $g(t)=t^*$ *and* $h(t)=t_*$ *for all* $t \in [a,b]$. *Then the following statements hold:*

(i) *The integral* $\int_a^b f(t)\,\Delta t$ *exists if and only if the integral* $\int_a^b \widetilde{f}(t)\,\mathrm{d}g(t)$ *exists; in this case, both integrals have the same value.*
(ii) *The integral* $\int_a^b f(t)\,\nabla t$ *exists if and only if the integral* $\int_a^b \widetilde{f}(t)\,\mathrm{d}h(t)$ *exists; in this case, both integrals have the same value.*

Bibliographic remarks

More information about the time scale calculus can be found e.g. in [Bohner, Peterson (2001); Hilger (1990)]. The basic reference on Riemann Δ- and ∇-integrals (as well as Lebesgue integrals on time scales) is [Bohner, Guseinov (2003)]. The definition presented there is slightly different from our Definition 8.6.3: The conditions $\xi_i \in [\alpha_{i-1}, \rho(\alpha_i)]_{\mathbb{T}}$ and $\xi_i \in [\sigma(\alpha_{i-1}), \alpha_i]_{\mathbb{T}}$ are replaced by $\xi_i \in [\alpha_{i-1}, \alpha_i)_{\mathbb{T}}$ and $\xi_i \in (\alpha_{i-1}, \alpha_i]_{\mathbb{T}}$, respectively. Nevertheless, the two definitions are equivalent; see Remark 5.17 in [Bohner, Guseinov (2003)].

Kurzweil Δ- and ∇-integrals were introduced in [Peterson, Thompson (2006)], and subsequently discussed in [Thomson (2008)]. In [Peterson, Thompson (2006)], the definition of a Δ-gauge and a ∇-gauge is slightly less restrictive as it allows the possibility that $\delta_L(a)=0$ and $\delta_R(b)=0$. However, one can easily see that this modification has no influence on the definition of a δ-fine partition of $[a,b]_{\mathbb{T}}$.

Riemann-Stieltjes Δ- and ∇-integrals were treated in [Mozyrska et al. (2010)]. As far as we are aware, there is no literature dealing with the Kurzweil-Stieltjes Δ- and ∇-integrals. Special cases of Corollary 8.6.9 with $\widetilde{f}(t)=f^*(t)$ were obtained in [Federson, Mesquita, Slavík (2013); Slavík (2012)].

8.7 Dynamic equations on time scales

The time scale calculus introduced in the previous section makes it possible to unify the theories of differential and difference equation by considering the so-called *dynamic equations*, where classical derivatives or differences are replaced by Δ- or ∇-derivatives. The general form of a Δ-dynamic equation is

$$x^{\Delta}(t) = f(x(t), t), \tag{8.7.1}$$

while ∇-dynamic equations have the form

$$x^{\nabla}(t) = f(x(t), t), \tag{8.7.2}$$

where in both cases $x : \mathbb{T} \to \mathbb{R}^n$ and $f : \mathbb{R}^n \times \mathbb{T} \to \mathbb{R}^n$. Integration leads to the equations

$$x(t) = x(t_0) + \int_{t_0}^{t} f(x(s), s)\,\Delta s, \tag{8.7.3}$$

$$x(t) = x(t_0) + \int_{t_0}^{t} f(x(s), s)\,\nabla s, \tag{8.7.4}$$

where $t_0, t \in \mathbb{T}$, and the integrals on the right-hand sides are the Kurzweil Δ- and ∇-integrals introduced in the previous section.

Equations (8.7.3) and (8.7.4) are the main subject of this section. Since the indefinite Kurzweil integrals are always continuous, each solution x of (8.7.3) or (8.7.4) necessarily has the same property. Moreover, if $s \mapsto f(x(s), s)$ is rd-continuous, then the integral on the right-hand side of (8.7.3) exists as a Riemann integral, it is Δ-differentiable, and the equation reduces back to (8.7.1); similarly, if $s \mapsto f(x(s), s)$ is ld-continuous, then the integral on the right-hand side of (8.7.4) is ∇-differentiable, and we get (8.7.2). Nevertheless, we focus on the more general integral equations (8.7.3) and (8.7.4) without imposing any continuity conditions on f.

Since we know that Kurzweil Δ- and ∇-integrals can be rewritten as Kurzweil-Stieltjes integrals, it comes as no surprise that equations (8.7.3) and (8.7.4) are equivalent to certain Kurzweil-Stieltjes integral equations. This relation is described in the next theorem, where we use the notation (8.6.2) and (8.6.3) from the previous section.

8.7.1 Theorem. *Suppose that $a, b, t_0 \in \mathbb{T}$, $a \le t_0 \le b$, and consider a function $f : B \times [a, b]_{\mathbb{T}} \to \mathbb{R}^n$, where $B \subset \mathbb{R}^n$. Let $\widetilde{f} : B \times [a, b] \to \mathbb{R}^n$ be an arbitrary function such that $\widetilde{f}(x, t) = f(x, t)$ for all $x \in B$, $t \in [a, b]_{\mathbb{T}}$. Also, let $g(t) = t^*$ and $h(t) = t_*$ for all $t \in [a, b]$. Then the following statements hold:*

(i) *If a function $x : [a, b]_{\mathbb{T}} \to B$ satisfies*

$$x(t) = x(t_0) + \int_{t_0}^{t} f(x(s), s)\,\Delta s, \quad t \in [a, b]_{\mathbb{T}}, \tag{8.7.5}$$

then the function $y:[a,b]\to B$ *given by* $y=x^*$ *satisfies*

$$y(t)=y(t_0)+\int_{t_0}^{t}\widetilde{f}(y(s),s)\,\mathrm{d}g(s),\quad t\in[a,b].\tag{8.7.6}$$

Conversely, each function $y:[a,b]\to B$ *satisfying* (8.7.6) *has the form* $y=x^*$, *where* $x:[a,b]_{\mathbb{T}}\to B$ *satisfies* (8.7.5).

(ii) *If a function* $x:[a,b]_{\mathbb{T}}\to B$ *satisfies*

$$x(t)=x(t_0)+\int_{t_0}^{t}f(x(s),s)\,\nabla s,\quad t\in[a,b]_{\mathbb{T}},\tag{8.7.7}$$

then the function $y:[a,b]\to B$ *given by* $y=x_*$ *satisfies*

$$y(t)=y(t_0)+\int_{t_0}^{t}\widetilde{f}(y(s),s)\,\mathrm{d}h(s),\quad t\in[a,b].\tag{8.7.8}$$

Conversely, each function $y:[a,b]\to B$ *satisfying* (8.7.8) *has the form* $y=x_*$, *where* $x:[a,b]_{\mathbb{T}}\to B$ *satisfies* (8.7.7).

Proof. Suppose that $x:[a,b]_{\mathbb{T}}\to B$ satisfies (8.7.5) and $y=x^*$. For each $t\in[a,b]$, Corollary 8.6.9 implies

$$\begin{aligned}y(t)=x(t^*)=x(t_0)+\int_{t_0}^{t^*}f(x(s),s)\,\Delta s\\=x^*(t_0)+\int_{t_0}^{t^*}\widetilde{f}(x^*(s),s)\,\mathrm{d}g(s)=y(t_0)+\int_{t_0}^{t^*}\widetilde{f}(y(s),s)\,\mathrm{d}g(s).\end{aligned}$$

Since the function g is constant on $[t,t^*]$, we have $\int_t^{t^*}\widetilde{f}(y(s),s)\,\mathrm{d}g(s)=0$, and therefore (8.7.6) holds.

Conversely, assume that $y:[a,b]\to B$ satisfies (8.7.6). Since the function g is constant on each interval $[t,t^*]$ with $t\in[a,b]$, it follows that y has the same property. Hence, $y=x^*$, where $x:[a,b]_{\mathbb{T}}\to B$ is the restriction of y to $[a,b]_{\mathbb{T}}$. For each $t\in[a,b]_{\mathbb{T}}$, Corollary 8.6.9 implies

$$\begin{aligned}x(t)=y(t)=y(t_0)+\int_{t_0}^{t}\widetilde{f}(y(s),s)\,\mathrm{d}g(s)\\=x^*(t_0)+\int_{t_0}^{t}\widetilde{f}(x^*(s),s)\,\mathrm{d}g(s)=x(t_0)+\int_{t_0}^{t}f(x(s),s)\,\Delta s.\end{aligned}$$

The proof of the second statement is similar and is left to the reader. □

A consequence of the previous theorem is that all results on generalized differential equations are directly applicable to dynamic equations on time scales. We illustrate this fact in the context of linear equations, and show how the theory developed in Chapter 7 leads to results on dynamic equations.

First, we establish the following useful lemma; it is formulated for scalar functions, but remains valid for vector and matrix functions.

8.7.2 Lemma. *Consider a function* $p\colon [a,b]_{\mathbb{T}} \to \mathbb{R}$, *and let* $\widetilde{p}\colon [a,b] \to \mathbb{R}$ *be an arbitrary function such that* $\widetilde{p}(t) = p(t)$ *for all* $t \in [a,b]_{\mathbb{T}}$.

(i) *Suppose that* p *is Kurzweil* Δ*-integrable, and there exists a Kurzweil* Δ*-integrable function* $m\colon [a,b]_{\mathbb{T}} \to \mathbb{R}$ *such that*

$$\Big|\int_u^v p(s)\,\Delta s\Big| \le \int_u^v m(s)\,\Delta s \quad \textit{whenever } u,v \in [a,b]_{\mathbb{T}},\ u \le v.$$

Let $P\colon [a,b] \to \mathbb{R}$ *be given by*

$$P(t) = \int_{t_0}^t \widetilde{p}(s)\,\mathrm{d}g(s) \quad \textit{for} \ \ t \in [a,b], \tag{8.7.9}$$

where $g(t) = t^*$ *for* $t \in [a,b]$. *Then* P *is left-continuous,*

$$\Delta^+ P(t) = \begin{cases} p(t)\,\mu(t) & \textit{if} \ \ t \in [a,b)_{\mathbb{T}}, \\ 0 & \textit{if} \ \ t \in [a,b) \backslash \mathbb{T} \end{cases}$$

and

$$\operatorname{var}_a^b P \le \int_a^b m(s)\,\Delta s.$$

(ii) *Suppose that* p *is Kurzweil* ∇*-integrable, and there exists a Kurzweil* ∇*-integrable function* $m\colon [a,b]_{\mathbb{T}} \to \mathbb{R}$ *such that*

$$\Big|\int_u^v p(s)\,\nabla s\Big| \le \int_u^v m(s)\,\nabla s \quad \textit{whenever } u,v \in [a,b]_{\mathbb{T}},\ u \le v.$$

Let $P\colon [a,b] \to \mathbb{R}$ *be given by*

$$P(t) = \int_{t_0}^t \widetilde{p}(s)\,\mathrm{d}h(s), \quad t \in [a,b], \tag{8.7.10}$$

where $h(t) = t_*$ *for all* $t \in [a,b]$. *Then* P *is right-continuous,*

$$\Delta^- P(t) = \begin{cases} p(t)\,\nu(t) & \text{if } t \in [a,b)_{\mathbb{T}}, \\ 0 & \text{if } t \in [a,b) \setminus \mathbb{T} \end{cases}$$

and

$$\operatorname{var}_a^b P \le \int_a^b m(s)\,\nabla s.$$

Proof. We prove only the first statement and leave the second to the reader. The function g is regulated (because it is nondecreasing) and left-continuous. According to Corollary 6.5.5, the function P given by (8.7.9) is regulated and left-continuous. Using the same corollary and the definition of g, we see that $\Delta^+ P(t) = p(t)\,\mu(t)$ for all $t \in [a,b)_{\mathbb{T}}$, and $\Delta^+ P(t) = 0$ otherwise.

Note that $P(t^*) = P(t)$ for each $t \in [a,b]$, since g is constant on $[t, t^*]$. Hence, if $\boldsymbol{\alpha}$ is an arbitrary division of $[a,b]$, we get

$$\begin{aligned} V(P,\alpha) &= \sum_{j=1}^{\nu(\boldsymbol{\alpha})} |P(\alpha_j) - P(\alpha_{j-1})| = \sum_{j=1}^{\nu(\boldsymbol{\alpha})} |P(\alpha_j^*) - P(\alpha_{j-1}^*)| \\ &= \sum_{j=1}^{\nu(\boldsymbol{\alpha})} \Big| \int_{\alpha_{j-1}^*}^{\alpha_j^*} \widetilde{p}(s)\,\mathrm{d}g(s) \Big| = \sum_{j=1}^{\nu(\boldsymbol{\alpha})} \Big| \int_{\alpha_{j-1}^*}^{\alpha_j^*} p(s)\,\Delta s \Big| \\ &\le \sum_{j=1}^{\nu(\boldsymbol{\alpha})} \int_{\alpha_{j-1}^*}^{\alpha_j^*} m(s)\,\Delta s = \int_a^b m(s)\,\Delta s, \end{aligned}$$

which shows that $\operatorname{var}_a^b P \le \int_a^b m(s)\,\Delta s$. □

Now, let us return back to dynamic equations on time scales. Let arbitrary functions $p\colon [a,b]_{\mathbb{T}} \to \mathscr{L}(\mathbb{R}^n)$ and $q\colon [a,b]_{\mathbb{T}} \to \mathbb{R}^n$ be given. If we let $f(x,t) = p(t)\,x + q(t)$ for all $x \in \mathbb{R}^n$ and $t \in [a,b]_{\mathbb{T}}$, equations (8.7.5) and (8.7.7) become

$$x(t) = x(t_0) + \int_{t_0}^t (p(s)\,x(s) + q(s))\,\Delta s \quad \text{for } t \in [a,b]_{\mathbb{T}}, \tag{8.7.11}$$

$$x(t) = x(t_0) + \int_{t_0}^t (p(s)\,x(s) + q(s))\,\nabla s \quad \text{for } t \in [a,b]_{\mathbb{T}}. \tag{8.7.12}$$

Thus, if $\widetilde{p}\colon [\alpha,\beta] \to \mathscr{L}(\mathbb{R}^n)$ and $\widetilde{q}\colon [\alpha,\beta] \to \mathbb{R}^n$ are arbitrary functions such that $\widetilde{p}(t) = p(t)$ and $\widetilde{q}(t) = q(t)$ for all $t \in [a,b]_{\mathbb{T}}$, Theorem 8.7.1 implies that equations (8.7.11), (8.7.12) are respectively equivalent to the Kurzweil-Stieltjes integral equations

$$y(t) = y(t_0) + \int_{t_0}^{t} (\widetilde{p}(s)\, y(s) + \widetilde{q}(s))\, \mathrm{d}g(s) \quad \text{for} \quad t \in [a,b], \tag{8.7.13}$$

$$y(t) = y(t_0) + \int_{t_0}^{t} (\widetilde{p}(s)\, y(s) + \widetilde{q}(s))\, \mathrm{d}h(s) \quad \text{for} \quad t \in [a,b], \tag{8.7.14}$$

with $g(t) = t^*$ and $h(t) = t_*$ for all $t \in [a,b]$. Using Theorem 6.6.1 (substitution theorem), we see that the last two equations can be rewritten as generalized linear differential equations

$$y(t) = y(t_0) + \int_{t_0}^{t} \mathrm{d}A_1(s)\, y(s) + B_1(t) - B_1(t_0) \quad \text{for} \quad t \in [a,b], \tag{8.7.15}$$

$$y(t) = y(t_0) + \int_{t_0}^{t} \mathrm{d}A_2(s)\, y(s) + B_2(t) - B_2(t_0) \quad \text{for} \quad t \in [a,b], \tag{8.7.16}$$

where $A_1, A_2 : [a,b] \to \mathscr{L}(\mathbb{R}^n)$ and $B_1, B_2 : [a,b] \to \mathbb{R}^n$ are given by

$$A_1(t) = \int_{t_0}^{t} \widetilde{p}(s)\, \mathrm{d}g(s), \quad B_1(t) = \int_{t_0}^{t} \widetilde{q}(s)\, \mathrm{d}g(s), \tag{8.7.17}$$

$$A_2(t) = \int_{t_0}^{t} \widetilde{p}(s)\, \mathrm{d}h(s), \quad B_2(t) = \int_{t_0}^{t} \widetilde{q}(s)\, \mathrm{d}h(s) \tag{8.7.18}$$

for all $t \in [a,b]$.

These considerations lead to the following existence and uniqueness theorem for linear dynamic equations.

8.7.3 Theorem. *The following statements hold:*

(i) *Suppose that* $p : [a,b]_{\mathbb{T}} \to \mathscr{L}(\mathbb{R}^n)$, $q : [a,b]_{\mathbb{T}} \to \mathbb{R}^n$ *are Kurzweil* Δ-*integrable,* $I + p(t)\mu(t)$ *is invertible for each* $t \in [a, t_0)_{\mathbb{T}}$, *and there exists a Kurzweil* Δ-*integrable function* $m : [a,b]_{\mathbb{T}} \to \mathbb{R}$ *such that*

$$\left| \int_u^v p(s)\, \Delta s \right| \le \int_u^v m(s)\, \Delta s \quad \text{if } u, v \in [a,b]_{\mathbb{T}} \text{ and } u \le v. \tag{8.7.19}$$

Then equation (8.7.11) *has a unique solution* $x : [a,b]_{\mathbb{T}} \to \mathbb{R}^n$, *which is obtained as the restriction to* $[a,b]_{\mathbb{T}}$ *of the unique solution to equation* (8.7.15) *with* A_1, B_1 *being given by* (8.7.17).

(ii) *Suppose that* $p\colon [a,b]_{\mathbb{T}} \to \mathscr{L}(\mathbb{R}^n)$, $q\colon [a,b]_{\mathbb{T}} \to \mathbb{R}^n$ *are Kurzweil* ∇-*integrable,* $I - p(t)\,\nu(t)$ *is invertible for each* $t \in (t_0, b]_{\mathbb{T}}$, *and there exists a Kurzweil* ∇-*integrable function* $m\colon [a,b]_{\mathbb{T}} \to \mathbb{R}$ *such that*

$$\Big|\int_u^v p(s)\,\nabla s\Big| \leq \int_u^v m(s)\,\nabla s \quad \text{if } u,v \in [a,b]_{\mathbb{T}} \text{ and } u \leq v. \tag{8.7.20}$$

Then equation (8.7.12) *has a unique solution* $x\colon [a,b]_{\mathbb{T}} \to \mathbb{R}^n$, *which is obtained as the restriction to* $[a,b]_{\mathbb{T}}$ *of the unique solution to equation* (8.7.16) *with* A_2, B_2 *being given by* (8.7.18).

Proof. We prove the first statement; the proof of the second one is similar and is left to the reader. By Lemma 8.7.2, the functions $A_1 : [a,b] \to \mathscr{L}(\mathbb{R}^n)$ and $B_1\colon [a,b] \to \mathbb{R}^n$ given by (8.7.17) have bounded variation and are left-continuous. Moreover, if $t \in [a, t_0)_{\mathbb{T}}$, then $I + \Delta^+ A_1(t) = I + p(t)\,\mu(t)$. On the other hand, if $t \in [a, t_0) \setminus \mathbb{T}$, then $I + \Delta^+ A_1(t) = I$. In both cases, $I + \Delta^+ A_1(t)$ is an invertible matrix.

Thus, all assumptions of Theorem 7.4.5 are satisfied, and equation (8.7.15) has a unique solution $y\colon [a,b] \to \mathbb{R}^n$. By Theorem 8.7.1, equation (8.7.11) has a unique solution $x\colon [a,b]_{\mathbb{T}} \to \mathbb{R}^n$, which is obtained as the restriction of y to $[a,b]_{\mathbb{T}}$. □

8.7.4 Remark. In the theory of dynamic equations, the requirement that $I + p(t)\mu(t)$ is invertible for each $t \in [a, t_0)_{\mathbb{T}}$ is known as μ-regressivity (or simply regressivity), while the condition that $I - p(t)\nu(t)$ is invertible for each $t \in (t_0, b]_{\mathbb{T}}$ is called ν-regressivity. For Δ-dynamic equations, it is usually supposed (see Chapter 5 of [Bohner, Peterson (2001)]) that p and q are rd-continuous functions, while the common assumption for ∇-equations is ld-continuity of p and q (see [Anderson et al (2003)]). Under these assumptions, both (8.7.19) and (8.7.20) hold, because p is bounded. Moreover, equations (8.7.11) and (8.7.12) reduce to $x^\Delta(t) = p(t)\,x(t) + q(t)$ and $x^\nabla(t) = p(t)x(t) + q(t)$, respectively. Our Theorem 8.7.3 is more general – no continuity or boundedness of p or q is assumed.

The next result is a continuous dependence theorem for solutions of linear Δ-dynamic equations.

8.7.5 Theorem. *Let* $p\colon [a,b]_{\mathbb{T}} \to \mathscr{L}(\mathbb{R}^n)$, $q\colon [a,b]_{\mathbb{T}} \to \mathbb{R}^n$, $p_k\colon [a,b]_{\mathbb{T}} \to \mathscr{L}(\mathbb{R}^n)$, $q_k\colon [a,b]_{\mathbb{T}} \to \mathbb{R}^n$, $k \in \mathbb{N}$, *be Kurzweil* Δ-*integrable. Suppose that* $I + p(t)\,\mu(t)$ *is invertible for each* $t \in [a, t_0)_{\mathbb{T}}$, *and there exists a Kurzweil* Δ-*integrable* *function* $m:[a,b]_{\mathbb{T}} \to \mathbb{R}$ *such that*

$$\Big|\int_u^v p_k(s)\,\Delta s\Big| \le \int_u^v m(s)\,\Delta s \quad \textit{for} \;\; u,v\in[a,b]_{\mathbb{T}}, u\le v, k\in\mathbb{N}. \tag{8.7.21}$$

Furthermore, assume that

$$\lim_{k\to\infty}\sup_{t\in[a,b]_{\mathbb{T}}}\Big|\int_a^t p_k(s)\,\Delta s - \int_a^t p(s)\,\Delta s\Big| = 0, \tag{8.7.22}$$

$$\lim_{k\to\infty}\sup_{t\in[a,b]_{\mathbb{T}}}\Big|\int_a^t q_k(s)\,\Delta s - \int_a^t q(s)\,\Delta s\Big| = 0. \tag{8.7.23}$$

Finally, let $x_k^0\in\mathbb{R}^n$, $k\in\mathbb{N}$, *be a sequence satisfying*

$$\lim_{k\to\infty} x_k^0 = x^0.$$

Then there exists a $k_0\in\mathbb{N}$ *such that the equations*

$$x_k(t) = x_k^0 + \int_{t_0}^t (p_k(s)\,x_k(s) + q_k(s))\,\Delta s \quad \textit{for} \;\; t\in[a,b]_{\mathbb{T}} \tag{8.7.24}$$

have unique solutions $x_k:[a,b]_{\mathbb{T}}\to\mathbb{R}^n$ *for* $k\ge k_0$. *Moreover,* $x_k \rightrightarrows x$, *where* $x:[a,b]_{\mathbb{T}}\to\mathbb{R}^n$ *is the unique solution of the equation*

$$x(t) = x^0 + \int_{t_0}^t (p(s)\,x(s) + q(s))\,\Delta s \quad \textit{for} \;\; t\in[a,b]_{\mathbb{T}}. \tag{8.7.25}$$

Proof. Consider the functions

$$A(t) = \int_{t_0}^t \widetilde{p}(s)\,\mathrm{d}g(s), \quad B(t) = \int_{t_0}^t \widetilde{q}(s)\,\mathrm{d}g(s),$$

and

$$A_k(t) = \int_{t_0}^t \widetilde{p}_k(s)\,\mathrm{d}g(s), \quad B_k(t) = \int_{t_0}^t \widetilde{q}_k(s)\,\mathrm{d}g(s) \quad \text{for} \;\; k\in\mathbb{N},$$

where $\widetilde{p}:[a,b]\to\mathscr{L}(\mathbb{R}^n)$, $\widetilde{q}:[a,b]\to\mathbb{R}^n$, $\widetilde{p}_k:[a,b]\to\mathscr{L}(\mathbb{R}^n)$, $\widetilde{q}_k:[a,b]\to\mathbb{R}^n$, $k\in\mathbb{N}$, are arbitrary functions satisfying

$$\widetilde{p}(t)=p(t),\ \widetilde{q}(t)=q(t),\ \widetilde{p}_k(t)=p_k(t),\ \widetilde{q}_k(t)=q_k(t)$$

for all $t\in[a,b]_{\mathbb{T}}$ and $k\in\mathbb{N}$. From Lemma 8.7.2, we know that the functions A, A_k, B, B_k are regulated, left-continuous, $I+\Delta^+A(t)$ is invertible for each $t\in[a,t_0)$, and

$$\operatorname{var}_a^b A_k \le \int_a^b m(s)\,\Delta s \quad \text{for} \;\; k\in\mathbb{N}.$$

Corollary 8.6.9 and the assumptions (8.7.22) and (8.7.23) imply that $A_k \rightrightarrows A$ and $B_k \rightrightarrows B$ on $[a,b]_{\mathbb{T}}$. Since A, A_k, B, B_k are constant on each interval $[t,t^*]$ with $t\in[a,b]$, we conclude that $A_k \rightrightarrows A$ and $B_k \rightrightarrows B$ on $[a,b]$.

Thus, by Theorem 7.6.3, there exists a $k_0\in\mathbb{N}$ such that for every $k\geq k_0$, the equation

$$y_k(t)=x_k^0+\int_{t_0}^t \mathrm{d}A_k(s)\,y_k(s)+B_k(t)-B_k(t_0) \quad \text{for } t\in[a,b], \qquad (8.7.26)$$

has a unique solution $y_k:[a,b]\to\mathbb{R}^n$. Moreover, $y_k\rightrightarrows y$, where $y:[a,b]\to\mathbb{R}^n$ is the unique solution of the equation

$$y(t)=x^0+\int_{t_0}^t \mathrm{d}A(s)\,y(s)+B(t)-B(t_0) \quad \text{for } t\in[a,b].$$

By Theorem 8.7.1, the restrictions of y and y_k to $[a,b]_{\mathbb{T}}$ are the unique solutions of the equations (8.7.25) and (8.7.24), respectively; this completes the proof. □

8.7.6 Exercise. Formulate and verify the counterpart of Theorem 8.7.5 for linear ∇-dynamic equations.

We conclude this section with a brief overview of the time scale exponential functions.

8.7.7 Theorem. *Consider a function $p:[a,b]_{\mathbb{T}}\to\mathbb{R}$, and let $\widetilde{p}:[a,b]\to\mathbb{R}$ be an arbitrary function such that $\widetilde{p}(t)=p(t)$ for all $t\in[a,b]$.*

(i) *Suppose that p is Kurzweil Δ-integrable, $1+p(t)\,\mu(t)\neq 0$ for every $t\in[a,t_0)_{\mathbb{T}}$, and there exists a Kurzweil Δ-integrable function $m:[a,b]_{\mathbb{T}}\to\mathbb{R}$ such that*

$$\left|\int_u^v p(s)\,\Delta s\right|\leq\int_u^v m(s)\,\Delta s \quad \textit{whenever } u,v\in[a,b]_{\mathbb{T}},\ u\leq v.$$

Then, the equation

$$x(t)=1+\int_{t_0}^t p(s)\,x(s)\,\Delta s \quad \textit{for } t\in[a,b]_{\mathbb{T}} \qquad (8.7.27)$$

has a unique solution $x:[a,b]_{\mathbb{T}}\to\mathbb{R}$, given by $x(t)=e_{\mathrm{d}P}(t,t_0)$ for $t\in[a,b]_{\mathbb{T}}$, where $P(t)=\int_{t_0}^t\widetilde{p}(s)\,\mathrm{d}g(s)$ and $g(t)=t^*$ for all $t\in[a,b]$.

(ii) *Suppose that* p *is Kurzweil* ∇*-integrable,* $1-p(t)\,\nu(t)\neq 0$ *for every* $t\in(t_0,b]_{\mathbb{T}}$*, and there exists a Kurzweil* ∇*-integrable function* $m\colon[a,b]_{\mathbb{T}}\to\mathbb{R}$ *such that*

$$\Big|\int_u^v p(s)\,\nabla s\Big|\leq\int_u^v m(s)\,\nabla s \quad \textit{whenever } u,v\in[a,b]_{\mathbb{T}},\ u\leq v.$$

Then, the equation

$$x(t)=1+\int_{t_0}^t p(s)\,x(s)\,\nabla s \quad \textit{for } t\in[a,b]_{\mathbb{T}} \tag{8.7.28}$$

has a unique solution $x\colon[a,b]_{\mathbb{T}}\to\mathbb{R}$*, given by* $x(t)=e_{\mathrm{d}\widehat{P}}(t,t_0)$ *for* $t\in[a,b]_{\mathbb{T}}$*, where* $\widehat{P}(t)=\int_{t_0}^t\widetilde{p}(s)\,\mathrm{d}h(s)$ *and* $h(t)=t_*$ *for all* $t\in[a,b]$*.*

Proof. The unique solvability of equations (8.7.27) and (8.7.28) follows from Theorem 8.7.3. By Lemma 8.7.2, the functions P, $\widehat{P}$ have bounded variation, P is left-continuous, $1+\Delta^+P(t)\neq 0$ for each $t\in[a,t_0)$, $\widehat{P}$ is right-continuous, and $1-\Delta^-\widehat{P}(t)\neq 0$ for each $t\in(t_0,b]$. Thus, the formulas $x(t)=e_{\mathrm{d}P}(t,t_0)$ and $x(t)=e_{\mathrm{d}\widehat{P}}(t,t_0)$, respectively, follow from Definition 8.5.2. □

The solutions of equations (8.7.27) and (8.7.28) are referred to as the Δ- and ∇-exponential functions; they are denoted by $t\mapsto e_p(t,t_0)$ and $t\mapsto\widehat{e}_p(t,t_0)$, $t\in[a,b]_{\mathbb{T}}$. In [Bohner, Peterson (2001)] and [Anderson et al (2003)], they are studied in the case when p is rd-continuous or ld-continuous, respectively. The previous theorem shows that the time scale exponential functions can be introduced under weaker hypotheses on p.

The properties of the time scale exponential functions can be obtained from the properties of the generalized exponential function. This fact is illustrated by the next result. We use the convention that $\sigma(b)=b$ and $\mu(b)=0$.

8.7.8 Theorem. *If all the exponentials below are well defined, then the following relations hold:*

(i) $e_0(t,t_0)=1$ *for every* $t\in[a,b]_{\mathbb{T}}$.

(ii) $e_p(t,t)=1$ *for every* $t\in[a,b]_{\mathbb{T}}$.

(iii) $e_p(t,s)\,e_p(s,r)=e_p(t,r)$ *for all* $t,s,r\in[a,b]_{\mathbb{T}}$.

(iv) $e_p(t,s)=e_p(s,t)^{-1}$ *for all* $t,s\in[a,b]_{\mathbb{T}}$.

(v) $e_p(\sigma(t),t_0)=(1+\mu(t)\,p(t))\,e_p(t,t_0)$ *for every* $t\in[a,b]_{\mathbb{T}}$.

(vi) $e_p(t,t_0)\,e_q(t,t_0)=e_r(t,t_0)$ *for every* $t\in[a,b]_{\mathbb{T}}$, *where* $r:[a,b]_{\mathbb{T}}\to\mathbb{R}$ *is given by* $r(t)=p(t)+q(t)+\mu(t)\,p(t)\,q(t)$.

(vii) $e_p(t,t_0)^{-1}=e_u(t,t_0)$ *for every* $t\in[a,b]_{\mathbb{T}}$, *where* $u:[a,b]_{\mathbb{T}}\to\mathbb{R}$ *is given by* $u(t)=-p(t)(1+p(t)\,\mu(t))^{-1}$.

(viii) *If* $t>t_0$, *then*

$$e_p(t,t_0)=\frac{\exp\Big(\int_{t_0}^t p(s)\,\Delta s\Big)\prod_{\tau\in[t_0,t)_{\mathbb{T}}}(1+p(\tau)\,\mu(\tau))}{\exp\Big(\sum_{\tau\in[t_0,t)_{\mathbb{T}}}p(\tau)\,\mu(\tau)\Big)}.$$

(ix) *If* $1+p(s)\,\mu(s)>0$ *for all* $s\in[t_0,t)_{\mathbb{T}}$, *then*

$$0<e_p(t,t_0)\le\exp\Big(\int_{t_0}^t p(s)\,\Delta s\Big).$$

Proof. Throughout this proof, we use the following convention: If f is a function defined on $[a,b]_{\mathbb{T}}$, then $\widetilde{f}$ is a function defined on $[a,b]$ such that $\widetilde{f}(t)=f(t)$ for all $t\in[a,b]_{\mathbb{T}}$. Also, we let $g(t)=t^*$ for all $t\in[a,b]$.

The first four statements are immediate consequences of Theorem 8.5.3 and the relation $e_p(t,s)=e_{\mathrm{d}P}(t,s)$, where $P(t)=\int_{t_0}^t\widetilde{p}(s)\,\mathrm{d}g(s)$.

To prove the fifth statement, note that P is constant on each interval of the form $(t,\sigma(t)]$, where $t\in[a,b]_{\mathbb{T}}$. Thus, $t\mapsto e_{\mathrm{d}P}(t,t_0)$ has the same property. For each $t\in[a,b]_{\mathbb{T}}$, Theorem 8.5.3 implies

$$\begin{aligned}e_p(\sigma(t),t_0)&=e_{\mathrm{d}P}(\sigma(t),t_0)=e_{\mathrm{d}P}(t+,t_0)\\&=(1+\Delta^+P(t))\,e_p(t,t_0)=(1+\mu(t)\,p(t))\,e_p(t,t_0).\end{aligned}$$

To prove the sixth statement, we use Theorem 8.5.6 to get

$$e_p(t,t_0)\,e_q(t,t_0)=e_{\mathrm{d}P}(t,t_0)\,e_{\mathrm{d}Q}(t,t_0)=e_{\mathrm{d}(P\oplus Q)}(t,t_0),$$

where $P(t)=\int_{t_0}^t\widetilde{p}(s)\,\mathrm{d}g(s)$ and $Q(t)=\int_{t_0}^t\widetilde{q}(s)\,\mathrm{d}g(s)$. Using left-continuity of P and Theorem 6.6.1 (substitution theorem), we see that the formula for $P\oplus Q$ from Theorem 8.5.6 reduces to

$$\begin{aligned}(P\oplus Q)(t) &= P(t)+Q(t)+\int_{t_0}^{t}\Delta^+Q(s)\,\mathrm{d}P(s)\\ &= P(t)+Q(t)+\int_{t_0}^{t}\Delta^+Q(s)\,\widetilde{p}(s)\,\mathrm{d}g(s)\\ &= \int_{t_0}^{t}(\widetilde{p}(s)+\widetilde{q}(s)+\widetilde{q}(s)\,\Delta^+g(s)\,\widetilde{p}(s))\,\mathrm{d}g(s)=\int_{t_0}^{t}\widetilde{r}(s)\,\mathrm{d}g(s),\end{aligned}$$

and it follows that $e_{\mathrm{d}(P\oplus Q)}(t,t_0)=e_r(t,t_0)$.

The proof of the seventh statement begins with an application of Theorem 8.5.8, which yields

$$e_p(t,t_0)^{-1}=e_{\mathrm{d}P}(t,t_0)^{-1}=e_{\mathrm{d}(\ominus P)}(t,t_0).$$

Using subsequently the formula for $\ominus P$ from Theorem 8.5.8, left-continuity of P, Lemma 6.3.15 and Theorem 6.6.1 (substitution theorem), we get

$$\begin{aligned}(\ominus P)(t) &= -P(t)+\sum_{s\in[t_0,t)}\frac{(\Delta^+P(s))^2}{1+\Delta^+P(s)}\\ &= -P(t)+\sum_{s\in[t_0,t)}\frac{\Delta^+P(s)}{1+\Delta^+P(s)}\,\Delta P(s)\\ &= -P(t)+\int_{t_0}^{t}\frac{\Delta^+P(s)}{1+\Delta^+P(s)}\,\mathrm{d}P(s)\\ &= -\int_{t_0}^{t}\widetilde{p}(s)\,\mathrm{d}g(s)+\int_{t_0}^{t}\frac{\widetilde{p}(s)\,\widetilde{\mu}(s)}{1+\widetilde{p}(s)\,\widetilde{\mu}(s)}\,\mathrm{d}P(s)\\ &= -\int_{t_0}^{t}\widetilde{p}(s)\,\mathrm{d}g(s)+\int_{t_0}^{t}\frac{\widetilde{p}(s)^2\,\widetilde{\mu}(s)}{1+\widetilde{p}(s)\,\widetilde{\mu}(s)}\,\mathrm{d}g(s)\\ &= -\int_{t_0}^{t}\widetilde{p}(s)\Big(1-\frac{\widetilde{p}(s)\,\widetilde{\mu}(s)}{1+\widetilde{p}(s)\,\widetilde{\mu}(s)}\Big)\,\mathrm{d}g(s)\\ &= -\int_{t_0}^{t}\frac{\widetilde{p}(s)}{1+\widetilde{p}(s)\,\widetilde{\mu}(s)}\,\mathrm{d}g(s)=\int_{t_0}^{t}\widetilde{u}(s)\,\mathrm{d}g(s),\end{aligned}$$

and hence $e_{\mathrm{d}(\ominus P)}(t,t_0)=e_u(t,t_0)$ for $t\in[a,b]_{\mathbb{T}}$.

Using Theorem 8.5.4 and left-continuity of P, we get the formula

$$\begin{aligned} e_p(t,t_0) &= \frac{e^{P(t)-P(t_0+)} \prod_{\tau\in[t_0,t)}(1+\Delta^+ P(\tau))}{\exp\left(\sum_{\tau\in(t_0,t)} \Delta^+ P(\tau)\right)} \\ &= \frac{e^{P(t)-P(t_0)} \prod_{\tau\in[t_0,t)}(1+\Delta^+ P(\tau))}{\exp\left(\sum_{\tau\in[t_0,t)} \Delta^+ P(\tau)\right)} \\ &= \frac{\exp\left(\int_{t_0}^t \widetilde{p}(s)\,\mathrm{d}g(s)\right) \prod_{\tau\in[t_0,t)_{\mathbb{T}}}(1+p(\tau)\,\mu(\tau))}{\exp\left(\sum_{\tau\in[t_0,t)_{\mathbb{T}}} p(\tau)\,\mu(\tau)\right)}, \end{aligned}$$

which proves the eighth statement.

If $1+p(\tau)\,\mu(\tau)>0$ for all $\tau\in[t_0,t)_{\mathbb{T}}$, then $1+\Delta^+P(\tau)>0$ for all $\tau\in[t_0,t)$, and the ninth statement follows from Corollary 8.5.5. □

8.7.9 Exercise. Formulate and verify the counterpart of Theorem 8.7.8 for the ∇-exponential function.

Bibliographic remarks

A special case of the relation between dynamic equations on time scales and Kurzweil-Stieltjes integral equations was described in [Slavík (2012)]. Integral equations having the form

$$y(t)=y(t_0)+\int_{t_0}^t f(y(s),s)\,\mathrm{d}g(s),$$

where g is a nondecreasing functions, are sometimes called *measure differential equations.* They include not only dynamic equations on time scales, but also differential equations with impulses; see e.g. [Federson, Mesquita, Slavík (2013); Monteiro, Slavík (2016)]. Basic results concerning the existence, uniqueness and continuous dependence of solutions to nonlinear measure differential equations are available in [Monteiro, Slavík (2016); Slavík (2015)]. The applicability of the theory of linear generalized differential equations in the context of linear dynamic equations was demonstrated in [Monteiro, Tvrdý (2013)], which contains a special case of Theorem 8.7.5. The concept of *Stieltjes differential equations* is closely related to measure differential equations, and is based on Stieltjes derivatives instead of Stieltjes integrals; more details can be found in [Frigon, Pouso (2017)]. The fact that the Δ-exponential function is a special case of the generalized exponential function was observed in [Monteiro, Slavík (2014)]; similarly, one can introduce the time scale hyperbolic and trigonometric functions as special cases of the generalized hyperbolic and trigonometric functions.

Bibliography

[Aleksandrov (1924)]
Aleksandrov, P. S., *Über die Äquivalenz des Perronschen und des Denjoyschen Integralbegriffes.* Math. Z. 20 (1924), 213-222.

[Anderson et al. (2003)]
Anderson, D., Bullock, J., Erbe, L., Peterson, A., and Tran, H., *Nabla dynamic equations.* In Bohner, M., Peterson, A. (eds.), Advances in dynamic equations on time scales. Birkhäuser, 2003, 47–83.

[Appell, Banás, Merentes (2014)]
Appell, J., Banás, J., and Merentes, N., *Bounded variation and around.* De Gruyter, 2014.

[Apostol (1974)]
Apostol, T. M., *Mathematical analysis. 2nd edition.* Addison-Wesley, 1974.

[Arzelà (1885)]
Arzelà, C., *Sulla integrazione per serie. I, II.* Rom. Acc. L. Rend. (4) I (1885), 532–537, 566–569.

[Ashordia (2016)]
Ashordia, M., *On the Opial type criterium for the well-posedness.* Math. Bohem. 141 (2016), 183–215.

[Aumann (1954)]
Aumann, G., *Reelle Funktionen.* Springer, 1954.

[Bartle (2001)]
Bartle, R. G., *A modern theory of integration.* American Mathematical Society, 2001.

[Bendová, Malý (2011)]
Bendová, H., and Malý, J., *An elementary way to introduce a Perron-like integral.* Ann. Acad. Sci. Fenn. Math. 36 (2011), 153–164.

[Bohner, Peterson (2001)]
Bohner, M., and Peterson, A., *Dynamic equations on time scales. An introduction with applications.* Birkhäuser, 2001.

[Bohner, Guseinov (2003)]
Bohner, M., and Guseinov, G., *Riemann and Lebesgue integration.* In Bohner, M., Peterson, A. (eds.), Advances in dynamic equations on time scales. Birkhäuser, 2003, 117–163.

[Bressoud (2008)]
Bressoud,D. M., *A radical approach to Lebesgue's theory of integration.* Cambridge University Press, 2008.

[Brokate, Krejčí (2003)]
Brokate, M., and Krejčí, P., *Duality in the space of regulated functions and the play operator.* Math. Z. 245 (2003), 667–688.

[Chen (1998/99a)]
Chen, H., *A pair of adjoint classes of Riemann-Stieltjes integrable functions.* Real Anal. Exch. 23 (1998/99), 235–240.

[Chen (1998/99b)]
Chen, H., *Adjoint classes of generalized-Stieltjes integrable functions.* Real Anal. Exch. 24 (1998/99), 139–148.

[Chen (2000/01)]
Chen, H., *Adjoint classes of Lebesgue-Stieltjes integrable functions.* Real Anal. Exch. 26 (2000/01), 421–428.

[Chistyakov (1997)]
Chistyakov, V. V., *On mappings of bounded variation.* J. Dynam. Control Systems 3 (1997), 261–289.

[Chistyakov (1998)]
Chistyakov, V. V., *On the theory of set-valued maps of bounded variation of one real variable.* Sb. Math. 189 (1998), 153–176.

[Cousin (1895)]
Cousin, P., *Sur les fonctions de n variables complexes.* Acta Math. 19 (1895), 1–62.

[Darboux (1875)]
Darboux, G., *Memoire sur les fonctions discontinues.* Ann. de l'Éc. Norm. (2) 4 (1875), 57–112.

[Doetsch (1935)]
Doetsch, G., *Über die Existenz der Darbouxschen Integrale und verwandter Begriffe.* J. Reine Angew. Math. 172 (1935), 183–192.

[Dudley, Norvaiša (1999)]
Dudley, R. M., and Norvaiša, R., *Differentiability of six operators on nonsmooth functions and p-variation.* Springer, 1999.

[Dudley, Norvaiša (2011)]
Dudley, R. M., and Norvaiša, R., *Concrete functional calculus.* Springer, 2011.

[Dunford, Schwartz (1958,1963)]
Dunford, N., and Schwartz, J.T., *Linear operators I, II.* Interscience Publishers, 1958 and 1963.

[Dushnik (1931)]
Dushnik, B., *On the Stieltjes integral.* Ph.D. thesis, University of Michigan, 1931.

[Federson, Mesquita, Slavík (2013)]
Federson, M., Mesquita, J. G., and Slavík, A., *Basic results for functional differential and dynamic equations involving impulses.* Math. Nachr. 286 (2013), 181–204.

[Fraňková (1989)]
Fraňková, D., *Continuous dependence on a parameter of solutions of generalized differential equations.* Čas. Pěst. Mat. 114 (1989), 230–261.

[Fraňková (1991)]
Fraňková, D., *Regulated functions.* Math. Bohem. 116 (1991) 20–59.

[Friedlander, Joshi (1999)]
Friedlander, F. G., and Joshi, M., *Introduction to the theory of distributions.* Cambridge University Press, 1999.

[Frigon, Pouso, (2017)]
Frigon, M., and Pouso, R. L., *Theory and applications of first-order systems of Stieltjes differential equations.* Adv. Nonlinear Anal. 6 (2017), 13–36.

[Garg (1992)]
Garg, K. M., *Relativization of some aspects of the theory of functions of bounded variation.* Diss. Math. 320 (1992), 123 pp.

[Getchell (1935)]
Getchell, B. C., *On the equivalence of two methods of defining Stieltjes integrals.* Bull. Amer. Math. Soc. 41 (1935), 413–418.

[Gordon (1994)]
Gordon, R. A., *The integrals of Lebesgue, Denjoy, Perron, and Henstock.* American Mathematical Society, 1994.

[Gordon (1998)]
Gordon, R. A., *The use of tagged partitions in elementary real analysis.* Amer. Math. Monthly 105 (1998), 107–117.

[Gowurin (1936)]
Gowurin, M., *Über die Stieltjessche Integration abstrakter Funktionen.* Fundam. Math. 27 (1936), 255–268.

[Groh (1980)]
Groh, J., *A nonlinear Volterra-Stieltjes integral equation and a Gronwall inequality in one dimension.* Illinois J. Math. 24 (1980), 244–263.

[Hake (1921)]
Hake, H.,, *Über de la Vallée-Poussins Ober- und Unterfunktionen einfacher Integrale und die Integraldefinition von Perron.* Math. Ann. 83 (1921), 119-142.

[Hakl et al. (2004)]
Hakl, R., Lomtatidze, A., and Stavrolaukis, I. P., *On a boundary value problem for scalar linear functional differential equations.* Abstr. Appl. Anal. 2004 (2004), 45–67.

[Halas, Monteiro, Tvrdý (2011)]
Halas, Z., Monteiro, G. A., and Tvrdý, M., *Emphatic convergence and sequential solutions of generalized linear differential equations.* Mem. Differ. Equ. Math. Phys. 54 (2011), 27–49.

[Halperin (1952)]
Halperin, I., *Introduction to the theory of distributions.* University of Toronto Press, 1952.

[Hanung, Monteiro, Tvrdý (2016)]
Hanung, U. M., Monteiro, G. A., and Tvrdý, M., *Bounded convergence theorem for abstract KurzweilStieltjes integral.* Monatsh. Math. 180 (2016), 409–434.

[Hawkins (2001)]
Hawkins, T., *Lebesgue's theory of integration. Its origins and development.* American Mathematical Society, 2001.

[Heikkilä (2012)]
Heikkilä, S. , *On the solvability of distributional and impulsive systems of Cauchy problems.* Nonlinear Anal. 75 (2012), 4604–4631.

[Helly (1912)]
Helly, E., *Über lineare Funktionaloperationen.* Wien. Ber. 121 (1912), 265–297.

[Henstock (1946)]
Henstock, R., *On interval functions and their integrals.* J. Lond. Math. Soc. 21 (1946), 204–209.

[Henstock (1968)]
Henstock, R., *A Riemann-type integral of Lebesgue power.* Can. J. Math. 20 (1968), 79–88.

[Henstock (1988)]
Henstock, R., *Lectures on the theory of integration.* World Scientific, 1988.

[Hildebrandt (1917–18)]
Hildebrandt, T. H., *On integrals related to and extensions of the Lebesgue integrals.* Bull. Amer. Math. Soc. 24 (1917), 113–144; 24 (1918), 177–202.

[Hildebrandt (1938)]
Hildebrandt, T. H., *Definitions of Stieltjes integrals of the Riemann type.* Amer. Math. Monthly 45 (1938), 265–278.

[Hildebrandt (1959)]
Hildebrandt, T. H., *On systems of linear differentio-Stieltjes-integral equations.* Illinois J. Math. 3 (1959), 352–373.

[Hildebrandt (1963)]
Hildebrandt, T. H., *Introduction to the theory of integration.* Academic Press, 1963.

[Hildebrandt (1966)]
Hildebrandt, T. H., *Linear continuous functionals on the space (BV) with weak topologies.* Proc. Amer. Math. Soc. 17 (1966), 658–664.

[Hilger (1990)]
Hilger, S., *Analysis on measure chains – a unified approach to continuous and discrete calculus.* Results Math. 18 (1990), 18–56.

[Hönig (1973)]
Hönig, Ch. S., *The abstract Riemann-Stieltjes integral and its applications to linear differential equations with generalized boundary conditions.* Notas do Instituto de Matemática e Estatística da Universidade de Sao Paulo, Série Matemática, No. 1. Universidade de Sao Paulo, 1973, 135 pp.

[Hönig (1975)]
Hönig, Ch. S., *Volterra-Stieltjes integral equations. Functional analytic methods; Linear constraints.* North Holland & American Elsevier, 1975.

[Hönig (1980)]
Hönig, Ch. S., *Volterra-Stieltjes integral equations.* In Izé, A. F. (ed.), Functional differential equations and bifurcation, Proceedings of a conference held at the Instituto de Ciências Matemáticas de São Carlos, Universidade de São Paulo, São Carlos, July 2–7, 1979, Springer, 1980, 173–216.

[Jarník (1976)]
Jarník, V., *Integrální počet II.* Academia, 1976.

[Jordan (1881)]
Jordan, C., *Sur la série de Fourier.* C. R. XCII (1881), 228–230.

[Kaczor, Nowak (2003)]
Kaczor, W. J., and Nowak, M. T., *Problems in mathematical analysis III. Integration.* American Mathematical Society, 2003.

[Kaltenborn (1934)]
Kaltenborn, H. S., *Linear functional operations on functions having discontinuities of the first kind.* Bull. Amer. Math. Soc. 40 (1934), 702–708.

[Kelley, Peterson (2010)]
Kelley, W. and Peterson, A., *The theory of differential equations. Classical and qualitative. Second edition.* Springer, 2010.

[Kiguradze (1988)]
Kiguradze, I. T., *Boundary-value problems for systems of ordinary differential equations.* J. Sov. Math. 43 (1988), 2259–2339.

[Kolmogorov, Fomin (2012)]
Kolmogorov, A. N., and Fomin, S. V., *Elements of the theory of functions and functional analysis.* Martino Fine Books, 2012.

[Krabbe (1965)]
Krabbe, G.L., *A Helly convergence theorem for Stieltjes integrals.* Indag. Math. 27 (1965), 52–65.

[Krejčí (2003)]
Krejčí, P., *The Kurzweil integral with exclusion of negligible sets.* Math. Bohem. 128 (2003), 277–292.

[Krejčí (2006)]
Krejčí, P., *The Kurzweil integral and hysteresis.* In: Proceedings of the International Workshop on Multi-Rate Processes and Hysteresis (Cork, 3.4.2006–8.4.2006, eds: M. Mortell, R. O'Malley, A. Pokrovskii, V. Sobolev). Journal of Physics: Conference Series, 55 (2006) 144–154.

[Krejčí, Kurzweil (2002)]
Krejčí, P., and Kurzweil, J., *A nonexistence result for the Kurzweil integral.* Math. Bohem., 127 (2002) 571–580.

[Krejčí, Lamba, Melnik, Rachinskii (2015)]
Krejčí, P., Lamba, H. Melnik, S., and Rachinskii, D., *Kurzweil integral representation of interacting Prandtl-Ishlinskii operators.* Discrete Contin. Dyn. Syst. Ser. B 20 (2015), 2949–2965.

[Krejčí, Lamba, Monteiro, Rachinskii (2016)]
Krejčí, P., Lamba, H., Monteiro G. A., and Rachinskii, D., *The Kurzweil integral in financial market modeling.* Math. Bohem. 141, 2016, 261–286.

[Krejčí, Laurençot (2002)]
Krejčí, P., and Laurençot, P., *Generalized variational inequalities.* J. Convex Anal. 9 (2002), 159–183.

[Krejčí, Liero (2009)]
Krejčí, P., and Liero, M., *Rate independent Kurzweil processes.* Appl. Math. 54 (2009), 117–145.

[Kurzweil (1957)]
Kurzweil, J., *Generalized ordinary differential equations and continuous dependence on a parameter.* Czech. Math. J. 7 (82) (1957), 418–449.

[Kurzweil (1958a)]
Kurzweil, J., *Generalized ordinary differential equations.* Czech. Math. J. 8 (83) (1958), 360–387.

[Kurzweil (1986)]
Kurzweil, J., *Ordinary differential equations. Introduction to the theory of ordinary differential equations in the real domain.* Elsevier, 1986.

[Kurzweil (1980)]
Kurzweil, J., *Nichtabsolut konvergente Integrale.* Teubner, 1980.

[Kurzweil (2000)]
Kurzweil, J., *Henstock-Kurzweil integration. Its relation to topological vector spaces.* World Scientific, 2000.

[Kurzweil (2002)]
Kurzweil, J., *Integration between the Lebesgue integral and the Henstock-Kurzweil integral. Its relation to local convex vector spaces.* World Scientific, 2002.

[Kurzweil (2012)]
Kurzweil, J., *Generalized ordinary differential equations. Not absolutely continuous solutions.* World Scientific, 2012.

[Lebesgue (1910)]
Lebesgue, H., *Sur l'intégrale de Stieltjes et sur les opérations fonctionnelles linéaires.* C. R. 150 (1910), 86–88.

[Lee, Výborný (2000)]
Lee, P. Y., and Výborný, R., *The integral: An easy approach after Kurzweil and Henstock.* Cambridge University Press, 2000.

[Lewin (1986)]
Lewin, J. W., *A truly elementary approach to the bounded convergence theorem.* Amer. Math. Monthly 93 (1986), 395–397.

[Liu, Krejčí, Ye (2017)]
Liu, W., Krejčí, P., and Ye, G., *Continuity properties of Prandtl-Ishlinskii operators in the space of regulated functions.* Discrete Contin. Dyn. Syst. Ser. B 22 (2017), 3783–3795.

[Looman (1935)]
Looman, H., *Über die Perronsche Integraldefinition.* Math. Ann. 93 (1935), 153-156.

[Lukeš, Malý (2017)]
Lukeš, J., and Malý, J., *Measure and integral.* MatfyzPress, 2017.

[Luxemburg (1971)]
Luxemburg, W. A. J., *Arzela's dominated convergence theorem for the Riemann integral.* Amer. Math. Monthly 78 (1971), 970–979.

[Malý (2014)]
Malý, J., *Non-absolutely convergent integrals with respect to distributions.* Ann. Mat. Pura Appl. 193 (2014), 1457–1484.

[Mawhin (1992)]
Mawhin, J., *Analyse. Fondements, techniques, évolution.* De Boeck Université, 1992.

[Mawhin (2001)]
Mawhin, J., *L'eternal retour des sommes de Riemann-Stieltjes dans l'evolution du calcul intégral.* Bull. Soc. R. Sci. Liège 70 (2001), 345–364.

[McLeod (1980)]
McLeod, R. M., *The generalized Riemann integral.* Mathematical Association of America, 1980.

[MacNerney (1963)]
MacNerney, J. S., *An integration by parts formula.* Bull. Amer. Math. Soc. 69 (1963), 803–805.

[Medvedev (1991)]
Medvedev, F. A., *Scenes from the history of real functions.* Birkhäuser, 1991.

[Monteiro (2015)]
Monteiro, G. A., *On functions of bounded semivariation.* Real Anal. Exch. 40 (2015), 233–276.

[Monteiro (2017)]
Monteiro, G. A., *Adjoint classes of Kurzweil-Stieltjes integrable functions.* Monatsh. Math. 186 (2018), 175–188.

[Monteiro, Satco (2016)]
Monteiro, G. A., and Satco, B., *Distributional, differential and integral problems: equivalence and existence results.* Electron. J. Qual. Theory Differ. Equ. (2017), paper no. 7, 1–26.

[Monteiro, Slavík (2014)]
Monteiro, G. A., and Slavík, A., *Generalized elementary functions.* J. Math. Anal. Appl. 411 (2014), 838–852.

[Monteiro, Slavík (2016)]
Monteiro, G. A., and Slavík, A., *Extremal solutions of measure differential equations.* J. Math. Anal. Appl. 444 (2016), 568–597.

[Monteiro, Tvrdý (2012)]
Monteiro, G. A., and Tvrdý, M., *On Kurzweil-Stieltjes integral in Banach space.* Math. Bohem. 137 (2012), 365–381.

[Monteiro, Tvrdý (2013)]
Monteiro, G. A., and Tvrdý, M., *Generalized linear differential equations in a Banach space: Continuous dependence on a parameter.* Discrete Contin. Dyn. Syst. 33 (2013), 283–303.

[Monteiro, Tvrdý (2014)]
Monteiro, G. A., and Tvrdý, M., *Continuous dependence of solutions of abstract generalized linear differential equations with potential converging uniformly with a weight.* Boundary Value Problems 2014:71 (2014), 1–18.

[Mozyrska et al. (2010)]
Mozyrska, D., Pawluszewicz, E., and Torres, D. F. M., *The Riemann-Stieltjes integral on time scales.* Aust. J. Math. Anal. Appl. 7 (2010), 1–14.

[Opial (1967)]
Opial, Z., *Continuous parameter dependence in linear systems of differential equations.* J. Differential Equations 3 (1967), 571–579.

[Osgood (1897)]
Osgood, W. F., *Non-uniform convergence and the integration of series term by term.* American J. Math. 19 (1897), 155–190.

[Pelant, Tvrdý (1993)]
Pelant, M., and Tvrdý, M., *Linear distributional differential equations in the space of regulated functions.* Math. Bohem. 118 (1993), 379–400.

[Pesin (1970)]
Pesin, I. N., *Classical and modern integration theories.* Academic Press, 1970.

[Peterson, Thompson (2006)]
Peterson, A. and Thompson, B., *Henstock-Kurzweil delta and nabla integrals.* J. Math. Anal. Appl. 323 (2006), 162–178.

[Pollard (1920)]
Pollard, S., *The Stieltjes' integral and its generalizations.* Quart. J. 49 (1923), 73–138.

[Radon (1913)]
Radon, J., *Theorie und Anwendungen der absolut additiven Mengenfunktionen.* Wien. Ber. 122 (1913), 1295–1438.

[Rana (2002)]
Rana, I. K., *An introduction to measure and integration. Second edition.* American Mathematical Society, 2002.

[Riesz (1909)]
Riesz, F., *Sur les opérations functionelles linéaires.* C. R. 149 (1909), 974–977.

[Riesz (1911)]
Riesz, F., *Sur certains systémes singuliers d'équations intégrales.* Ann. de l'Éc. Norm. 28 (1911), 33–62.

[Rosentrater (2015)]
Rosentrater, C. R., *Varieties of integration.* Mathematical Association of America, 2015.

[Rudin (1973)]
Rudin, W., *Functional analysis.* McGraw-Hill, 1973.

[Saks (1927)]
Saks, S., *Sur les fonctions d'intervalle.* Fund. Math. 10 (1927), 211–224.

[Saks (1937)]
Saks, S., *Theory of the integral. 2nd revised edition.* G. E. Stechert & Co., 1937.

[Schechter E. (1997)]
Schechter, E., *Handbook of analysis and its foundations.* Academic Press, 1997.

[Schechter M. (2001)]
Schechter, M., *Principles of functional analysis. Second edition.* American Mathematical Society, 2001.

[Schwabik (1971)]
Schwabik, Š., *Verallgemeinerte lineare Differentialgleichungssysteme.* Čas. Pěst. Mat. 96 (1971), 183–211.

[Schwabik (1973a)]
Schwabik, Š., *On the relation between Young's and Kurzweil's concept of Stieltjes integral.* Čas. Pěst. Mat. 98 (1973), 237–251.

[Schwabik (1973b)]
Schwabik, Š., *On a modified sum integral of Stieltjes type.* Čas. Pěst. Mat. 98 (1973), 274–277.

[Schwabik (1988)]
Schwabik, Š., *A survey of some new results for regulated functions.* Seminário Brasileiro de Análise 28 (1988), 201–209.

[Schwabik (1992)]
Schwabik, Š., *Generalized ordinary differential equations.* World Scientific, 1992.

[Schwabik (1996)]
Schwabik, Š., *Abstract Perron-Stieltjes integral.* Math. Bohem. 121 (1996), 425–447.

[Schwabik (1999a)]
Schwabik, Š., *Integrace v $\mathbb{R}$ (Kurzweilova teorie).* Karolinum, 1999.

[Schwabik (1999b)]
Schwabik, Š., *Linear Stieltjes integral equations in Banach spaces.* Math. Bohem. 124 (1999), 433–457.

[Schwabik (2000)]
Schwabik, Š., *Linear Stieltjes integral equations in Banach spaces II: operator valued solutions.* Math. Bohem. 125 (2000), 431–454.

[Schwabik (2001)]
Schwabik, Š., *A note on integration by parts for abstract Perron-Stieltjes integrals.* Math. Bohem. 126 (2001), 613–629.

[Schwabik, Tvrdý (1979)]
Schwabik, Š., and Tvrdý, M., *Boundary value problems for systems of generalized linear differential equations.* Czech. Math. J. 29 (1979), 451–477.

[Schwabik, Tvrdý, Vejvoda (1979)]
Schwabik, Š., Tvrdý, M., and Vejvoda, O., *Differential and integral equations. Boundary value problems and adjoints.* Academia and Reidel, 1979.

[Schwabik, Ye (2005)]
Schwabik, Š., and Ye, G., *Topics in Banach space integration.* World Scientific, 2005.

[Schwartz (1998)]
Schwartz, L., *Théorie des distributions I, II.* Hermann, 1998.

[Shilov (1996)]
Shilov, G. E., *Elementary real and complex analysis.* Dover Publications, 1996.

[Shilov, Gurevich (1977)]
Shilov, G. E., and Gurevich, B. L., *Integral, measure and derivative: a unified approach.* Dover Publications, 1977.

[Slavík (2012)]
Slavík, A., *Dynamic equations on time scales and generalized ordinary differential equations.* J. Math. Anal. Appl. 385 (2012), 534–550.

[Slavík (2013)]
Slavík, A., *Generalized differential equations: Differentiability of solutions with respect to initial conditions and parameters.* J. Math. Anal. Appl. 402 (2013), 261–274.

[Slavík (2015)]
Slavík, A., *Well-posedness results for abstract generalized differential equations and measure functional differential equations.* J. Differential Equations 259 (2015), 666–707.

[Stieltjes (1894)]
Stieltjes, T.J., *Recherches sur les fractions continues.* Toulouse Ann. 8 (1894), J1–J122, and 9 (1895), A1–A47. English translation: *Investigations on continued fractions.* In van Dijk, G. (ed.), Œuvres compltes. Collected papers, vol. 2. Springer, 1993, 609–745.

[Stroock (1994)]
Stroock, D.A., *A concise introduction to the theory of integration. 2nd edition.* Birkhäuser, 1994.

[Swartz (1998/99)]
Swartz, C., *Norm convergence and uniform integrability for the Henstock-Kurzweil integral.* Real Anal. Exch. 24 (1998/99), 423–426.

[Swartz (2001)]
Swartz, C., *Introduction to gauge integrals.* World Scientific, 2001.

[Sworowski (2007)]
Sworowski, P., *Adjoint classes of functions in the H_1 sense.* Czech. Math. J. 57 (132) (2007), 505–522.

[Talvila (2008)]
Talvila, E., *The distributional Denjoy integral.* Real Anal. Exch. 33 (2008), 51–82.

[Talvila (2009)]
Talvila, E., *The regulated primitive integral.* Illinois J. Math. 53 (2009), 1187–1219.

[Taylor, Lay (1980)]
Taylor, A. E., and Lay, D. C., *Introduction to functional analysis. 2nd edition.* John Wiley, 1980.

[Thomson (2008)]
Thomson, B. S., *Henstock-Kurzweil integrals on time scales.* PanAmer. Math. J. 18 (2008), 1–19.

[Tvrdý (1989)]
Tvrdý, M., *Regulated functions and the Perron-Stieltjes integral.* Čas. Pěst. Mat. 114 (1989), 187–209.

[Tvrdý (1991)]
Tvrdý, M., *Generalized differential equations in the space of regulated functions (boundary value problems and controllability).* Math. Bohem. 116 (1991), 225–244.

[Tvrdý (1994)]
Tvrdý, M. *Linear distributional differential equations of the second order.* Math. Bohem. 119 (1994), 415–436.

[Tvrdý (2002)]
Tvrdý, M., *Differential and integral equations in the space of regulated functions.* Mem. Differ. Equ. Math. Phys. 25 (2002), 1–104.

[Tvrdý (2012)]
Tvrdý, M., *Stieltjesův integrál. Kurzweilova teorie.* Univerzita Palackého v Olomouci, Olomouc, 2012.

[de la Vallée Poussin (1916)]
de la Vallée Poussin, C. J. *Intégrales de Lebesgue. Fonctions d'ensemble. Classes de Baire.* Gauthier-Villars, 1916.

[van Vleck (1917)]
van Vleck, E. B., *Haskin's momental theorem and its connection with Stieltjes's problem of moments.* Trans. Amer. Math. Soc. 18 (1917), 326–330.

[Ward (1936)]
Ward, A. J., *The Perron-Stieltjes integral.* Math. Z. 41 (1936), 578–604.

[Ye, Liu (2016)]
Ye, G., and Liu, W., *The distributional Henstock-Kurzweil integral and applications.* Monatsh. Math. 181 (2016), 975–989.

[Yeh (2014)]
Yeh, J., *Real analysis. Theory of measure and integration. 3rd edition.* World Scientific, 2014.

[L.C. Young (1927)]
Young, L. C., *The theory of integration.* Cambridge University Press, 1927.

[L.C. Young (1936)]
Young, L. C., *An inequality of the Hölder type, connected with Stieltjes integration.* Acta Math. 67 (1936), 251–282.

[W.H. Young (1914)]
Young, W. H., *Integration with respect to a function of bounded variation.* Proc. Lond. Math. Soc. (2) 13 (1914), 109–150.

Subject index

Symbol index

Sets

Spaces

www.ingramcontent.com/pod-product-compliance
Ingram Content Group UK Ltd.
Pitfield, Milton Keynes, MK11 3LW, UK
UKHW020401150926
8624IPUK00004B/281

9 789814 641777